Supplements to the 2nd Edition of

RODD'S CHEMISTRY OF CARBON COMPOUNDS

ELSEVIER SCIENTIFIC PUBLISHING COMPANY
335 Jan van Galenstraat
P.O. Box 211, Amsterdam, The Netherlands

AMERICAN ELSEVIER PUBLISHING COMPANY, INC.
52 Vanderbilt Avenue
New York, New York, 10017

Library of Congress Card Number: 64-4605

ISBN 0-444-41133-X

Printed in The Netherlands

Supplements to the 2nd Edition of

RODD'S CHEMISTRY OF CARBON COMPOUNDS

───────────

ADVISORS

Professor Sir ROBERT ROBINSON, O.M., M.A.(Oxon.), D.Sc.(Manc.),
Hon.D.Sc. (Lond., Liv., Wales, Dunelm, Sheff., Belfast, Bris., Oxon., Nott.,
Strath., Delhi, Sydney, Zagreb), Hon.Sc.D.(Cantab.), Hon.LL.D.(Manc., Edin.,
Birm., St. Andrews, Glas., Liv.), Hon.D.Pharm.(Madrid and Paris),
Hon.F.R.S.E., F.R.S., *London*

Professor Sir DEREK BARTON, Ph.D., D.Sc. (Lond.), F.R.I.C., F.R.S., Hon.D.Sc.
(Montpelier, Dublin, St. Andrews, Columbia, Univ. NYC, Coimbra), *London*

Professor R. A. RAPHAEL, Ph.D., D.Sc.(Lond.), F.R.S., F.R.S.E., *Cambridge*

Supplements to the 2nd Edition of

RODD'S CHEMISTRY OF CARBON COMPOUNDS

VOLUME I

ALIPHATIC COMPOUNDS

★

VOLUME II

ALICYCLIC COMPOUNDS

★

VOLUME III

AROMATIC COMPOUNDS

★

VOLUME IV

HETEROCYCLIC COMPOUNDS

★

VOLUME V

MISCELLANEOUS

GENERAL INDEX

★

Supplements to the 2nd Edition (Editor S. Coffey) of

RODD'S CHEMISTRY OF CARBON COMPOUNDS

A modern comprehensive treatise

Edited by

M. F. ANSELL
Ph.D., D.Sc. (London), F.R.I.C.
Department of Chemistry, Queen Mary College,
University of London (Great Britain)

Supplement to
VOLUME II   ALICYCLIC COMPOUNDS

Part A: Monocarbocyclic compounds to and including five ring atoms
Part B: Six- and higher-membered monocyclic compounds

ELSEVIER SCIENTIFIC PUBLISHING COMPANY
AMSTERDAM
1974

CONTRIBUTORS TO THIS VOLUME

M. F. ANSELL, Ph.D., D.Sc.(London), F.R.I.C.
Department of Chemistry, Queen Mary College, Mile End Road, London, E.1

A. J. BELLAMY, B. Sc., Ph.D.
Department of Chemistry, University of Edinburgh, Edinburgh

P. H. BOYLE, M.A., Ph.D.
University of Dublin, Chemical Laboratory, Trinity College, Dublin 2

J. B. DAVIS, B.Sc., Ph.D., D.I.C.
Tropical Products Institute, 56–62 Gray's Inn Road, London WC1X 8LU

IAN FLEMING, M.A., Ph.D.
University Chemical Laboratory, Lensfield Road, Cambridge CB2 1EW

S. H. GRAHAM, B.Sc., Ph.D.
Department of Chemistry, University College of Wales, Aberystwyth SY23 1NE

S. H. HARPER, Ph.D., D.Sc., D.I.C., F.R.I.C.
Department of Chemistry, University of Rhodesia, Salisbury (Rhodesia)

P. W. HICKMOTT, M.Sc., Ph.D., F.R.I.C.
Department of Chemistry, University of Salford, Salford M5 4WT

R. E. FAIRBAIRN, B.Sc., Ph.D., F.R.I.C.
formerly of Research Department, Dyestuffs Division, I.C.I. Ltd., Manchester 9 (*Index*)

PREFACE TO SUPPLEMENT II AB

The appearance of this volume is in continuance of the policy, detailed in the preface to the Supplement I CD, of publishing supplements to the Second Edition of *Rodd's Chemistry of Carbon Compounds*. The eight chapters of this volume parallel those of Volumes II A and II B of the Second Edition, with the exception of the introductory first chapter which mainly deals with the basic examples of pericyclic reactions. Many examples of these reactions will be found in this volume throughout which an attempt has been made to integrate descriptive material with the more theoretical and mechanistic aspect of the subject.

In this volume the basic chemistry of the mono-cyclic alicyclic systems is surveyed and, in addition, Chapters 6 and 7 respectively, advances in the chemistry of monoterpenoids and carotenoids are discussed. In both these chapters the related acyclic and cyclic compounds are dealt with together.

I am most grateful to all the contributors to this volume, for the careful and critical way in which each of them has selected material from the extensive literature available and in every case condensed it into a very readable review. Each chapter, by supplementing the chapters in the Second Edition, provides a valuable guide to the advances in the field reviewed.

As editor I am pleased to acknowledge the help I have received from the advisors.

March, 1974   M. F. ANSELL

CONTENTS

VOLUME II AB

Alicyclic Compounds; Monocarbocyclic Compounds to and including Five Rings Atoms; Six- and Higher-Membered Monocyclic Compounds.

Chapter 1. Introduction
by MARTIN F. ANSELL

Chapter 2. The Cyclopropane Group
by P. H. BOYLE

Chapter 5. The Cyclohexane Group
by A. J. BELLAMY

Chapter 8. The Cycloheptane, Cyclo-octane and Macrocyclic Groups
by S. H. GRAHAM

OFFICIAL PUBLICATIONS

B.P.	British (United Kingdom) Patent
F.P.	French Patent
G.P.	German Patent
Sw.P.	Swiss Patent
U.S.P.	United States Patent
U.S.S.R.P.	Russian Patent
B.I.O.S.	British Intelligence Objectives Sub-Committee Reports, H.M. Stationery Office, London.
C.I.O.S.	Combined Intelligence Objectives Sub-Committee Reports
F.I.A.T.	Field Information Agency, Technical Reports of U.S. Group Control Council for Germany
B.S.	British Standards Specification
A.S.T.M.	American Society for Testing and Materials
A.P.I.	American Petroleum Institute Projects
C.I.	Colour Index Number of Dyestuffs and Pigments

SCIENTIFIC JOURNALS AND PERIODICALS

With few obvious and self-explanatory modifications the abbreviations used in references to journals and periodicals comprising the extensive literature on organic chemistry, are those used in the World List of Scientific Periodicals.

LIST OF COMMON ABBREVIATIONS AND SYMBOLS USED

A	acid
Å	Ångström units
Ac	acetyl
a	axial; antarafacial
as, asymm.	asymmetrical
at.	atmosphere
B	base
Bu	butyl
b.p.	boiling point
C, mC and μC	curie, millicurie and microcurie
c, C	concentration
C.D.	circular dichroism
conc.	concentrated
crit.	critical
D	Debye unit, 1×10^{-18} e.s.u.
D	dissociation energy
D	dextro-rotatory; dextro configuration
DL	optically inactive (externally compensated)
d	density
dec. or decomp.	with decomposition
deriv.	derivative
E	energy; extinction; electromeric effect; Entgegen (opposite) configuration
E1, E2	uni- and bi-molecular elimination mechanisms
E1cB	unimolecular elimination in conjugate base
e.s.r.	electron spin resonance
Et	ethyl
e	nuclear charge; equatorial
f	oscillator strength
f.p.	freezing point
G	free energy
g.l.c.	gas liquid chromatography
g	spectroscopic splitting factor, 2.0023
H	applied magnetic field; heat content
h	Planck's constant
Hz	hertz
I	spin quantum number; intensity; inductive effect
i.r.	infrared
J	coupling constant in n.m.r. spectra; joule
K	dissociation constant
kJ	kilojoule
k	Boltzmann constant; velocity constant
kcal.	kilocalories
L	laevorotatory; laevo configuration
M	molecular weight; molar; mesomeric effect
Me	methyl
m	mass; mole; molecule; *meta-*

ml	millilitre
m.p.	melting point
Ms	mesyl (methanesulphonyl)
[M]	molecular rotation
N	Avogadro number; normal
nm	nanometre (10^{-9} metre)
n.m.r.	nuclear magnetic resonance
n	normal; refractive index; principal quantum number
o	*ortho-*
o.r.d.	optical rotatory dispersion
P	polarisation, probability; orbital state
Pr	propyl
Ph	phenyl
p	*para-;* orbital
p.m.r.	proton magnetic resonance
R	clockwise configuration
S	counterclockwise config.; entropy; net spin of incompleted electronic shells; orbital state
S_N1, S_N2	uni- and bi-molecular nucleophilic substitution mechanisms
S_Ni	internal nucleophilic substitution mechanisms
s	symmetrical; orbital; suprafacial
sec	secondary
soln.	solution
symm.	symmetrical
T	absolute temperature
Tosyl	*p*-toluenesulphonyl
Trityl	triphenylmethyl
t	time
temp.	temperature (in degrees centigrade)
tert	tertiary
U	potential energy
u.v.	ultraviolet
v	velocity
Z	zusammen (together) configuration
α	optical rotation (in water unless otherwise stated)
$[\alpha]$	specific optical rotation
α_A	atomic susceptibility
α_E	electronic susceptibility
ϵ	dielectric constant; extinction coefficient
μ	microns (10^{-4} cm); dipole moment; magnetic moment
μ_B	Bohr magneton
μg	microgram (10^{-6} g)
λ	wavelength
ν	frequency; wave number
χ, χ_d, χ_μ	magnetic, diamagnetic and paramagnetic susceptibilities
~	about
(+)	dextrorotatory
(−)	laevorotatory
⊖	negative charge
⊕	positive charge

Chapter 1

Introduction

MARTIN F. ANSELL

1. Pericyclic reactions

An outstanding advance in theoretical chemistry, which has had a major impact on alicyclic chemistry, is the rationalisation of concerted reactions by Woodward and Hoffmann ("Conservation of Orbital Symmetry" by R. B. Woodward and R. Hoffmann, Academic Press, New York, 1970; Accts. chem. Res., 1968, 1:17; "Organic Reactions and Orbital Symmetry", by T. L. Gilchrist and R. C. Storr, Cambridge University Press, 1972; "The Conservation of Orbital Symmetry", G. B. Gill and M. R. Willis, Methuen, London, 1969; see also Gill, Quart. Review, 1968, 22: 238).

Examples of concerted reactions and their rationalisation will be found in the following chapters. The purpose of this chapter is to provide definitions of the terms used together with a summary of the general "Woodward–Hoffmann Rules" governing such reactions. It is beyond the scope of this chapter to present fully the various alternative theoretical bases of the rules.

Pericyclic reactions are concerted reactions which involve a cyclic transition state, and are subdivided into electrocyclic reactions, cycloaddition reactions, sigmatropic rearrangements and chelotropic reactions.

Electrocyclic reactions involve the formation of a single σ-bond between the termini of a fully-conjugated linear π-bond system, or the reverse reaction. Thus if the linear precursor contains k π-electrons the cyclic product contains $(k - 2)$ π-electrons. Such reactions may be illustrated by:

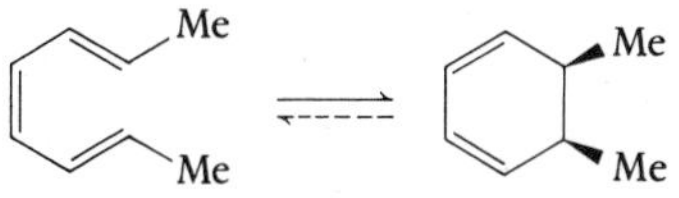

(a)

[1]

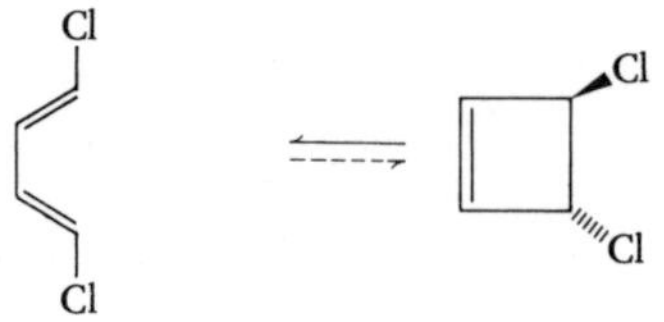

(b)

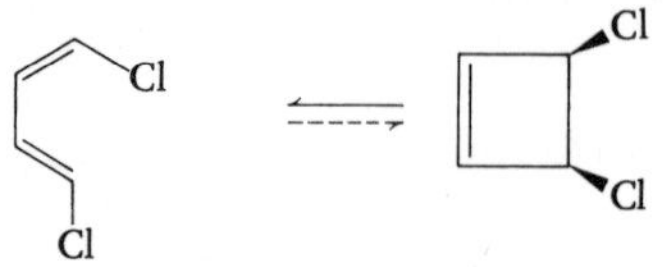

(b)

(a) E. Vogel, W. Grimme and E. Dinne, Tetrahedron Letters, 1965, 391.
(b) R. Criegee, D. Seebach, R. E. Williams and B. Borretzen, Ber., 1965, 98: 2339.

Such reactions are usually highly stereospecific, and there are four possible steric courses for such reactions:

Conrotatory

Enantiomers

Disrotatory

Enantiomers

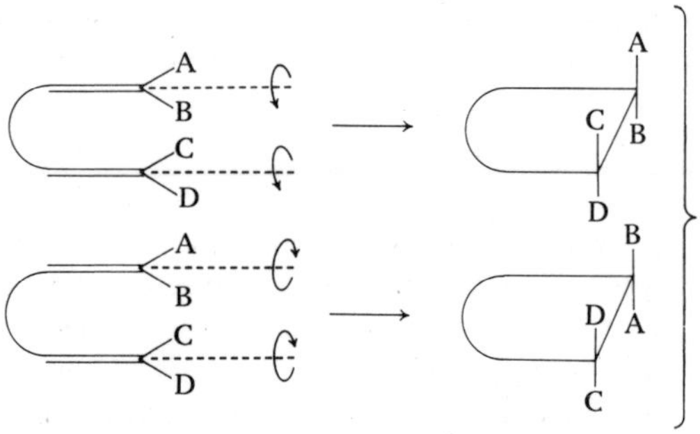

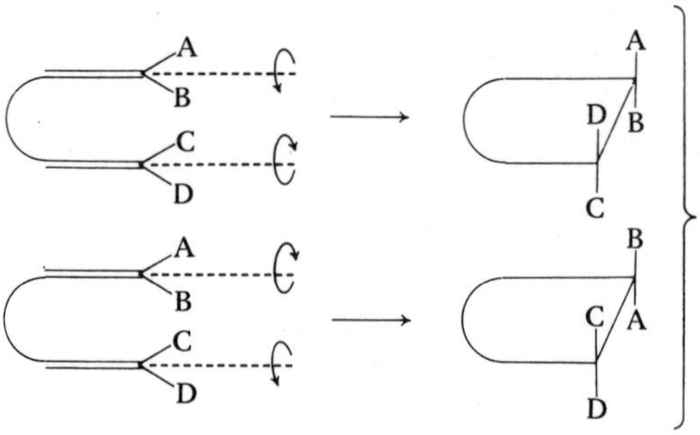

The cyclisation (or ring-opening) reaction may proceed in either a disrotatory or a conrotatory fashion. The four possible modes of cyclisation cannot always be distinguished and the "rules" consider only disrotatory and conrotatory modes of reaction. Predictions within a cyclisation mode depend on secondary steric factors. The preferred steric course of the reaction depends on the method of initiation, thermal or photochemical, and the number of π-electrons in the linear partner of the reaction as shown below:

STERIC COURSE OF ELECTROCYCLIC REACTIONS

Number of participating	Mode of initiation	
π-electrons	Thermal	Photochemical
$4n$	Conrotatory	Disrotatory
$4n + 2$	Disrotatory	Conrotatory

Cycloaddition reactions occur when two or more reactants combine additively to form a cyclic compound, no small fragments are eliminated, and σ-bonds are formed but not broken. The reverse reactions are cycloreversion reactions.

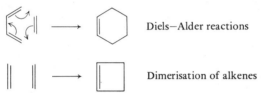

Diels—Alder reactions

Dimerisation of alkenes

Such reactions are classified on either the number of ring atoms or the number of π-electrons provided by each component. Thus the Diels—Alder reaction may be referred to as either a 4 + 2 or a $(4\pi + 2\pi)$ cycloaddition reaction. In these reactions the two new σ-bonds which arise can, in principle, be formed either on the same face (suprafacial process) or on opposite faces (antarafacial process) of each reactant:

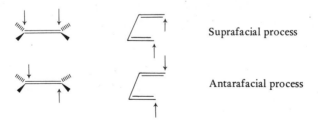

Suprafacial process

Antarafacial process

The Diels–Alder reaction proceeds suprafacially with respect to each component and may be designated [4s + 2s] reaction. Alternative styles of expressing this are: $[\pi_s^4 + \pi_s^2]$ and $[\pi 4_s + \pi 2_s]$.

For thermally initiated reactions concerted s,s or a,a cycloadditions are allowed (preferred) when $(4n + 2)$ π-electrons participate and s,a additions when $4n$ π-electrons are involved. These rules are reversed for photochemically induced cycloadditions:

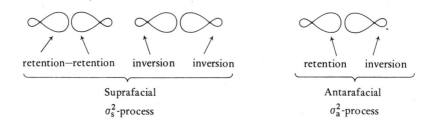

The terms antarafacial and suprafacial also apply to σ-bonds participating in pericyclic reactions. Reactions which proceed with either inversion or retention at both termini of the breaking σ-bond are suprafacial, those which proceed with retention at one terminus and inversion at the other are antarafacial:

A non-bonding p-orbital (designated ω) which participates in a pericyclic reaction can form bonds to both flanking groups from the same lobe (suprafacially) or opposite lobes (antarafacially)

The suprafacial participation of an empty p-orbital is designated $^{\omega}0_s$. *Sigmatropic rearrangements* are concerted thermally or photochemically induced reactions in which a σ-bond, flanked by one or more π-electron systems, migrates. Typical examples are:

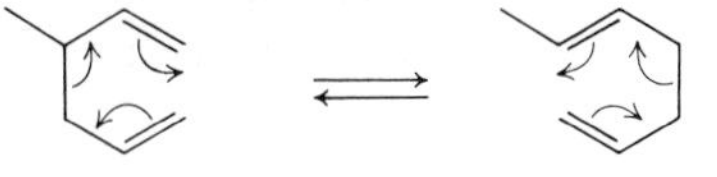

Cope rearrangement

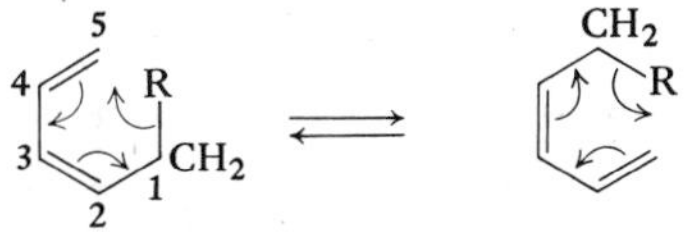

Hydrogen or alkyl shift

a 1,5-shift

Nomenclature. If the atoms at the termini of the migrating σ-bond are numbered, 1 and 1′ in the initial position and i and j in the final position, the reaction is said to be of the class [i,j]. Thus the Cope rearrangement is a [3,3] sigmatropic rearrangement:

$$
\begin{array}{ccc}
1' & 2' & 3' \\
CH_2 & -CH= & CH_2 \\
| & & \\
CH_2 & -CH= & CH_2 \\
1 & 2 & 3
\end{array}
\qquad
\begin{array}{ccc}
1' & 2' & 3' \\
CH_2= & CH- & CH_2 \\
 & & | \\
CH_2= & CH- & CH_2 \\
1 & 2 & 3
\end{array}
$$

The following reaction is a [1,5] sigmatropic rearrangement:

$$
\begin{array}{l}
1' \\
H \\
| \\
CH_2-CH=CH-CH=CH_2 \\
\;1\quad 2\quad 3\quad 4\quad 5
\end{array}
\longrightarrow
\begin{array}{l}
1' \\
H \\
| \\
CH_2=CH-CH=CH-CH_2 \\
\;1\quad 2\quad 3\quad 4\quad 5
\end{array}
$$

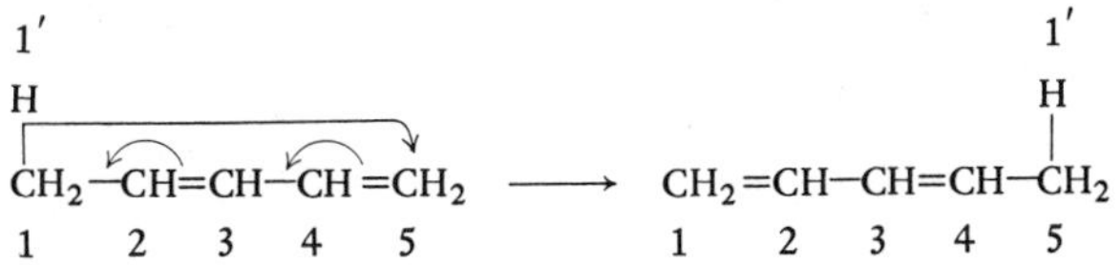

Such rearrangements may involve either a suprafacial or antarafacial process

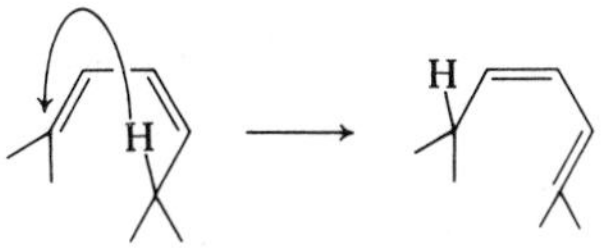

suprafacial

antarafacial

The selection rules for sigmatropic reactions involving neutral molecules are given below:

Order of reaction i + j	Method of initiation	
	Thermal	Photochemical
4n	a + s	s + s
	s + a	a + a
4n + 2	s + s	a + s
	a + a	s + a

For a discussion of sigmatropic reactions involving charged species, see Gilchrist and Storr, op. cit., p. 215.

Chelotropic reactions are those reactions in which two σ-bonds which terminate at a single atom are made or broken in concert. Such reactions are a special case of cycloaddition reactions.

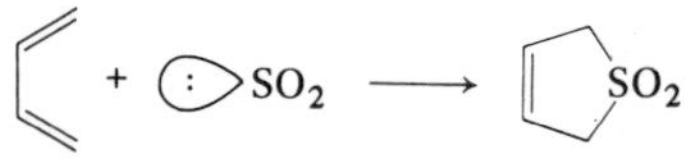

The addition of sulphur dioxide to butadiene is comparable to the Diels–Alder reaction, the lone pair on sulphur replacing the π-electrons of the dieneophile.

Extrusion of carbon monoxide often occurs readily, especially when an aromatic product is formed:

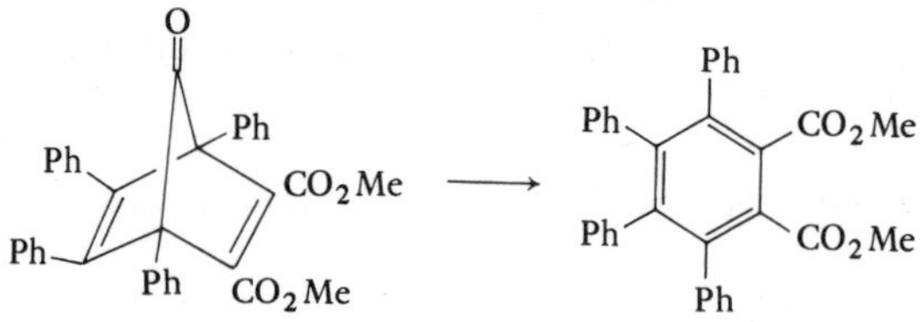

(a) Generalised Woodward–Hoffmann rule for pericyclic reaction

A thermally induced pericyclic reaction in a neutral molecule is allowed if the total number of suprafacial regroupings of the participating bonds is an odd number. For a photochemically induced pericyclic reaction to be allowed, the number of suprafacial regroupings must be even (J. B. Hendrickson, D. J. Cram, G. S. Hammond, "Organic Chemistry", 3rd Edn., McGraw Hill, New York, 1970, p. 871).

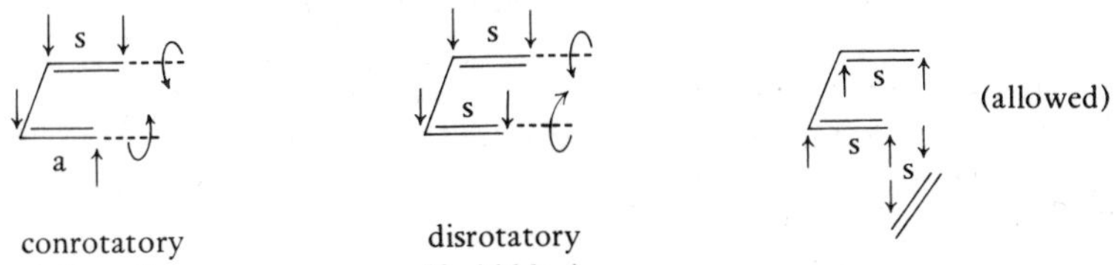

conrotatory
(allowed)

disrotatory
(forbidden)

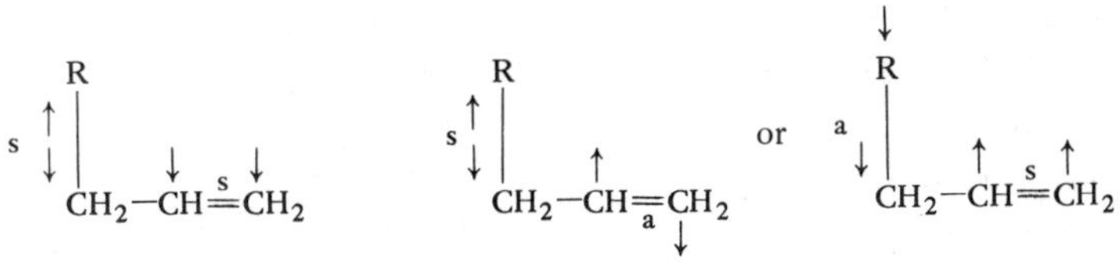

(allowed)

forbidden

permitted

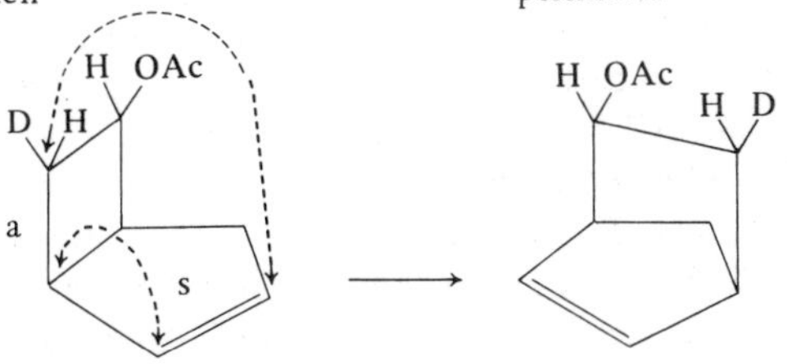

The above [1,3] sigmatropic rearrangement occurs by the allowed route, rearrangement occurs with inversion at the migrating group.

(b) Theoretical basis of Woodward–Hoffmann rules

The rationalisation of the Woodward–Hoffmann rules may be made:

(a) on the basis of conservation of orbital symmetry between the reactants and the products (Woodward and Hoffmann, op. cit);

(b) by considering the symmetry of the highest occupied molecular orbital (HOMO) of one reactant and of the lowest unoccupied molecular orbital (LUMO) of the other (frontier orbital analysis), (K. Fukui, Accts. chem. Res., 1971, 4: 57);

(c) using the aromatic transition state concept in which $(4n + 2)$ π-electrons constitute a Hückel state and a $4n$ π-system constitutes a Mobius system (a Mobius polyene forms a continuous orbital system with a sign discontinuity and a Hückel system two separate π-clouds of the normal cyclic π-system), (M. J. S. Dewar, Angew. Chem., Intern. Edn., 1971, 10: 761; H. E. Zimmerman, Accts. chem. Res., 1971, 4: 272).

2. Stereochemistry and conformational analysis

Conformational or topographical analysis of alicyclic systems has been reviewed by V. Prelog (Pure and Applied Chem., 1971, 25: 465). Other leading review articles are: "Conformational Analysis", ed. G. Chiurdoglu, Academic Press, London, 1971; "Stereochemistry of Many-membered Rings", by J. Sicher in "Progress in Stereochemistry", eds. P. B. D. de la Mare and W. Klyne, Butterworths, London, 1962; "Conformations of Medium Rings", by J. D. Dunnitz in "Perspectives in Structural Chemistry", Vol. 2, 1968, p. 1; "Conformational Equilibria and n.m.r. studies", J. D. Roberts, Chem. in Britain, 1966, 2: 529; Study of Ring Inversion by n.m.r., J. E. Anderson, Quart. Review, 1965, 19: 426. Conformational Analysis, E. L. Eliel, Angew. Chem., Intern. Edn., 1965, 4: 761; "Catenanes, Rotenanes and Knots", G. Schill, Academic Press, New York, 1971.

It has been found possible to calculate the geometries and energies of a wide variety of hydrocarbons by a semi-empirical method referred to as "molecular mechanics" or the "Westheimer Method" (F. H. Westheimer in "Steric Effects in Organic Chemistry", ed. M. S. Newman, Wiley, New York, 1956; S. E. Williams, P. J. Strang and P. von Schleyer, Ann. rev. Phys. Chem., 1968, 19: 531). A review of ab initio calculations in this field has been given by J.-M. Lehn in Conformational Analysis, ed. G. Chiurdoglu (loc. cit.). An illustration of the accuracy of such calculations is shown in the following table which compares the calculated and the observed thermodynamic data for perhydroanthracenes (see N. L. Allinger and M. T. Wuesthoff, J. Org. Chem., 1971, 36: 2051. Allinger, M. T. Trebble, M. A. Miller and D. H. Wertz, J. Amer. Chem. Soc., 1971, 93: 1637; Allinger and Wu, Tetrahedron, 1971, 27: 5093).

THERMODYNAMIC DATA FOR PERHYDROANTHRACENES

	Experimental				Calculated			
Isomer	ΔH°	ΔS°	$T\Delta S^\circ$	ΔG°	ΔH°	ΔS°	$T\Delta S^\circ$	ΔG° (544°K)
tst	0	0	0	0	0	0	0	0
ct	+2.76	+2.1	+1.4	+1.62	+2.62	+2.8	+1.52	+1.10
tat	+4.15	−1.6	−0.87	+5.02	+5.86	0	0	+5.86
cac	+5.58	+0.3	+0.16	+5.42	+5.56	+1.4	+0.76	+4.80
csc	+8.74	+4.0	+2.18	+6.56	+8.13	+2.2	+1.20	+6.93

Chapter 2

The Cyclopropane Group

P. H. BOYLE

1. Preparative methods

The past few years have seen many important advances in our understanding of the chemistry of the cyclopropane group of compounds. In this introductory section on preparative methods, new approaches to the construction of the three-membered ring are reviewed. New synthetic methods for individual compounds, and preparative methods which utilise cyclopropane precursors, are described in later sections under the appropriate functional group heading.

(a) Methods based on 1,3-cyclisation

A variety of cyclopropane compounds may now be synthesised by the use of sulphonium ylids, which have been developed into extremely versatile and useful reagents for this purpose (S. R. Landor and N. Punja, J. chem. Soc., C, 1967, 2495). When the ylid is treated with an α,β-unsaturated ester, aldehyde, ketone or nitrile, its nucleophilic carbon atom adds to the β-carbon atom of the Michael acceptor, and the cyclopropane ring is formed by 1,3-elimination of sulphide from the intermediate. The stereochemistry of the product is usually *trans*:

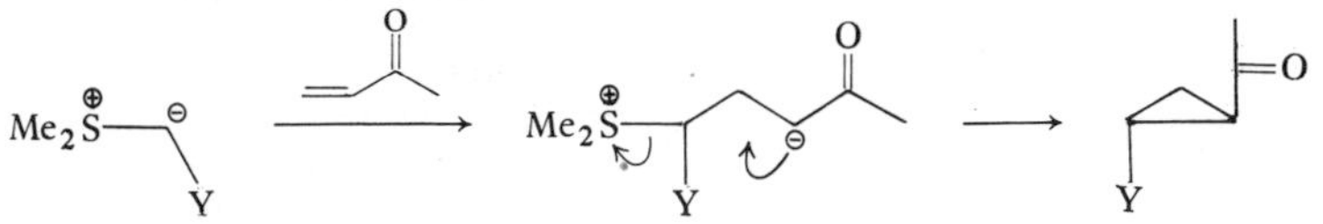

Y = CO$_2$R, CO·R

The electron-attracting group Y reduces the nucleophilicity of the ylid carbon atom and thus prevents attack on the carbonyl group of the Michael acceptor. It also stabilises the ylid, which in some cases may be isolated e.g.

[9]

$Me_2\overset{\oplus}{S}-\overset{\ominus}{CH}\cdot CO_2Et$, $Me_2\overset{\oplus}{S}-\overset{\ominus}{CH}\cdot CO\cdot Ph$ (J. Adams, L. Hoffman Jr. and B. M. Trost, J. org. Chem., 1970, 35: 1600; A. W. Johnson and R. T. Amel, ibid., 1969, 34: 1240; G. B. Payne, ibid., 1967, 32: 3351). The sulphoxonium ylid $PhSO(NMe_2)-\overset{\ominus}{CH}_2$ has been obtained in optically active form, giving optically active cyclopropanes on addition to the double bond of Michael acceptors (C. R. Johnson and C. W. Schroeck, J. Amer. chem. Soc., 1968, 90: 6852). Dimethylsulphoxonium methylid, $Me_2\overset{\oplus}{S}O-\overset{\ominus}{CH}_2$, is an effective reagent for the conversion of conjugated nitro-olefins into substituted nitrocyclopropanes (J. Asunskis and H. Shechter, J. org. Chem., 1968, 33: 1164) while dimethylsulphonium methylid reacts with α,β-unsaturated sulphonic acid esters or amides to give the corresponding cyclopropanesulphonates or cyclopropanesulphonamides (W. E. Truce and C. T. Goralski, ibid., 1968, 33: 3849). A direct approach to the *gem*-dimethylcyclopropane system is afforded by the reaction of diphenylsulphonium isopropylide, $Ph_2\overset{\oplus}{S}-\overset{\ominus}{C}Me_2$, with conjugated carbonyl compounds (E. J. Corey and M. Jautelat, J. Amer. chem. Soc., 1967, 89: 3912). A novel synthesis of cyclopropylcarbonyl compounds involves the treatment of an α-haloketone or ester with three moles of carbonyl-stabilised ylid. In this reaction two of the methylene groups of the new cyclopropane ring are derived from the ylid (P. Bravo et al., Tetrahedron, 1971, 27: 3563).

When the conjugate base of a C–H acid is added to a vinyl sulphonium compound, an ylid is formed which readily undergoes proton migration followed by 1,3-elimination. The resulting cyclopropane derivative is usually the *trans*-isomer (J. Gosselck and G. Schmidt, Tetrahedron Letters, 1969, 2623, 2615):

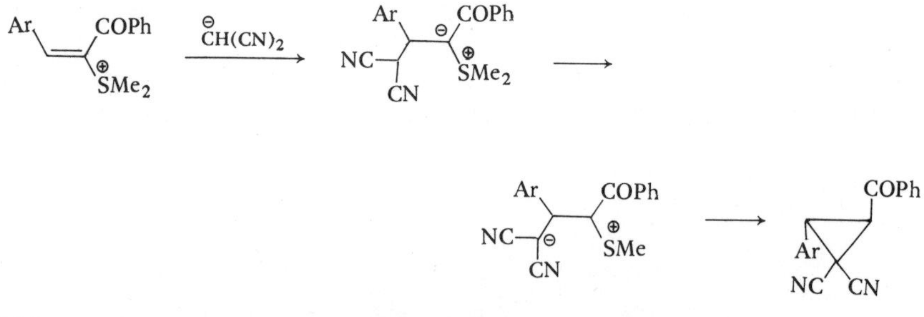

This reaction can be compared with a new method which has been described for the preparation of cyclopropane carboxylic acid esters. Michael addition of an active methylene group compound to ethyl 2-(*O,O*-diethylphosphonooxy)acrylate is followed by 1,3-elimination, and here too the thermodynamically more stable *trans* product is formed preferentially (U. Schmidt, R. Schröer and A. Hochrainer, Ann., 1970, 733: 180):

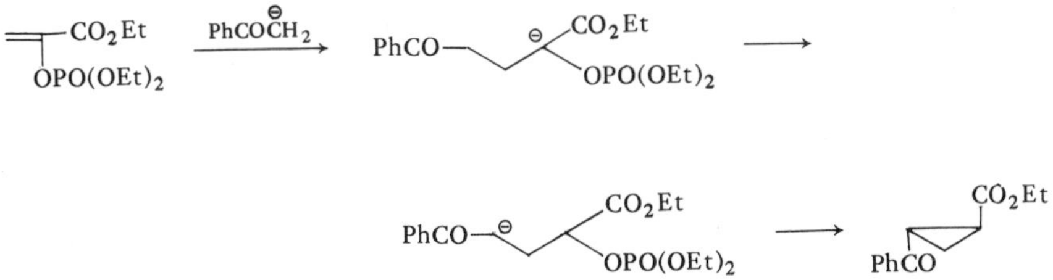

J. Gosselck and co-workers have exploited the elimination of the sulphonium group in a general route to cyclopropanes which are geminally substituted by electron-attracting groups (Angew. Chem., intern. Edn., 1968, 7: 456):

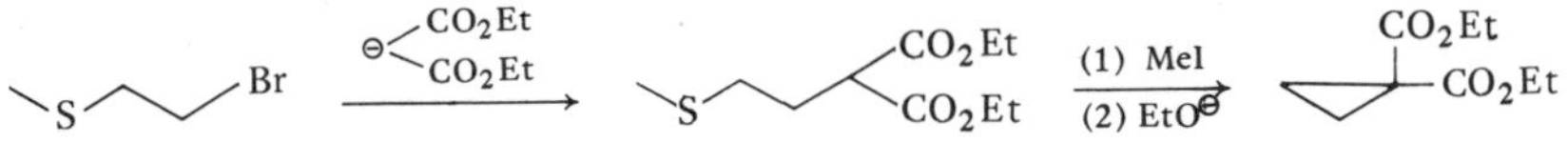

1,1-Disubstituted-2-methylcyclopropanes can also be readily prepared by this reaction sequence if $CH_3 \cdot S \cdot CH_2 \cdot CHBr \cdot CH_3$, obtainable by the addition of HBr to $CH_3 \cdot S \cdot CH_2 \cdot CH=CH_2$, is used as the starting material (Tetrahedron Letters, 1970, 2437).

An alternative approach to the synthesis of 1,1-dicyanocyclopropanes utilises the free-radical addition of bromomalononitrile to an alkene, followed by 1,3-elimination of hydrogen bromide under the influence of base (P. Boldt, L. Schulz and J. Etzemüller, Ber., 1967, 100: 1281):

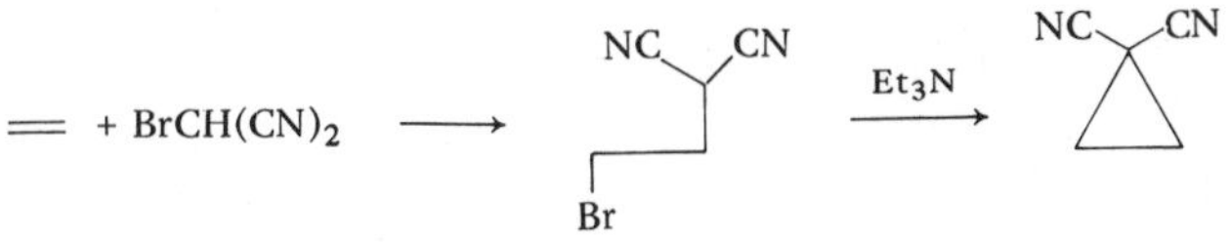

A similar scheme affords 1,1,2,2-tetracyanocyclopropanes or 1,2,2-tricyano-cyclopropanecarboxylic acid esters when bromomalononitrile is added to an ylidenemalononitrile or an ylidenecyanoacetate, respectively (Y. C. Kim and H. Hart, J. chem. Soc., C, 1969, 2409).

γ-Haloalkynes undergo 1,3-elimination when treated with lithium diethyl-amide in tetrahydrofuran and this offers a route to cyclopropyl acetylenes (J. K. Crandall and D. J. Keyton, Chem. Comm., 1968, 1069).

H. C. Brown and S. P. Rhodes describe a useful new entry into the cyclopropane series using 9-borabicyclo[3,3,1]nonane (9-BBN). This reagent readily adds to the double bond of allyl chlorides, and facile cyclisation of the adduct occurs on treatment with base (J. Amer. chem.

Soc., 1969, 91: 2149). The advantage of 9-BBN lies in the high selectivity with which it adds to the double bond in the required direction:

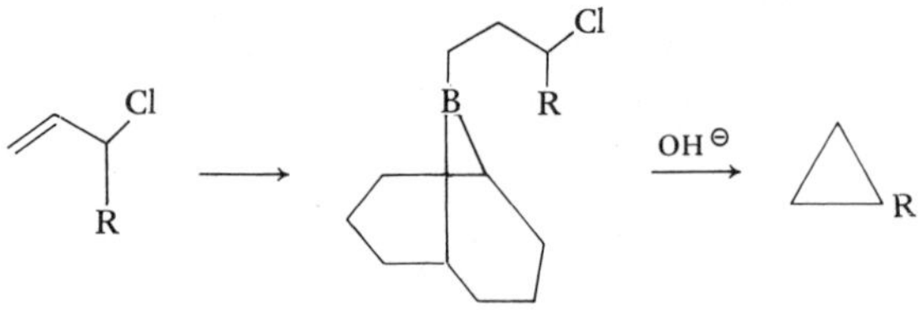

If two moles of 9-BBN are added to α-bromoacetylene, sodium hydroxide treatment of the adduct gives a cyclopropyl borane which can be readily oxidised to cyclopropanol in 65% yield (idem, ibid., 1969, 91: 4306).

A possible alternative to the classical Wurtz coupling reaction is reported by J. K. Kochi and D. M. Singleton who found that 1,3-dihalides and 1,3-halotosylates are reductively cyclised to cyclopropanes in excellent yield using ethylenediaminechromium(II) reagent in N,N-dimethylformamide at room temperature (J. org. Chem., 1968, 33: 1027). 1,3-Dibromides have also been cyclised to cyclopropanes by electrolytic reduction (R. Gerdil, Helv., 1970, 53: 2100; M. R. Rifi, J. Amer. chem. Soc., 1967, 89: 4442), while 1,3-diiodopropane was quantitatively converted into cyclopropane in a radical induced 1,3-elimination promoted by benzoyl or tert-butyl peroxide (L. Kaplan, ibid., 1967, 89: 1753).

Molecules possessing the di-π-methane moiety, in which two π systems are bonded to a single sp³ carbon, undergo a general photochemical trans-formation to vinylcyclopropanes. The direction of the rearrangement follows the demand for maximum electron delocalisation in the intermediates and proceeds with retention of the stereochemistry of the double bonds (H. E. Zimmerman et al., J. Amer. chem. Soc., 1971, 93: 3646; 1970, 92: 6267, 6259):

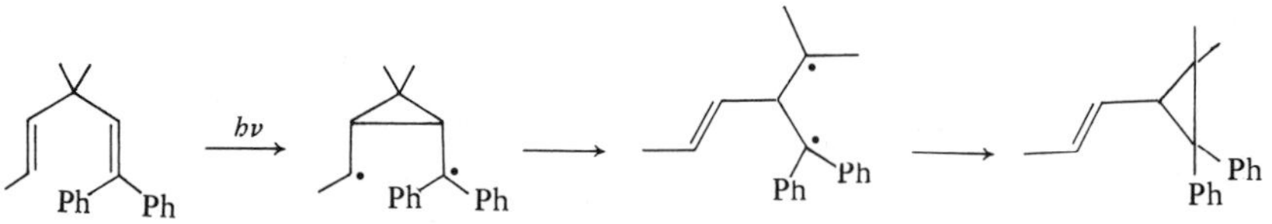

(b) Methods based on ring contraction

The formation of a cyclopropane ring by the elimination of a neutral molecule from a higher membered ring remains a very important technique with pyrazolines, but no analogous methods of comparable utility have emerged for other ring systems. Photolytic elimination of carbon monoxide from cyclobutanones may in certain cases lead to cyclopropane formation. Thus H. U. Hostettler reported that photolysis of 2,2,4,4-tetramethylcyclo-butanones carrying a variety of substituents in the 3-position gave cyclo-propane products, in addition to tetrahydrofuran derivatives (Helv., 1966, 49: 2417):

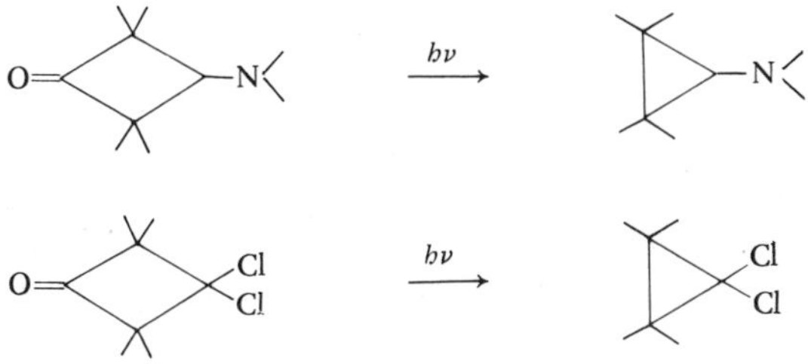

N. J. Turro and W. B. Hammond examined the photolysis of cyclobutane-1,3-diones as a possible general route to cyclopropanones but found it to be of limited value (Tetrahedron, 1968, 24: 6017). Thietans are not good substrates for desulphurisation. Thietan sulphones, however, lose sulphur dioxide on either photolysis (A. Padwa and R. Gruber, J. org. Chem., 1970, 35: 1781) or pyrolysis, affording cyclopropanes in a non-stereospecific re-action. The pyrolysis product also contains isomeric olefins. In contrast, the decomposition of thietanonium salts with butyl lithium proceeds with high stereospecificity, giving lower yields of cyclopropanes but with fewer olefinic contaminants (B. M. Trost et al., J. Amer. chem. Soc., 1971, 93: 676; 1969, 91: 4320):

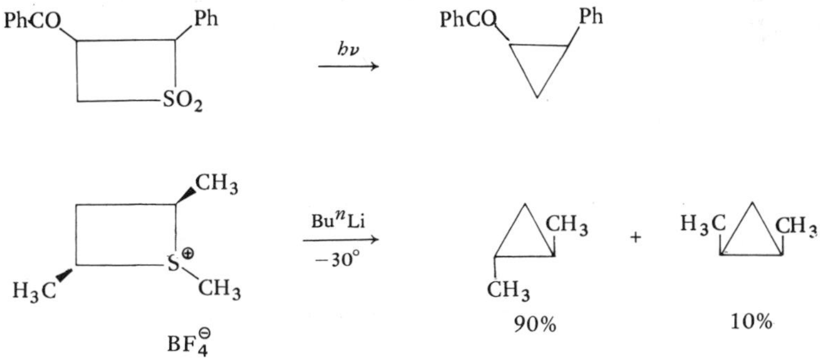

M. Franck-Neumann has described a general synthesis of gem-dimethylcyclo-propanes by the addition of 2-diazopropane to electrophilic alkenes, followed by benzophenone-sensitised photolytic decomposition of the resulting pyrazoline (Angew. Chem., intern. Edn., 1968, 7: 65). 2-Diazopropane also adds to electrophilic alkynes in a useful new synthesis of cyclopropenes carrying an electron-attracting group attached to the double bond (Franck-Neumann and C. Buchecker, Tetrahedron Letters, 1969, 15). Trifluoromethylcyclopropanes are readily formed when 2,2,2-trifluorodiazoethane is photolysed with suitable alkenes, an interesting reaction in that it involves facile diazo addition to a non-activated double bond (J. H. Atherton and R. Fields, J. chem. Soc., C, 1968, 1507).

Favorskii rearrangement of 2-halocyclobutanones takes place readily on treatment of the latter with a variety of nucleophiles, and the corresponding cyclopropane carboxylic acid derivatives are obtained in good yield. A semibenzilic mechanism is suggested (J.-L. Ripoll, Ann. Chim. (France), 1967, 233). J. Ciabattoni and A. E. Feiring report a novel ring-contraction reaction which occurs when 1,2,3-tri-*tert*-butyl-3,4-dichlorocyclobutene is solvolysed in aqueous dioxane containing sodium bicarbonate. 1,2-Di-*tert*-butyl-3-pivaloylcyclopropene is formed in nearly quantitative yield and the following mechanism is suggested (J. Amer. chem. Soc., 1972, 94: 5113):

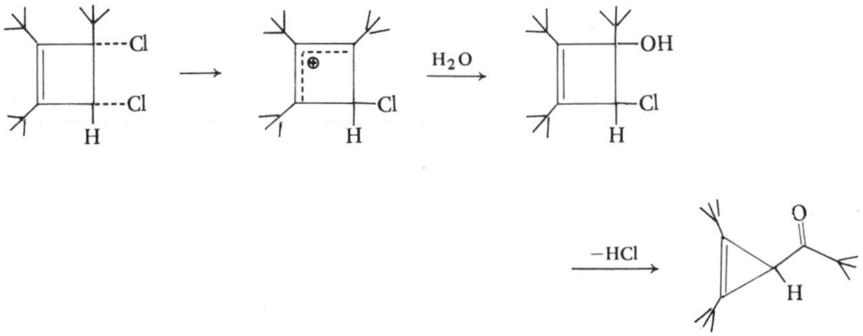

(c) Methods based on carbene addition

(i) Halocarbenes

The most important method of preparation of halocyclopropanes is by the addition of a halocarbene to an olefin, and Seyferth has extensively pioneered the use of mercurials as effective divalent carbon-transfer agents. Some of the reagents which he has developed are shown in Table 1 (D. Seyferth et al., J. org. Chem., 1970, 35: 1297; J. organometal. Chem., 1970, 25: 293; J. Amer. chem. Soc., 1971, 93: 3714; 1969, 91: 6536).

When an olefin is treated with one of these reagents in a suitable solvent, a halocyclopropane is formed by carbene addition to the double bond.

Table 1

MERCURIALS AS DIVALENT CARBON-TRANSFER AGENTS

Mercurial reagent	Species transferred	Mercurial reagent	Species transferred
$PhHgCF_3$	CF_2		
$PhHgCCl_2F$	$CFCl$	$PhHgCBr_2H$	$CHBr$
$PhHgCCl_2Br$	CCl_2	$PhHgCCl_2Ph$	$CPhCl$
$PhHgCCl_3$	CCl_2	$PhHgCCl_2CO_2Me$	$CClCO_2Me$
$PhHgCClBr_2$	$CClBr$	$PhHgCBr_2CO_2Me$	$CBrCO_2Me$
$PhHgCBr_3$	CBr_2	$PhHgCClBrCF_3$	$CClCF_3$
$PhHgCCl_2H$	$CHCl$	$PhHgCCl_2\overset{O}{\underset{O}{C}}R$	$CCl\overset{O}{\underset{O}{C}}R$
$PhHgCClBrH$	$CHCl$	(R = H, Me, Ph)	

Carbene release from some of the mercurials may be slow but it can be greatly accelerated by carrying out the reaction in presence of sodium iodide. The iodide ion displaces $\overset{\ominus}{C}X_3$ from the mercurial and the $\overset{\ominus}{C}X_3$ anion then readily generates the carbene by loss of $\overset{\ominus}{X}$ (Seyferth et al., loc. cit.; J. Amer. chem. Soc., 1967, 89: 959).

Difluorocarbene has been known for some years but it is only recently that convenient methods for its generation have become available. Phenyl-(trifluoromethyl)mercury, listed in Table 1, is easily prepared, and in presence of sodium iodide it releases $\overset{..}{C}F_2$ under mild neutral conditions. Seyferth had earlier reported that the action of sodium iodide on trimethyl-(trifluoromethyl)tin, Me_3SnCF_3, in 1,2-dimethoxyethane, also affords an extremely mild source of $\overset{..}{C}F_2$ (J. org. Chem., 1967, 32: 2980). Another useful new reagent for effecting the transfer of difluoromethylene is hexafluoropropylene oxide (P. B. Sargeant, J. org. Chem., 1970, 35: 678).

A convenient procedure has been described for the generation of dichloro-carbene from chloroform, aqueous sodium hydroxide, and a catalytic amount of triethylbenzylammonium chloride. This method allowed the preparation of dichlorocyclopropanes from olefins which yielded little or no product with potassium tertiary butoxide and chloroform (E. V. Dehmlow and J. Schönefeld, Ann., 1971, 744: 42; M. Makosza and W. Wawrzyniewicz, Tetrahedron Letters, 1969, 4659). A quaternary ammonium salt is also used as catalyst in another simple preparation of 1,1-dihalocyclopropanes in which an olefin is treated with trihalomethane and ethylene oxide in presence of tetraethylammonium bromide (P. Weyerstahl et al., Ber., 1967, 100: 1858). Depending on the degree of substitution of the double bond, however, ring-opened products sometimes predominate in this reaction (F. Nerdel et al., Ann., 1971, 746: 6).

(*ii*) *Other carbenoid reactions*

One of the most widely used methods for the methylenation of olefins utilises the Simmons—Smith reaction. A modification of this has been reported by R. J. Rawson and I. T. Harrison, who found that the use of a mixture of zinc dust and cuprous halide obviates the necessity for separate preparation of the zinc—copper couple (J. org. Chem., 1970, 35: 2057). Another modification replaces the zinc—copper couple with diethylzinc or diethylcadmium, either of which in the presence of methylene iodide gives good yields of cyclopropanes from olefins, in a stereospecific addition of methylene (J. Furukawa et al., Tetrahedron, 1970, 26: 243; 1968, 24: 53). The use of ethylidene iodide in place of methylene iodide allows the preparation of methylcyclopropanes (idem, ibid., 1969, 25: 2647).

Asymmetric Simmons—Smith synthesis has been achieved by treatment of the (−)menthyl esters of α,β- and β,γ-unsaturated carboxylic acids with methylene iodide and a zinc—copper couple (O. Cervinka and O. Kriz, Z. Chem., 1971, 11: 63; S. Sawada, K. Takehana and Y. Inouye, J. org. Chem., 1968, 33: 1767). Achiral olefins afforded optically active cyclopropanes when the Simmons—Smith reaction was carried out in presence of free (−)-menthol (S. Sawada, J. Oda and Y. Inouye, ibid., 1968, 33: 2141). The optical purity of the products in both these reactions was low. Methylenation using diethylzinc and methylene iodide yielded optically active cyclopropanes when carried out in presence of L-leucine (J. Furukawa, N. Kawabata and J. Nishimura, Tetrahedron Letters, 1968, 3495).

A general method for the preparation of aryl cyclopropanes from olefins is based on the zinc halide catalysed carbenoid decomposition of aryldiazomethanes. This reaction is particularly valuable for the preparation of the thermodynamically less stable *syn*-isomers, which are the predominant products (S. H. Goh, L. E. Closs and G. L. Closs, J. org. Chem., 1969, 34: 25).

Cyclopropanes have been formed by the methylenation of olefins by treatment with methylene halide and an organolithium reagent. Yields are modest but the reaction is of importance theoretically. Evidence indicates that the divalent carbon species involved here is a carbenoid, i.e. a complexed rather than a free carbene (L. Friedman, R. J. Honour and J. G. Berger, J. Amer. chem. Soc., 1970, 92: 4640; U. Burger and R. Huisgen, Tetrahedron Letters, 1970, 3049):

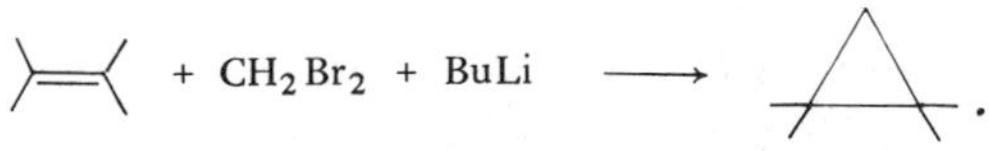

Disubstituted methylenecyclopropanes may be prepared by treatment of 5,5-dialkyl-*N*-nitrosooxazolidones with lithium alkoxides in presence of an olefin. The reaction proceeds by formation of an unsaturated carbene, dialkyl ethylidene (M. S. Newman and T. B. Patrick, J. Amer. chem. Soc., 1969, 91: 6461):

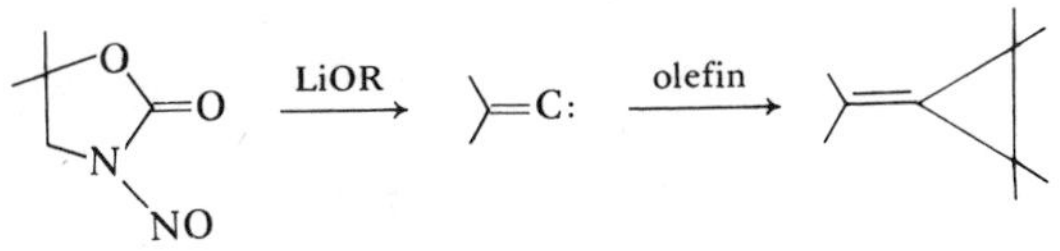

Cyclopropanols can be prepared from olefins by a very useful new general method which initially involves the reaction of the olefin with dichloro-methyl β-chloroethyl ether and methyl-lithium. A β-chloroethyl cyclopropyl ether is formed which may be readily converted into the corresponding cyclopropanol (U. Schöllkopf et al., Ber., 1966, 99: 3391):

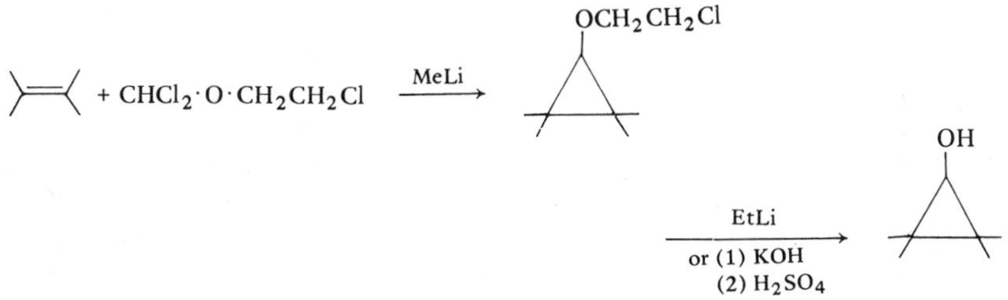

A novel technique for the conversion of 1-alkynes into cyclopropyl deriva-tives has been reported by G. Zweifel et al. (J. Amer. chem. Soc., 1971, 93: 1305). The 1-alkyne is treated with a dialkylaluminium hydride, forming a 1-alkenylalane which is converted into a cyclopropylalane using the Simmons–Smith reaction. The cyclopropylalane may be hydrolysed to give an alkyl cyclopropane or halogenated to give a trans-1-halo-2-alkylcyclo-propane:

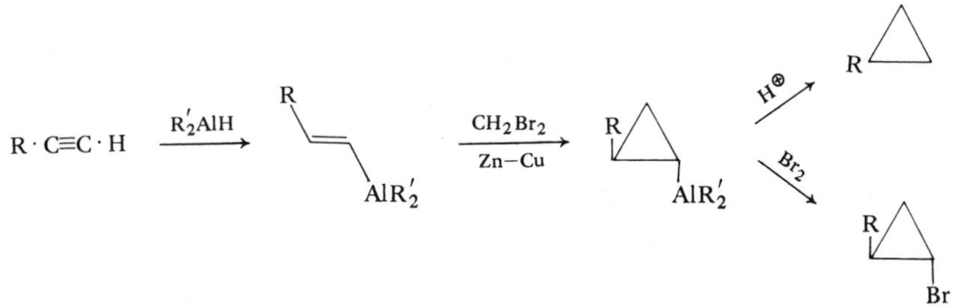

Cyclopropenes are formed by photolysis of the sodium salts of the tosylhydrazones of α,β-unsaturated ketones (H. Dürr, Angew. Chem., intern. Edn., 1967, 6: 1084).

Two unusual diazo-compounds, which are precursors of new carbenes, have been described. *Diazoacetaldehyde* is a yellow liquid which detonates violently if overheated. It decomposes smoothly to give formyl carbene, however, when treated with copper acetylacetonate (Z. Arnold, Chem. Comm., 1967, 299). *Ethyl diazoiodoacetate*, the first halogenodiazoalkane to be characterised, is a red oil and generates ethoxycarbonyliodocarbene on photolysis (F. Gerhart, U. Schöllkopf and H. Schumacher, Angew. Chem., intern. Edn., 1967, 6: 74).

2. Saturated hydrocarbons

The susceptibility of the cyclopropane ring to attack by electrophiles has been known for a long time but it is only during the past few years that the intermediacy of the protonated cyclopropane species has been recognised. Evidence for the existence of protonated cyclopropanes has been derived largely from isotope scrambling experiments, where the observed scrambling pattern is such as to exclude any mechanism invoking solely classical carbonium ions. Protonated cyclopropanes have now been implicated as intermediates in reactions not only on cyclopropyl substrates, when treated with acidic or electrophilic reagents, but also on open-chain substrates such as 1-amino-, 1-chloro-, or 1-hydroxy-propane (see G. J. Karabatsos, C. Zioudrou and S. Meyerson, J. Amer. chem. Soc., 1970, 92: 5996; C. C. Lee and D. J. Woodcock, ibid., 1970, 92: 5992; N. C. Deno et al., ibid., 1970, 92: 3700, for recent refs.; and C. J. Collins, Chem. Reviews, 1969, 69: 543; C. C. Lee in "Progress in Physical Organic Chemistry", Vol. 7, A Streitwieser Jr. and R. W. Taft (eds.), Interscience, New York, 1970, p. 129, for useful reviews).

Three common representations of the structure of the protonated cyclopropane cation are the face-, corner- (or methyl-bridged), and edge-protonated species:

The face-protonated structure may be excluded from further consideration both on the basis of incompatibility with empirical data and also on theoretical considerations. It is more difficult to decide between the corner- and edge-protonated systems. Many experimental results can be explained by invoking either of these species as intermediates and they are both of comparable stability. Most experimental data in the literature have been discussed in terms of edge-protonated cyclopropanes which are equilibrating via the corner-protonated species. The latter may occur either as a transition state or as a discrete intermediate between two edge-protonated species. This is illustrated below for deuterated cyclopropane:

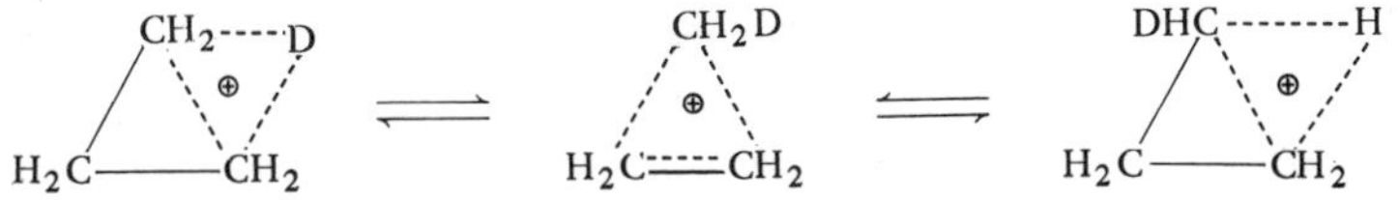

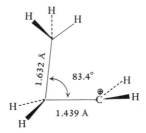

However, J. A. Pople, P. von R. Schleyer and co-workers (J. Amer. chem. Soc., 1971, 93: 1813) reported, on the basis of new molecular orbital calculations, that neither of these structures occurs as an energy minimum and they describe a new species of highly unusual structure as a probable intermediate. It is the methyl-eclipsed 1-propyl cation, and is predicted to be the most stable form of $C_3H_7^\oplus$, after the 2-propyl cation, which is the most stable:

Not surprisingly, in view of the above discussion, it has now been conclusively shown that the cyclopropane ring can act as a weak proton acceptor in hydrogen bonding. This may be intramolecular as in cyclopropyl-carbinols, which may show two O—H stretching frequencies in the infrared (L. Joris, Schleyer and R. Gleiter, J. Amer. chem. Soc., 1968, 90: 327; M. Oki et al., Bull. chem. Soc., Japan, 1969, 42: 1986; see also T. Suga et al., ibid., 1971, 44: 204), or intermolecular, as seen in the infrared spectra of p-fluorophenol in carbon tetrachloride solution in presence of 1,1-dimethyl-cyclopropane or dicyclopropylmethane (Joris et al., loc. cit.). In agreement with the studies on protonated cyclopropanes, the preferred site for proton—donor interaction in hydrogen bonding is the edge of the cyclopropane ring.

In n.m.r. spectra of cyclopropanes, an alkyl substituent shields the cis-vicinal cyclopropyl proton by about 0.25 p.p.m. The trans-vicinal proton is deshielded, but to a lesser extent. The cis-proton coupling constants in cyclopropanes are generally larger than the *trans*-proton coupling constants (D. T. Longone and A. H. Miller, Chem. Comm., 1967, 447). G. E. Schenck and F. A. L. Anet report that cyclopropane derivatives cause a downfield solvent shift in the n.m.r., opposite in sign to that caused by benzene. They relate this to the anisotropy of cyclopropane both in its magnetic susceptibility and in its basicity (Tetrahedron Letters, 1971, 2779).

3. Unsaturated hydrocarbons

Several convenient methods have been reported for the preparation of simple cyclopropenes. For example, *cyclopropene* itself may be very easily made by the low-pressure Hofmann elimination of *N,N*-dimethylcyclopropylamine (D. A. Archer, J. chem. Soc., C, 1971, 1327). 1-*Methylcyclopropene* is readily available by aqueous treatment of 1-lithio-2-methylcyclopropene, easily prepared by reaction of methallyl chloride with specially purified phenyllithium. The lithiocyclopropene can be stored for up to three months at low temperatures and serves as an instant source of 1-methylcyclopropene (R. M. Magid, T. C. Clarke and C. D. Duncan, J. org. Chem., 1971, 36: 1320):

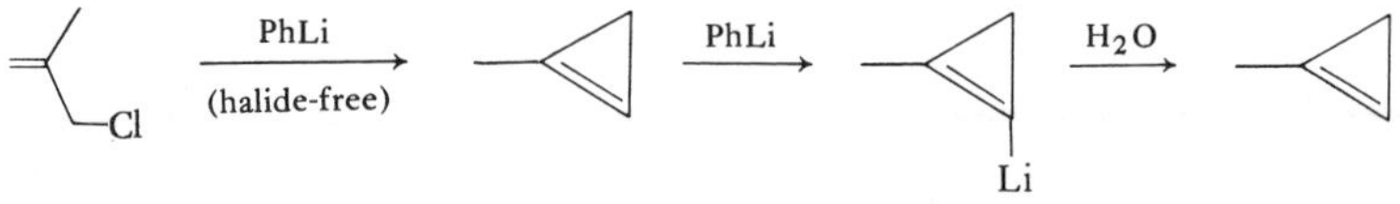

In an improved method for the preparation of 3-*methylcyclopropene*, crotylchloride is treated with lithium amide in boiling dioxane (R. Köster, S. Arora and P. Binger, Angew. Chem. intern. Edn., 1970, 9: 810). The first preparation of a 1,2-*diarylcyclopropene* was reported by D. T. Longone and D. M. Stehouwer (Tetrahedron Letters, 1970, 1017) who reduced 1,2-diphenylcyclopropenylium fluoroborate with sodium borohydride. The success of the procedure demands inverse addition because the initially formed 1,2-diphenylcyclopropene readily reacts with the starting cyclopropenylium salt to give 1,2,4,5-tetraphenylbenzene.

Molecular-orbital calculations show that in cyclopropene the vinylic protons are more acidic than the methylene protons (D. T. Clarke, Chem. Comm., 1969, 637). They are also more acidic than normal vinyl protons. This property is clearly demonstrated by the preparation of 1,2-dideuteriocyclopropene by refluxing cyclopropene with potassium *tert*-butoxide in *tert*-butanol-*d* (E. A. Dorko and R. W. Mitchell, Tetrahedron Letters, 1968, 341) and also in the ready formation of 1-lithio-2-methylcyclopropene mentioned above (Magid et al., loc. cit.). Grignard reagents, however, add stereospecifically *cis* to the double bond of 1-methylcyclopropenes (M. Y. Lukina, T. Y. Rudashevskaya and O. A. Nesmeyanova, Doklady Akad. Nauk, S.S.S.R., 1970, 190: 1109; C.A., 1970, 72: 132113), and phenyllithium adds *cis* to the double bond of cyclopropene (Magid and J. G. Welch, J. Amer. chem. Soc., 1968, 90: 5211).

Bicyclic cyclopropylpyrazolines have been obtained by the cycloaddition of diazoalkanes to the double bond of cyclopropenes (M. Franck-Neumann and C. Buchecker, Tetrahedron Letters, 1969, 2659). The pyrazolines may decompose to bicyclobutanes on pyrolysis (M. I. Komendantov and R. R. Bekmukhametov, Zhur. org. Khim., 1971, 7: 423; C.A., 1971, 74: 141051). With alkylidene cyclopropenes, diazoalkanes add to the cyclopropene double bond rather than to the exocyclic double bond, giving ultimately ring-opened products (T. Eicher and E. von Angerer, Ber., 1970, 103: 339; R. S. Pyron and W. M. Jones, J. org. Chem., 1967, 32: 4048).

Methylenecyclopropane can be readily prepared from methallyl chloride by treatment with potassamide in boiling tetrahydrofuran (R. Köster, S. Arora and P. Binger, Angew. Chem., intern. Edn., 1969, 8: 205) or by the near quantitative base-catalysed isomerisation of 1-methylcyclopropene (Köster et al., loc. cit.; I. S. Krull and D. R. Arnold, Org. Prep. Proced., 1969, 1: 283). A route to other alkylidene cyclopropanes is opened up by the report that cyclopropyltriphenylphosphonium bromide undergoes Wittig olefination reactions upon treatment with strong non-hydroxylic base and carbonyl compounds (E. E. Schweizer, C. J. Berninger and J. G. Thompson, J. org. Chem., 1968, 33: 336).

2-Phenyl-1-methylenecyclopropane and 2,2-diphenyl-1-methylenecyclo-propane form five-membered cycloadducts when treated with tetracyano-ethylene (R. Noyori, N. Hayashi and M. Katô, J. Amer. chem. Soc., 1971, 93: 4948):

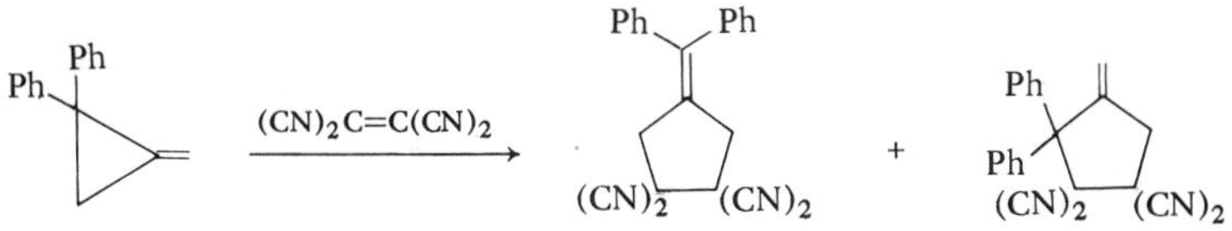

This reaction is interesting in that it represents, formally at any rate, a $[\sigma^2 + \pi^2 + \pi^2]$ cycloaddition. Tetracyanoethylene was the only olefin found to undergo this cycloaddition with the phenylmethylenecyclopropanes, how-ever, and an investigation of other methylenecyclopropanes when treated with 4-phenyl-1,2,4-triazoline-3,5-dione failed to reveal any further examples of the reaction (D. J. Pasto and A. Chen, Tetrahedron Letters, 1972, 2995).

With *trans*-2,3-dimethylmethylenecyclopropane, for example, a $[\pi^2 + \pi^2]$ cycloaddition was observed instead:

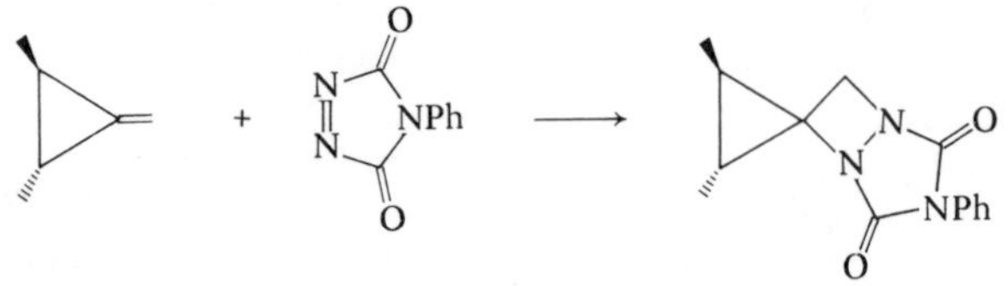

In contrast, alkenylidene cyclopropanes very readily undergo cycloaddition reactions with 4-phenyl-1,2,4-triazoline-3,5-dione (Pasto and Chen, J. Amer. chem. Soc., 1971, 93: 2562):

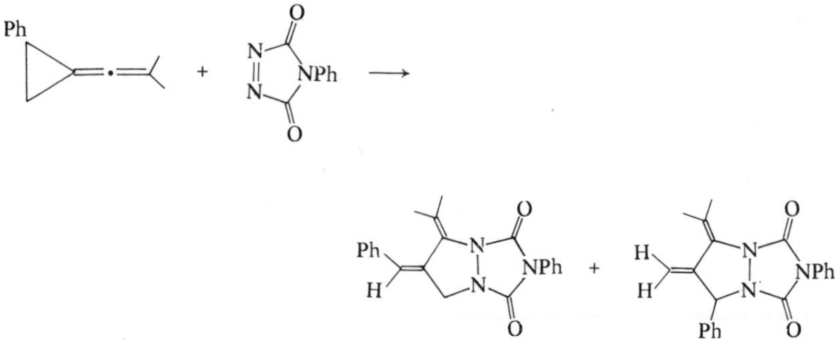

Noyori and co-workers have reported novel cycloaddition reactions of methylene cyclopropanes with olefins, which take place under the influence of a nickel(O) catalyst (ibid., 1972, 94: 4018; 1970, 92: 5780):

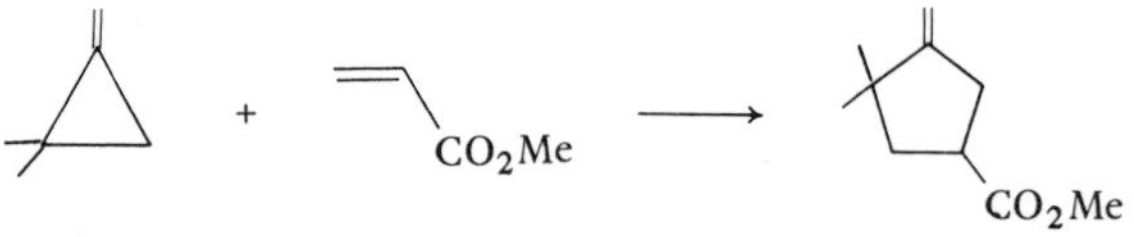

Methylenecyclopropane and its substituted derivatives undergo a thermally induced rearrangement, illustrated here for a simple case:

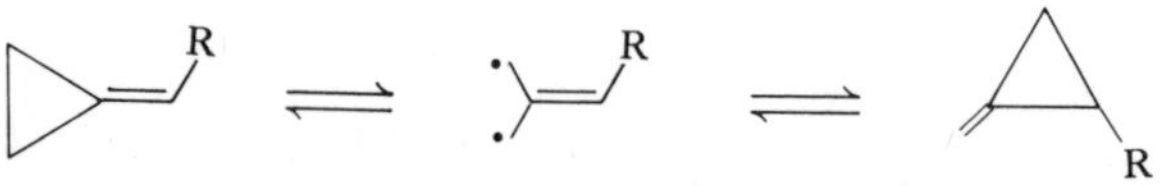

Optically active substrates can retain their optical activity during this rearrangement. Therefore the intermediate cannot be a planar delocalised diradical, and a diradical, with one methylene orthogonal to the plane of the remaining atoms has been suggested (see W. von E. Doering and H. D. Roth, Tetrahedron, 1970, 26: 2825, for references and for description of a new "pivot" mechanism):

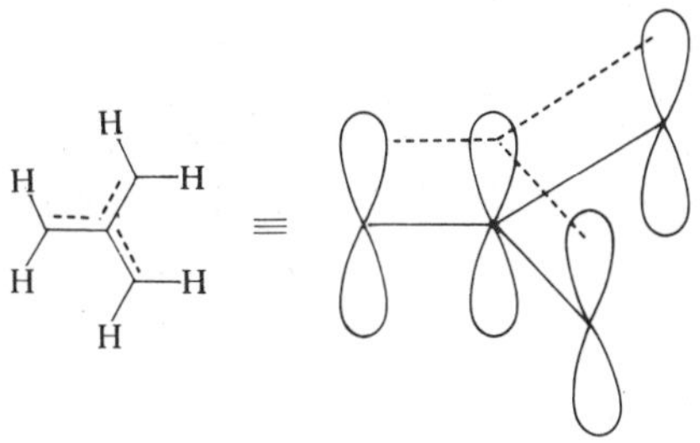

planar

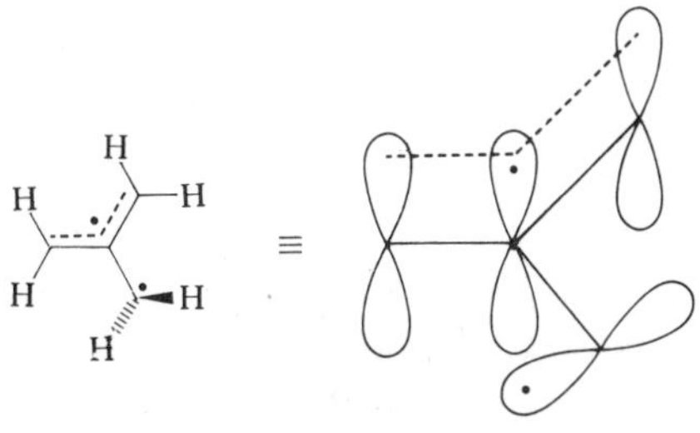

orthogonal

This conclusion is consistent with recent calculations reported by M. J. S. Dewar and J. S. Wasson (J. Amer. chem. Soc., 1971, 93: 3081) who showed that while triplet trimethylenemethane is most stable when planar, the ground state of the singlet form is the open shell, orthogonal structure.

In an analogous rearrangement alkenylidene cyclopropanes undergo smooth conversion to dimethylene cyclopropanes on pyrolysis (D. R. Paulson, J. K. Crandall and C. A. Bunnell, J. org. Chem., 1970, 35: 3708):

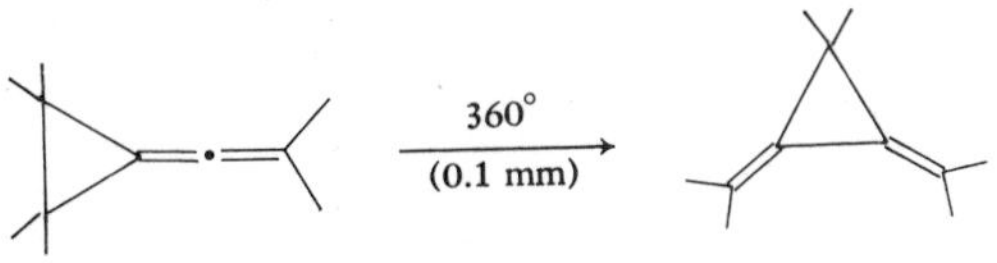

The parent compound, *dimethylenecyclopropane*, was prepared for the first time by this method (R. Bloch, P. Le Perchec and J.-M. Conia, Angew. Chem. intern. Edn., 1970, 9: 798).

Inconsistencies in earlier reports on the preparation and properties of *trimethylenemethane* have been resolved by K. H. Rhee and F. A. Miller (Spectrochim. Acta, 1971, 27A: 1). Raman and infrared spectra, and also electron-diffraction data, indicate a planar trigonal structure of D_{3h} symmetry, with a short C—C bond length and some electron delocalisation. Trimethylenemethane is extremely sensitive to oxygen but is remarkably stable to dilute mineral acids and to strong bases.

Methylenecyclopropene has not yet been synthesised. Stabilised derivatives are known however, in which the exocyclic negative charge can be delocalised over an aromatic ring (B. Föhlisch and P. Bürgle, Ann., 1967, 701: 67) or onto carbonyl groups (T. Eicher and N. Pelz, Ber., 1970, 103: 2647):

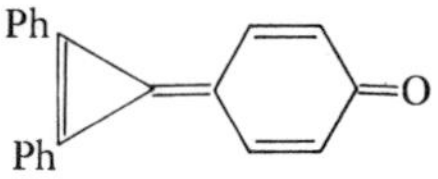

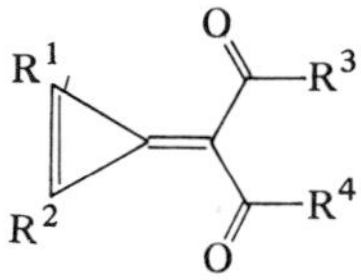

West and Zecher report the derivative shown below, which can be oxidised to a substituted trimethylenemethane of unprecedented stability, being a dark blue powder, stable in air to above 250° (J. Amer. chem. Soc., 1967, 89: 152):

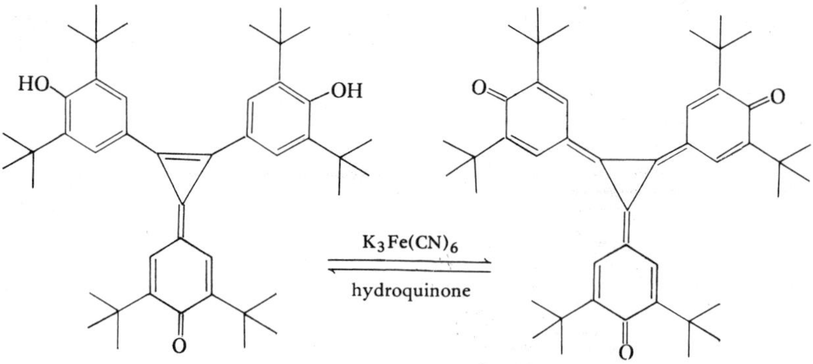

Vinylcyclopropanes are formed in the Wittig reaction using cyclopropyl-methyltriphenylphosphonium bromide. No ring-opened products have been observed, showing that no equilibration exists between the ylidic Wittig reagent and an acyclic isomer (E. E. Schweizer, J. G. Thompson and T. A. Ulrich, J. org. Chem., 1968, 33: 3082). This contrasts with earlier demonstrations that cyclopropylcarbinyl anions may exist in reversible equilibrium with the isomeric acyclic homoallylic carbanions (A. Maercker, Angew. Chem., intern. Edn., 1967., 6: 557).

Selective reduction of the double bond of a vinyl cyclopropane may be achieved by the use of diimide as reducing agent (J. B. Pierce and H. M. Walborsky, J. org. Chem., 1968, 33: 1962). Catalytic hydrogenation usually results in appreciable hydrogenolysis of the ring, although Raney nickel and tristriphenylphosphine rhodium chloride have been used with moderate success (M. T. Wuesthoff and B. Rickborn, J. org. Chem., 1968, 33: 1311; C. H. Heathcock and S. R. Poulter, Tetrahedron Letters, 1969, 2755). Z. N. Parnes et al. have reported the interesting ionic hydrogenation using Et_3SiH in trifluoroacetic acid, of isopropenyl- and vinyl-cyclopropane, giving isopropyl- and ethyl-cyclopropane, respectively (Doklady Akad. Nauk, S.S.S.R., 1968, 178: 620; C.A., 1968, 69: 2535). While a cyclopropane ring is normally inert under the usual hydroboration conditions, hydroboration of a vinyl cyclopropane may lead to cleavage of the ring:

75%   12%

A stepwise mechanism has been suggested for this reaction involving an initial hydroboration of the double bond followed by a homoallylic rearrangement of the α-cyclopropylcarbinylborane to an allylcarbinylborane derivative. The latter may then undergo further hydroboration (E. Breuer et al., J. org. Chem., 1972, 37: 2242).

In a study of the ultraviolet spectra of monocyclic vinylcyclopropanes, the cyclopropyl group was found to impart a bathochromic shift of from 8 to 15 nm to the olefinic $\pi \to \pi^*$ transition, when compared with isopropyl models (Heathcock and Poulter, J. Amer. chem. Soc., 1968, 90: 3766). A similar study has been reported on β-cyclopropyl-α,β-unsaturated esters (M. J. Jorgenson et al., ibid., 1968, 90: 3769; Helv., 1970, 53: 1421). The spectra of the esters showed that maximum cyclopropyl conjugation occurs when the plane of the double bond bisects the ring, i.e. the "bisected" conformation. Heathcock and Poulter (loc. cit.) suggest that there is no analogous conformational requirement for vinylcyclopropanes, although a similar suggestion for arylcyclopropanes has been questioned (R. C. Hahn et al., J. Amer. chem. Soc., 1969, 91: 3558; L. Martinelli et al., J. org. Chem., 1972, 37: 2278). J.-P. Pete has reviewed the ultraviolet spectra of cyclopropane derivatives (Bull. Soc. chim. Fr., 1967, 357).

The cyclopropane ring can *transmit*, as well as *extend*, the conjugation of contiguous unsaturated groups. This property, though not generally accepted in the past, has now been confirmed by n.m.r. studies on 1-aryl-2-benzoyl-cyclopropanes and on fluorine-substituted 1,2-diarylcyclopropanes (A. B. Turner et al., J. org. Chem., 1971, 36: 1107; R. G. Pews and N. D. Ojha, J. Amer. chem. Soc., 1969, 91: 5769), by Raman studies on 1-aryl-2-vinyl-cyclopropanes (E. G. Treshchova et al., Zhur. org. Khim., 1970, 6: 2358; C. A., 1971, 74: 41732) and also by chemical evidence (J. M. Stewart and G. K. Pagenkopf, J. org. Chem., 1969, 34: 7). The cyclopropene ring is also able to transmit conjugation (A. B. Thigpen Jr. and R. Fuchs, ibid., 1969, 34: 505).

The three-membered ring of an aryl cyclopropane is readily cleaved by alkali metal in liquid ammonia. The preferred direction of ring opening is controlled by a subtle interplay of steric and electronic factors (H. M. Walborsky, M. S. Aronoff and M. F. Schulman, ibid., 1971, 36: 1036; S. W. Staley and J. J. Rocchio, J. Amer. chem. Soc., 1969, 91: 1565).

Synthesis of the parent *cyclopropenylium ion*[*] has been achieved by mixing 3-chlorocyclopropene with either antimony pentachloride, aluminium trichloride, or silver fluoroborate (R. Breslow and J. T. Groves, ibid., 1970, 92: 984). The hexachloroantimonate is a white solid, stable for several days at room temperature and indefinitely at $-20°$. It exhibits a sharp singlet in the n.m.r. at δ 11.1–11.2, and its infrared spectrum indicates a symmetrical delocalised structure for the cation. The ^{13}C–H coupling constant is very large (265 Hz) and Breslow (loc. cit.) describes a hybridisation scheme consistent with this in which each carbon uses an sp orbital to hydrogen, two sp^3 orbitals in the plane for the bent single bonds and a p orbital for the π system:

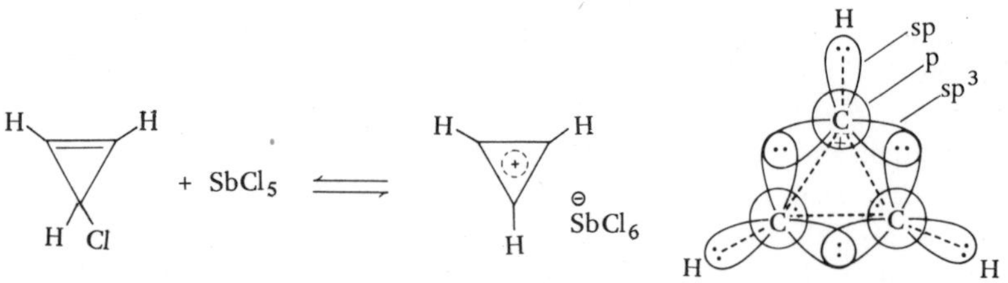

The parent cyclopropenylium ion has also been generated in solution by a process involving ester decarbonylation (D. G. Farnum, G. Mehta and R. G. Silberman, ibid., 1967, 89: 5048). The triphenylcyclopropenylium ion has been observed directly in the mass spectrometer in the spectra of triphenyl-

---

[*] Also termed the cyclopropenium or the cyclopropenyl cation.

cyclopropenyl bromide, chloride and fluoroborate (M. A. Battiste and B. Halton, Chem. Comm., 1968, 1368). The equivalence of the three carbon atoms of the three-membered ring in the triphenylcyclopropenylium ion has been demonstrated by ^{14}C-labelling experiments (I. A. D'yakonov, R. R. Kostikov and A. P. Molchanov, Zhur. org. Khim., 1970, 6: 316; C.A., 1970, 72: 110857).

Triphenylcyclopropenyl bromide electrophilically substitutes into activated aromatic compounds, such as phenols or aromatic amines, giving 1-aryl-1,2,3-triphenylcyclopropene (B. Föhlisch and P. Bürgle, Ann., 1967, 701: 58). The 3-ethoxy-1,2-diphenylcyclopropenylium ion may suffer nucleophilic attack at either the 3- or the 1-position, by anions derived from active methylene group compounds (T. Eicher et al., Ber., 1971, 104: 605; 1969, 102: 319).

The 1-cyclopropylvinyl cation is generated under solvolytic conditions from 1-cyclopropyl-1-haloalkenes (R. G. Bergman and co-workers, J. Amer. chem. Soc., 1971, 93: 1953, 1941, 1925; M. Hanack and T. Bässler, ibid., 1969, 91: 2117).

4. Halogen derivatives of cyclopropane

The reduction of halocyclopropanes may usually be achieved with retention of configuration but the degree of stereospecificity depends critically on the reaction conditions used. T. Ando et al. described the reduction with tri-*n*-butyltin hydride of a series of *gem*-halofluorocyclopropanes, leading to fluorocyclopropanes with retention of configuration. They postulated the intermediacy of a pyramidal 1-fluorocyclopropyl radical, which abstracted hydrogen faster than it underwent inversion (J. org. Chem., 1970, 35: 33). Configuration was also retained in the reduction of *gem*-bromofluorocyclopropanes to fluorocyclopropanes with lithium aluminium hydride (H. Yamanaka et al., Chem. Comm., 1971, 380), and in the reduction of chlorocyclopropanes with sodium in liquid ammonia (H. M. Walborsky, F. P. Johnson and J. B. Pierce, J. Amer. chem. Soc., 1968, 90: 5222; D. B. Ledlie and S. MacLean, J. org. Chem., 1969, 34: 1123). L. J. Altman and B. W. Nelson have observed net inversion in the reduction of optically active cyclopropylbromide using triphenyltin hydride (J. Amer. chem. Soc., 1969, 91: 5163) and net retention using di-*n*-butyltin dihydride (Tetrahedron Letters, 1970, 4891). They interpret these results in terms of cage reduction of a rapidly inverting pyramidal cyclopropyl radical. Selective removal of chlorine from *gem*-chlorofluorocyclopropanes with sodium in liquid ammonia offers a useful route to fluorocyclopropanes (M. Schlosser, G. Heinz and Le Van Chau, Ber., 1971, 104: 1921).

Silver ion assisted solvolysis of cyclopropyl halides leads to open-chain allyl derivatives. This is a disrotatory process in which the $C_{(2)}$ and $C_{(3)}$ substituents which are *trans* to the departing halide ion, move outwards. This explains why *trans*-1-chloro-*cis*-2,3-di-*n*-propylcyclopropane reacts very much faster than its all-*cis* isomer; there is appreciable steric interaction between the two propyl groups in the allyl cation derived from the latter:

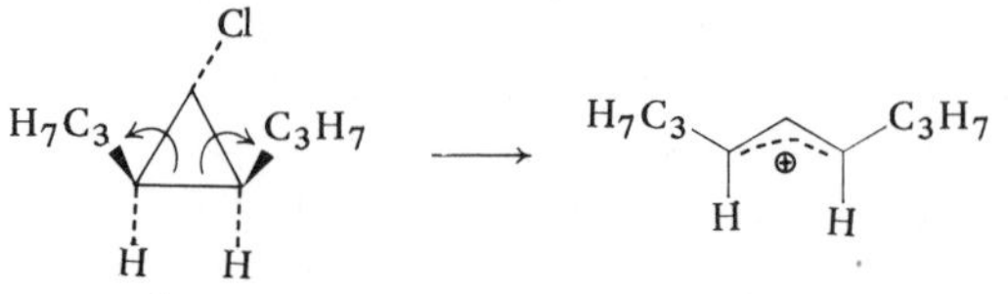

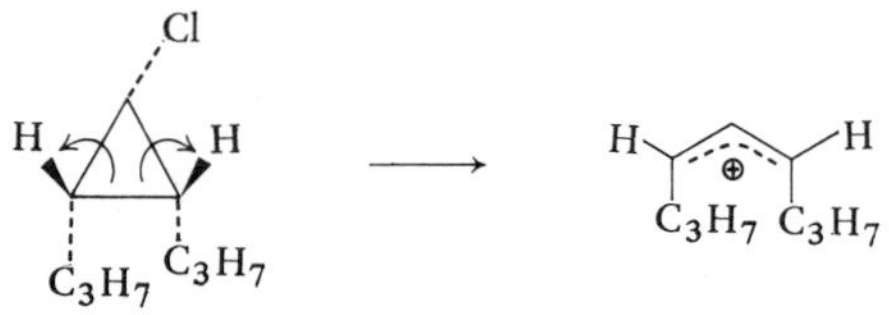

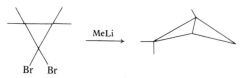

For similar reasons, 1,1-dihalo-2,3-*cis*-disubstituted cyclopropanes react faster than their *trans*-isomers. In the *cis*-isomer it is the halogen atom *cis* to the two hydrogen atoms at $C_{(2)}$ and $C_{(3)}$ which is lost preferentially. The overall reaction need not be stereospecific, for either the *cis* or the *trans* starting material may lead to the same final product (W. E. Parham and K. S. Yong, J. org. Chem., 1970, 35: 683; S. R. Sandler, ibid., 1967, 32: 3876). Direct evidence that isomeric cyclopropyl chlorides lead to different allyl cations has been obtained by observing the n.m.r. spectra of the latter in a $SbF_5 SO_2 ClF$ system at $-100°$ (P. von R. Schleyer et al., J. Amer. chem. Soc., 1969, 91: 5174).

Mono-, di- and tri-substituted 1,1-dihalocyclopropanes undergo ring-opening to allenes when treated with methyl-lithium. Tetra-substituted 1,1-dihalocyclopropanes, however, preferentially undergo a bond-insertion reaction to give bicyclobutane derivatives (W. R. Moore et al., Tetrahedron Letters, 1970, 4553, 4343, 2365; L. Skattebøl, ibid., 1970, 2361):

An optically active allene was obtained on treatment of *trans*-1,1-dibromo-2-methyl-3-phenylcyclopropane with butyl-lithium in presence of (−)-sparteine (H. Nozaki et al., Tetrahedron, 1971, 27: 905).

Chlorocyclopropanes can be dehydrochlorinated to the corresponding cyclopropenes by treatment with base. Only in special cases can the cyclopropene be isolated, however; it usually isomerises, or else reacts further by addition of the nucleophilic base across the double bond (T. C. Shields et al., J. Amer. chem. Soc., 1967, 89: 5425; Chem. Comm., 1967, 556; Chem. and Ind., 1967, 1999).

Tetrafluorocyclopropene, a flammable explosive gas, has been described by three groups of workers (G. Camaggi and F. Gozzo, J. chem. Soc., C, 1970, 178; P. B. Sargeant and C. G. Krespan, J. Amer. chem. Soc., 1969, 91: 415; W. Stuckey and J. Heicklen, ibid., 1968, 90: 3952). It is susceptible to attack by nucleophiles and, like all tetrahalocyclopropenes (D. C. F. Law and S. W. Tobey, ibid., 1968, 90: 2376), readily undergoes Diels–Alder addition to 1,3-diene systems.

The chemistry of the trichlorocyclopropenylium ion has been extensively investigated (see R. West, Accounts chem. Res., 1970, 3: 130, for review and refs.). In reactions with aromatic hydrocarbons, the chlorine atoms are successively replaced by one, two or three aryl groups, although introduction of the third aryl group requires an activated aromatic ring:

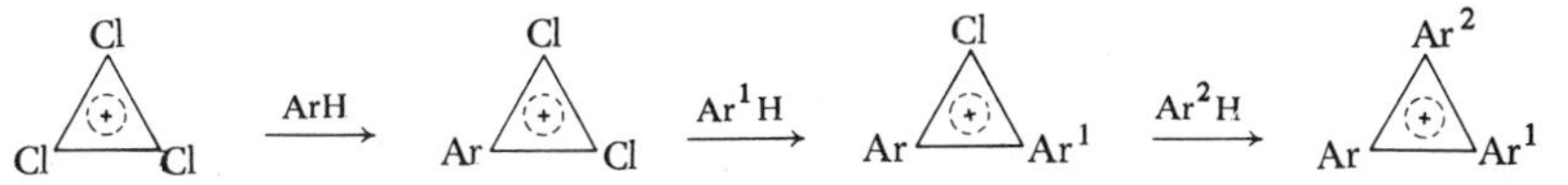

If one of the aryl residues is phenolic the products are *p*-hydroxyarylcyclopropenylium salts, which readily lose a proton to give aryl quinocyclopropenes:

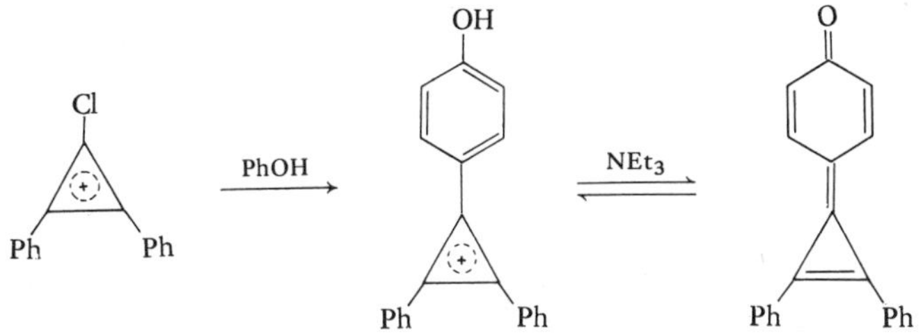

A tri(*p*-hydroxyaryl)cyclopropenylium ion gives, on loss of proton, a di(*p*-hydroxyaryl)quinocyclopropene. This on oxidation gives a triquinocyclopropane, which is a trimethylenemethane or radialene type of compound (see p. 24 for structure). The diaryl chlorocyclopropenylium ion can be hydrolysed by water to a 1,2-diaryl-3,3-dichlorocyclopropene, which in turn may be hydrolysed to a 2,3-diarylcyclopropenone:

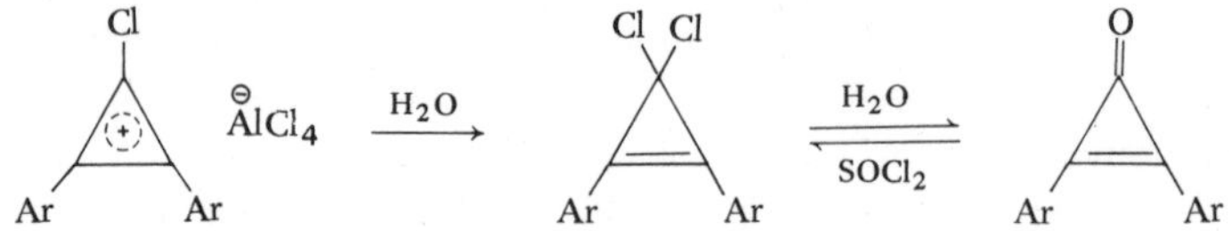

A 2,3-di(p-hydroxyaryl)cyclopropenone gives an unstable diquinocyclo-propanone on oxidation:

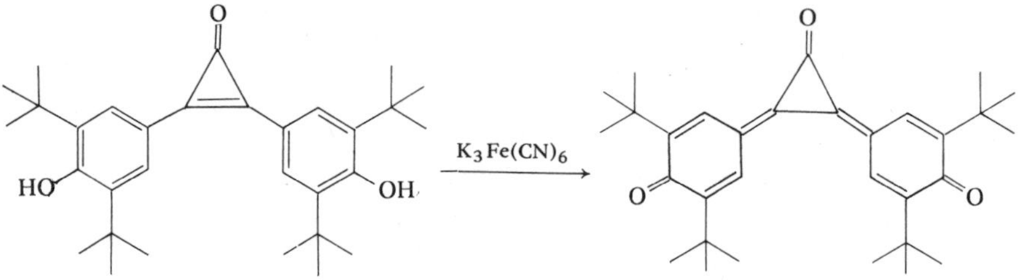

5. Organometallic compounds derived from cyclopropanes

Vinylcyclopropanes, methylenecyclopropanes and 1,2,3-triphenylcyclo-propene all form stable π-allylic palladium complexes (T. Shono et al., J. org. Chem., 1968, 33: 876; R. Noyori and H. Takaya, Chem. Comm., 1969, 525; P. Mushak and M. A. Battiste, J. organometal. Chem., 1969, 17: 46).

Platinum chloride complexes of substituted cyclopropanes may be pre-pared by a displacement reaction:

$$(C_2H_4PtCl_2)_2 + \text{cyclopropane} \rightarrow C_2H_4 + (\text{cyclopropane} \cdot PtCl_2)_n$$

Properties of these platinum complexes suggest that they contain an intact cyclopropane ring, edge-coordinated to the metal. 1,1-Diphenylcyclopro-pane, however, gave a π-allylic ring-opened complex (W. J. Irwin and F. J. McQuillin, Tetrahedron Letters, 1968, 1937). The parent complex $[PtCl_2(C_3H_6)]_n$, formed from cyclopropane and chloroplatinic acid, has been shown to be a tetramer in which the cyclopropane ring has been cleaved (R. D. Gillard and co-workers, J. chem. Soc., A, 1969, 1227).

Very stable cyclopropylcarbonyl complexes are formed by manganese, rhenium, iron and iridium. The manganese complex gives a π-allylic deriva-tive on decarbonylation (M. I. Bruce, M. Z. Iqbal and F. G. A. Stone, J. organometal. Chem., 1969, 20: 161). K. Öfele describes a remarkably stable

complex,   pentacarbonyl(2,3-diphenylcyclopropenylidene)chromium(O),
which is interesting in that it contains a carbene ligand but no heteroatom
(Angew. Chem., intern. Edn., 1968, 7: 950). D. L. Weaver and co-workers
described the properties of $[(\pi\text{-}Ph_3 C_3)NiCl(py)_2]\cdot$ py which is the first
example of a transition-metal π-complex with a three-membered aromatic
ring system, and also the mixed sandwich complex, $(\pi\text{-}C_5 H_5)Ni(\pi\text{-}C_3 Ph_3)$, in
which the nickel atom is symmetrically bonded to both five- and three-
membered unsaturated carbocyclic rings (J. Amer. chem. Soc., 1969, 91:
6506; 1970, 92: 4981).

6. Alcohols derived from cyclopropanes

Cyclopropanols may be conveniently prepared from cyclopropyl bromides
by treatment of the latter with lithium followed by oxidation with oxygen
(D. T. Longone and W. D. Wright, Tetrahedron Letters, 1969, 2859).

Cyclopropanols suffer ring cleavage when treated with either bases or
electrophilic reagents. The course of the reaction varies with the reagent and
with the substituents on the three-membered ring, and is usually highly
stereospecific. Thus in the acid-catalysed ring-opening reaction of *trans*-1-
methyl-2-phenylcyclopropanol, the configuration of the benzylic carbon is
fully retained. On the other hand, complete inversion occurs with base, or
with mercuric acetate, or with electrophilic bromine reagents (e.g. *N*-bromo-
succinimide or *tert*-butylhypochlorite). Ring-opening with base occurs towards the carbon which can better stabilise a negative charge. Electrophilic
halogen reagents, and sometimes mercuric acetate, also open the ring
selectively in one direction. In contrast, acid treatment leads to a mixture of
products resulting from both $C_{(1)}-C_{(2)}$ and $C_{(1)}-C_{(3)}$ bond cleavage (C. H.
DePuy et al., J. Amer. chem. Soc., 1970, 92: 4008; 1968, 90: 1830):

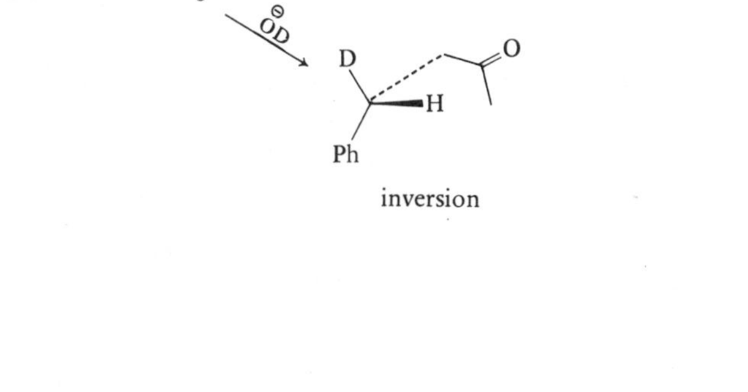

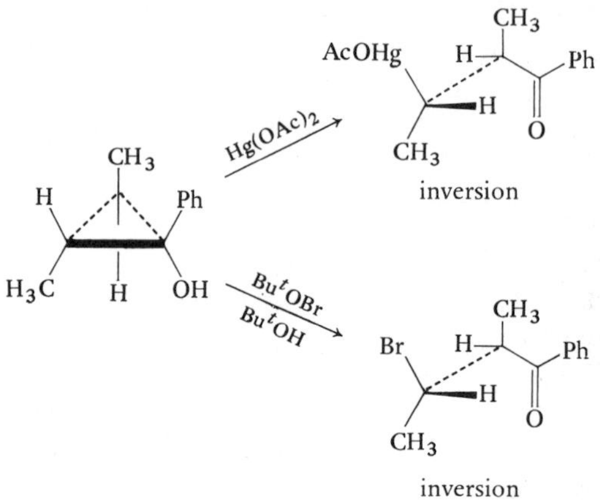

Solvolysis of cyclopropyl tosylates leads to allyl derivatives, and as with cyclopropyl halides, the ring-opening is a disrotatory process with the $C_{(2)}$ and $C_{(3)}$ substituents *trans* to the departing tosyl group rotating outwards (see U. Schöllkopf, Angew. Chem., intern. Edn., 1968, 7: 588, for review; also DePuy, Accounts chem. Res., 1968, 1: 33, for review of this and other aspects of cyclopropanol chemistry).

The methyl ethers of cyclopropanols may be obtained either by treating the cyclopropanol with diazomethane and boron trifluoride or by treating the cyclopropyl acetate with methyl-lithium followed by trimethyloxonium fluoroborate (A. DeBoer and C. H. DePuy, J. Amer. chem. Soc., 1970, 92: 4008).

The nature of the cyclopropylcarbinyl cation which is formed during the solvolysis of cyclopropylcarbinol derivatives, such as tosylates, has attracted much attention. Such solvolysis proceeds with rate enhancement and gives rise to a mixture of cyclopropylcarbinyl, cyclobutyl, and allylcarbinyl products, with substantial isotopic scrambling. These results necessitate an electronic interaction between the three-membered ring and the electron-deficient carbinyl carbon, and a number of representations of the structure of the intermediate ion have been considered. They include rapidly equi-librating unsymmetrical bicyclobutonium ions such as I, the "bisected" cation II, its perpendicular conformer III, and the tricyclobutonium ions IV and V. Of these, only structures I and II have retained support:

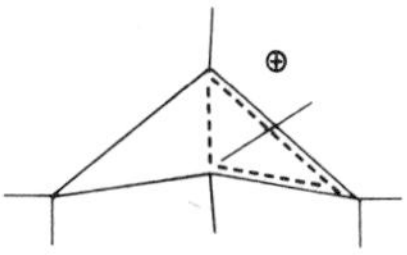

The tricyclobutonium ions IV and V are deemed unlikely for theoretical reasons, and ion III may be eliminated in favour of the "bisected" conformer II. In the latter, maximum overlap is achieved between the ring and the empty p orbital of the carbinyl carbon, and its stability as compared with III can be demonstrated by molecular-orbital calculations. Furthermore, n.m.r. spectra of the dimethylcyclopropylcarbinyl cation clearly indicate two non-equivalent methyl groups. In systems where steric factors preclude attainment of the "bisected" conformation, the cyclopropane ring loses its electron-releasing and rate-enhancing properties. The "bisected"structure II is at present the most popular representation although the available evidence cannot rigorously distinguish between it and rapidly equilibrating bicyclo-butonium ions of type I. Isotopic scrambling and other observations require a facile degenerate interconversion of one cyclopropylcarbinyl cation into another, and it has been suggested that this rearrangement proceeds through a C_s symmetric ion of structure VI (see D. S. Kabakoff and E. Namanworth, J. Amer. chem. Soc., 1970, 92: 3234; G. A. Olah et al., ibid., 1970, 92: 2544; and B. R. Ree and J. C. Martin, ibid., 1970, 92: 1660, for leading refs.).

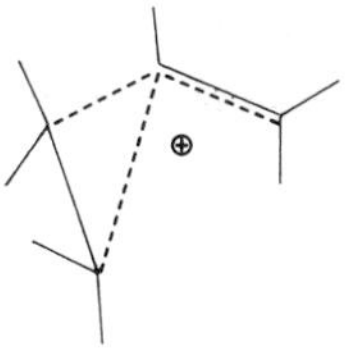

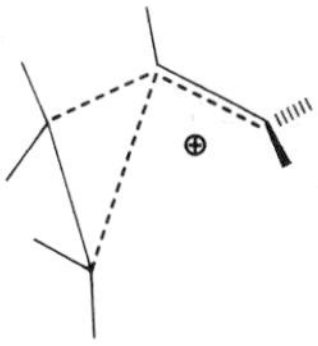

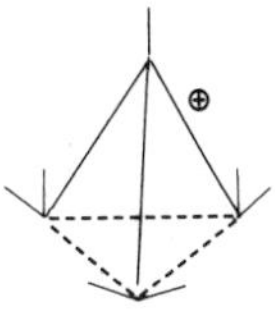

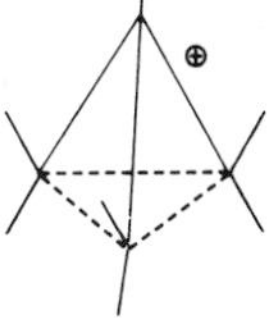

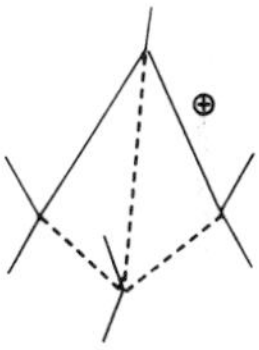

7. Sulphur derivatives of cyclopropanes

Diphenylcyclopropenethione has been prepared by treatment of diphenyl-cyclopropenone with P_4S_{10} (G. Laban, J. Fabian and R. Mayer, Z. Chem., 1968, 8: 414) or with hydrogen sulphide and hydrogen chloride (P. Metzner and J. Vialle, Bull. Soc. chim. Fr., 1970, 3739) or with thionyl chloride followed by thioacetic acid (J. W. Lown and T. W. Maloney, J. org. Chem.,

1970, 35: 1716). It affords the unstable diphenylcyclopropenethione-S-oxide on oxidation with monoperphthalic acid (Lown and Maloney, loc. cit.):

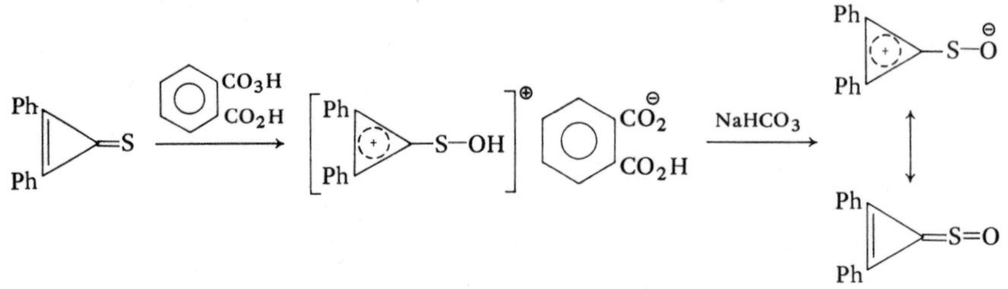

8. Nitrocyclopropanes

Reduction of nitrocyclopropane either electrolytically or with zinc gives N-cyclopropylhydroxylamine. This, without isolation, has been oxidised to nitrosocyclopropane with diethylazodicarboxylate (R. Stammer, J. B. F. N. Engberts and Th. J. de Boer, Rec. trav. Chim., 1970, 89: 169), and to *cyclopropylazoxycyclopropane* by electrolysis at a mercury electrode. This is the first report of an azoxy compound in the cyclopropane series (P. E. Iversen, Ber., 1971, 104: 2195):

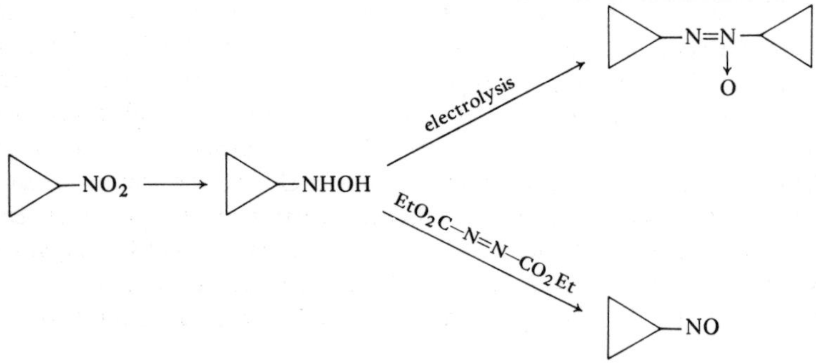

9. Amines and related compounds derived from cyclopropanes

N-Alkylcyclopropylamines may be obtained by reduction of the appropriate imine, formed by the reaction of a ketone with cyclopropylamine. The reduction may be effected either by hydrogenation or by sodium borohydride (J. R. Boissier and R. Ratouis, C. A., 1968, 68: 114154; I. G. Bolesov et al., Zhur. org. Khim., 1969, 5: 1510; C.A., 1969, 71: 112464):

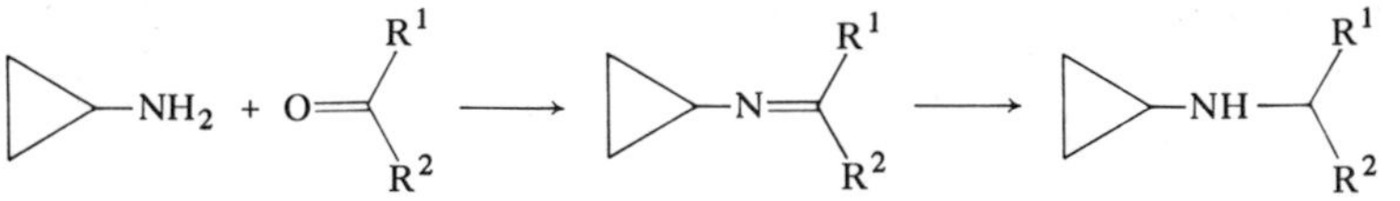

Cyclopropylcarbinyl amines carrying one, two or three cyclopropyl groups are formed by ammonolysis of the benzoate or *p*-nitrobenzoate of the appropriate cyclopropyl carbinol. Oxidative coupling of the cyclopropyl-carbinol amine with iodine pentafluoride affords the azo compound (J. W. Timberlake and J. C. Martin, J. org. Chem., 1968, **33**: 4054):

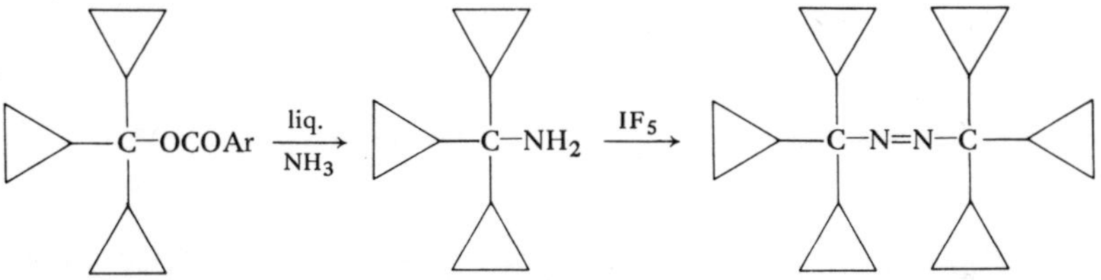

Deamination of a primary cyclopropylamine with nitrous acid leads to ring-opened products, such as the corresponding allyl alcohol. The inter-mediacy of a free cyclopropyl cation is uncertain. When *N*-nitroso-*N*-cyclo-propylureas are treated with base the corresponding cyclopropyldiazonium ion is generated. This may lose a proton to give the diazocyclopropane which decomposes stereospecifically to an allene. Under very weakly basic con-ditions, however, and in presence of a nucleophile, diazocyclopropane formation is suppressed, and the cyclopropyldiazonium ion decomposes to give, in addition to ring-opened products, low yields of cyclopropane derivatives. This decomposition is non-stereospecific and the cyclopropyl cation is suggested as an intermediate here (W. Kirmse and H. Schütte, Ber., 1968, **101**: 1674; J. Amer. chem. Soc., 1967, **89**: 1284):

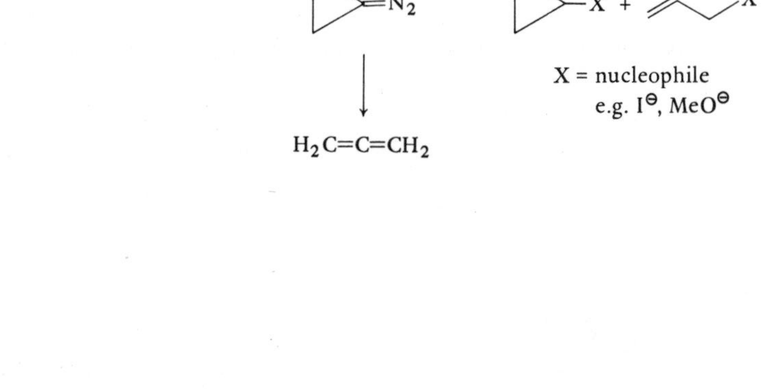

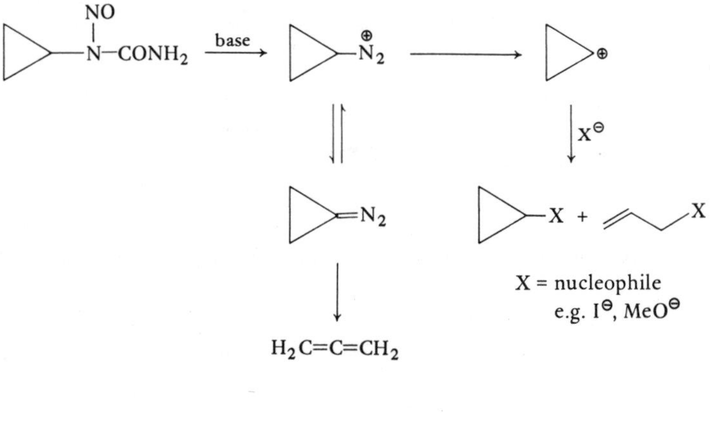

The stereospecific decomposition of diazocyclopropanes to allenes is exploited in a useful new general synthesis of optically active allenes (J. M. Walbrick, J. W. Wilson Jr. and W. M. Jones, ibid., 1968, 90: 2895) as shown below:

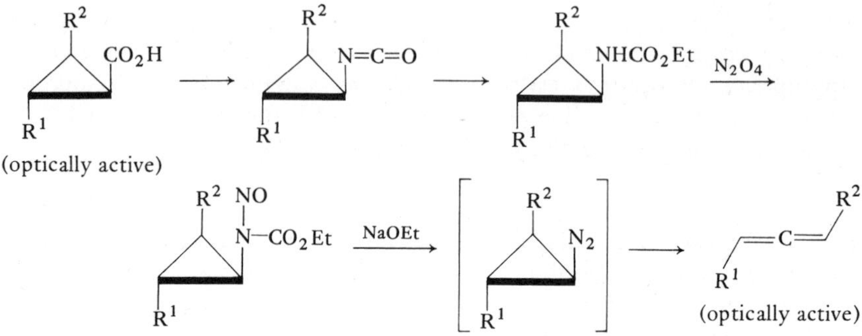

(optically active)

(optically active)

Both cyclopropyldiazomethane and the cyclopropylmethyldiazotate anion have been generated from N-cyclopropylmethyl-N-nitrosourethan. On aqueous hydrolysis, or treatment with benzoic acid or benzoyl chloride, both give a mixture of cyclopropylcarbinyl, cyclobutyl and allylcarbinyl derivatives (R. A. Moss and F. C. Shulman, Tetrahedron, 1968, 24: 2881). Cyclopropyldiazomethane has also been prepared from N-nitroso-N-(1,1-dimethyl-3-oxobutyl)-N-cyclopropylmethylamine (P. B. Shevlin and A. P. Wolf, J. Amer. chem. Soc., 1966, 88: 4735).

The first preparation of *cyclopropyl azide* was achieved by treating N-nitroso-N-cyclopropylurea with lithium azide (Kirmse and Schütte, loc. cit.). A general synthesis of 2,2-dichlorocyclopropylazides was described by A. B. Levy and A. Hassner (J. Amer. chem. Soc., 1971, 93: 2051) by addition of dichlorocarbene to vinyl azides. These dichlorocyclopropyl azides explode if overheated but decompose smoothly and regiospecifically at 105—125° to give the corresponding 3,3-dichloroazetines:

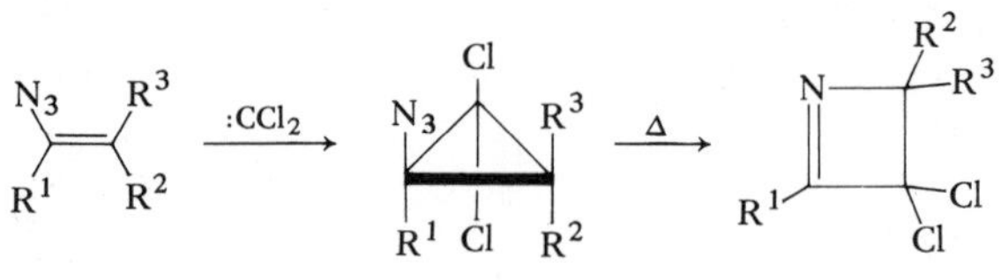

10. Aldehydes and ketones derived from cyclopropanes

Two new methods of synthesis of *cyclopropanecarboxaldehyde* make this compound now readily available. The first utilises ceric ammonium nitrate oxidation of cyclopropyl carbinol (L. B. Young and W. S. Trahanovsky, J. org. Chem., 1967, 32: 2349) and the second, which has the advantage of using common starting materials, is shown below (J. M. Stewart et al., ibid., 1970, 35: 2040):

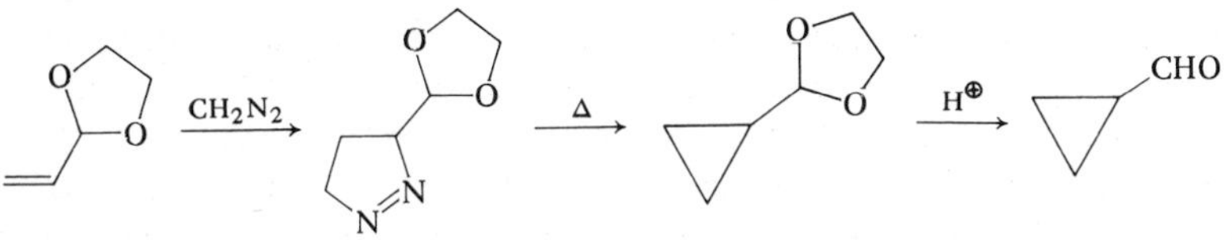

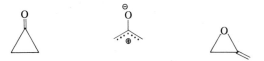

cis-*Cyclopropane*-1,2-*dicarboxaldehyde* has been prepared by addition of dichlorocarbene to 2,5-dimethoxy-2,5-dihydrofuran, reductive removal of the chlorine atoms by sodium in liquid ammonia, and then acid hydrolysis. The dialdehyde could not be obtained by oxidation of *cis*-1,2-dihydroxy-methylcyclopropane (G. Maier and T. Sayrac, Ber., 1968, 101: 1354). *cis*-2-Vinylcyclopropanecarboxaldehyde undergoes a bis-allylic type thermal rearrangement to its valence isomer, 2,5-dihydrooxepin. This is analogous to the Cope rearrangement of *cis*-1,2-divinylcyclopropane (S. J. Rhoads and R. D. Cockroft, J. Amer. chem. Soc., 1969, 91: 2815):

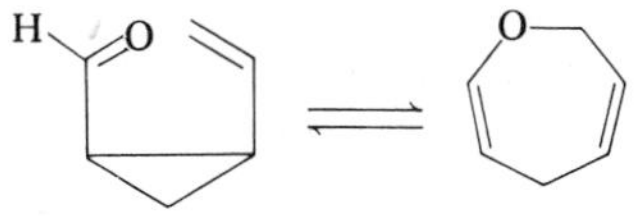

The chemistry of cyclopropanones has been reviewed by N. J. Turro (Accounts chem. Res., 1969, 2: 25) who lists three approaches to their synthesis: (a) by the addition of a diazo compound to a ketene; (b) by the photoelimination of CO from cyclobutane-1,3-diones; (c) by the extraction of cyclopropanone from its equilibrium with a labile adduct, such as 1-hydroxy-1-methoxycyclopropane. Of these, method (a) is the most useful.

Theoretical considerations indicate that the following three structures should be of comparable energy for the cyclopropanone molecule:

Accumulated spectral data, however, together with a detailed microwave analysis (J. M. Pochan, J. E. Baldwin and W. H. Flygare, J. Amer. chem. Soc., 1969, 91: 1896) have shown unequivocally that the molecule is best described as the closed three-ring ketone. The $C_{(2)}-C_{(3)}$ bond length (1.575 Å) is one of the longest ever measured spectroscopically and indicates a weak $C_{(2)}-C_{(3)}$ bond, consistent with the observed chemical properties of cyclopropanone. These considerations do not preclude the possibility of a tautomeric equilibrium between the closed three-ring ketone and the open-ring dipolar structure, and in some of its cycloaddition reactions, cyclopropanone may initially tautomerise to the 1,3-dipole. For instance, kinetic measurements of the cycloaddition of cyclopropanones with 1,3-dienes suggest that in the transition state of this reaction the cyclopropanone has assumed a large degree of zwitterionic character (S. S. Edelson and Turro, ibid., 1970, 92: 2770).

The cyclopropanone ring system is highly unusual in that it can undergo both $3 + 4 \rightarrow 7$ and $3 + 2 \rightarrow 5^*$ symmetry allowed cyclo-addition reactions. In these, substituted cyclopropanones are much more reactive than cyclopropanone itself (Turro et al., ibid., 1969, 91: 2283; J. org. Chem., 1970, 35: 2058):

$$3 + 4 \rightarrow 7$$

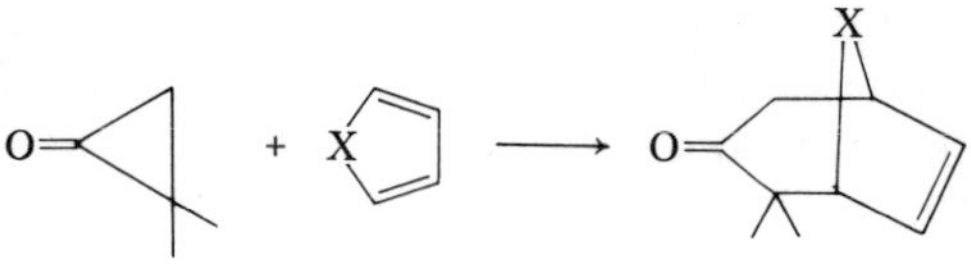

$$X = CH_2, CH_2CH_2, NCH_3, O, C=C(CH_3)_2$$

$$3 + 2 \rightarrow 5$$

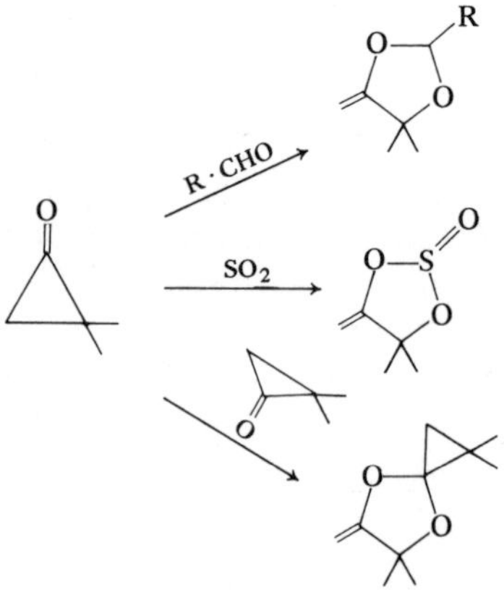

Dimethylketene and 1,1-dimethoxyethylene undergo $2 + 2 \rightarrow 4$ addition to the carbonyl group of 2,2-dimethylcyclopropanone to give the products shown below (Turro et al., loc. cit.):

* Huisgen's nomenclature (Angew. Chem., intern. Edn., 1963, 2: 565).

Although substituted cyclopropanones yield only ring-opened products when treated with acidic reagents, with cyclopropanone itself addition to the carbonyl group occurs with a wide variety of compounds of the type HX (Turro and W. B. Hammond, Tetrahedron, 1968, 24: 6029):

$$X = OCH_3, Cl, OH, CH_3CO,$$
$$PhNH, PhNCH_3$$

The cyclopropanone ring is cleaved by base. The direction of ring-opening is determined both by steric factors and by the stability of the intermediate carbanion (C. Rappe et al., J. Amer. chem. Soc., 1970, 92: 2032). Cyclopropanones may be ring-expanded to cyclobutanones by treatment with diazomethane (Turro and R. B. Gagosian, ibid., 1970, 92: 2036).

In contrast to the extreme instability of simpler members of the family, cyclopropanones which carry very bulky substituents appear to be relatively stable. Thus 2,2-*di*-tert-*butylcyclopropanone* has been obtained as a waxy solid by peracid oxidation of 1,1-di-*tert*-butylallene (J. K. Crandall and W. H. Machleder, ibid., 1968, 90: 7347). Similar oxidation of 1,3-di-*tert*-butylallene leads to trans-2,3-*di*-tert-*butylcyclopropanone*. Here, the highly unusual allene oxide could be isolated as an intermediate (R. L. Camp and F. D. Greene, ibid., 1968, 90: 7349):

R = *tert*-butyl

Greene and co-workers have also prepared trans-2,3-*di*-tert-*butylcyclopropanone* by the dehydrobromination of α-bromodineopentyl ketone, and have partially resolved it by asymmetric destruction with *d*-amphetamine. The optically active compound racemised on heating (ibid., 1970, 92: 7488; 1967, 89: 1030).

Di-tert-*butylcyclopropenone* has been prepared in an analogous manner by base treatment of α,α'-dibromodineopentyl ketone (J. Ciabattoni and E. C. Nathan, J. Amer: chem. Soc., 1969, 91: 4766). The parent compound, *cyclopropenone*, has been prepared from tetrachlorocyclopropene, by reduction with tri-*n*-butyltin hydride followed by hydrolysis with water. First obtainable only in solution, it has more recently been prepared as a pure solid, m.p. −28 to −29°, b.p. 30°. The pure compound is stable for many weeks below its melting point but it rapidly polymerises at room temperature. Spectral evidence indicates that it exists as the free ketone rather than as a *gem*-diol, even in aqueous solution (R. Breslow et al., ibid., 1972, 94: 4787; 1967, 89: 3073).

Cyclopropenones are susceptible to attack at either the carbonyl carbon or at a double-bond carbon, by a variety of nucleophilic reagents. Cleavage of the ring usually results (E. V. Dehmlow, Ber., 1969, 102: 3863; Ann., 1969, 729: 64) although *diphenylcyclopropenone* readily forms a 2,4-dinitrophenylhydrazone (E. D. Bergmann and I. Agranat, Tetrahedron, 1970, 26: 4859). A 1:1 adduct, which has been shown to be a 2,3-diphenyl-4-pyridone, is formed when diphenylcyclopropenone is treated with a 1-azirine. The suggested mechanism involves nucleophilic attack of the weakly basic azirine nitrogen on the electrophilic cyclopropenone ring, followed by an intramolecular Cope cyclisation (A. Hassner and A. Kascheres, J. org. Chem., 1972, 37: 2328):

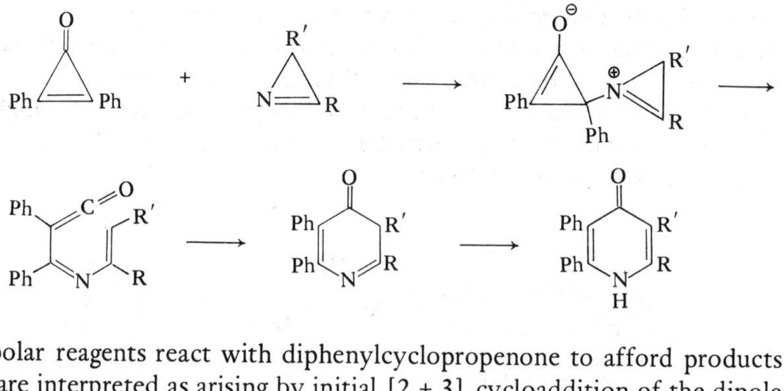

1,3-Dipolar reagents react with diphenylcyclopropenone to afford products which are interpreted as arising by initial [2 + 3] cycloaddition of the dipole to the cyclopropenone. Depending on the structure of the 1,3-dipole, exclusive addition occurs to either the C=C or the C=O of the cyclopropenone (J. W. Lown, T. W. Maloney and G. Dallas, Canad. J. Chem., 1970,

48: 584). Nitration or bromination of the benzene rings in diphenylcyclo-propenone gives *meta*-substituted products. This is presumably because under the acidic conditions used for the reaction, the species undergoing substitution is the protonated cyclopropenone i.e. the hydroxycyclo-propenylium ion (Agranat and M. R. Pick, Tetrahedron Letters, 1972, 3111; C. W. Bird and A. F. Harmer, Org. Prep. Proced., 1970, 2: 79):

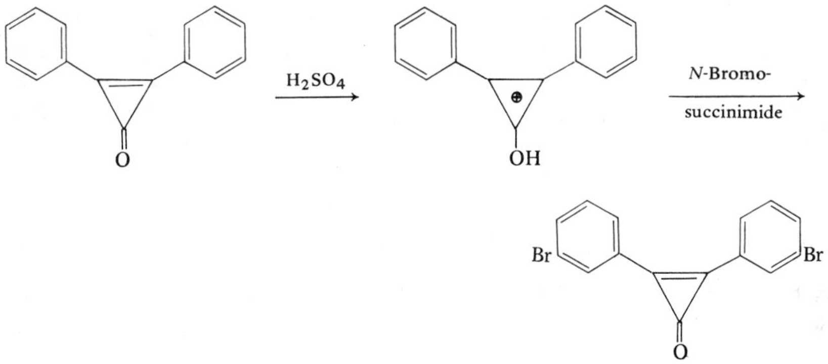

H_2SO_4

N-Bromo-succinimide

Diphenylcyclopropenone forms metal complexes in which coordination occurs through the carbonyl oxygen (Bird and E. M. Briggs, J. chem. Soc., A, 1967, 1004).

Cyclopropyl methyl ketones exist predominantly in the symmetrical "bisected" *s-cis* conformation (J.-L. Pierre and P. Arnaud, Bull. Soc. chim. Fr., 1966, 1690). In contrast, cyclopropylcarboxaldehyde exists mainly as the *s-trans* conformer (G. J. Karabatsos and N. Hsi, J. Amer. chem. Soc., 1965, 87: 2864):

s-cis   *s-trans*

The "bisected" conformation is preferred because it allows maximum electronic interaction between the three-membered ring and the adjacent p-orbital of the carbonyl group and it represents a conformational energy minimum not only for cyclopropyl aldehydes and ketones but also for the cyclopropylcarbinyl cation, cyclopropyl carboxylic acid chloride, phenyl-cyclopropane, vinylcyclopropane, and cyclopropylsemidiones (see A. H. Cowley and T. A. Furtsch, J. Amer. chem. Soc., 1969, 91: 39 for refs.).

The three-membered ring of a cyclopropyl ketone is reductively cleaved by solutions of metals in liquid ammonia. The bond which cleaves preferentially is that which has the greater overlap with the carbonyl π-system. This is readily demonstrated in bicyclic systems but is also important in monocyclic cyclopropyl ketones which carry a *cis*-substituent in the 2-position of the ring. Such a substituent precludes free rotation of the carbonyl side-chain, resulting in maximum overlap with, and therefore preferential cleavage of, the $C_{(1)}-C_{(2)}$ bond. In the absence of such steric effects, as in *trans*-substituted cyclopropyl ketones, the bond that cleaves is the one that gives the more thermodynamically stable carbanion intermediate (W. G. Dauben and R. E. Wolf, J. org. Chem., 1970, 35: 2361, 374):

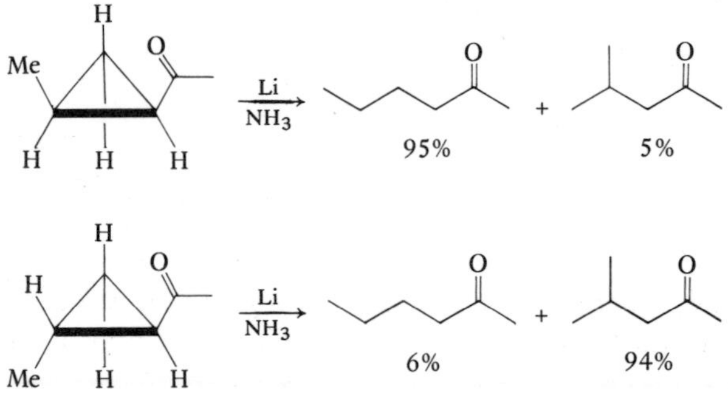

Photolysis of cyclopropyl ketones may lead to reaction via two distinct mechanistic pathways. The first operates when the three-membered ring carries an alkyl substituent *cis* to the ketone side-chain, offering a γ-hydrogen atom which can be abstracted by the excited carbonyl group (Norrish type II reaction):

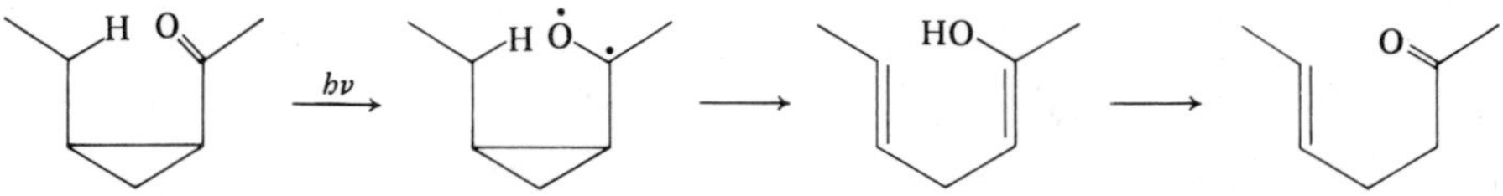

In the absence of an abstractible γ-hydrogen atom, the primary photochemical process is cleavage of one of the ring bonds adjacent to the carbonyl group (Norrish type I reaction). In sterically constrained cyclopropyl ketones, e.g. fused-ring systems, selective cleavage occurs of that ring bond which enjoys greater overlap with the carbonyl electrons. Ring cleavage generates a diradical which in the vapour phase or in inert solvent undergoes 1,2-hydrogen migration to give a conjugated enone (W. G. Dauben et al.,

ibid., 1969, 34: 2512, 2301, 1849; J. Amer. chem. Soc., 1970, 92: 6273;
D. G. Marsh et al., ibid., 1971, 93: 333):

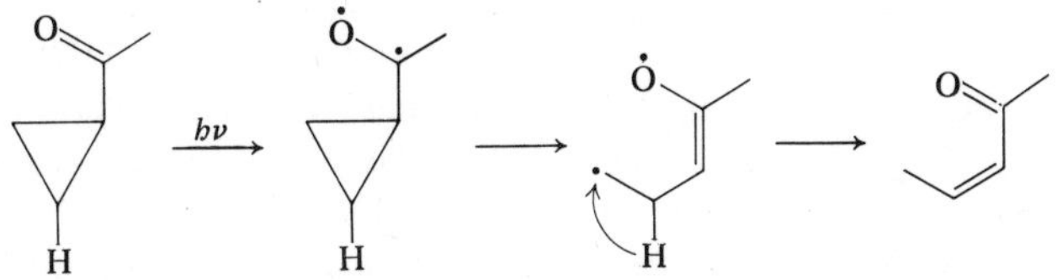

Dauben and co-workers have also described the novel formation of a
cyclopropyl ketone by acetophenone-sensitised photolysis of a β,γ-un-
saturated ketone, in a reaction directly analogous to the di-π-methane
rearrangement of 1,4-dienes (ibid., 1970, 92: 1786):

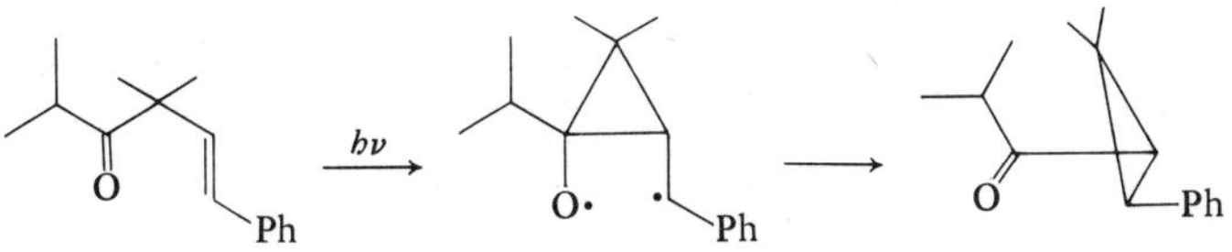

Photolysis of benzoyl aryl cyclopropanes leads to a facile *cis—trans* isomerisation reaction (H. E. Zimmerman et al., ibid., 1970, 92: 6931, 2023,
2000):

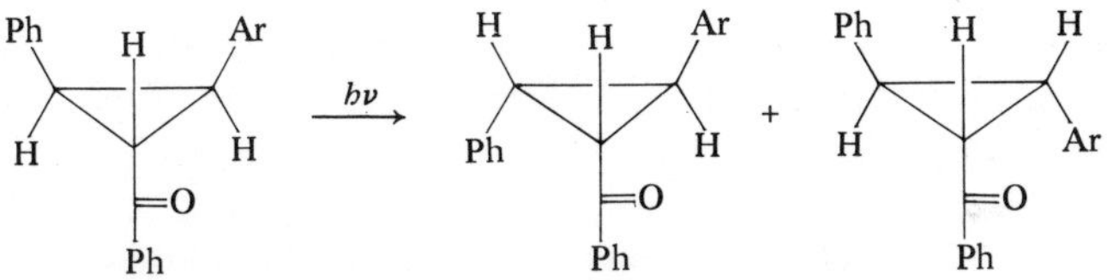

Pyrolysis of cyclopropyl ketones which carry a *cis*-2-alkyl substituent on the
ring causes a 1,5-hydrogen shift followed by ring-opening, to give a
γ,δ-unsaturated ketone (R. M. Roberts et al., ibid., 1967, 89: 1404):

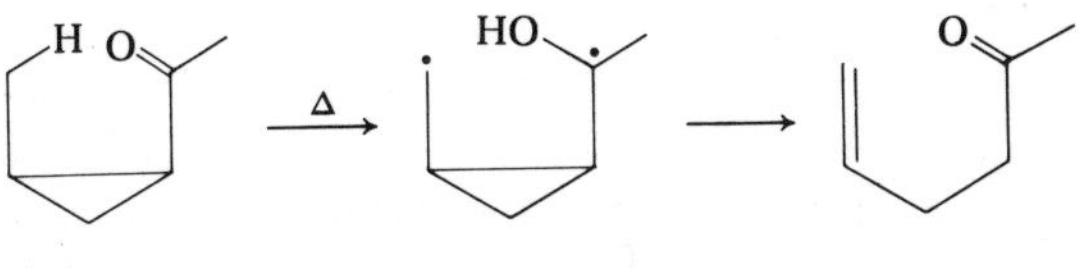

11. Cyclopropane- and cyclopropene-carboxylic acids

Methyl cyclopropanecarboxylates carrying a *cis*-2-alkyl substituent undergo
ring cleavage with concomitant 1,5-hydrogen shift when heated, affording
γ,δ-unsaturated esters. This reaction offers a method for determining the

stereochemistry of *cis–trans* isomers (D. E. McGreer and N. W. K. Chiu, Canad. J. Chem., 1968, 46: 2217). Hydrogenolysis of cyclopropanecarboxylic acids and esters, and also of other cyclopropanes having an adjacent carbonyl group, with hydrogen and a palladium-on-carbon catalyst, leads to predominant ring cleavage at the $C_{(1)}-C_{(2)}$ ring bond (A. L. Schultz, J. org. Chem., 1971, 36: 383). An ethyl 1,2-disubstituted cyclopropene-3-carboxylate is formed as the initial product when ethoxy carbonyl carbene is added to an alkyne in presence of copper sulphate. Depending on the reaction conditions, the ester may isomerise to the 2,3-disubstituted-5-ethoxyfuran (I. A. D'yakonov, M. I. Komendantov and T. S. Smirnova, Zhur. org. Khim., 1969, 5: 1742; C.A., 1970, 72: 12455):

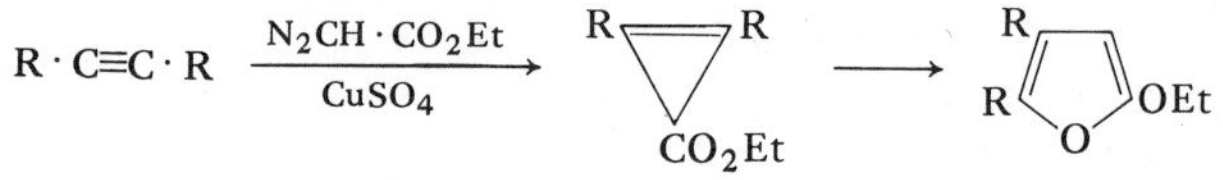

The double bond in *Feist's ester* (dimethyl 3-methylenecyclopropane-*trans*-1,2-dicarboxylate) undergoes normal peroxide-catalysed free radical addition of thiophenol and bromotrichloromethane. No evidence could be found for participation of a nonclassical cyclopropylmethyl radical. The double bond could not be epoxidised but it did react readily with ethyl diazoacetate to give the spiropentane ester (T. L. Gilchrist and C. W. Rees, J. chem. Soc., C, 1968, 776). R. C. Cookson and co-workers found that the double bond of Feist's ester gave normal Diels–Alder adducts with furan and cyclopentadiene, though not with anthracene (ibid., 1967, 928).

12. Cyclopropyl and cyclopropenyl carbanions

Cyclopropyl ketones undergo slow deuterium exchange of the cyclopropyl methine hydrogen, indicating a reluctance to form the planar enolate anion (H. W. Amburn, K. C. Kauffman and H. Shechter, J. Amer. chem. Soc., 1969, 91: 530; C. Rappe and W. H. Sachs, Tetrahedron, 1968, 24: 6287). The latter is disfavoured by the development of double-bond character towards the three-membered ring. H. M. Walborsky and J. M. Motes have investigated the stereochemical fate of the cyclopropyl carbanionic centre (J. Amer. chem. Soc., 1970, 92: 3697, 2445). In protic (though not in aprotic) solvents, the 1-cyano-2,2-diphenylcyclopropyl anion exhibits a high degree (99%) of retention of configuration. The 1-benzoyl-2,2-diphenylcyclopropyl anion in contrast exhibits only a moderate degree (27%) of retention. It is suggested that the delocalisation energy associated with the carbonyl group, being larger than that of the cyano group, is better able to overcome the barrier to planarity (I strain) of the carbanion. On the basis of molecular-orbital calculations, the cyclopropyl anion is predicted to be nonplanar with the $C_{(1)}$–H bond at 68° to the plane of the ring, with a barrier to inversion of 20.85 kcal/mole (D. T. Clark and D. R. Armstrong, Chem. Comm., 1969, 850).

The cyclopropenyl anion is theoretically more interesting in that it contains 4π-electrons and this does not satisfy the $4n + 2$ Hückel requirement for aromaticity. Not only is the cyclopropenyl anion non-aromatic, however, but evidence has been adduced which indicates it to be "anti-aromatic", i.e. it is actually destabilised by resonance (R. Breslow et al., Angew. Chem., intern. Edn., 1968, 7: 565; J. Amer. chem. Soc., 1968, 90: 2698; 1967, 89: 4383). For example, kinetic studies show that the base-catalysed deuterium exchange of the methine hydrogen in 3-benzoyl-1,2-diphenylcyclopropene is 6000 times slower than that of the corresponding hydrogen in cis-2,3-diphenyl-trans-1-benzoylcyclopropane. D. T. Clark has carried out molecular-orbital calculations on the cyclopropenyl anion and predicts it to be non-planar, with the $C_{(1)}$–H bond at 68° to the plane of the ring with a barrier to inversion of 52.3 kcal/mole (Chem. Comm., 1969, 637). The anti-aromatic cyclopropenyl anion derived from tert-butyl 1,2-diphenylcyclopropene-3-carboxylate is reported to be stable (I. N. Domnin, I. A. D'yakonov and N. I. Komendantov, Zhur. org. Khim., 1967, 3: 2076; C.A., 1968, 69: 2538).

13. Cyclopropyl radicals

Cyclopropyl radicals have been generated by thermal decomposition of cyclopropane carbonyl peroxides (H. M. Walborsky and J.-C. Chen, J. Amer. chem. Soc., 1971, 93: 671; 1970, 92: 7573; A. P. Stefani, L.-Y. Y. Chuang and H. E. Todd, ibid., 1970, 92: 4168), and of tert-butylcyclopropaneper-carboxylates (T. Shono, M. Akashi and R. Oda, Tetrahedron Letters, 1968, 1507); by the decarboxylation of cyclopropanecarboxylic acids with lead tetraacetate (T. Aratani, Y. Nakanisi and H. Nozaki, Tetrahedron, 1970, 26: 1675); and by the reaction of cyclopropyl bromides with lithium (M. J. S. Dewar and J. M. Harris, J. Amer. chem. Soc., 1969, 91: 3652) or with sodium dihydronaphthylide (J. Jacobus and D. Pensak, Chem. Comm., 1969, 400).

Configuration at the radical site is not usually preserved; optically active precursors generally afford racemic products if the cyclopropyl radical is generated at the chiral centre. While this could be interpreted as implying a planar structure, present evidence points to a rapidly inverting pyramidal structure for the cyclopropyl free radical. A few reactions have been described where partial retention of configuration occurs at the radical site. These reactions either occur in a solvent cage (Walborsky and Chen, loc cit.) or else involve a low inversion rate for the radical so that the latter is trapped before completely losing its stereochemical integrity (Jacobus and Pensak, loc. cit.; Dewar and Harris, loc. cit.) (see also p. 27). Evidence from a study of the addition of cyclopropane radicals to a series of simple olefins suggests that these radicals are nucleophilic in character (Stefani, Chuang and Todd, loc. cit.).

Much interest has centred on the question of whether or not the cyclopropylcarbinyl free radical is stabilised by a conjugative interaction between the three-membered ring and the adjacent radical site. J. C. Martin and J. W. Timberlake present evidence supporting such stabilisation (J. Amer. chem. Soc., 1970, 92: 978); earlier workers had ascribed the observed effect of the cyclopropane ring solely to ring strain (D. C. Neckers and A. P. Schaap, J. org. Chem., 1967, 32: 22). The cyclopropylcarbinyl radical rapidly rearranges to the corresponding allylcarbinyl radical and a rapid reversible equilibrium may exist between the two (J. K. Kochi, P. J. Krusic and D. R. Eaton, J. Amer. chem. Soc., 1969, 91: 1877; T. A. Halgren et al., ibid., 1967, 89: 3051). The e.s.r. spectrum of the cyclopropylcarbinyl radical indicates a preferred "bisected" conformation, analogous to that of the cyclopropylcarbinyl cation (see p. 33), and shows that the three methylene groups are not equivalent (Kochi, Krusic and Eaton, loc. cit.). There is no evidence for a non-classical structure.

14. Natural products containing a cyclopropyl ring

New syntheses of (±)-trans-*chrysanthemic acid* have been reported by several groups (R. W. Mills, R. D. H. Murray and R. A. Raphael, Chem. Comm., 1971, 555; E. J. Corey and M. Jautelat, J. Amer. chem. Soc., 1967, 89: 3912; M. Julia and A. Guy-Rouault, Bull. Soc. chim. Fr., 1967, 1411; S. Julia, M. Julia and G. Linstrumelle, ibid., 1966, 3499). (+)-trans-*Chrysanthemic acid* has been stereospecifically synthesised from (+)-car-3-ene (M. Matsui et al., Agr. Biol. Chem., Tokyo, 1967, 31: 33; C.A., 1967, 67: 32834), while L. Crombie, C. F. Doherty and G. Pattenden describe the preparation of ^{14}C-labelled (+)-*trans*-chrysanthemum mono- and di-carboxylic acids and related compounds (J. chem. Soc., C, 1970, 1076). The n.m.r. spectra of the natural pyrethrins, and of the derived alcohols and acids are presented and discussed by A. F. Bramwell et al. (Tetrahedron, 1969, 25: 1727). T. A. King and H. M. Paisley describe the corresponding mass spectra, and also discuss methods used for the isolation of the pyrethroids (J. chem. Soc., C, 1969, 870).

Syntheses of *methyl malvalate* and of *methyl sterculate* have been achieved by W. J. Gensler and co-workers (J. org. Chem., 1970, 35: 2301; J. Amer. chem. Soc., 1970, 92: 2472). The proposed biosynthetic route to malvalic acid, by a biological α-oxidation of sterculic acid, is strongly supported by new evidence from L. J. Morris and S. W. Hall, who have isolated the most likely intermediate, D-2-hydroxysterculic acid, from *Pachira* and *Bombacopsis* seed oils (Chem. and Ind., 1967, 32). Related acids which have been isolated are *dihydrosterculic acid* from *Euphoria longana* seed oil (R. Kleiman, F. R. Earle and I. A. Wolff, Lipids, 1969, 4: 317) and *sterculynic acid* (8,9-methyleneoctadec-8-en-17-ynoic acid) from the seed oil of *Sterculia alata* (A. W. Jevans and C. Y. Hopkins, Tetrahedron Letters, 1968, 2167). cis-11,12-*Methylene-2-hydroxyoctadecanoic acid* has been isolated from *Thiobacillus thiooxidans* (H. W. Knoche and J. M. Shively, J. biol. Chem., 1969, 244: 4773).

The methyl esters of mycolic acids isolated from human and avian strains of tubercle bacilli may possess extremely long chains of carbon atoms, incorporating one or two cyclopropane rings. For example D. E. Minnikin and N. Polgar suggest the following structure for the avian mycolic ester (Chem. Comm., 1967, 916: 1172):

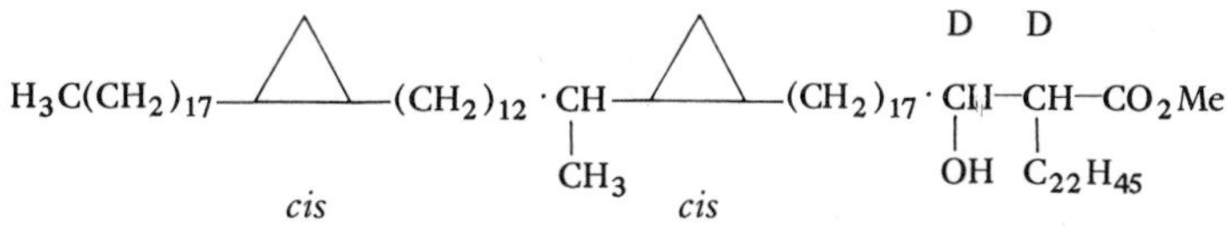

Minnikin has also discussed the problem of locating the position of the ring in cyclopropane fatty acids and has described a new method in which the cyclopropane fatty acid ester is treated with a boron trifluoride—methanol reagent. The methoxy derivatives which are formed by cleavage of the ring are then examined by mass spectrometry (Lipids, 1972, 7: 398).

Chapter 3

The Cyclobutane Group

IAN FLEMING

1. Occurrence, formation and general properties

Cyclobutanes, long regarded as rather difficult to make, have recently received a lot of attention. In the five years from 1966 to 1970 there have been published approximately six hundred papers in which the formation of a cyclobutane ring is described. While the generalisation that cyclobutanes are more difficult to prepare than their three- and five-membered ring counterparts remains true, it is now obvious that the difficulty has been over-emphasised.

Two or three dozen natural products containing a cyclobutane ring are now known. Cyclobutenes (R. Criegee, Angew. Chem., intern. Edn., 1968, 7: 559) and cyclobutadienes (M. P. Cava and M. J. Mitchell, "Cyclobutadiene and Related Compounds", Academic Press, New York 1967) have been reviewed, and vol. IV/4 of Houben-Weyl is devoted to cyclobutanes.

(a) Syntheses and preparative methods

For the purpose of this review, the methods of cyclobutane synthesis will be divided into eight categories according to the mechanism of the carbon—carbon bond-forming step.

(1) The best-established method involves the combination of a nucleophilic carbon with an electrophilic carbon in a ring-forming reaction.

A study (A. C. Knipe and C. J. M. Stirling, J. chem. Soc., B, 1968, 67) of the rates of ring-forming reactions of ω-halogenoalkyl malonic esters reveals that cyclobutane formation is only a little slower than cyclohexane formation, but that both are considerably slower than three- and five-membered ring formation. Much attention has recently been paid to the synthesis of bicyclic ketones, both those syntheses yielding cyclobutanones, as in the reaction of I with base (F. Nerdel, D. Frank and H. Marschall, Ber., 1967,

100: 720), and those yielding cyclobutyl ketones, most elegantly illustrated in the key step (II → III) in the synthesis of *copaene* (IV) (C. H. Heathcock, R. A. Badger and J. W. Patterson, J. Amer. chem. Soc., 1967, 89: 4133):

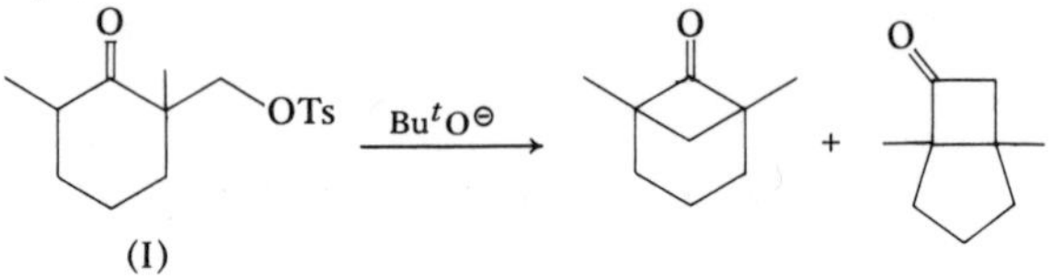

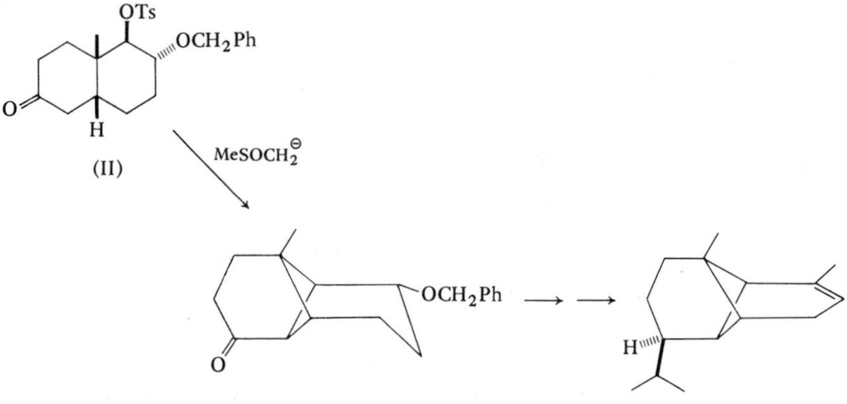

1,4-Dibromobutane gives cyclobutane itself with lithium amalgam in higher yield than usual for the Wurtz-type of reaction (D. S. Connor and E. R. Wilson, Tetrahedron Letters, 1967, 4925). The 1,4-dibromide (V) gives bicyclo[1,1,1]pentane with sodium and naphthalene (K. B. Wiberg and V. Z. Williams, J. org. Chem., 1970, 35: 369):

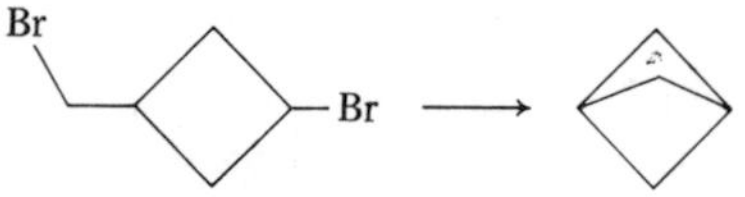

A number of cycloaddition reactions have been observed in which cyclobutane formation involves a stepwise ionic process. The reaction of enamines with electrophilic olefins giving, in the first instance, cyclobutanes, has been reviewed (A. G. Cook, "Enamines", Dekker, New York, 1969); and the cycloaddition of tetracyanoethylene to styrene and other olefins has also been reviewed (P. D. Bartlett, Quart. Review, 1970, 24: 473).

In reactions which extend the well-known homoallyl participation, it has been found that solvolysis of homopropargyl tosylates leads to three- and four-membered ring products (M. Hanack, Accounts chem. Res., 1970, 3: 209), and that allenes too can occasionally give four-membered ring products (e.g. M. Santelli and M. Bertrand, Tetrahedron Letters, 1969, 3699):

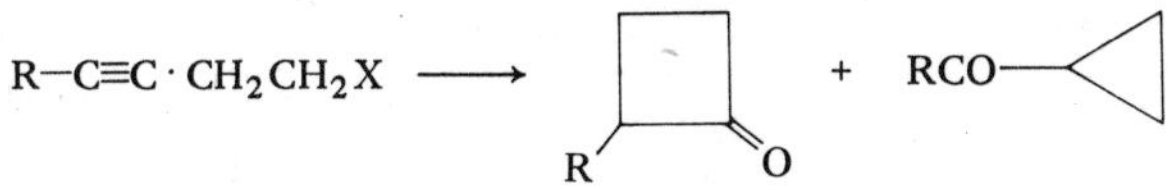

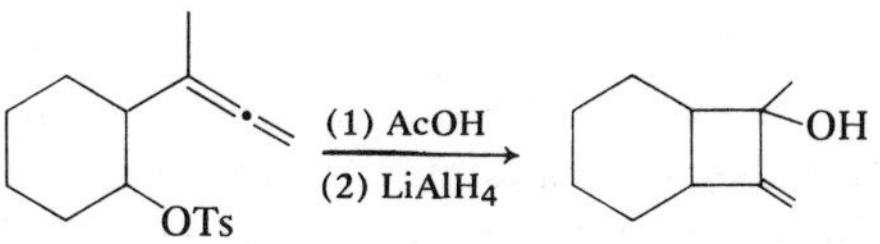

The Ramberg–Bäcklund reaction has been used to give cyclobutenes (L. A. Paquette and J. C. Philips, J. Amer. chem. Soc., 1969, 91: 3973):

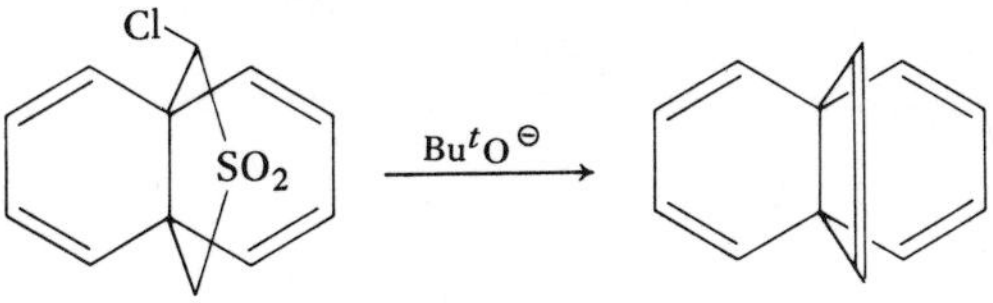

(2) A second route to cyclobutanes involves ring contraction or ring expansion. The former is not common, because it is usually unfavourable thermodynamically, but it has been observed (a) occasionally in carbonium ion chemistry (e.g. H. Tanida et al., J. Amer. chem. Soc., 1967, 89: 2928), and (b) in some photo-Wolff rearrangements (J. Meinwald and J. K. Crandall, J. Amer. chem. Soc., 1966, 88: 1292):

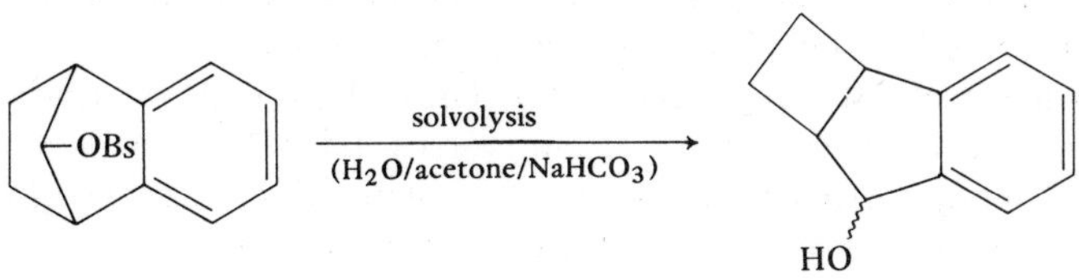

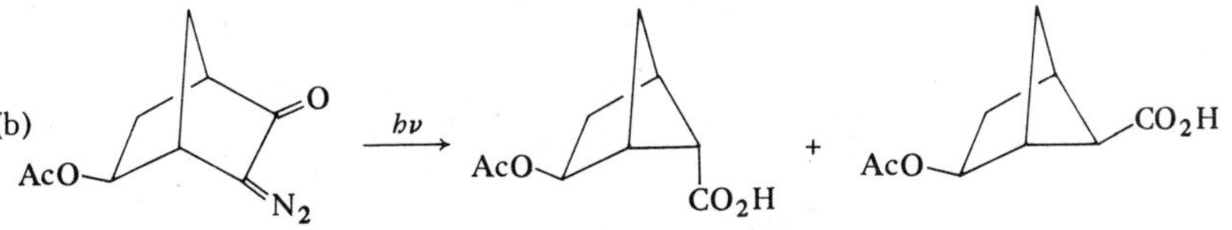

Ring expansion is more usual, and examples can be found in the many cyclopropylcarbinyl–cyclobutyl cation equilibria which have been studied, for example (K. B. Wiberg and J. E. Hiatt, Tetrahedron Letters, 1968, 3009):

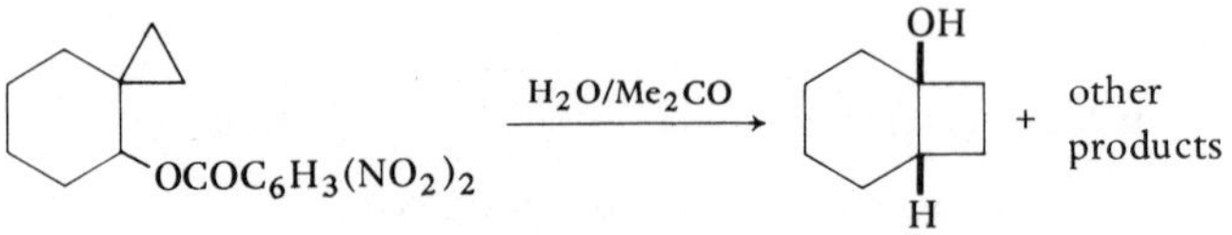

A versatile synthesis (a) of cyclobutanones belongs to this class (H. H. Wasserman, R. E. Cochoy and M. S. Baird, J. Amer. chem. Soc., 1969, 91: 2375): as does a one-step procedure (b) for the synthesis of spirocyclobutanones (B. M. Trost, R. LaRochelle and M. J. Bogdanowicz, Tetrahedron Letters, 1970, 3449, see also J. Amer. chem. Soc., 1972, 94: 4777):

(a)

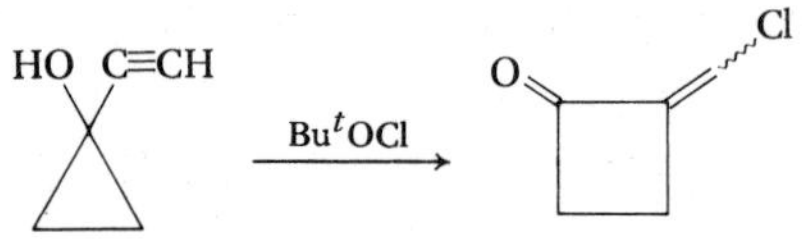

(b)

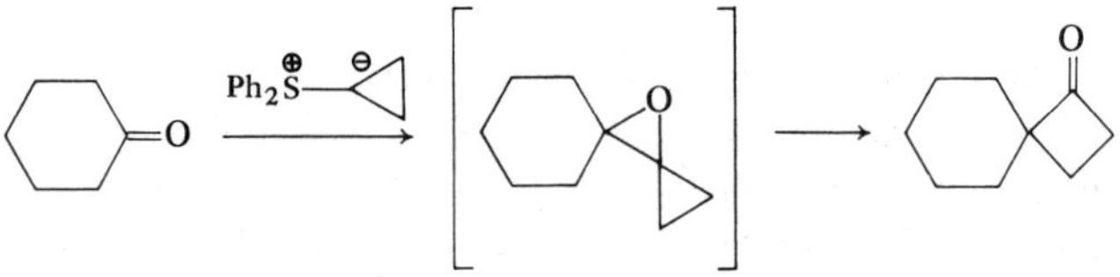

The formation of cyclobutenes from cyclopropanecarboxaldehyde and ketone tosylhydrazones is not clearly ionic, carbene or radical, but the reaction has been used quite frequently. Thus VI gives VII on irradiation (W. Kirmse and K. H. Pook, Angew. Chem., intern. Edn., 1966, 5: 594) and VIII gives IX, amongst other products (S. Masamune et al., J. Amer. chem. Soc., 1967, 89: 4804; M. Jones and S. D. Reich, ibid., p. 3935):

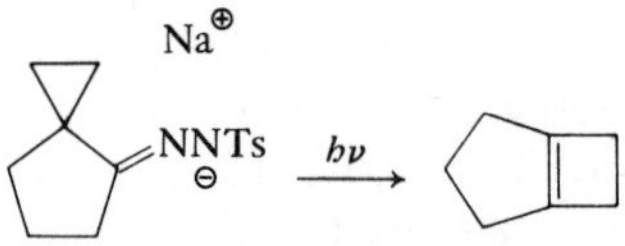

(VI)   (VII)

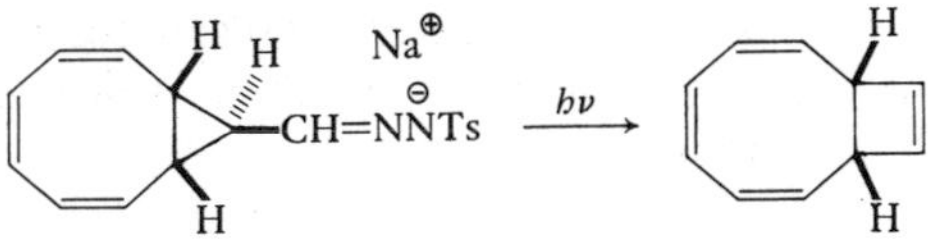

(VIII)   (IX)

(3) A number of carbene-like reactions involving insertion into C–H and C=C bonds and giving cyclobutanes have been found recently. Thus diazoketones give various amounts of cyclobutanones (X) when decomposed with silver or copper salts (E. Wenkert, B. L. Mylari and L. L. Davis, J. Amer. chem. Soc., 1968, 90: 3870):

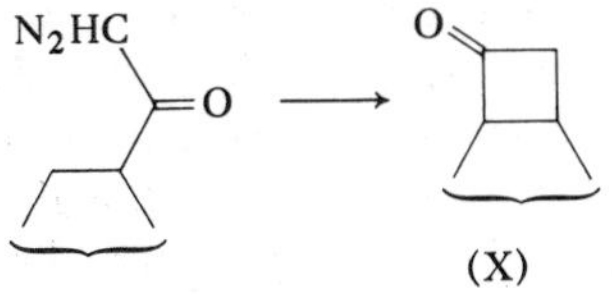

(X)

Bicyclobutylketones such as XI have been prepared by the photolysis of cyclopropenyl diazoketones (Masamune et al., J. Amer. chem. Soc., 1967, 89: 2792):

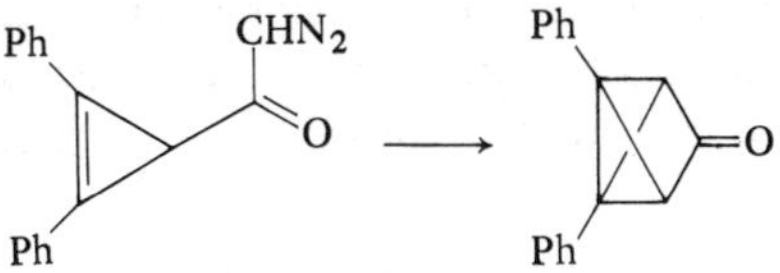

(XI)

(4) The intramolecular collapse of 1,4-diradicals is a well-established route to cyclobutanes. Recently an adaptation of the acyloin reaction, in which silylation is used to trap the intermediate enediol, has made the reaction suitable for the synthesis of four-membered rings (J. J. Bloomfield, Tetrahedron Letters, 1968, 587).

The formation of cyclobutanols which accompanies the Norrish type II photolysis of ketones, is well-established. Recent work has indicated that the proportion of cyclobutanol is greatly increased when the intermediate diradical (e.g. XIII) cannot arrange its orbitals in a favourable conformation for the type II fragmentation. Thus the ketone (XII) gives 89% of the cyclobutanol (XIV) (F. D. Lewis and T. A. Hilliard, J. Amer. chem. Soc., 1970, 92: 6672), whereas there is usually only about 10% of cyclobutanols. This factor is only important in the triplet-state reaction; in the singlet state, the proportion of cyclobutanol is little affected by geometrical constraints acting upon the diradical (I. Fleming, A. V. Kemp-Jones and E. J. Thomas, Chem. Comm., 1971, 1158).

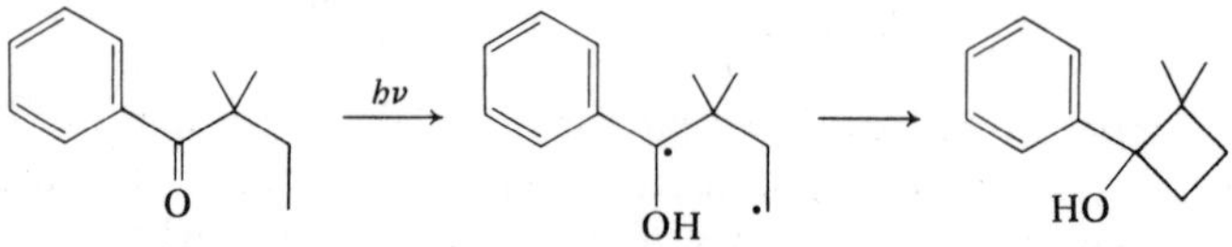

(XII)   (XIII)   (XIV)

The thermal coupling of halogenated olefins to give cyclobutanes has been established, in many cases, to be a stepwise process (P. D. Bartlett, Quart. Review, 1970, 24: 473; Science, 1968, 159: 833). The dimerisation of allenes, however, although possibly concerted (W. R. Moore et al., J. Amer. chem. Soc., 1969, 91: 5918), is currently thought to be a stepwise radical process (W. R. Dolbier and S.-H. Dai, J. Amer. chem. Soc., 1970, 92: 1774; Moore and W. R. Moser, ibid., p. 5469).

(5) Concerted, thermal [2 + 2] cycloadditions have been a particularly important subject recently because it is now clear that they must be [π2s + π2a] processes to be allowed. The geometrical constraints which this requirement places on the combining olefins is such that it is observed only in special cases. The most celebrated is the stereospecific dimerisation of the *trans*-olefin XV (K. Kraft and G. Koltzenburg, Tetrahedron Letters, 1967, 4357):

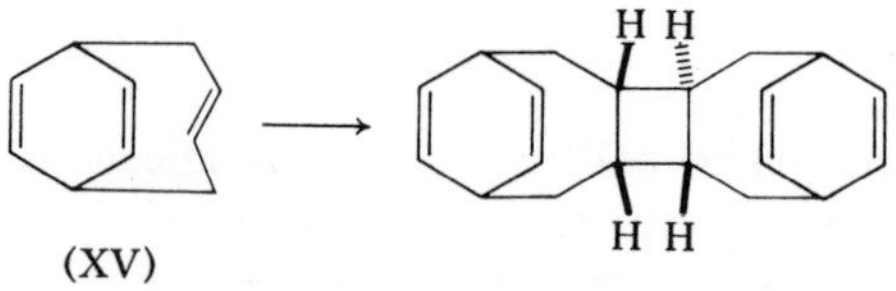

(XV)

Other closely related compounds have been found to combine non-stereo-specifically and, one presumes, non-concertedly (J. Leitich, Angew. Chem., intern. Edn., 1969, 8: 909; P. G. Gassman and H. P. Benecke, Tetrahedron Letters, 1969, 1089).

However, ketenes and vinyl cations, both of which have p-orbitals at right angles to the π-bond involved in cycloadditions, are capable of [π2s + π2a] cycloadditions (R. B. Woodward and R. Hoffmann, Angew. Chem., intern. Edn., 1969, 8: 781). This explains the Smirnov—Zamkov reaction, and also the extensive evidence (R. Huisgen et al., Ber., 1969, 102: 3391 et seq.; J. E.

Baldwin and J. A. Kapecki, J. Amer. chem. Soc., 1969, 91: 3106; H. M. Frey and N. S. Isaacs, J. chem. Soc., B, 1970, 830) that ketenes add to olefins in one step. Most recently, unsymmetrical ketenes have been found to react stereoselectively with olefins like cyclopentadiene to give, usually, the cyclobutanone with the larger group *endo* (W. T. Brady and E. F. Hoff, J. org. Chem., 1970, 35: 3733; A. S. Dreiding et al., Helv., 1970, 53: 417; T. DoMinh and O. P. Strausz, J. Amer. chem. Soc., 1970, 92: 1766; P. R. Brook, A. J. Duke and J. R. C. Duke, Chem. Comm., 1970, 574).

(6) Photochemical cycloadditions, on the other hand, are allowed for the geometrically accessible [π2s + π2s] process. They are by far the most studied of cyclobutane forming reactions. Many of the cycloadditions are photosensitised, and it is usual to assume that these take a stepwise diradical pathway. In a few cases, the singlet-state cycloadditions (or cyclo-reversals) have been found to be stereospecific. Thus the dimerisation of *cis*- and *trans*-but-2-ene has been found (H. Yamazaki and R. J. Cvetanović, J. Amer. chem. Soc., 1969, 91: 520) to be largely stereospecific. But both *cis*- and *trans*-stilbene add to dihydropyran non-stereospecifically (H. M. Rosenberg, R. Rondeau and P. Servé, J. org. Chem., 1969, 34: 471), and the 1,2-dichloroethylenes add to cyclopentenone non-stereospecifically (W. L. Dilling et al., J. Amer. chem. Soc., 1970, 92: 1399). It seems likely that most photocycloadditions are two-step, diradical reactions, and that the concerted reaction is observed only when the potential radical centres are very little stabilised by substituents.

Current interest in the mechanism of the two-step reactions is not yet matched by any clear understanding of the process. Attention has been focussed particularly on the nature of the exited state (E. J. Corey et al., J. Amer., chem. Soc., 1964, 86: 5570; O. L. Chapman et al., ibid., 1968, 90: 1657; R. O. Loutfy, P. de Mayo and M. F. Tchir, ibid., 1969, 91: 3984; Chapman and R. D. Lura, ibid., 1970, 92: 6352; D. O. Cowan and R. L. E. Drisko, ibid., p. 6286; D. Bryce-Smith, A. Gilbert and J. Grzonka, Chem. Comm., 1970, 498; R. A. Crellin, M. C. Lambert and A. Ledwith, ibid., p. 682; Rosenberg, Rondeau and Servé, op. cit.; J. J. McCullough et al., Canad. J. Chem., 1969, 47: 757 and Chem. Comm., 1970, 948; D. I. Schuster and D. H. Sussman, Tetrahedron Letters, 1970, 1657; Dilling, Chem. Reviews, 1969, 69: 845), on the orientation and stereochemistry of addition (De Mayo, S. T. Reid and R. W. Yip, Canad. J. Chem., 1964, 42: 2828; P. Yates et al., ibid., 1967, 45: 2927, 2933; D. Valentine, N. J. Turro and G. S. Hammond, J. Amer. chem. Soc., 1964, 86: 5202; H. Ziffer et al., ibid., 1970, 92: 1597; Dilling et al., ibid., pp. 928, 1399; J. W. Hanifin and E. Cohen, ibid., 1969, 91: 4494; J. Carnduff et al., Chem. Comm., 1969, 1218; G. W. Griffin et al., Tetrahedron Letters, 1968, 6173; S. Sasson, I. Rosenthal and D. Elad, ibid.,

1970, 4513; C. H. Krauch, S. Farid and G. O. Schenck, Ber., 1966, 99: 625), on the effect of solvent (De Mayo et al., Chem. Comm., 1967, 704; 1968, 982; G. Mark, F. Mark and O. E. Polansky, Ann., 1968, 719: 151; H. Morrison and R. Kleopfer, J. Amer. chem. Soc., 1968, 90: 5037), and on the effect of the crystal lattice in solid-state photochemistry (M. Lahav and G. M. J. Schmidt, J. chem. Soc., B, 1967, 312). There are several reviews available (Dilling, Chem. Reviews, 1969, 69: 845; P. Eaton, Accounts chem. Res., 1968, 1: 50; G. J. Fonken in "Organic Photochemistry", Dekker, New York, Vol. I, and D. J. Trecker, ibid., Vol. II). A widely based account of cycloadditions, both thermal and photochemical ones, concludes that they are generally concerted reactions in which the substituents modify the otherwise strict rules (N. D. Epiotis, J. Amer. chem. Soc., 1972, 94: 1924 et seq.).

Photochemical cycloadditions have often been used in the synthesis of natural products (P. G. Sammes, Quart. Review, 1970, 24: 37). Particularly useful, from a synthetic point of view, have been cycloadditions of α,β-unsaturated ketones to isolated double bonds, to oxygenated double bonds and to allenes (E. J. Corey, R. B. Mitra and H. Uda, J. Amer. chem. Soc., 1964, 86: 485; Corey and S. Nozoe, ibid., p. 1652; J. D. White and D. N. Gupta, ibid., 1966, 88: 5364; J. B. Siddall et al., ibid., 1970, 92: 425; De Mayo et al., Chem. Comm., 1967, 704: K. Wiesner et al., Tetrahedron Letters, 1965, 2441; 1967, 1523; Z. Valenta, Canad. J. Chem., 1970, 48: 1436).

Other synthetically useful photochemical cycloadditions have been those which are intramolecular. Cubane (P. E. Eaton and T. W. Cole, J. Amer. chem. Soc., 1964, 86: 3157; J. C. Barborak, L. Watts and R. Pettit, ibid., 1966, 88: 1328; N. B. Chapman, J. M. Key and K. J. Toyne, J. org. Chem., 1970, 35: 3860), hexamethylprismane (D. M. Lemal and J. P. Lokensgard, J. Amer. chem. Soc., 1966, 88: 5934; R. Criegee et al., Angew. Chem., intern. Edn., 1967, 6: 78), Dewar benzenes (E. E. van Tamelen and S. P. Pappas, J. Amer. chem. Soc., 1962, 84: 3789; H. R. Ward and J. S. W. Wishnok, ibid., 1968, 90: 1085; M. G. Barlow, R. N. Haszeldine and R. Hubbard, J. chem. Soc., C, 1970, 1232), basketane (W. G. Dauben and D. L. Whalen, Tetrahedron Letters, 1966, 3743), bicyclo[1,1,0]butanes (Dauben and J. S. Ritscher, J. Amer. chem. Soc., 1970, 92: 2925), and a great many other cage compounds (e.g. H. Prinzbach et al., Pure appl. Chem., 1968, 16: 17; Helv., 1970, 53: 2201) have been prepared in this way.

(7) Transition-metal carbonyls and complex ions have been found to catalyse the dimerisation of olefins. Thus C. W. Bird, R. C. Cookson and J. Hudec (Chem. and Ind., 1960, 20) found that norbornadiene with nickel carbonyl gave the dimer XVI:

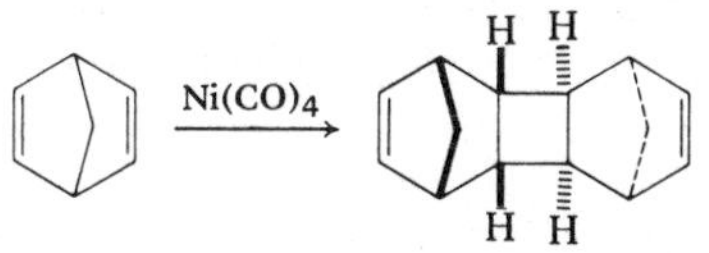

(XVI)

Several other workers have found various metal catalysts for both thermal and photochemical dimerisation of this and other olefins (W. Jennings and B. Hill, J. Amer. chem. Soc., 1970, 92: 3199 and refs. therein). It is too early yet to say what the mechanism of this reaction is, but F. D. Mango (Tetrahedron Letters, 1969, 4813; see also W. T. van der Lugt, ibid., 1970, 2281) has stated the case for the participation of the metal in such a way that it converts a disallowed into an allowed reaction, and T. J. Katz and S. A. Cerefice (J. Amer. chem. Soc., 1969, 91: 6519) have provided some evidence for a stepwise pathway for this kind of reaction.

The reversibility of the reaction (P. Heimbach and H. Hey, Angew. Chem., intern. Edn., 1970, 9: 528) has considerable potential usefulness in olefin metathesis (N. Calderon et al., J. Amer. chem. Soc., 1968, 90: 4133) and for making macrocycles (E. Wasserman, D. A. Ben-Efraim and R. Wolovsky, J. Amer. chem. Soc., 1968, 90: 3286).

(8) In a few miscellaneous reactions a four-membered ring is formed adventitiously. Thus dienophiles reacting with cyclobutadiene give bicyclo-[2,2,0]hexenes (Pettit et al., J. Amer. chem. Soc., 1966, 88: 623, 1328; Criegee and R. Huber, Ber., 1970, 103: 1855, 1862). The cycloadditions of cycloheptatriene (S. Ito, Y. Fujisi and M. C. Woods, Tetrahedron Letters, 1967, 1059; Ito, Fujisi and M. Sato, ibid., 1969, 691) give cage compounds, formed by successive [6 + 4] and [4 + 2] cycloadditions, in which one ring is a cyclobutane. And many valence tautomerisms, of cyclooctatrienes for example (W. J. Farrissey et al., Tetrahedron Letters, 1964, 3635; T. S. Cantrell, J. Amer. chem. Soc., 1970, 92: 5480) and of the divinylcyclo-propane (XVII) (M. S. Baird and C. B. Reese, Chem. Comm., 1970, 1519), give cyclobutanes:

(XVII)

(b) General properties of cyclobutanes

Most of the interesting reactions of cyclobutanes are ring-opening reactions, relieving ring strain, under conditions which would not normally be conducive to the breaking of carbon–carbon bonds. Thus phenylmagnesium bromide reacts with the ester XVIII to give the ketone XIX (L. Weintraub et al., J. org. Chem., 1965, 30: 1805):

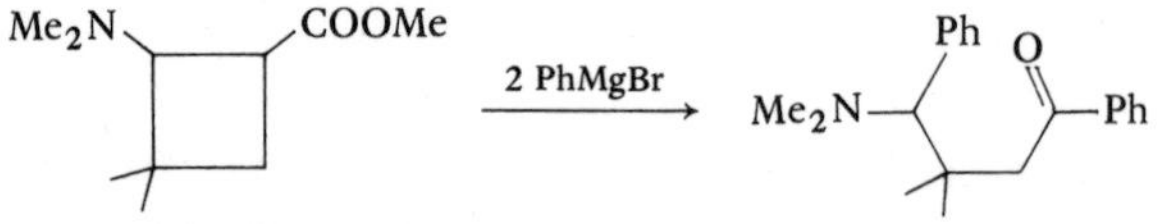

(XVIII)   (XIX)

The photochemistry of cyclobutanones is unusual in giving acetals (N. J. Turro and R. M. Southam, Tetrahedron Letters, 1967, 545; see also P. Dowd, A. Gold and K. Sachdev, J. Amer. chem. Soc., 1970, 92:5724):

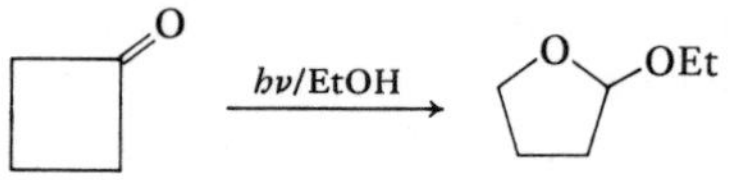

The reverse of this ring-expansion step has also been observed (A. M. Foster and W. C. Agosta, J. Amer. chem. Soc., 1972, 94: 5777):

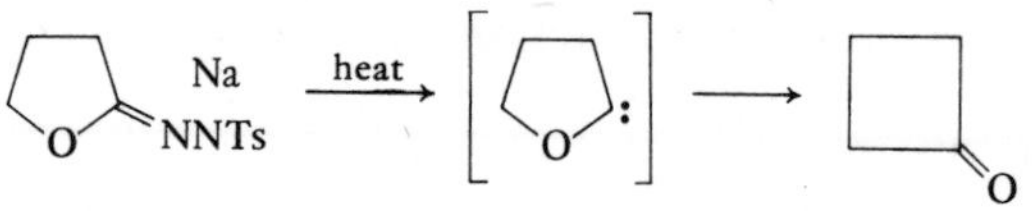

Cyclobutylmethyl sulphonates (e.g. XX) solvolyse with ring expansion (Dauben and J. L. Chitwood, J. Amer. chem. Soc., 1970, 92: 1624) and, in the case of XX, 10^4 times faster than neopentyl tosylate:

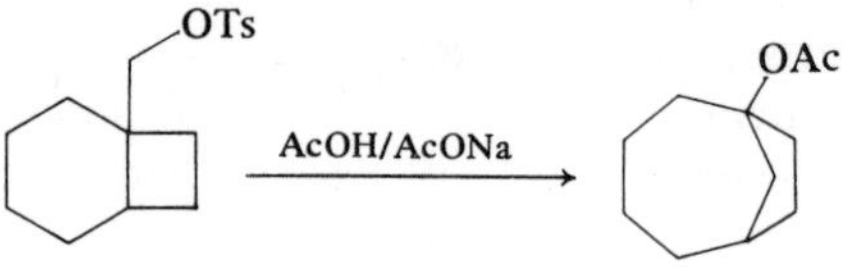

(XX)

3-Hydroxycyclobutyl ketones can be oxidised to give 1,4-diketones (Z. Valenta et al., Canad. J. Chem., 1970, 48: 1436):

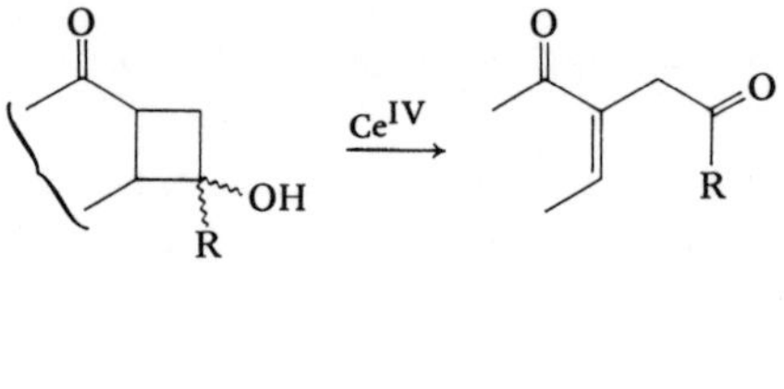

The exceptionally strained cyclobutanol XXI opens in base (A. Padwa and E. Alexander, J. Amer. chem. Soc., 1970, 92: 5674) as do cyclopropanols:

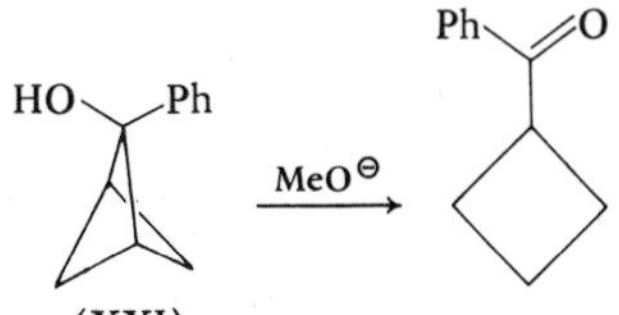

HO—Ph

$\xrightarrow{\text{MeO}^{\ominus}}$

Ph—O

(XXI)

The opening of the bicyclo[2,1,0]pentene (XXII) occurs by the [σ2s + σ2a] path, which is allowed, and gives the cyclopentadiene XXIII as the first-formed product (J. E. Baldwin and A. H. Andrist, Chem. Comm., 1970, 1561):

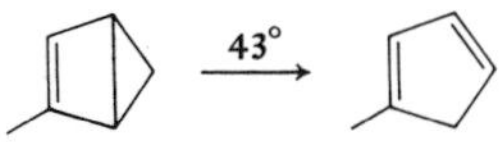

$\xrightarrow{43°}$

(XXII)   (XXIII)

This conclusion has been questioned (S. McLean, D. M. Findlay and G. I. Dmitrienko, J. Amer. chem. Soc., 1972, 92: 1380) and supported by further evidence (J. E. Baldwin and G. D. Andrews, J. Amer. chem. Soc., 1972, 92: 1775).

It is now well established that the conformation of cyclobutanes is puckered in most cases. Centrosymmetrically substituted cyclobutanes are sometimes, but not always, planar in the solid state (T. N. Margulis, Chem. Comm., 1969, 215, and refs. therein). The puckering causes the bonds to substituents to have axial and equatorial character. Thus it seems likely that the *cis-* and the *trans-*bicyclo[4,2,0]octanes (XXIV and XXV) will be found to be not very different in energy, because the otherwise strained *trans-*fused rings will have the ring residues equatorial in both rings (see K. B. Wiberg, J. E. Hiatt and K. Hseih, J. Amer. chem. Soc., 1970, 92: 544):

(XXIV)   (XXV)

2. Natural products containing a cyclobutane ring

Pinene, verbenone, myrtenal, caryophyllene, the *truxillic* and *truxinic acids,* the *lumicolchicines, chrysanthenone* (E. P. Blanchard, Chem. and Ind., 1958,

293), *lyconnitine* (K. Wiesner et al., Tetrahedron, 1958, 4: 87), *flavenso-mycenic acid* (L. Canonica, G. Jommi and F. Pelizzoni, Tetrahedron Letters, 1961, 537), and α-*longipinene* (H. Erdtman and L. Westfelt, Acta chem. Scand., 1963, 17: 2351) have been known for several years. More recently several sesquiterpenes with the copaene (IV) skeleton (p. 50) have been found: *copaene* (G. Büchi et al., Proc. chem. Soc., 1963, 214), *mustakone* (S. Dev, Tetrahedron Letters, 1963, 1933), β-*copaene* and β-*ylangene* (Westfelt, Acta chem. Scand., 1967, 21: 152), α-*ylangene* (Y. Ohta and Y. Hirose, Tetrahedron Letters, 1969, 1601), *copadiene* (Dev et al., Tetrahedron Letters, 1967, 4661), and the *brachylaenolones* (C. J. W. Brooks and M. M. Campbell, Chem. Comm., 1969, 630). There are two other sesquiterpene skeletons containing a cyclobutane ring: that of α-*bergamotene* (XXVI) (E. Kovats, Helv., 1963, 46: 2705; T. W. Gibson and W. F. Erman, J. Amer. chem. Soc., 1969, 91: 4771) and that of the *bourbonenes* (XXVII) (F. Sorm et al., Tetrahedron, 1967, Suppl. 8: 53; Tetrahedron Letters, 1966, 359, 3209; M. Brown, J. org. Chem., 1968, 33: 162; C. Gianotti and H. Schwang, Bull. Soc. chim. Fr., 1968, 2452):

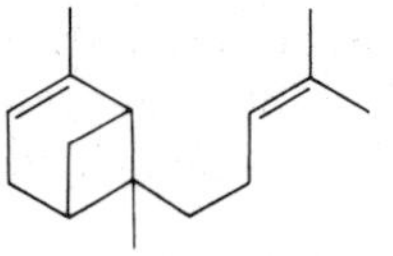

(XXVI) 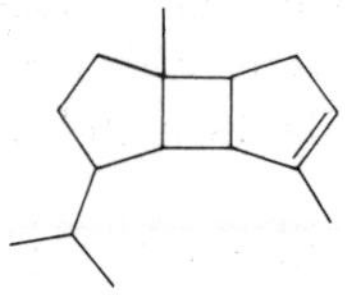   (XXVII)

Kobusone and *isokobusone* (H. Hikino, K. Aota and T. Takemoto, Chem. pharm. Bull., Japan, 1969, 17: 1390) are related to caryophyllene. Other natural products with cyclobutane rings are *fomannosin* (J. A. Kepler et al., J. Amer. chem. Soc., 1967, 89: 1260), *filifolone* (R. B. Bates et al., Chem. Comm., 1967, 1037), *illudol* (T. C. McMorris, M. S. R. Nair and M. Anchel, J. Amer. chem. Soc., 1967, 89: 4562), *paeoniflorin* (S. Shibata et al., Tetrahedron, 1969, 25: 1825), the *coriolins* (S. Takahashi et al., Tetrahedron Letters, 1969, 4663; 1970, 1637), *bicyclomahanimbicine* (S. P. Kureel, R. S. Kapil and S. P. Popli, Chem. and Ind., 1970, 958; but see L. Crombie et al., Chem. Comm., 1970, 1547, for a revision of the structure), and a component of the boll weevil pheromone (J. B. Siddall et al., J. Amer. chem. Soc., 1970, 92: 425).

Chapter 4

The Cyclopentane Group

P. W. HICKMOTT

1. Synthesis and spatial geometry of cyclopentane compounds

(a) *Preparative methods*

For the more important general methods of forming cyclopentyl compounds see Vol. II A, 1st and 2nd Edns., pp. 71 and 104 et seq., respectively. The following methods exemplify the formation of five-membered rings by ring-contraction, ring-expansion and some miscellaneous cyclizations. Further examples may be found in the text. Carbocyclic ring-expansion reactions in general have been reviewed (C. D. Gutsche and D. Redmore, Advances in Alicyclic Chemistry, Suppl. 1, eds. H. Hart and G. J. Karabatsos, Academic Press, New York, 1968).

(i) *Ring-contraction*

Photolysis of 2-diazo-5,5-dimethylcyclohexane-1,3-dione gives 3,3-dimethylcyclopentanone (18%). Reaction occurs via the intermediacy of a keten which can be trapped, in methanol, to give methyl 4,4-dimethyl-2-oxocyclopentane carboxylate (H. Veschambre and D. Vocelle, Canad. J. Chem., 1969, 47: 1981). Similarly photolysis of 2-diazo-4,5-diphenylcyclohexane-1,3-dione in carbon tetrachloride gives *trans*-3,4-diphenylcyclopentanone (W. D. Barker et al., Canad. J. Chem., 1969, 47: 2853). Photolysis of 6-acetoxy-3,5-bis-(chloromethyl)-2,4,6-trimethylcyclohexa-2,4-dien-1-one in ether gave the bicyclic intermediate I which rearranged to II on heating (M. R. Morris and A. J. Waring, Chem. Comm., 1969, 526):

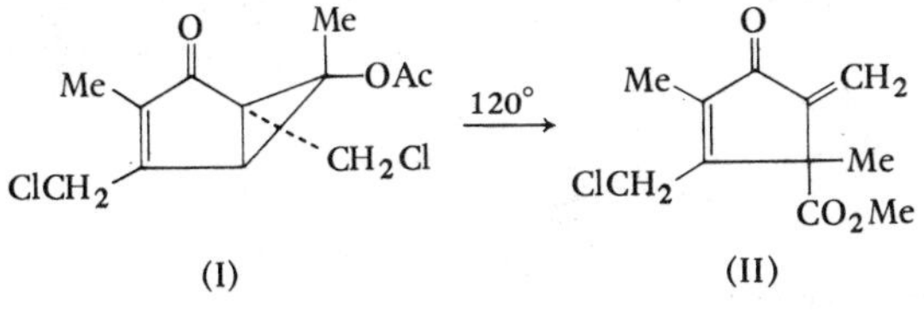

(I)   (II)

[61]

1-Alkoxycarbonyl-2,4-diphenyl-5-arylcyclopenta-1,4-dien-3-ones are formed by heating 6-hydroxy-6-alkoxy-1,3-diphenyl-4-arylcyclohex-3-ene-2,5-diones in acetic anhydride (W. Ried, W. Kunkel and P. B. Olschewski, Ann., 1969, 724: 199). Ring-contraction may also be effected by Clemmensen reduction of 2,2,4,4,6,6-hexamethylcyclohexa-1,3,5-trione (T. J. Curphey and R. C. McCartney, J. org. Chem., 1969, 34: 1964):

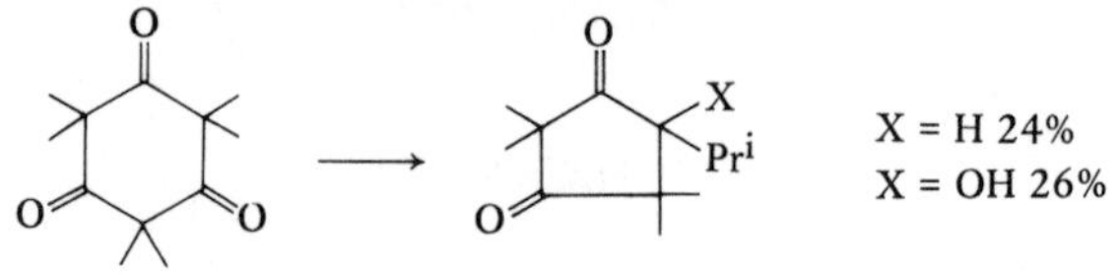

X = H 24%
X = OH 26%

Pyrolytic ring-contraction of *trans*-1-hydroxy-2-acetoxycyclohexanes gives cyclopentyl alkyl ketones (J. C. Leffingwell and R. E. Shackelford, Tetrahedron Letters, 1970, 2003):

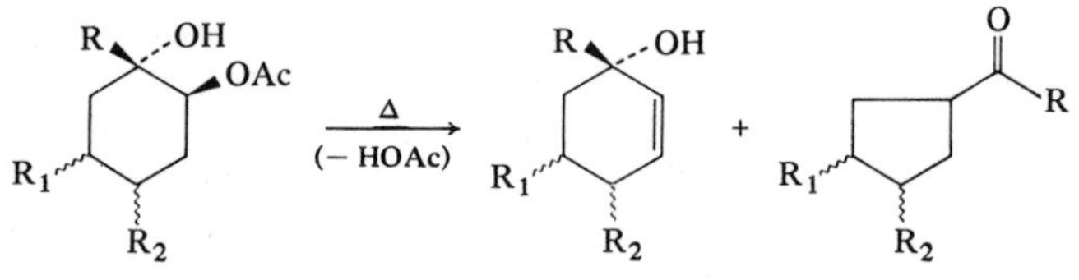

Treatment of 2,2,4,4,6,6-hexabromodimedone with sodium acetate gives 2,3,5,5-tetrabromo-4,4-dimethylcyclopenta-2-en-1-one (95%) (H. de Pooter and N. Schamp, Bull. Soc. chim. Belg., 1969, 78: 17).

Acid-induced ring-contraction of hexamethyldewarbenzene (HMDB) gives 1-(1-substituted ethyl)pentamethylcyclopentadienes (R. Criegee and H. Grüner, Angew. Chem., intern. Edn., 1968, 6: 467). To account for the fact that both moieties of the acid reagent become bonded to the same carbon atom, it has been proposed that protonation of HMDB occurs from its *endo*-surface with subsequent or concomitant migration of the central bond of the resulting carbonium ion (L. A. Paquette and G. R. Krow, Tetrahedron Letters, 1968, 2139; see also M. Kunz and W. Lüttle, Ber., 1970, 103: 315):

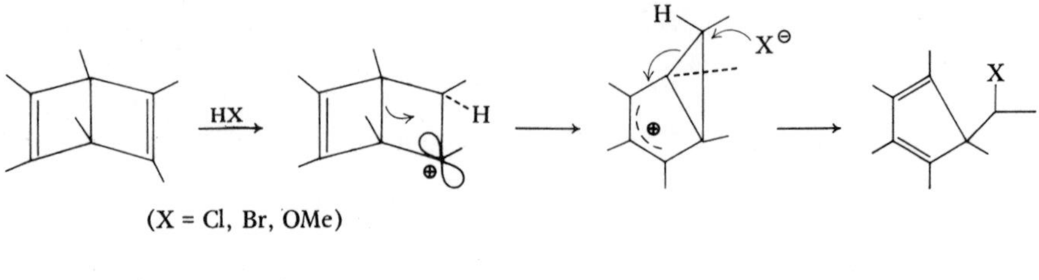

(X = Cl, Br, OMe)

Oxidation of HMDB with perbenzoic acid gives the *endo*-oxide which, on treatment with acid, rearranges to 1-acetyl-1,2,3,4,5-pentamethylcyclopenta-2,4-diene (H.-N. Junker, W. Schäfer and H. Niedenbrück, Ber., 1967, 100: 2508; Paquette and Krow, loc. cit.):

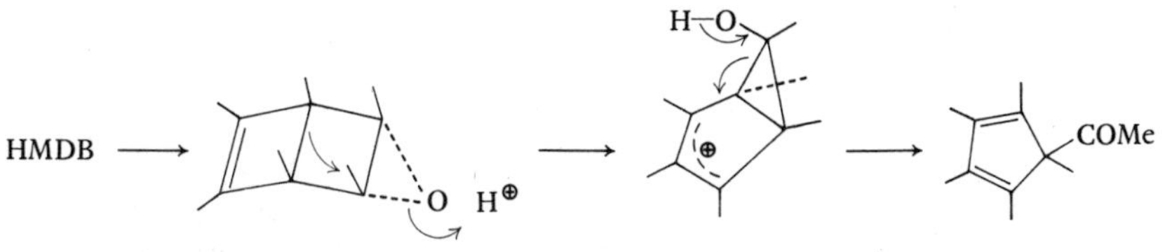

The intermediacy of bicyclo[3,1,0]hexenyl and bicyclo[2,1,1]hexenyl cations, in these rearrangements, is indicated by low temperature n.m.r. studies (Paquette et al., J. Amer. chem. Soc., 1968, 90: 7147). Treatment of hexamethyldewarbenzene with rhodium trichloride in methanol results in a novel ring-contraction to give a pentamethylcyclopentadienylrhodium complex (J. W. Kang and P. M. Maitlis, J. Amer. chem. Soc., 1968, 90: 3259; B. L. Booth, R. N. Haszeldine and M. Hill, J. chem. Soc., A, 1969, 1299):

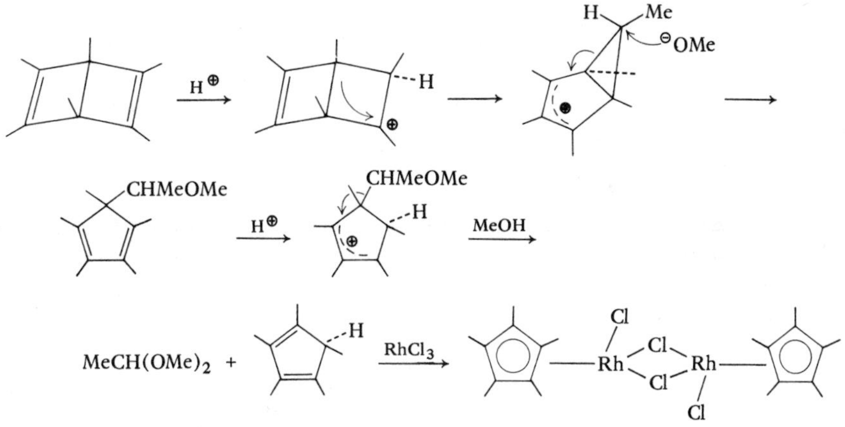

1-Acetyl-2-methylcyclopentene is obtained by a dehydrogenation—acetylation rearrangement of cyclohexane in the presence of acetyl chloride and aluminium chloride. Hydride abstraction by the acetyl chloride—aluminium chloride complex is postulated, followed by rearrangement of the cyclohexyl cation produced to a tertiary carbonium ion (I. Tabushi, K. Fujita and R. Oda, Tetrahedron Letters, 1968, 4247):

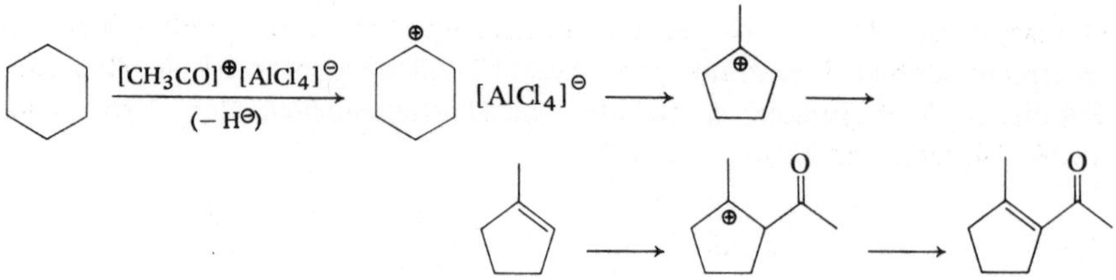

Branched hydrocarbons (viz. methylcyclohexane or methylcyclohexene) undergo acetylation without ring-contraction (idem, ibid., 1968, 5455). Cyclohex-2-en-1-one is converted quantitatively into 3-methylcyclopent-2-enone on treatment with HF–SbF$_5$ (H. Hageveen, Rec. Trav. chim., 1968, 87: 1295).

(ii) Ring-expansion

Cyclopentenes are obtained by thermal or photolytic rearrangement of cyclopropylacrylic esters (M. J. Jorgenson and C. H. Heathcock, J. Amer. chem. Soc., 1965, 87: 5264) and 2-cyclopropyl-1-phenylethylenes (P. H. Mazzocchi and R. C. Ladenson, Chem. Comm., 1970, 469, and refs. therein):

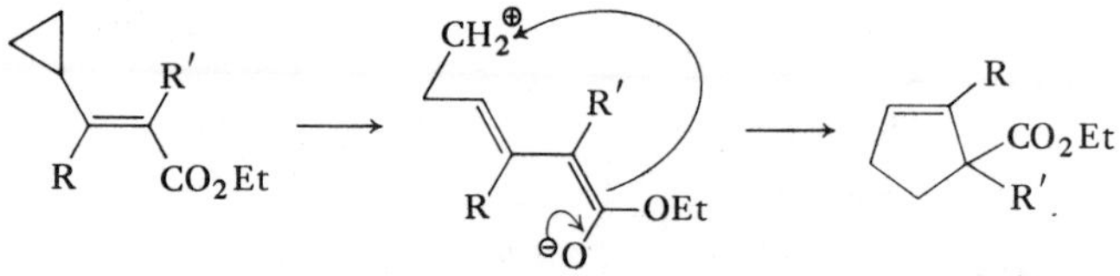

Thermolysis of bicyclo[2,1,0]pentanes as, for example, in the formation of 1-acetyl-1,2-dimethylcyclopent-2-ene and 1-acetyl-2,3-dimethylcyclopent-2-ene (Jorgenson and A. F. Thacher, Chem. Comm., 1969, 1030), also leads to cyclopentenes:

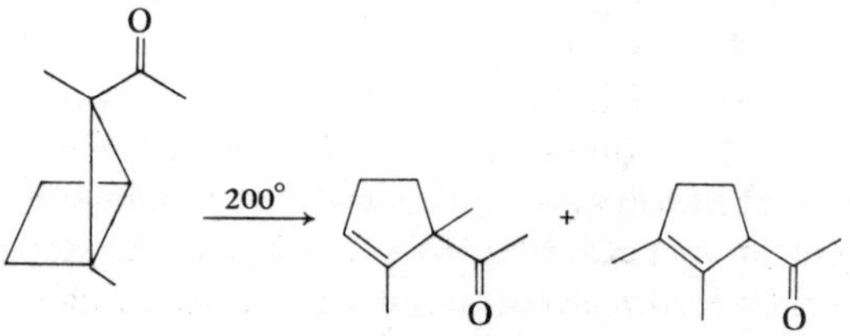

Photolysis of 1-cyclopropyl-1-alkylbutadienes gives alkylvinylcyclopentenes (Ital. P. 793,265/1967; C.A., 1969, 70: 37286u):

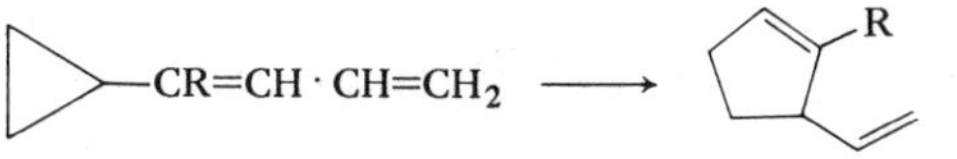

$$R = C_1 - C_5 \text{ alkyl}$$
$$C_3 - C_5 \text{ cycloalkyl}$$

Thermal rearrangement of 1-phenyl-2-vinylcyclopropanes gives 4-phenyl-cyclopentenes (I. G. Bolesov et al., Zhur. obshcheĭ Khim., 1969, 5: 1707; C.A., 1970, 72: 3127u).

(iii) Miscellaneous cyclizations

Radical cyclization of $\Delta^{6,7}$-unsaturated esters, nitriles and ketones occurs on treatment with dibenzoylperoxide (M. Julia and M. Maumy, Bull. Soc. chim. Fr., 1969, 2415, 2427) to give a mixture of cyclopentane (IV) and cyclohexane (V) derivatives:

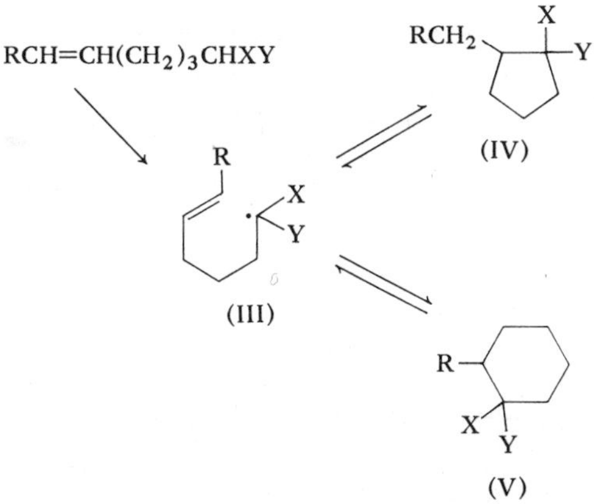

$$X = CO_2Et, Y = H \text{ or } CO_2Et$$
$$X = Cl, Y = CO_2Et$$
$$X = Ac, Y = H \text{ or } CO_2Et$$

The ratio of IV:V depends markedly on the nature of the substituents present (idem, ibid., 1966, 434) and on the experimental conditions. Under conditions of kinetic control the cyclopentane derivative (IV) is the major product; under conditions of thermodynamic control and with two electronegative substituents present (X and Y = CN or CO_2R) to stabilize the radical (III) homolytic cleavage of the cyclopentane occurs and the cyclohexane derivative (V) is the major product (Julia, Maumy and L. Mion, ibid., 1967, 2641). Radical cyclization also occurs during the addition of perfluoroalkyl or trichloromethyl radicals to ethyl diallylacetate or ethyl diallylmalonate, thus giving entry into partially halogenated acids (viz., ethyl 3-methyl-4-(perfluoropropyl)methylcyclopentyl-1-carboxylate) (N. O. Brace, J. org. Chem., 1969, 34, 2441). Acylcyclopentenes are obtained by thermal cyclization of ω-acetylenic ketones at 300° (J.-M. Conia et al., Bull. Soc. chim. Fr., 1967, 3554; Tetrahedron, 1968, 24: 5971; Fr. Addn. 91,043/1966 to Fr. P. 1,458,236):

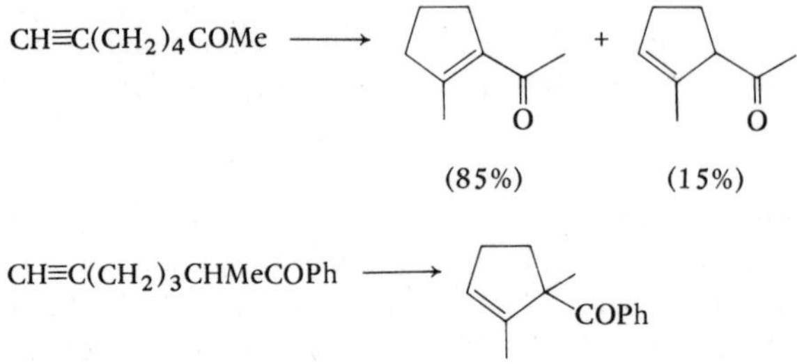

$$CH{\equiv}C(CH_2)_4COMe \longrightarrow$$

(85%)   (15%)

$$CH{\equiv}C(CH_2)_3CHMeCOPh \longrightarrow$$

Condensation of methyl ketones with dimethyl dimethylmaleate in the presence of sodium hydride gives methyl 4-acyl-2-methyl-3-oxocyclo-pentane-1-carboxylates (M. Elliott, N. F. Jones and K. A. Jeffs, J. chem. Soc., C, 1845, 1969):

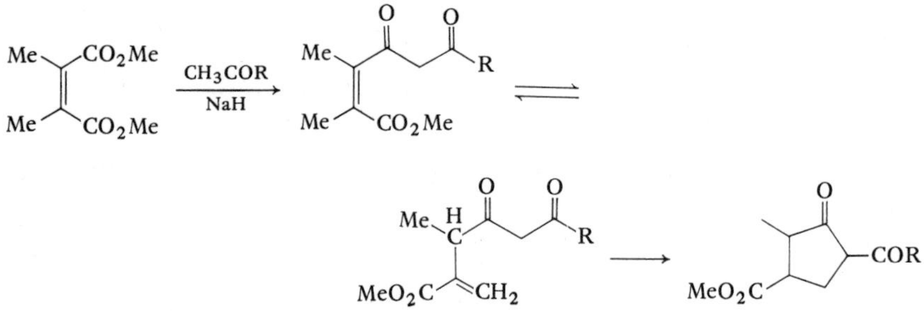

Epoxidation of substituted vinylallenes gives cyclopent-2-en-1-ones (J. Grimaldi and M. Bertrand, Tetrahedron Letters, 1969, 3269):

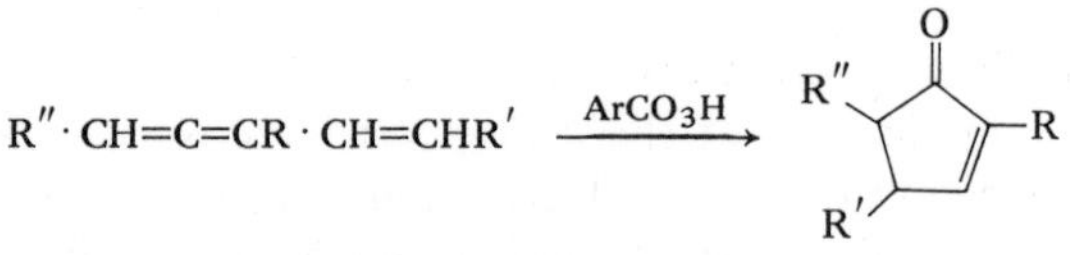

$$R''{\cdot}CH{=}C{=}CR{\cdot}CH{=}CHR' \xrightarrow{ArCO_3H}$$

(R = H, Me, Pr; R' = H, Me; R'' = H, Pr)

(b) Conformational analysis

Unlike six-membered rings, where the energy difference between possible conformations is sufficiently large for one conformation, the chair, to be strongly favoured in the majority of compounds, the precise conformation of five-membered rings depends on the number and nature of the substituents present. Non-bonded repulsions between adjacent substituents are minimized by puckering of the ring. However, the energy differences between the various puckered conformations is generally small and the ring is particularly mobile. The process whereby the angle of puckering moves round the ring (pseudo-rotation), a concept introduced by Pitzer et al. to account for the observed spectral and thermodynamic data for cyclopentane (J. E. Kilpatrick, K. S. Pitzer and R. Spitzer, J. Amer. chem. Soc., 1947, 69: 2483; Pitzer and W. E. Donath, ibid., 1959, 81: 3213) has been established (J. P. McCullough, J. chem. Phys., 1958, 29: 966; McCullough et al., J. Amer. chem. Soc., 1959, 81: 5880) and further studied by infrared spectroscopy (J. R. Durig and D. W. Wertz, J. chem. Phys., 1968, 49: 2118) and gas-phase electron diffraction (W. J. Adams, H. J. Geise and L. S. Bartell, J. Amer. chem. Soc., 1970, 92: 5013). The small difference in energy between the puckered conformations has been further supported by calculations (J. B. Hendrickson, ibid., 1961, 83: 4537; N. L. Allinger et al., ibid., 1968, 90: 1199; J. R. Hoyland, J. chem. Phys., 1969, 50: 2775). Two puckered conformations which often represent energy minima in substituted cyclopentanes are the envelope conformation, in which the substituent occupies the flap, and the half-chair conformation, favoured by methylenecyclopentane and cyclopentanone (Pitzer and Donath, loc. cit.; M. Scholz and H. J. Kohler, Tetrahedron, 1969, 25: 1863), and in which the substituents may be in equatorial or axial orientations:

envelope (C_5 symmetry)

half-chair (C_2 symmetry)

However, since the puckering in the cyclopentane ring is not as great as the complete staggering found in cyclohexane, the preference for a substituent to occupy an equatorial position is not as great either. In fact a conformational preference for the axial form in cyclopentyl fluoride and chloride has been demonstrated, assuming an envelope conformation, with the halogen substituent situated at the flap (I. O. C. Ekejiuba and H. E. Hallam, Spectrochim. Acta, 1970, 26A: 67) and cyclopentanol has been shown to be a mixture of conformers, by infrared spectroscopy (idem, J. chem. Soc., B,

1970, 209). 3,5-Disubstituted cyclopentenones and 3,5-disubstituted cyclo-pentenes are reported to have quasi-coplanar and envelope conformations, respectively (F. Cocu, Helv., 1970, 53: 739). The use of n.m.r. spectroscopy in the conformational analysis of cyclic compounds has been reviewed (H. Booth, Progr. Nucl. Magnetic Resonance Spectroscopy, eds. J. W. Emsley, J. Feeney and L. H. Sutcliffe, Pergamon, Oxford, 1969, 5: 149).

2. Hydrocarbons

(a) Saturated hydrocarbons

The cyclopentyl cation has been generated in highly acidic solvents ($F \cdot SO_3H{-}SbF_5$) by hydride abstraction from cyclopentane (G. A. Olah and J. Lukas, J. Amer. chem. Soc., 1968, 90: 933). The n.m.r. spectrum shows a single peak at $\tau 5.25$ indicating that the cation undergoes a series of 1,2-hydride shifts resulting in complete hydrogen degeneracy:

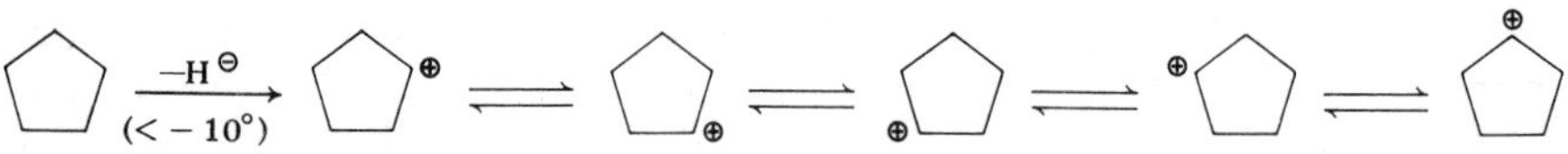

The 1-methylcyclopentyl cation, in which the positive charge is localized at the tertiary carbon atom, has been similarly generated from a variety of 1-methylcyclopentyl and cyclohexyl precursors (Olah et al., ibid., 1967, 89: 2692).

(b) Mono-unsaturated hydrocarbons

(i) Cyclopentenes

The base-catalysed double-bond migration of cyclopentene ^{14}C has been studied and the activation parameters reported (viz., $\Delta H^{\ddagger}$ 22.1 ± 1.6 kcal/mole and $\Delta S^{\ddagger}$ − 23.3 ± 2.6 e.u.) (S. B. Tjan, H. Steinberg and T. J. de Boer, Rec. Trav. chim., 1969, 88: 690). Selective degradation of [^{14}C]cyclopentene, whereby the specific activity of each ring-carbon atom can be derived, has also been described (idem, ibid., 1969, 88: 673). Addition of hydrogen bromide to cyclopentene is accompanied by isomeriz-ation so that mainly 1,2- and some 1,3-addition occurs (Yu. G. Bundel et al., Izvest. Akad. Nauk, S.S.S.R., Ser. Khim., 1969, 1403).

Sicher et al. have shown that in the Hoffmann elimination of cyclic and acyclic trimethylammonium hydroxides, trans-olefins are formed largely by a syn-elimination and cis-olefins largely by an anti-elimination mechanism

(M. Pánková, J. Sicher and J. Závada, Chem. Comm., 1967, 394, and refs. therein). Further work has shown that the *syn*-mode of elimination can also predominate in the formation of *cis*-olefins derived from small and medium ring compounds (four to eight membered), as in the case of N,N,N-trimethyl-3,3-dimethylcyclopentylammonium hydroxide (J. L. Coke and M. P. Cooke, Tetrahedron Letters, 1968, 2253).

The rearrangement of cyclohexenyl cations to cyclopentenyl cations (N. C. Deno and J. H. Houser, J. Amer. chem. Soc., 1964, 86: 1741) has been shown to be reversible and the mechanism discussed (T. S. Sorensen and K. Ranganayakulu, ibid., 1970, 92: 6539):

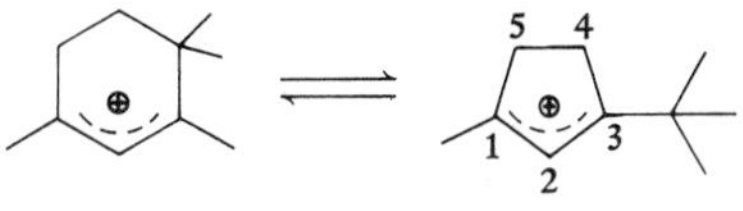

The greater stability of the cyclopentyl cation has been rationalized in terms of greater charge delocalization (Sorensen, ibid., 1969, 91: 6398) in that the smaller the ring the greater the extent of transannular 1,3-π-interaction, which places a charge at the 2-position and which has been postulated to account for the enhanced stability of cyclobutenyl cations (T. J. Katz and E. H. Gold, ibid., 1964, 86: 1600). Cyclic 6π-electron delocalization can also be envisaged in cyclopentenyl cations, by invoking both methylene groups (at $C_{(4)}$ and $C_{(5)}$) in a simultaneous hyperconjugative interaction.

Spirocyclopentenes have been prepared by ring-expansion of a cyclopentadienylcyclopropanol (H. H. Wasserman and D. C. Clagett, ibid., 1966, 88: 5368) and by malonic ester cyclization of 1,1-dimethylolcyclopent-3-ene ditoluene-*p*-sulphonate (E. J. Grubbs, D. J. Lee and A. G. Bellettini, J. org. Chem., 1966, 31: 4069):

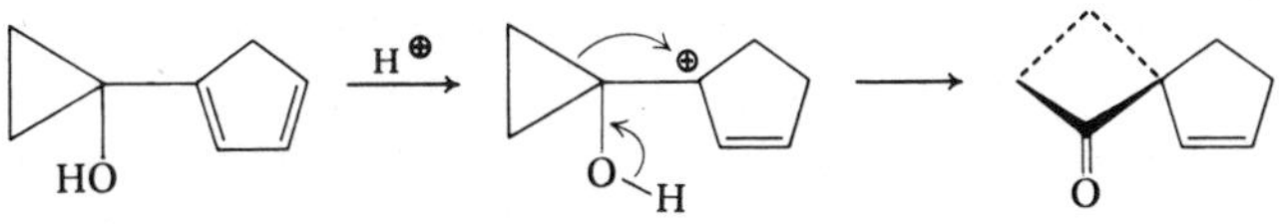

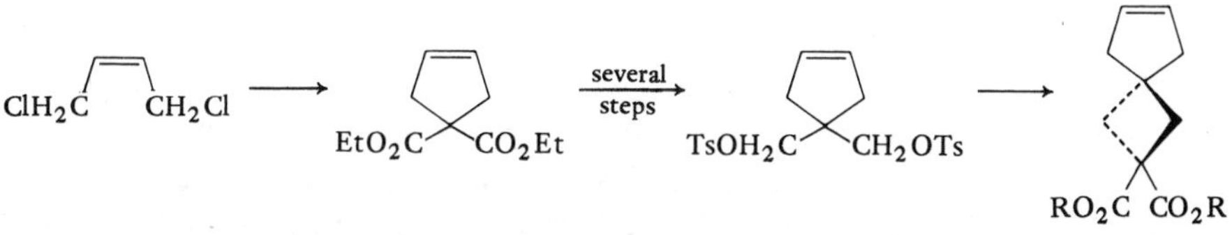

Deamination of spiro[4.5]dec-6-ylamine and spiro[4.4]non-1-ylamine gives a complex mixture of products resulting from ring-expansion or -contraction (H. Christol and J.-M. Bessière, Bull. Soc. chim. Fr., 1968, 2141, 2147). Fused bicyclic systems containing five-membered rings are obtained by cycloaddition of cyanoallenes to enamines (W. Ried and W. Käppeler, Ann., 1965, 687: 183):

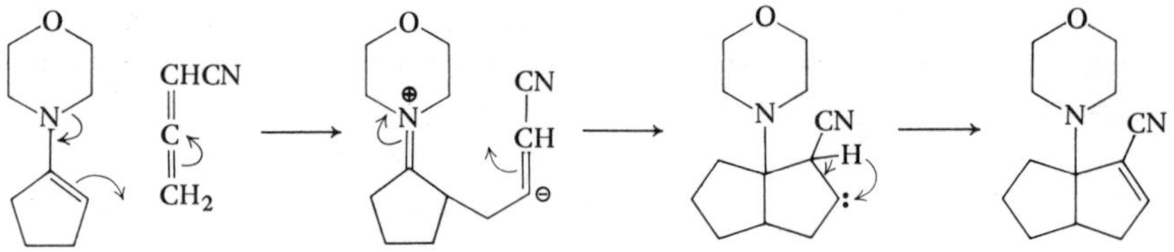

Two cyanogenetic glycosides recently shown to be cyclopentene derivatives are *gynocardin* (the β-D -glucopyranoside of 3σ-cyano-3ρ,4σ,5ρ-trihydroxy-cyclopentene) (L. Long et al., J. org. Chem., 1966, 31: 4312; Chem. Comm., 1970, 381) and *deidaclin* (the β-D -glucopyranoside of 2-cyclopenten-1-one cyanohydrin) (idem, J. Amer. chem. Soc., 1970, 92: 6378):

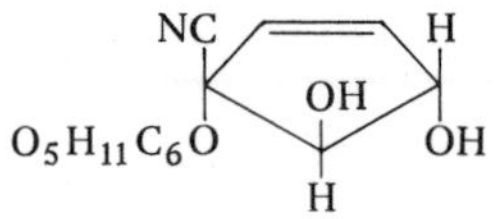

Gynocardin 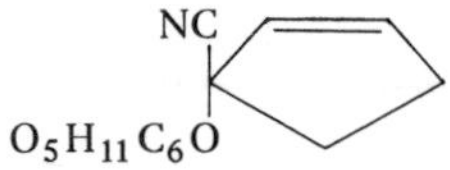   Deidaclin

(ii) Hydrocarbons with unsaturated side chains

Cyclodimerization of butadiene by nickel catalysts gives 1-vinyl-2-methyl-enecyclopentane (I) (J. Kiji, K. Masui and J. Furukawa, Tetrahedron Letters, 1970, 2561). Tetramerization of allene with halorhodium-triphenylphos-phine complexes gave 1,4,7-trimethylenespiro[4,4]nonane (II) (F. N. Jones and R. V. Lindsey, J. org. Chem., 1968, 33: 3838; S. Otsuka, A. Nakamura and H. Minamida, Chem. Comm., 1969, 191) and acid-catalysed rearrange-ment of hexamethyldewarbenzene gave III (R. Criegee and H. Grüner,

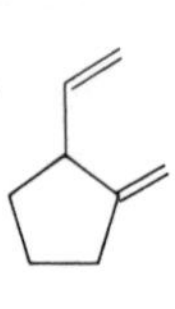

(I) 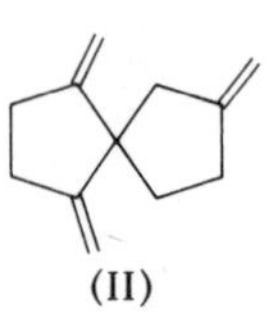   (II) 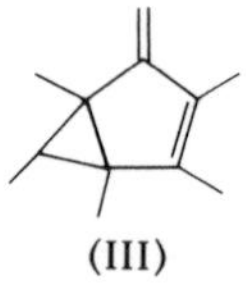   (III)

Angew. Chem., intern. Edn., 1968, 7: 467). Intramolecular cyclizations of acetylenic Grignard reagents have been shown to yield alkylidenecyclopentanes in preference to the more stable alkylcyclohexenes (H. G. Richey and A. M. Rothman, Tetrahedron Letters, 1968, 1457).

Methylenecyclopentane undergoes the ene-reaction with maleic anhydride, formaldehyde, and sulphur trioxide to give IV, V and VI respectively (R. T. Arnold, R. W. Amidon and R. M. Dodson, J. Amer. chem. Soc., 1950, 72: 2871). These reactions are related to the Diels—Alder reaction and can be regarded as intermolecular variants of the symmetry-allowed 1,5-hydrogen shift* (H. M. R. Hoffmann, Angew. Chem., intern. Edn., 1969, 8: 556):

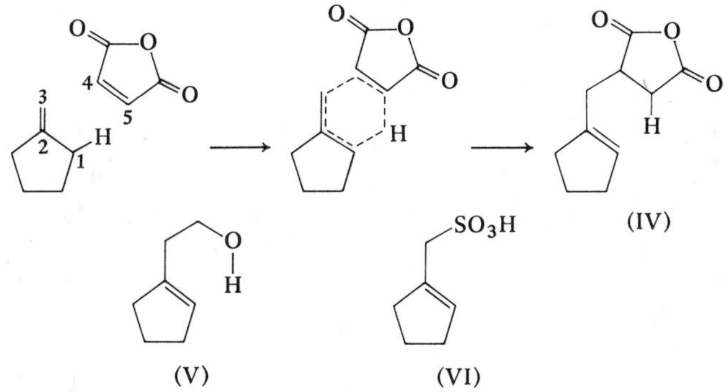

Amongst the intramolecular ene-reactions leading to 5-membered rings may be mentioned the cyclization of octa-1,6-diene to the *cis*-cyclopentane (VII) (W. D. Huntsman, V. C. Solomon and D. Eros, J. Amer. chem. Soc., 1958, 80: 5455), the conversion of 3,7-dimethylocta-1,6-diene to VIII (H. Pines, N. E. Hoffmann and V. N. Ipatieff, ibid., 1954, 76: 4412) and oct-6-en-1-yne to I (W. D. Huntsman and R. P. Hall, J. org. Chem., 1962, 27: 1988):

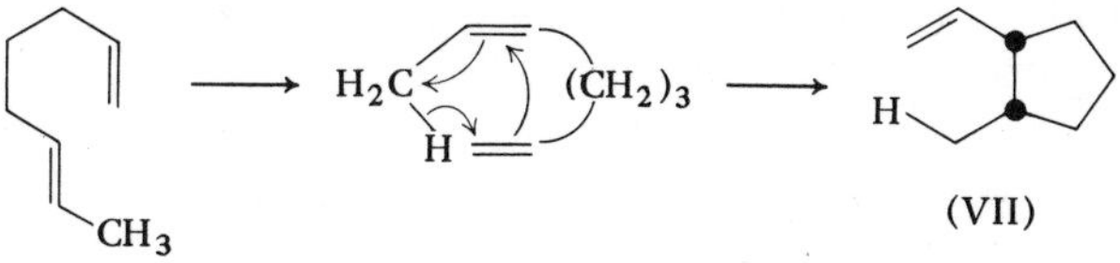

---

* For a comprehensive treatment of the principle of conservation of orbital symmetry in concerted reactions see R. B. Woodward and R. Hoffmann (Angew. Chem., intern. Edn., 1969, 8: 781).

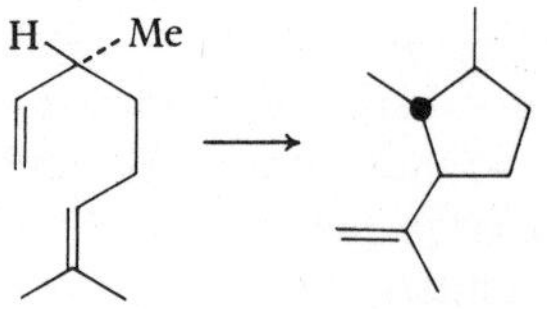

(VIII)

The cycloaddition of electrophilic olefins to enamines of cyclopentanealde-
hyde offers a convenient method for the formation of [4,3]spiro derivatives
(H. Christol, D. Lafont and F. Plénat, Bull. Soc. chim. Fr., 1966, 3947):

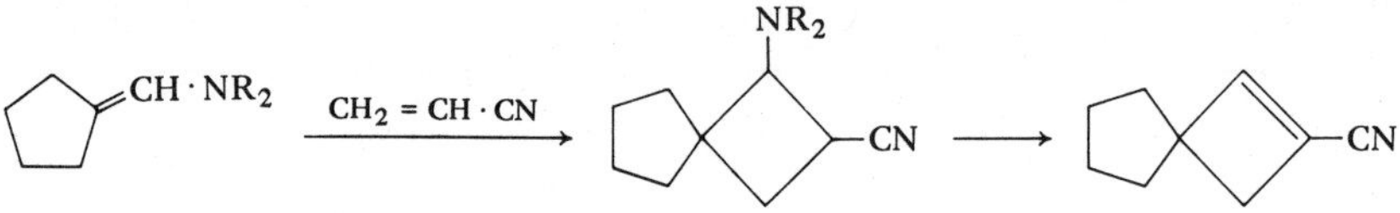

(iii) Cyclopentyne

Further evidence for the intermediacy of short-lived cycloalkynes formed
in the coupling reactions of 1-chlorocyclopentene, and higher homologues,
with phenyllithium has been reported (L. K. Montgomery and L. E.
Applegate, J. Amer. chem. Soc., 1967, 89: 2952). Cyclopentyne formed by
thermal decomposition of 1-lithio-2-bromocyclopentene has been trapped as
the adduct with 2,5-diphenyl-3,4-benzofuran (G. Wittig and J. Heyn, Ann.,
1969, 726: 57).

(c) Poly-unsaturated hydrocarbons

(i) Cyclopentadiene

The Diels—Alder reactions of cyclopentadiene have been comprehensively
reviewed (S. Seltzer, Advances in Alicyclic Chemistry, Vol. 2, eds. H. Hart
and G. J. Karabatsos, Academic Press, New York, 1968; J. Sauer, Angew.
Chem., intern. Edn., 1966, 5: 211; 1967, 6: 16) and the factors affecting the
endo/exo ratios of the products formed have been discussed further (R. A.
Grieger and C. A. Eckert, J. Amer. chem. Soc., 1970, 92: 2918; Y. Kobuke,
T. Fueno and J. Furukawa, ibid., 6548; W. C. Herndon and L. H. Hall,
Tetrahedron Letters, 1967, 3095; K. N. Houk, ibid., 1970, 2621, and refs.
therein). The heat and rate of dimerization of cyclopentadiene to endo-
dicyclopentadiene in the liquid phase at 25° has been found to be −9.22 ±
0.3 kcal/mole monomer and $4.99 \cdot 10^{-5} \mathrm{l} \cdot \mathrm{mole}^{-1} \cdot \mathrm{min}^{-1}$ respectively (A.
G. Turnbull and H. S. Hull, Austral. J. Chem., 1968, 21: 1789). In contrast
to the marked preference for 1,4-cycloaddition of dienophiles, 1,2-dichloro-
1,2-difluoroethylene undergoes a two-step biradical 1,2- and concerted

1,4-cycloaddition to cyclopentadiene (R. Wheland and P. D. Bartlett, J. Amer. chem. Soc., 1970, 92: 3822).

1,2-Cycloaddition of ketens to cyclopentadiene gives the less stable cyclobutanone with the large group (L) *endo*, as a direct consequence of antarafacial addition. The keten approaches with the minimum of steric repulsion (as IX) and then twists as the antarafacial process occurs (A. S. Dreiding et al., Helv., 1970, 53: 417; W. T. Brady et al., Tetrahedron Letters, 1970, 819; J. Amer. chem. Soc., 1970, 92: 4618; P. R. Brook, J. M. Harrison and A. J. Duke, Chem. Comm., 1970, 589):

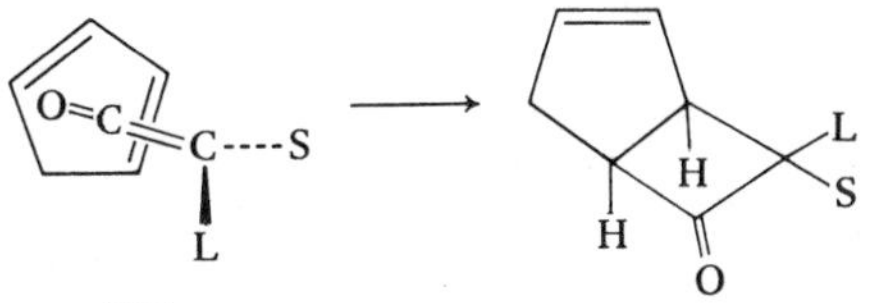

(IX)

Tropolone is obtained, in low yield, by the solvolysis of the dichloroketen—cyclopentadiene adduct (H. C. Stevens et al., J. Amer. chem. Soc., 1965, 87: 5257) and 4,5-benzotropolone is similarly obtained from indene (R. W. Turner and T. Seden, Chem. Comm., 1966, 399).

(ii) Alkyl- and aryl-cyclopentadienes

Trimerization of but-2-yne with palladium chloride gives a complex [formulated as X] which reacts with Lewis bases to give hexamethylbenzene, vinylpentamethylcyclopentadiene and 1-chlorovinylpentamethylcyclopentadiene (H. Reinheimer, J. Moffat and P. M. Maitlis, J. Amer. chem. Soc., 1970, 92: 2285):

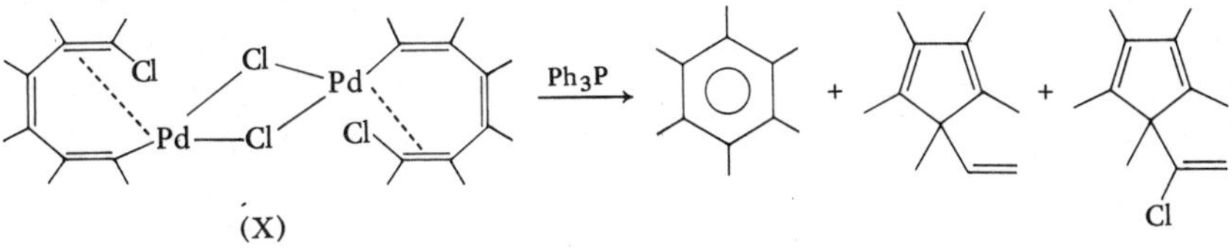

(X)

Geminal dimethylcyclopentadienes isomerize on heating to vicinal dimethylcyclopentadienes by a suprafacial [1,5]-sigmatropic shift of a methyl group. In this way 1,2,3- and 1,2,4-trimethylcyclopentadienes have been obtained from 1,5,5- and 2,5,5-trimethylcyclopentadienes (J. W. de Haan and H. Kloosterziel, Rec. Trav. chim., 1968, 87: 298). At higher temperatures (500°), 1,2,3-trimethylcyclopentadiene is converted into the 1,2,4-

isomer (V. A. Mironov, A. P. Ivanov and A. A. Akhrem, Izvest. Akad. Nauk, S.S.S.R., Ser. Khim., 1969, 1403). Thermal isomerization of 1,2,4,5,5-penta-methylcyclopentadiene into 1,2,3,4,5-pentamethylcyclopentadiene occurs through two sequential [1,5] sigmatropic methyl migrations (Mironov et al., ibid., 1968, 182; Tetrahedron Letters, 1968, 3997). Attempts to synthesize 1,2-diphenylcyclopentadiene led only to its Diels–Alder dimer, although 1,4-dimethyl-2,3-diphenylcyclopentadiene, 1,2,3-trimethyl-4,5-diphenyl-cyclopentadiene and 1,2-di(p-methoxyphenyl)cyclopentadiene exist as stable monomers (P. L. Pauson et al., J. chem. Soc., C, 1966, 306; M. Rosenblum et al., J. Amer. chem. Soc., 1962, 84: 2726).

(iii) Cyclopentadienyl ions and radicals

Photolysis of diazocyclopentadiene* readily affords the singlet carbena-cyclopentadiene XI (cyclopentadienylidene) (R. A. Moss and J. R. Przybyla, J. org. Chem., 1968, 33: 3816). Recent calculations and experimental evidence are consistent with a quasi-aromatic structure XII (R. Gleiter and R. Hoffmann, J. Amer. chem. Soc., 1968, 90: 5457; M. Jones, A. M. Harrison and K. R. Rettig, ibid., 1969, 91: 7462) resulting from partici-pation of the electron pair with the π-electron system and increased sp^2 character of the vacant orbital (XIIb and XIIc):

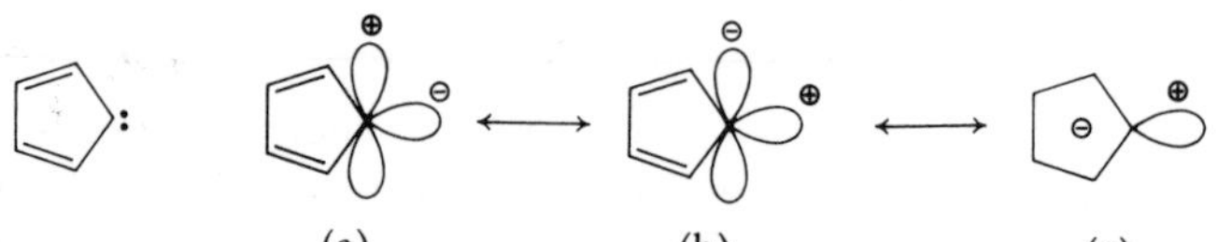

(a)   (b)   (c)

(XI)   (XII)

Photolysis of diazocyclopentadiene in saturated hydrocarbons leads to carbon hydrogen insertion (Moss, J. org. Chem., 1966, 31: 3296) and in the presence of olefins a mixture of spiro and alkenylcyclopentadienes is obtained (Moss, Chem. Comm., 1965, 622; H. Dürr and L. Schrader, Ber., 1969, 102: 2026):

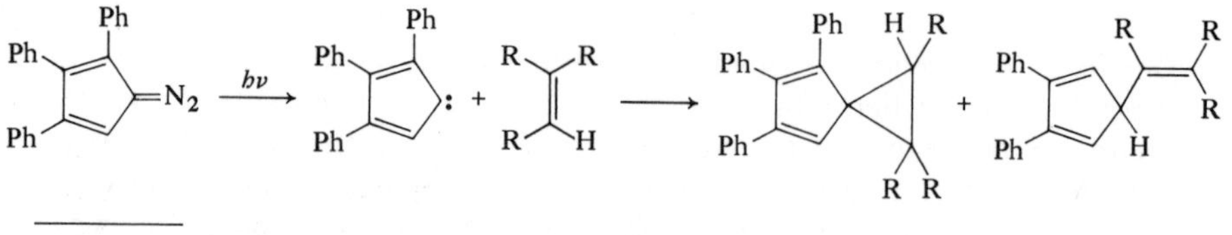

---

* Distillation of diazocyclopentadiene can be explosive (A. G. Wedd, Chem. and Ind., 1970, 109).

In the presence of norbornadiene or cyclo-octatetraene the spiro compounds XIII or XIV are formed, respectively (Dürr, G. Scheppers and Schrader, Chem. Comm., 1969, 257; Dürr and H. Kober, Ann., 1970, 740: 74):

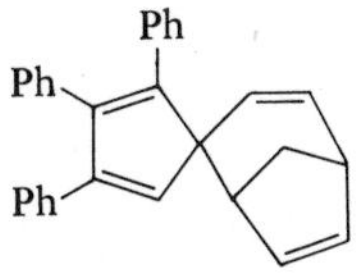

(XIII) 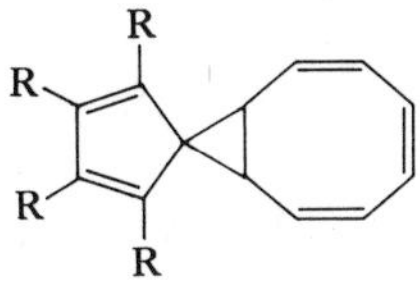   (XIV)

With cycloheptatriene the carbena-cyclopentadiene intermediates give spiro-homonorcaradienes and cyclopentadienylheptatrienes (Dürr, R. Sergio and Scheppers, ibid., 1970, 740: 63):

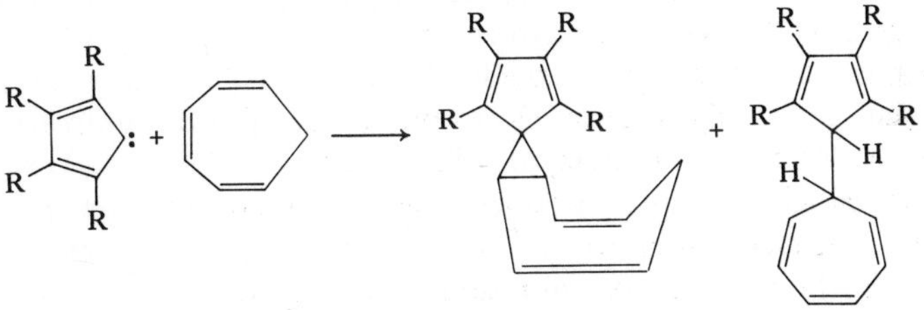

Photolysis of substituted diazocyclopentadienes in benzene leads, via carbene cycloaddition and ring expansion, to 7H- and 5H-benzocycloheptenes (idem, ibid., 1970, 734: 141) or homoazulenes (idem, Ber., 1970, 103: 380):

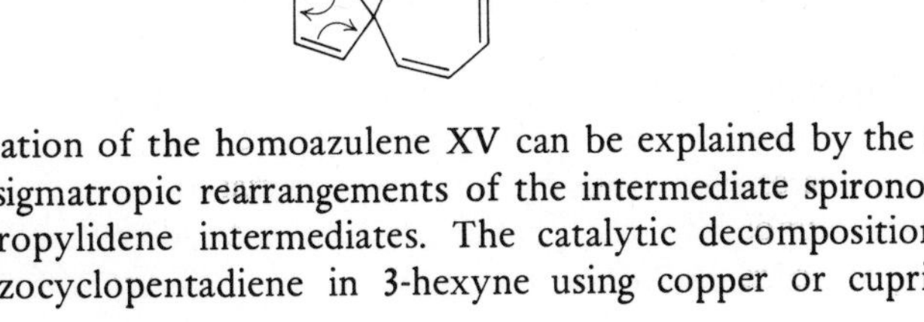

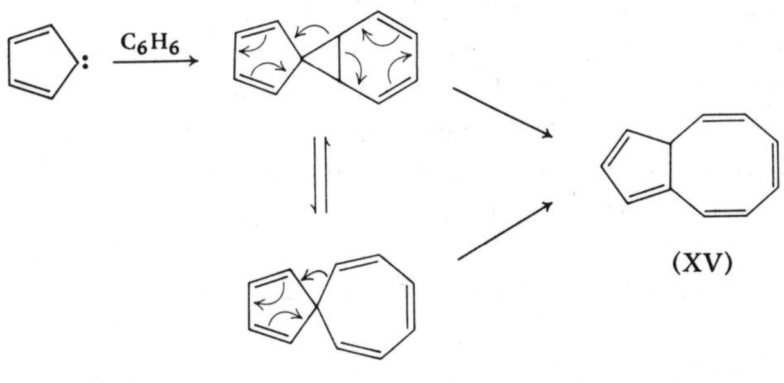

(XV)

The formation of the homoazulene XV can be explained by the occurrence of [1,5]-sigmatropic rearrangements of the intermediate spironorcaradienes or spirotropylidene intermediates. The catalytic decomposition of tetra-chlorodiazocyclopentadiene in 3-hexyne using copper or cupric sulphate gives the spiro-[2,4]-heptatriene XVI, whereas use of di-μ-chloro-di-π-allyl-palladium results in the formation of the tetrachlorotetraethylspiro-[4,4]-nonatetraene XVII (E. T. McBee, G. W. Calundann and T. Hodgins, J. org. Chem., 1966, 31: 4260). The latter rearranges thermally to the indene XVIII.

(XVI)   (XVII)   (XVIII)

The synthesis and reactions of diazotetracyanocyclopentadiene have also been reported (O. W. Webster, J. Amer. chem. Soc., 1966, 88: 4055).

The aromaticity of the cyclopentadienide anion has been demonstrated, in the classical sense, by typical electrophilic substitution in methylation reactions (S. McLean and P. Haynes, Tetrahedron, 1965, 21: 2313, 2343). Where the basicity of the cyclopentadienide anion is lowered, by the introduction of electronegative groups, more strongly acidic conditions can be used as in the bromination, nitration, and acetylation of the tetracyano-cyclopentadienide anion (R. C. Cookson and K. R. Friedrich, J. chem. Soc., C, 1966, 1641; O. W. Webster, J. org. Chem., 1967, 32: 39). Cyano groups can be introduced in a stepwise manner by treatment of the cyclopenta-dienide anion with cyanogen chloride (Webster, J. Amer. chem. Soc., 1966, 88: 3046). Photolysis of sodium cyclopentadienide gives *meso*- and *dl*-3-(3-cyclopentenyl)cyclopentene (E. E. van Tamelen, J. I. Brauman and L. E. Ellis, ibid., 1967, 89: 5073) and cycloaddition to benzyne gives 1,4-dihydro-1,4-methanonaphthalene (W. T. Ford, R. Radue and J. A. Walker, Chem. Comm., 1970, 966). The use of flash vacuum pyrolysis of nickelocene in the generation and characterization of the cyclopentadienyl radical has been reported (E. Hedaya, Accounts Chem. Research, 1969, 2: 367). Cyclopentadienyl radicals are also formed in the flash photolysis of substituted benzenes (G. Porter and B. Ward, Proc. Roy. Soc., A, 1968, 303: 139).

The Olah technique (G. A. Olah and P. von R. Schleyer, "Carbonium Ions", Vol. 1, Interscience, New York, 1968) for the generation and observation of carbonium ions by n.m.r. spectroscopy in fluorosulphonic acid has been applied to the pentamethylcyclopentadienylmethyl cation (R. F. Childs, M. Sakai and S. Winstein, J. Amer. chem. Soc., 1968, 90: 7144):

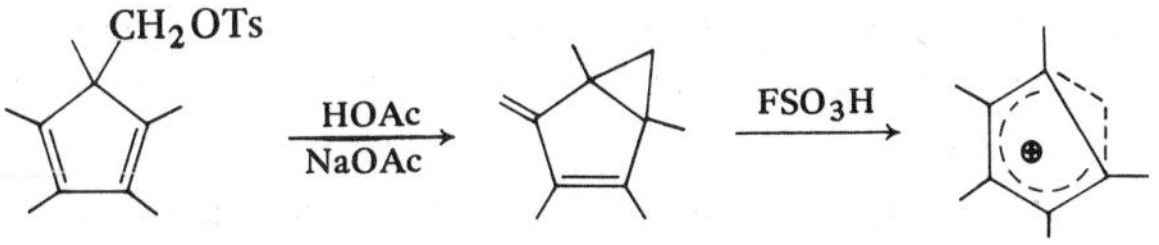

The e.s.r. spectra of some cyclopentadienyl cations have also been determined and the triplet state detected (R. Breslow et al., ibid., 1967, 89: 1112). The thermal rearrangement of pentadienylic cations into cyclopentenyl cations has been shown to be a conrotatory electrocyclic reaction (T. S. Sorensen et al., ibid., 1969, 91: 6404):

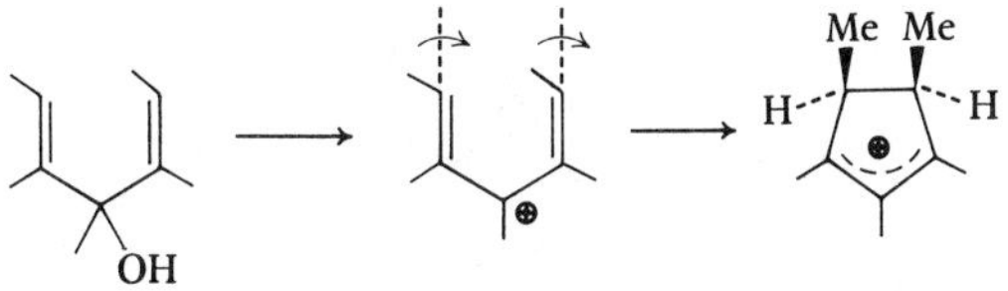

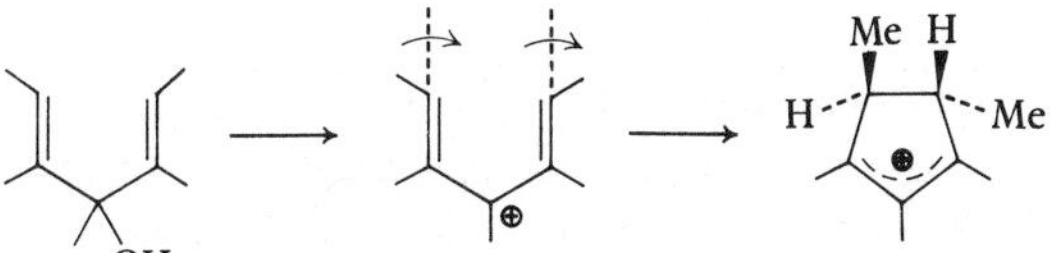

The antiaromatic (4 n π-electrons) highly reactive pentachlorocyclopentadiene cation XIX is most probably involved in the dimerization of hexachloropentadiene to "Prins dimer" (cf. p. 111). With aluminium chloride a partially ionic, chlorine-bridged complex XX appears to be formed (H. P. Fritz and L. Schäfer, J. organometal. Chem., 1964, 1: 318). E.s.r. signals for the triplet species have been observed in antimony pentafluoride (Breslow et al., loc. cit.).

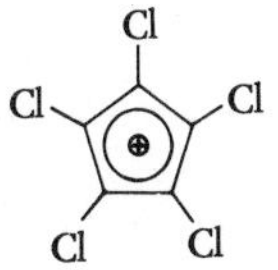

(XIX) 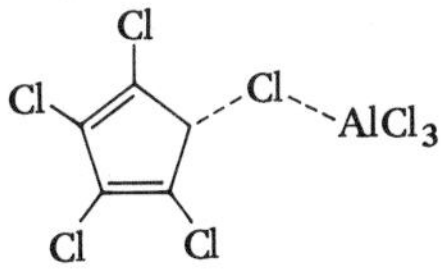   (XX)

(iv) Fulvenes and fulvalenes

The synthesis, spectroscopic and chemical properties, and theoretical aspects of fulvenes have been extensively reviewed (P. Yates, Advances in

Alicyclic Chemistry, eds. H. Hart and G. J. Karabatsos, Academic Press, New York, 1968, 2: 59; E. D. Bergmann, Chem. Reviews, 1968, 68: 41). The various fulvalenes have also been more briefly surveyed (Bergmann, loc. cit.).

Liquid- and vapour-phase photolysis of benzene causes partial isomerization to fulvene (H. J. F. Angus, J. M. Blair and D. Bryce-Smith, J. chem. Soc., 1960, 2003; H. R. Ward, J. S. Wishnok and P. D. Sherman, J. Amer. chem. Soc., 1967, 89: 162) and mechanistic aspects have been discussed (D. Bryce-Smith and H. C. Longuet-Higgins, Chem. Comm., 1966, 593; I. Jano and Y. Mori, Chem. Phys. Letters, 1968, 2: 185). Benzvalene (XXI) has also been isolated (K. E. Wilzbach, J. S. Ritscher and L. Kaplan, J. Amer. chem. Soc., 1967, 89: 1031).

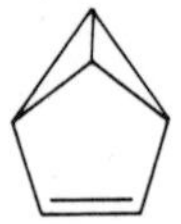

(XXI)

Various substituted 1H-azepines undergo thermal rearrangement to 6-amino-fulvene derivatives (R. F. Childs, R. Grigg and A. W. Johnson, J. chem. Soc., C, 1967, 201; M. Mahendran and Johnson, Chem. Comm., 1970, 10):

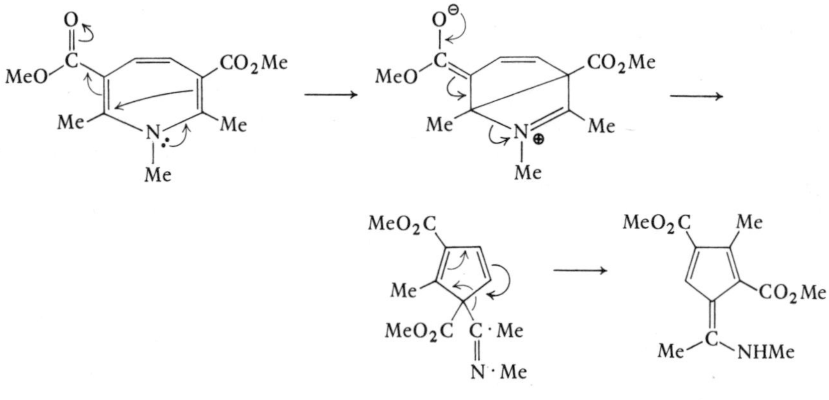

The driving force for this rearrangement is presumably supplied by the aromatic stability of the aminofulvene system (K. Hafner et al., Angew. Chem., intern. Edn., 1963, 2: 123; A. P. Downing, W. D. Ollis and I. O. Sutherland, Chem. Comm., 1968, 1053). 6,6-Diaminofulvenes are obtained by condensation of cyclopentadienyl salts with S-methylisothiouronium iodide (K. Hartke and G. Salamon, Ber., 1970, 103: 133, 147).

Diphenylfulvene undergoes 1,2-cycloaddition with dichloroketen (R. E. Harmon et al., Chem. Comm., 1970, 935).

In the case of 6,6-*dimethylfulvene*, ring expansion of the intermediate adduct gives 3-isopropenyltropolone (Y. Kitahara et al., Chem. Comm., 1970, 89). *m*-Chloro-α-alkylstyrenes are formed by a similar rearrangement of the adduct formed from dichlorocarbene (H. Hart et al., Tetrahedron Letters, 1969, 4933):

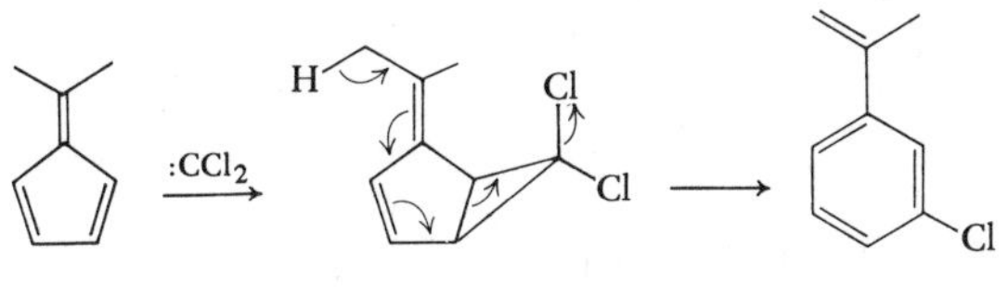

1,2-Cycloaddition of benzonitrile oxide (A. Quilico, P. Grunager and R. Mazzini, Gazz., 1952, 82: 349) and diazomethane (K. Alder, R. Braden and F. H. Flock, Ber., 1961, 94: 456) has also been claimed.

Fulvenes act as 6π-electron addends in the concerted cycloaddition of tropone. The reaction shows remarkable selectivity in that of 26 thermally allowed 1:1 adducts only two are formed; the reaction consists of two [6 + 4] cycloadditions and a [1,5]-hydrogen shift (K. N. Houk, L. J. Luskus and N. S. Bhacca, J. Amer. chem. Soc., 1970, 92: 6392):

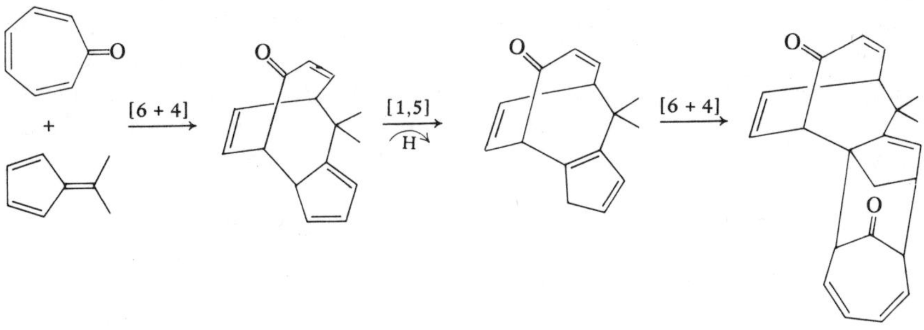

In most other examples 1,4-cycloaddition, or formation of bis-adducts, appears to be the preferred mode of reaction. Vilsmeier formylation of 1,2,3,4-tetrachloro-1,3-cyclopentadiene gives 2,3,4-trichloro-6-(dimethyl-amino)fulvene-1-carboxaldehyde (XXII; X = NMe₂) and 2,3,4-trichloro-6-hydroxyfulvene-1-carboxaldehyde (XXII; X = OH); the latter is readily converted to the trichlorocyclopentapyridazine XXIII (G. Seitz, Pharm. Zentralh., 1968, 107: 363; C.A., 1968, 69: 76702a):

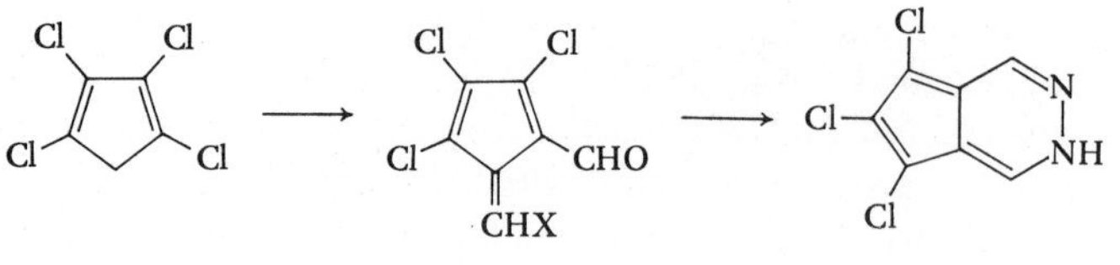

(XXII)   (XXIII)

6-(2-Dialkylaminovinyl)fulvenes can be cyclized to 3-dialkylamino-1,2-di-hydropentalenes (XXIV) (R. Kaiser and K. Hafner, Angew. Chem., intern. Edn., 1970, 9: 892):

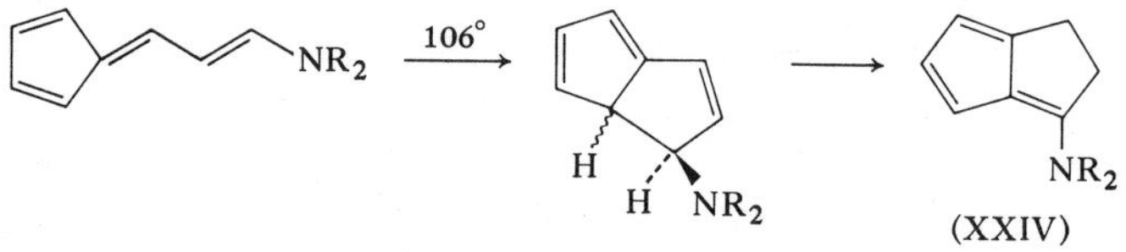

(XXIV)

Cyclopentadienylidenetriphenylphosphorane (XXV) is unreactive as a Wittig reagent (F. Ramirez and S. Levy, J. Amer. chem. Soc., 1957, 79: 67) and behaves as a non-benzenoid aromatic compound with mild electrophiles such as benzene diazonium chloride (idem, J. org. Chem., 1958, 23: 2035). The sulphur and selenium ylids (XXVI; Z = S, Se; n = 2) also failed to react with p-nitrobenzaldehyde (D. Lloyd and M. I. Singer, Chem. Comm., 1967, 390), but the corresponding arsonium and stibonium ylids (XXVI; Z = As, Sb; n = 3) react with aldehydes to give the corresponding fulvene (idem, Chem. and Ind., 1968, 1277). The increased nucleophilic reactivity is explained in terms of the effectiveness of the $p\pi-d\pi$ orbital overlap in the carbon—hetero-atom bond, which appears to be in the order S > P > As > Sb.

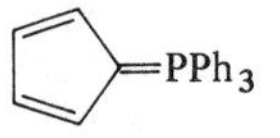

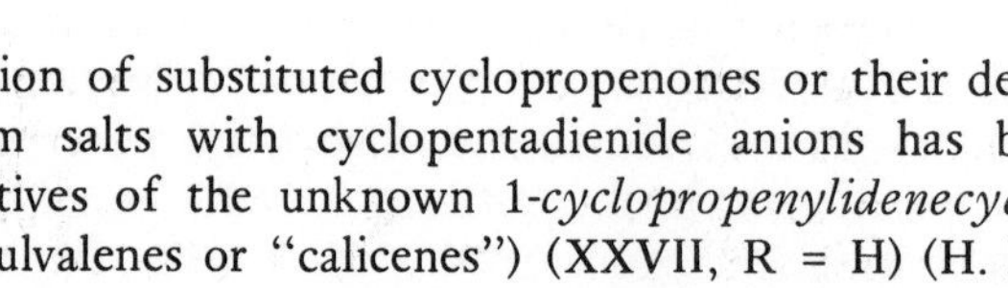

(XXV) 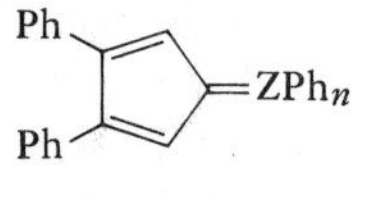   (XXVI)

The condensation of substituted cyclopropenones or their derived alkoxy-cyclopropenium salts with cyclopentadienide anions has been used to prepare derivatives of the unknown 1-*cyclopropenylidenecyclopentadiene* (i.e. pentatriafulvalenes or "calicenes") (XXVII, R = H) (H. Prinzbach, D. Seip and U. Fischer, Angew. Chem., intern. Edn., 1965, 4: 242; W. M. Jones and R. S. Pyron, J. Amer. chem. Soc., 1965, 87: 1608; A. S. Kende and P. T.

Izzo, ibid., 1609; Kende, Izzo and P. T. MacGregor, ibid., 1966, 3359; Y. Kitahara et al., Chem. Comm., 1966, 180). N.m.r. evidence indicates a considerable degree of cyclopropenium—cyclopentadienide aromatic character at least when the five-membered ring contains electronegative substituents (Kende, Izzo and MacGregor, loc. cit.; E. D. Bergmann and I. Agranat, Tetrahedron, 1966, 22: 1275).

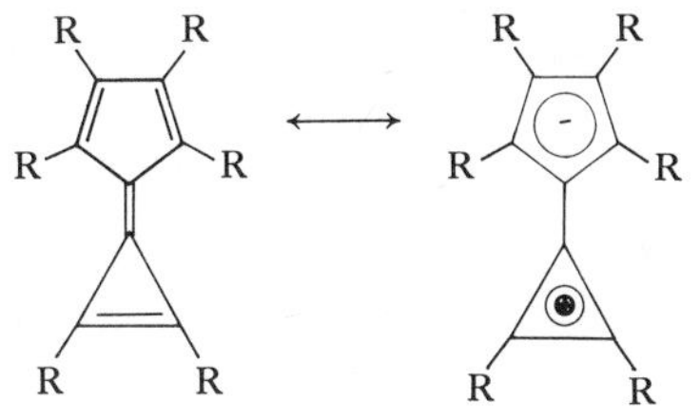

(XXVII) 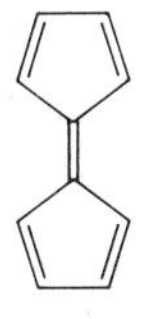　　　　(XXVIII) 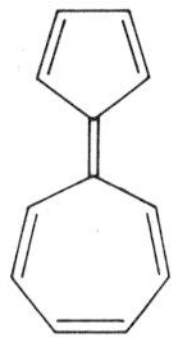　　　　(XXIX)

The hydrocarbon *pentafulvalene* ("*fulvalene*") (XXVIII) has been prepared by several methods but is unstable and has not been obtained pure. Other than fused-ring species the only other fulvalenes prepared are highly substituted phenyl or halogeno derivatives (P. T. Kwitowski and R. West, J. Amer. chem. Soc., 1966, 88: 4541, and refs. therein). There is no evidence for any aromatic character in these systems.

Similarly *heptapentafulvalene* ("*sesquifulvalene*") (XXIX) and its simple derivatives have the properties of reactive polyolefins (H. Prinzbach and D. Seip, Angew. Chem., 1961, 73: 169; Prinzbach and W. Rosswog, Tetrahedron Letters, 1963, 1217).

3. Metal cyclopentadienyls; metallocenes*

In view of the explosive growth in the literature on this topic, references quoted are generally the most recent in a series, by the same or different authors, particularly when these publications cite earlier relevant work in a comprehensive manner. Current areas of interest and annual surveys of organometallic chemistry, including cyclopentadienyl compounds, are reviewed in Organometallic Chemistry Reviews, 1966 et seq., Elsevier, Amsterdam; Advances in Organometallic Chemistry, 1964 et seq., Academic Press, New York; Progress in Inorganic Chemistry, 1959 et seq., Interscience, New York; Transition Metal Chemistry, 1965 et seq., Marcel Dekker, New York;

---

* See also Vol. II A, pp. 124 et seq.

olefin complexes of transition metals, including π-cyclopentadienyl compounds containing other olefinic ligands, have also been reviewed (H. W. Quinn and J. H. Tsai, Adv. Inorg. Chem. and Radiochem., 1969, 12: 217). Specific reference to specialized reviews and books are made in the text.

(a) Synthesis*

(1) *Ferrocene.* The mechanism for the formation of ferrocene has been investigated and the intermediacy of a σ-complex (tris-monohaptocyclopentadienyliron)† (I) is indicated:

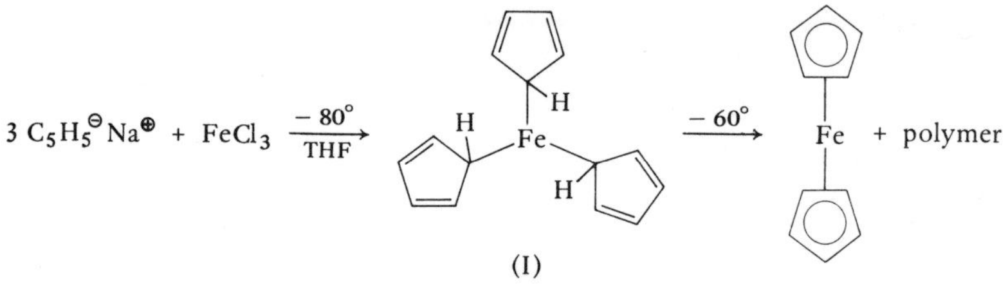

$$3\ C_5H_5^{\ominus}Na^{\oplus}\ +\ FeCl_3 \xrightarrow[THF]{-80°} \qquad \xrightarrow{-60°} \quad Fe\ +\ polymer$$

(I)

Use of a 2:1 mole ratio of sodium cyclopentadienide to ferric chloride gives di-monohaptocyclopentadienyliron chloride monotetrahydrofuranate (II) which rearranges to ferricenium chloride at higher temperatures (M. Tsutsui et al., J. Amer. chem. Soc., 1969, 91: 5233):

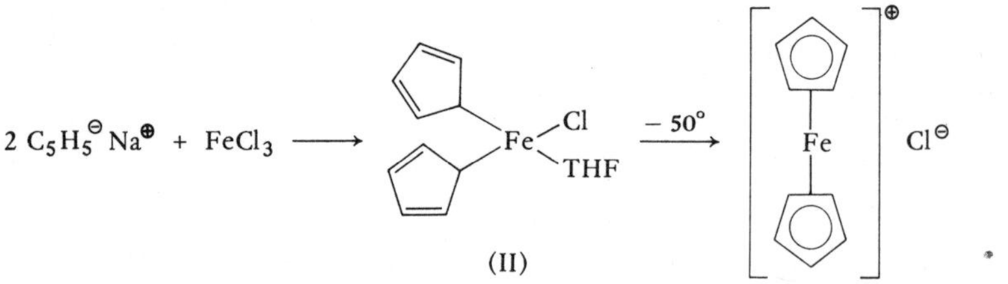

$$2\ C_5H_5^{\ominus}Na^{\oplus}\ +\ FeCl_3 \longrightarrow \qquad \xrightarrow{-50°}$$

(II)

(2) *Cyclopentadienyl metal complexes. Bis(cyclopentadienyl) mercury* is obtained in high yield from mercuric oxide, cyclopentadiene, and a primary amine (S. Lenzer, Austral. J. Chem., 1969, 22: 1303). New syntheses of *di-π-cyclopentadienyldihalogenotitanium*(IV), *zirconium*(IV), and

---

* For experimental techniques see R. B. King, Organometallic Syntheses, 1965, Academic Press, New York, and J. M. Birmingham, Adv. Organometal. Chem., 1964, 2: 365.
† The term *n*-hapto has been proposed for a ligand attached to the metal through *n* atoms (F. A. Cotton, J. Amer. chem. Soc., 1968, 90: 6230).

hafnium(IV) *complexes* are reported (P. M. Druce et al., J. chem. Soc., A, 1969, 2106; A. N. Nesmeyanov et al., Izvest. Akad. Nauk, S.S.S.R., Ser. Khim., 1969, 1323), and *dicyclopentadienidesamarium*(II)-1-*tetrahydrofuranate* (G. W. Watt and E. W. Gillow, J. Amer. chem. Soc., 1969, 91: 775).

The syntheses of *bis(cyclopentadienyl)titanium* (Watt, L. J. Baye and F. O. Drummond, J. Amer. chem. Soc., 1966, 88: 1138), *zirconium* (Watt and Drummond, ibid., 1966, 88: 5926) and *hafnium* (idem, ibid., 1970, 92: 826) have been reported, but in the former case (i.e. titanium) spectral and chemical evidence show the product to be a dimer containing a metal–metal bond and π- and σ-bonded cyclopentadienyl rings (J.-J. Salzmann and P. Mosimann, Helv., 1967, 50: 1831).

Cyclopentadienylcycloheptatrienyltitanium, prepared from cyclopentadienyltitanium trichloride, is thermally stable but sensitive to moisture and is reported to have a sandwich structure analogous to $C_5H_5VC_7H_7$ (H. O. van Oven and H. J. de Liefde Meijer, J. organometal. Chem., 1970, 23: 159). The corresponding π-complex of molybdenum and tungsten ($C_5H_5MC_7H_7$, M = Mo, W) have also been described (H. W. Wehner, E. O. Fischer and J. Muller, Ber., 1970, 103: 2258). Metal tricyclopentadienyls of arsenic and antimony have been isolated for the first time, together with the known bismuth derivative [$(C_5H_5)_3M$, M = As, Sb, Bi] from sodium pentadienide and MCl_3 (B. Deubzer et al., Ber., 1970, 103: 799).

The latest member of the mixed sandwich complexes containing a π-cyclopentadienyl metal moiety to be reported is *π-cyclopentadienyl π-triphenylcyclopropenylnickel* (M. D. Rausch, R. M. Tuggle and D. L. Weaver, J. Amer. chem. Soc., 1970, 92: 4981). Ring-expansion of 5-*exo*-halomethyl derivatives of cyclopentadienyl(cyclopentadiene) cobalt [Co-$(C_5H_5)(C_5H_5CH_2X)$] occurs at 0–60°, to give *cyclopentadienyl(cyclohexadienyl) cobalt* salts [$Co(C_5H_5)(C_6H_7)$]$^\oplus$ $X^\ominus$ (G. E. Herberich and J. Schwarzer, Ber., 1970, 103: 2016; Angew. Chem., intern. Edn., 1969, 8: 143) and a remarkable ring-contraction of hexamethyl Dewar-benzene occurs, on treatment with rhodium trichloride in methanol, to give *dichloro-(pentamethylcyclopentadienyl)rhodium dimer* (J. W. Kang, K. Moseley and P. M. Maitlis, J. Amer. chem. Soc., 1969, 91: 5970).

(3) Metal complexes of unstable cyclobutadiene and cyclopentadiene derivatives. Transition-metal complexes of otherwise unisolable or unstable unsaturated cyclic hydrocarbons such as fulvalene (III), pentalene (IV), calicene (V), and sesquifulvalene (VI) have been prepared [viz. VII (M. D. Rausch, R. F. Kovar and C. S. Kraihanzel, ibid., 1969, 91: 1259), VIII (T. J. Katz and M. Rosenberger, ibid., 1963, 85: 2030), IX and X (M. Cais and A. Eisenstadt, ibid., 1967, 89: 5468) respectively].

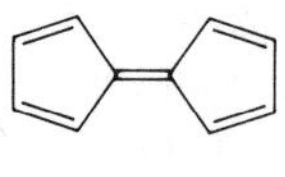

(III) 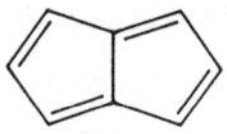　　　　　(IV) 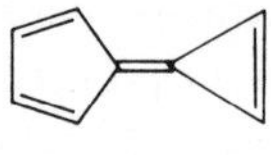　　　　　(V) 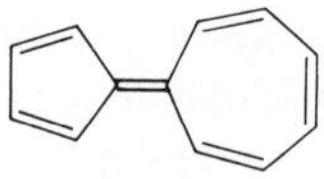　　　　　(VI)

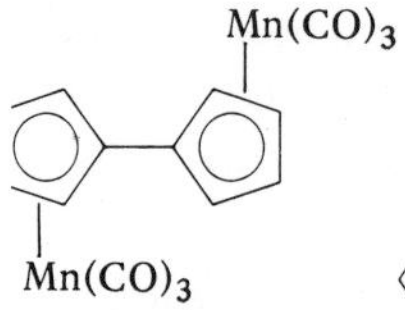

(VII) 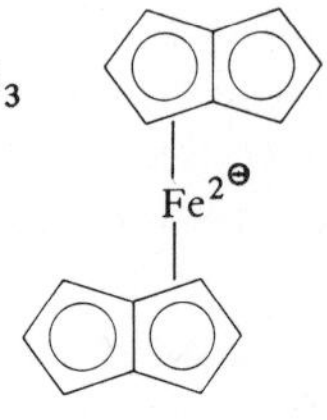　　　　　(VIII) 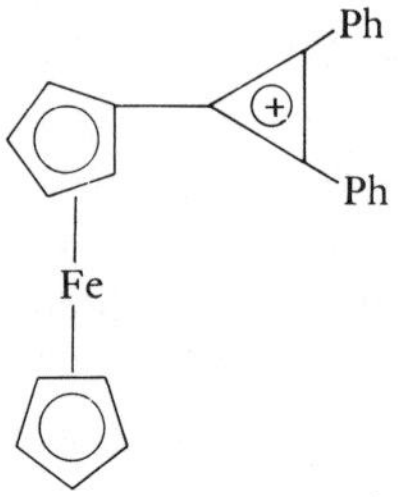　　　　　(IX) 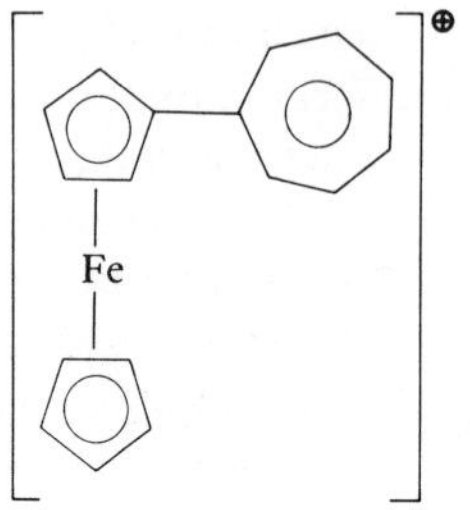　　　　　(X)

Complexes IX and X were obtained by direct substitution into the ferrocene nucleus. The synthesis of *cyclobutadiene* (*π-cyclopentadienyl*) *cobalt*, which is isoelectronic with ferrocene, has been effected as indicated below (R. G. Amiet and R. Pettit, ibid., 1968, 90: 1059) or by photolysis of photo-α-pyrone in the presence of cyclopentadienylcobalt dicarbonyl (M. Rosenblum and B. North, ibid., 1060):

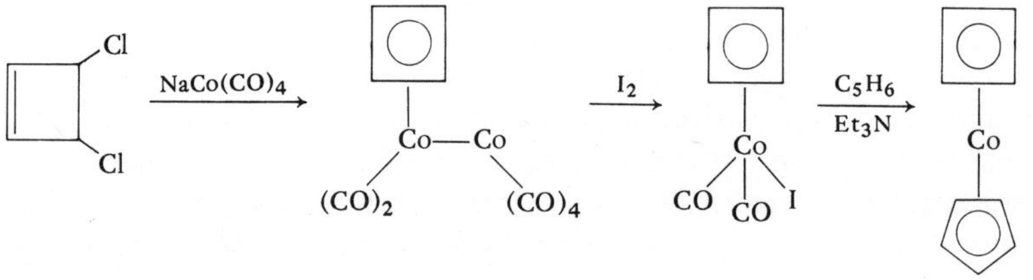

The cyclobutadiene ring appears to be the more reactive to electrophilic reagents, presumably due to the stability of the intermediate π-allyl complex.

(b) Chemical properties

(i) Electrophilic substitution

For a review of electrophilic and nucleophilic reactions of organo-transition metal compounds, including metallocenes, see D. A. White, Organometal Chem. Reviews, 1968, 3A: 497.

The mechanism of electrophilic substitution, formerly regarded as involving initial attack at the metal followed by migration to the ligand, thus leading to *endo*-orientation of the substituent in the resulting σ-complex, has recently been questioned. The rate of cyclization of the epimeric acids (XI and XII) to the homoannular cyclic ketone has been shown to be greater for the *exo*-isomer (XI) (M. Rosenblum and F. W. Abbate, J. Amer. chem. Soc., 1966, 88: 4178) indicating that direct attack on the ligand from the side remote from the metal is energetically favoured. Furthermore benzoylation of cobaltocene has been shown, by X-ray analysis, to yield the *exo*-orientated benzoyl derivative (M. R. Churchill, J. organometal. Chem., 1965, 4: 258). However, the factors which determine whether electrophilic (or nucleophilic) reagents attack the ligand or the metal are not clearly understood and could well depend on the nature of the ligand, the metal, the reagent, and on whether kinetic or thermodynamic control is operative. Directive influences of substituents are discernible in metallocene systems.

(XI)   (XII)

The activating or deactivating influence of a substituent in the ferrocene nucleus is felt not only by the ring bearing the substituent but also to a lesser extent by the unsubstituted ring. When the substituent is deactivating further electrophilic substitution takes place in the unsubstituted ring. When activating, and in the absence of strong resonance or *ortho* effects (i.e. for alkyl substituents), steric effects predominate and substitution occurs at the 3-position (β to the substituent). When non-bonding electron pairs are available on a hetero-atom attached to the ferrocene nucleus, electrophilic substitution at the 2-position (α to the substituent) is favoured (G. R. Knox et al., J. chem. Soc., C, 1967, 1853, and refs. therein). From an investigation into the acylation of ferrocenophanes it has been suggested that positive charge which accumulates in one ring, in the transition state, is transmitted stereospecifically to the other ring in a *pseudo-trans* manner by polarization of the metal—carbon bonds (H. L. Lentzner and W. E. Watts, Chem. Comm., 1970, 906). The interannular transmission coefficient of an electronic effect through the ferrocene nucleus has been estimated to be 0.28 (M. Sato et al., Bull. chem. Soc., Japan, 1970, 43: 1142). The effect of ligands on the electron density at the metal atom in $C_5H_5Mn(CO)_2L$ systems has been investigated by the effect on the carbonyl-stretching vibration and chemical shift of the cyclopentadienyl protons (M. Herberhold and C. R. Joblonski, Ber., 1969, 102: 778). Metalation of metallocenes has been reviewed (D. W. Slocum et al., J. chem. Educ., 1969, 46: 144). Mono- or hetero-annular di-metalation can be achieved and offers a convenient method for the introduction of a wide variety of substituents. Substituents containing a lone pair of electrons direct lithiation to the 2-position by means of a five- or six-membered chelate ring, as illustrated in the formation of 2-substituted vinylferrocenes (Slocum et al., Austral. J. Chem., 1968, 21: 2319), 1,1'- and 1,2-*diformylferrocenes* (C. Moise, J. Tirouflet and H. Singer (Bull. Soc. Chim. Fr., 1969, 1182), 2-*substituted alkoxymethylferrocenes* (Slocum and B. P. Koonsvitzky, Chem. Comm., 1969, 846 and refs. therein):

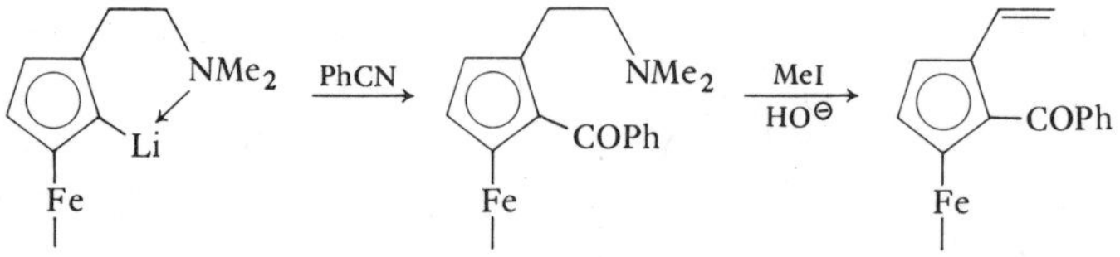

As a consequence of the cyclic mechanism lithiation of chiral α-ferrocenyl-alkylamines proceeds with asymmetric induction (I. Ugi, Rec. chem. Progr., 1969, 30: 289; Ugi et al., J. Amer. chem. Soc., 1970, 92: 5389; Tetrahedron Letters, 1970, 1771; D. Marguarding et al., Angew. Chem., intern. Edn., 1970, 9: 371). Monolithiation of alkylferrocenes occurs exclusively in the unsubstituted ring (A. N. Nesmeyanov et al., Izvest. Akad. Nauk, S.S.S.R., Ser. Khim., 1966, 1938), but sodiation occurs at the 3-position as well (R. A. Benkeser and J. L. Bach, J. Amer. chem. Soc., 1964, 86: 890).

(ii) Nucleophilic addition and substitution

The addition of nucleophiles to cyclopentadienyl metal cations or metal-locenes substituted by carbonyl or other electrophilic groups, occurs stereo-specifically from the *exo*-side. Thus cobaltocene gives XIII (R = CCl_3, CH_2CN, C:CR or CH_2COR) (H. Kojima et al., Bull. chem. Soc., Japan, 1970, 43: 2272), reduction of ferroceno[2,3:2',3'][1]indanones gives the *endo*-alcohol XIV (M. Le Plouzennec and R. Dabard, Compt. rend., 1969, 268: 1721), and borohydride reduction or lithium-alkyl alkylation of arenecyclopentadienyliron cations gives XV (R = H, CH_3 or Ph) (I. U. Khard, P. L. Pauson and W. E. Watts, J. chem. Soc., C, 1968, 2257, 2261; 1969, 116, 2024).

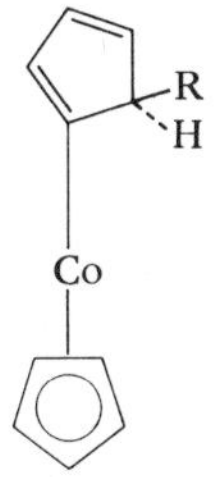

(XIII) 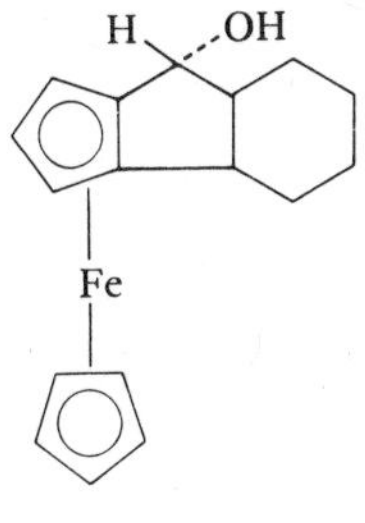   (XIV) 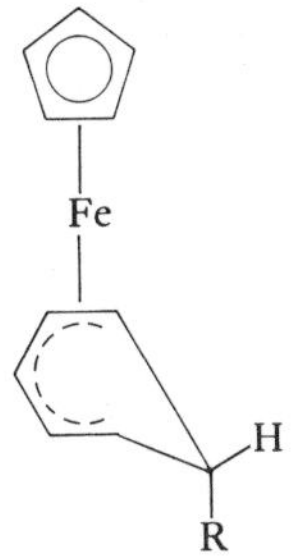   (XV) 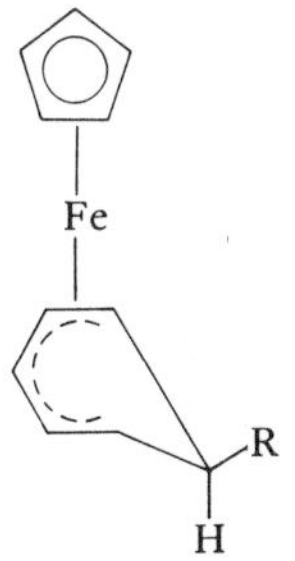   (XVI)

The *exo–endo* isomers (XV and XVI) may be differentiated by infrared (the *exo*-carbon–hydrogen stretching absorption occurs at lower frequency, possibly due to greater hyperconjugative electron drain in this isomer) and n.m.r. spectroscopy (*exo*-substituents give signals at higher field). The stereospecificity of *exo*-nucleophilic addition is attributed to steric shielding of the *endo*-face by the metal atom and attached groups, and to a concentration of electron density on the *endo*-face adjacent to the metal (Khard, Pauson and Watts, loc. cit.). However, reaction of μ-dichloro-dichloro-bis(pentamethylcyclopentadienyl)dirhodium (XVII) with cyclopentadiene in the presence of sodium carbonate leads stereospecifically to the *endo-H*(cyclopentodienyl)(pentamethylcyclopentadiene)rhodium (XVIII). This is attributed to the intermediacy of a complex in which the hydrogen being transferred, from the cyclopentadiene to the pentamethylcyclopentadienyl ring, is bound to the metal atom. The *endo*-isomer was shown to be less reactive than the corresponding *exo*-isomer (K. Moseley and P. M. Maitlis, Chem. Comm., 1969, 616).

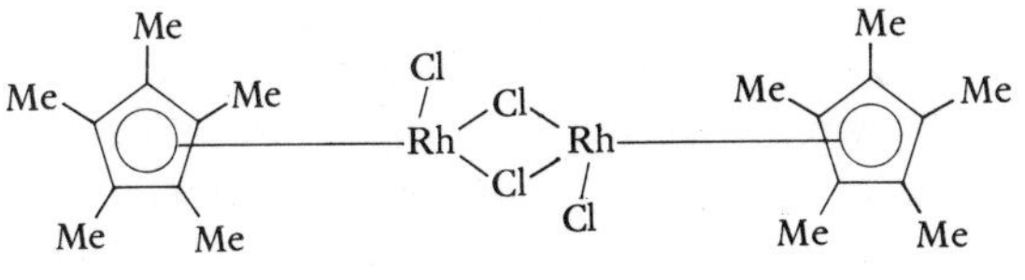

(XVII) 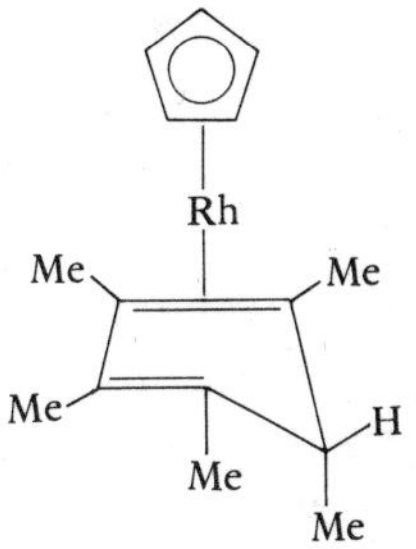   (XVIII)

Endo-substituents may be more reactive when participation by the interannular electrons can facilitate the reaction [whether these electrons be the non-bonding electrons of the metal (i.e. metal participation) (vide infra) or the d-π electrons involved in the metal–carbon bonding (i.e. metal hyper-conjugation) (See p. 94)] (M. J. Nugent et al., J. Amer. chem. Soc., 1969, 91: 6138, 6141, 6145):

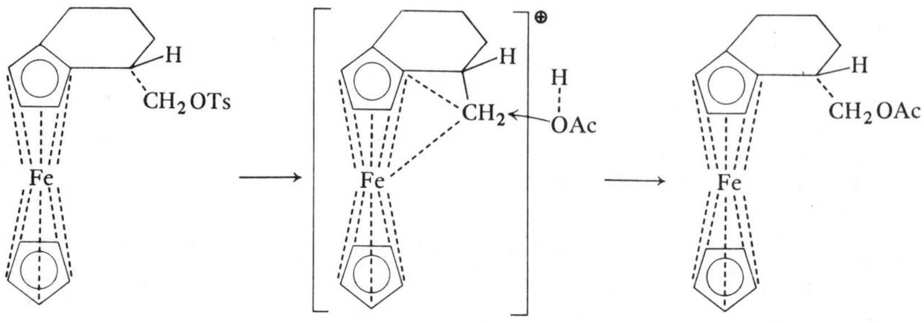

Photolysis of ferrocene in mixed solvents of ethanol and carbon tetra-chloride, chloroform, or methylene chloride results in substitution of the ferrocene nucleus by ethoxycarbonyl, formyl, or ethoxymethyl groups respectively (Y. Hoshi, T. Akiyama and A. Sugimori, Tetrahedron Letters, 1970, 1485; for a review of the photochemistry of ferrocene and derivatives see E. K. van Gustorf and F. W. Grenels, Fortschr. Chem. Forsch. 1969, 13: 366). Either 1,2- or 1,3-cycloaddition occurs on treatment of nickelocene with unsaturated electrophiles such as diethyl acetylenedicarboxylate or azodicarboxylate (M. Dubeck, J. Amer. chem. Soc., 1960, 82: 6193; M. Green, R. B. L. Osborn and F. G. A. Stone, J. chem. Soc., A, 1968, 3083 and refs. therein).

(iii) Reduction and oxidation

Reductive cleavage of ferrocene has been accomplished by: lithium in ethylamine (D. S. Trifan and L. Nicholas, J. Amer. chem. Soc., 1957, 79: 2746) or propylamine (A. D. Brown and H. Reich, J. org. Chem., 1970, 35: 1191); catalytic hydrogenation (A. N. Nesmeyanov et al., Izvest. Akad. Nauk, S.S.S.R., Otdel. Khim. Nauk, 1956, 749); potassium in ammonia (G. W. Watt and L. J. Boye, J. inorg. nucl. Chem., 1964, 26: 2099). Ferro-cenylalkylammonium salts are reduced to alkylferrocenes with sodium in ammonia without cleavage of the ferrocene ring system (D. W. Slocum and W. E. Jones, J. organometal. Chem., 1968, 15: 262; Slocum et al., J. org. Chem., 1969, 34: 1973). Raney-nickel hydrogenation of nickelocene gives (π-C_5H_5)Ni-(π-C_5H_7) and occurs stereospecifically, the hydrogen entering from the most hindered side of the cyclopentadienyl ring, suggesting that the nickel nucleus participates in the hydrogen-transfer step (K. W. Barnett, F. D. Mango and C. A. Reilly, J. Amer. chem. Soc., 1969, 91: 3387).

Resistance of ferrocene towards chemical oxidation is enhanced by electron-withdrawing substituents and the effect of two such substituents is roughly additive whether these be homoannular (A. N. Nesmeyanov et al., Izvest. Akad. Nauk, S.S.S.R., 1965, 909) or heteroannular (E. G. Perevalova et al., Doklady Akad. Nauk, S.S.S.R., 1964, 155: 857). Decachloroferrocene, prepared from 1,1'-dichloroferrocene by a stepwise dilithiation and chlorination process, is resistant to hot concentrated nitric and sulphuric acids (F. L. Hedberg and H. Rosenberg, J. Amer. chem. Soc., 1970, 92: 32). Oxidation of ferrocene Mannich bases to the aldehyde or carboxylic acid may be effected with silver oxide or manganese dioxide (K. Schlögl and M. Walser, Tetrahedron Letters, 1968, 5885; Monatsh., 1969, 100: 840). Direct oxygenation of the ferrocene ring has been reported (J. J. McDonnell, Tetrahedron Letters, 1969, 2039).

(iv) Ligand exchange and insertion

The chemical behaviour of co-ordinated ligands is an area of considerable current interest in view of the usually different geometry and reactivity of the co-ordinated ligand, compared to that of the free molecule, which is often reminiscent of the excited state. For example co-ordinated nitrogen is more easily reduced than free nitrogen (vide infra), and co-ordinated oxygen is considered to be in an excited singlet state, bonding proceeding by donation from an oxygen π-orbital to a vacant orbital on the metal atom with simultaneous back-donation from the metal d-orbitals to the antibonding π^* orbital of the oxygen ligand. The greater the back-donation the greater is the O—O bond length (J. A. Ibers et al., Science, 1967, 155: 709). The hydrocarbon structure of acenaphthylene is drastically changed in its pentacarbonyl di-iron complex, the aromaticity of the molecule being virtually destroyed. One iron atom is bonded symmetrically to the five-membered ring, making a substituted π-cyclopentadienyl system, and the other iron atom is bonded to three atoms of the acenaphthylene by a π-allyl type linkage (M. R. Churchill and J. Wormald, Chem. Comm., 1968, 1597). It appears that the product of the reaction between $[\pi\text{-}C_5H_5Fe(CO)_2]^\ominus$ and propargyl bromide which was originally formulated as $\pi\text{-}C_5H_5\text{-}Fe(CO)_2CH_2C\equiv CH$ and later reformulated as $\pi\text{-}C_5H_5Fe(CO)_2C\equiv C \cdot CH_3$, is in fact an allene $[\pi\text{-}C_5H_5Fe(CO)_2CH=C=CH_2]$ (M. D. Johnson and C. Mayle, Chem. Comm., 1969, 192). The X-ray analysis of rhodium allene complexes shows the allene to have a bent geometry analogous to that of the first excited state of the free ligand (T. G. Hewitt, K. Anzenhofer and J. J. de Boer, ibid., 312; T. Kashiwagi et al., ibid., 317). Strong Lewis acids promote cleavage of the ring–metal bond in ferrocene and its derivatives (E. W. Neuse, R. K. Crossland and K. Koda, J. org. Chem., 1966, 31: 2409, and refs. therein) and, in the presence of a suitable benzenoid compound, one π-C_5H_5 ring may be exchanged to give the arenecyclopentadienyliron cation (XX) (A. N. Nesmeyanov, N. A. Vol'kenau and I. N. Bolesova, Tetrahedron Letters, 1963, 1725). The reverse process (XX → XIX) occurs on u.v. irradiation (A. N. Nesmeyanov et al., Doklady Akad. Nauk, S.S.S.R., 1970, 190: 859):

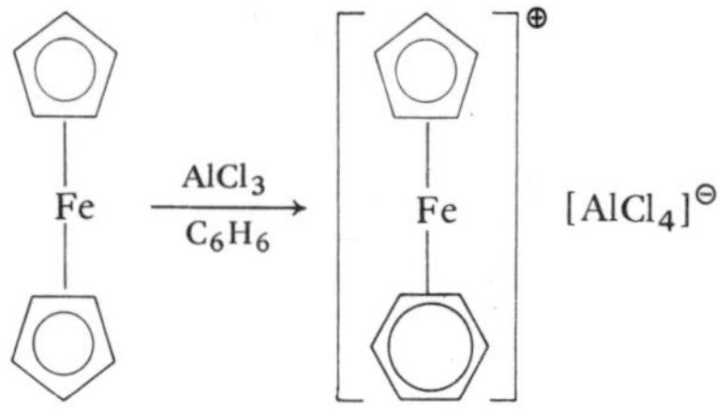

(XIX)   (XX)

Treatment of metallocenes with N-bromosuccinimide (D. Dell, A. Modiano and M. Cais, Israel J. Chem., 1969, 7: 779) and ferrocinium salts with chelating agents (O. N. Suvorova, G. A. Domracher and G. A. Razuvaev, Doklady Akad. Nauk, S.S.S.R., 1968, 183: 850) results in rapid cleavage of the metal–carbon bond. Ligand-exchange occurs on heating ferrocene with ruthenium trichloride at 250°C to give ruthenocene. Polyalkylated ferrocenes have been prepared by disproportionation and ring-migration [(π-ligand)–(π-ligand) exchange] of alkylferrocenes in the presence of aluminium chloride (D. E. Bublitz, J. organometal. Chem., 1969, 16: 149). Di-π-cyclopentadienylmolybdenum and tungsten dihalides undergo ligand exchange with azides, cyanides and thiocyanates to give (π-C_5H_5)$_2$MXY (X and Y = Cl, NCS, SCN, CN) (M = Mo or W) (M. L. H. Green and W. E. Linsell, J. chem. Soc., A, 1969, 2150), and with pentafluorophenyl lithium XXI is formed by substitution at the metal- and ring-carbon atoms (idem, ibid., p.

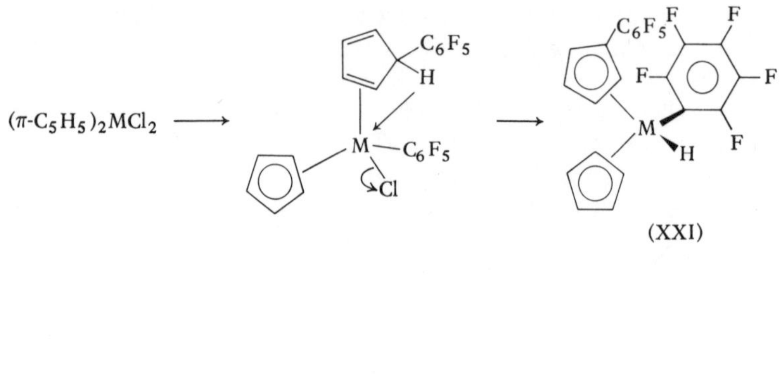

(XXI)

2215). Complexes of dicyclopentadienylvanadium $[(C_5H_5)_2VR]$ (R = C_6F_5, C_5H_5, allyl, etc.) are similarly obtained from the halide (F. W. Siegert and H. J. de Liefde Meijer, J. organometal. Chem., 1968, 15: 131). The cyclic derivatives (XXII–XXV) are obtained by treatment of the corresponding metal dichlorides with polysulphides (H. Köpf and B. Block, Ber., 1969, 102: 1504; Köpf, Angew. Chem., intern. Edn., 1969, 8: 375), or bismercaptans (M. G. Harris, M. L. Green and W. E. Lindsell, J. chem. Soc., A, 1969, 1453; R. B. King and C. A. Eggers, Inorg. Chem., 1968, 7: 340).

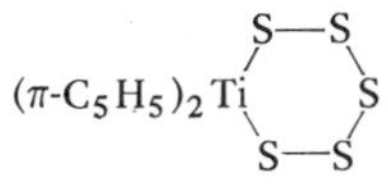

(XXII) 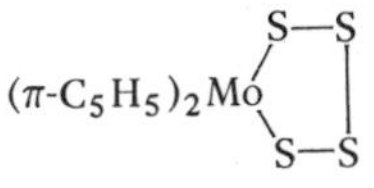   (XXIII) 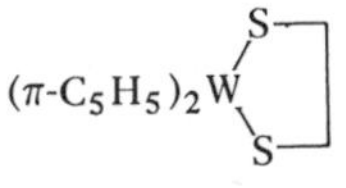   (XXIV) 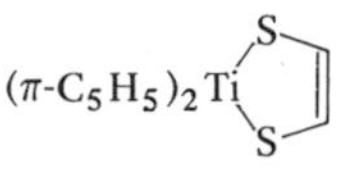   (XXV)

The selenium analogue of XXII has been prepared and the p.m.r. spectrum shows a similar temperature dependence (Köpf, Block and M. Schmidt, Ber., 1968, 101: 272). Hydrazine reacts with the cyclopentadienylirontricarbonyl cation to form a carbazoyl intermediate which loses ammonia to give an isocyanate complex (R. J. Angelici and L. Busetto, J. Amer. chem. Soc., 1969, 91: 3197):

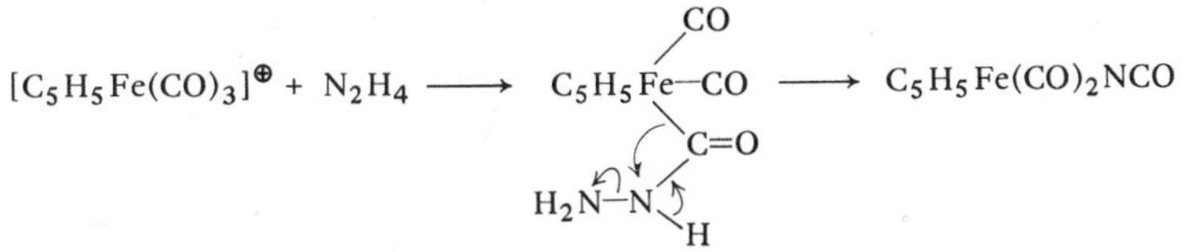

The lone electron pair enables isocyanides to act as carbon ligands and an insertion reaction into nickel–carbon sigma bonds has been reported (Y. Yamamoto, H. Yamazaki and N. Hagihara, J. organometal. Chem., 1969, 18: 189):

$$(\pi\text{-}C_5H_5)Ni(PPh_3)R \rightarrow (\pi\text{-}C_5H_5)Ni(C\equiv N \cdot C_6H_{11})(C(R)=N \cdot C_6H_{11})$$

Reaction of $\pi\text{-}C_5H_5Fe(CO)_2CH_2C\equiv CCH_3$ with SO_2 results in cycloaddition and metal–carbon bond migration to give the heterocyclic derivative

$$[\pi\text{-}C_5H_5Fe(CO)_2\overline{C=C(Me)\text{-}SO\text{-}OCH_2}]$$

rather than the allenyl-O-sulphinate derivative (metal-O–SO–CR=C=CH₂) as previously reported in similar reactions (M. R. Churchill et al., J. Amer.

chem. Soc., 1970, 92: 1795. [For a review of sulphur dioxide and sulphur trioxide insertion see W. Kitching and C. W. Fong (Organometal. Chem. Reviews, 1970, 5: 281).] Treatment of $Ti(C_5H_5)_2(C_6H_5)_2$ with carbon monoxide gives benzophenone (H. Masai, K. Sonogashira and N. Hagihara, Bull. chem. Soc., Japan, 1968, 41: 750). Metal—metal bonds have been formed by insertion into a metal—carbon σ-bond [(a) S. A. Keppie and M. F. Lappert, J. organometal. Chem., 1969, 19: 5; (b) A. N. Nesmeyanov, L. G. Makarova and V. N. Vinogradova, Izvest. Akad. Nauk, S.S.S.R., Ser. Khim., 1969, 1398)] :

(a) $(MeCN)_3M(CO)_3 + Me_3M'(\sigma\text{-}C_5H_5) \rightarrow \pi\text{-}C_5H_5(CO)_3M\text{—}M'Me_3$

(b) $(C_5H_5)Fe(CO)_2Br + Hg + h\upsilon \rightarrow (C_5H_5)Fe(CO)_2HgBr$

The importance of hydride complexes (for a review see J. Chatt, Science, 1968, 160: 723, and for recent refs. P. C. Wailes and H. Weigold, J. organometal. Chem., 1970, 24: 405) has been demonstrated in their use in the isomerization and reduction of alkenes and alkynes (see also G. A. Razuvaev and V. N. Latyaeva, Organometal. Chem. Reviews, 1967, 2: 349 for hydrogenation and polymerization by titanium complexes). The mechanism proposed for reduction envisages the complex serving as a template to which hydrogen and the unsaturated ligand are briefly co-ordinated (and thereby activated) before transfer of one to the other takes place. A concerted hydrogen transfer involving synchronous metal—H bond fission and C—H bond formation by two three-centre interactions has been proposed (G. Wilkinson et al., J. chem. Soc., A, 1966, 1711):

$$RhCl(PPh_3)_3 \rightleftharpoons RhCl(PPh_3)_2 \rightleftharpoons RhCl(PPh_3)_2H_2 \text{ (solvent)} \longrightarrow$$

With carbonyl hydrides of metal complexes, olefin insertion into the metal—hydrogen bond is followed by alkyl migration to a carbonyl carbon (i.e. carbonyl insertion) (K. Noack and F. Calderazzo, J. organometal. Chem., 1967, 10: 101) and results in the formation of aldehydes, ketones or alcohols (the oxo or hydroformylation process). This alkyl migration has been shown to occur with retention of configuration at the migrating carbon centre (G. M. Whitesides and D. J. Boschetto, J. Amer. chem. Soc., 1969, 91: 4313). The equilibrium may be displaced in favour of the migration product by the entry of other ligands such as phosphines or phosphites (P. J. Craig and M. Green, J. chem. Soc., A, 1968, 1978):

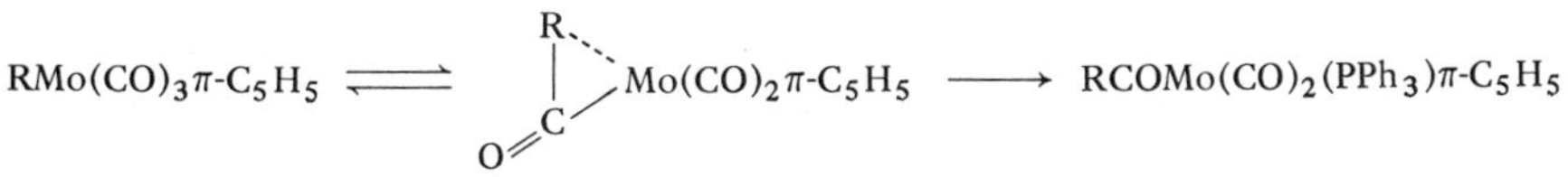

$$RMo(CO)_3\pi\text{-}C_5H_5 \rightleftharpoons \underset{O}{\overset{R}{\underset{\diagdown\!\!\diagdown}{\big|}}}\!\!\!\underset{\diagup}{C}\!\!-Mo(CO)_2\pi\text{-}C_5H_5 \longrightarrow RCOMo(CO)_2(PPh_3)\pi\text{-}C_5H_5$$

The reverse process, decarboxylation of aldehydes by rhodium complexes, has also been reported (M. C. Baird, C. J. Nyman and G. Wilkinson, J. chem. Soc., A, 1968, 348). The possibility of these or related processes, involving intramolecular nucleophilic or electrophilic attack on co-ordinated ligands favourably positioned to react in a template-type mechanism occurring in biosynthetic pathways seems likely [see J. Harris (Ann. N.Y. Acad. Sci. 1969, 153: 675) and S. J. Lippard (ibid., p. 677)] and for the possible uses of charge-transfer complexes formed between metalocinium cations and biological macromolecules see W. A. Korricker and B. L. Vallee (ibid., 1969, 153: 689). Several new hydrides of zirconium have been prepared, by lithium aluminium hydride reduction of bis(cyclopentadienyl)zirconium chlorides (B. Kautzner, P. C. Wailes and H. Weigold, Chem. Comm., 1969, 1105) and the first titanium hydride complex has been isolated as the dimer XXVI [di-μ-hydrido-bis(di-π-cyclopentadienyl)titanium] (J. E. Bercaw and H. H. Brintzinger, J. Amer. chem. Soc., 1969, 91: 7301). The metal–hydride bond in $(\pi\text{-}C_5H_5)_2MoH_2$ undergoes insertion or substitution on reaction with acetylenes or azo-compounds (viz. to give XXVII or XXVIII respectively), depending upon the π-acidity of the attacking reagent (S. Otsuka, A. Nakamura and H. Minamida, Chem. Comm., 1969, 1148).

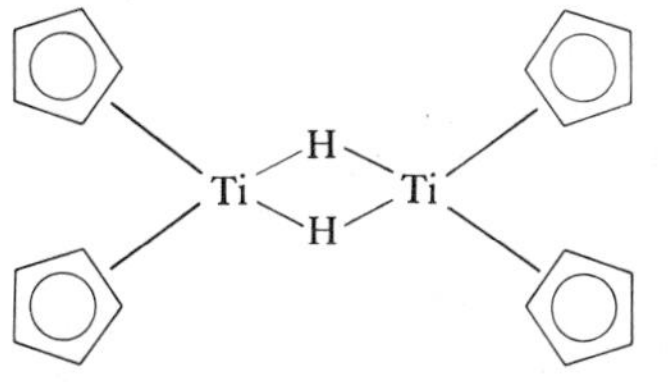

(XXVI) 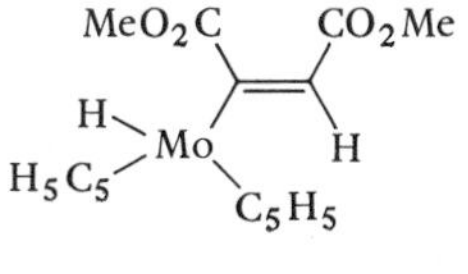   (XXVII) 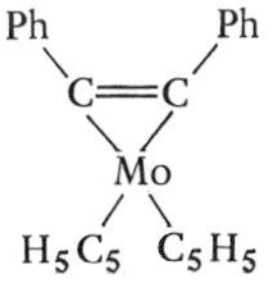   (XXVIII)

The protonation of cyclopentadienyl–metal complexes, and their reaction with a variety of Lewis acids, has been reviewed (M. A. Haas, Organometal. Chem. Reviews, 1969, 4A: 307; J. C. Katz and D. G. Pedrotty, ibid., 1969,

4: 479). Nitrogen acts as a ligand towards transition-metal centres and, in the presence of a reductant and proton source, the complexed nitrogen can be reduced to ammonia (M. E. Vol'pin and V. B. Shur, Nature, 1966, 209: 1236; E. E. van Tamelen et al., J. Amer. chem. Soc., 1969, 91: 1551, and refs. cited). The direct synthesis of amines from molecular nitrogen and aldehydes or ketones in the presence of dicyclopentadienyltitanium dichloride and sodium naphthalide has been reported (Van Tamelen and H. Rudler, ibid., 1970, 92: 5253). The elusive, carbene-like titanocene monomer [(π-C$_5$H$_5$)$_2$Ti] (see H. H. Brintzinger and L. S. Bartell, ibid., 1970, 92: 1105 for theoretical discussion) is believed to be the active entity in the fixation process (Van Tamelen et al., ibid., 1970, 92: 5251). An unstable molecular nitrogen complex is formed when isopropylmagnesium chloride is added to a solution of bis(cyclopentadienyl)titanium monochloride in ether under nitrogen, *at low temperatures*, which on solvolysis with hydrogen chloride at −40 to −100° gives a quantitative yield of hydrazine (A. Shilov et al., unpublished work).

This nitrogen complex must differ from that isolated by Vol'pin and Shur (loc. cit.) since it gives hydrazine and not ammonia on solvolysis and, at room temperature, molecular nitrogen is evolved. However, catalytic reduction of molecular nitrogen to hydrazine has now been accomplished with purely inorganic reagents (viz., molybdenum oxide trichloride in the presence of magnesium chloride and titanium trichloride) in aqueous media under alkaline conditions (A. Shilov et al., Nature, 1971, 231: 460).

(v) *Stability and reactions of the α-metallocenyl carbonium ion*

The stability of *α-ferrocenylcarbonium ions*, recently shown to be greater than originally indicated (E. A. Hill and R. Wiesner, J. Amer. chem. Soc., 1969, 91: 509; M. J. A. Habib and W. E. Watts, J. organometal. Chem., 1969, 18: 361), together with the stereoselectivity of its reactions (G. Gokel et al., Angew. Chem., intern. Edn., 1970, 9: 64, and refs. therein), has given rise to considerable speculation and controversy regarding the structural and electronic features responsible for these properties. The cation XXXI may be stabilized by direct metal participation resulting in displacement of the metal atom towards the α-carbon atom (as XXIX) or tilting or distortion of the cyclopentadienyl ring (as XXX) (Hill and J. H. Richards, J. Amer. chem. Soc., 1961, 83: 3840; M. Cais, Organometal. Chem. Reviews, 1966, 1: 435), or by a mesomeric effect (iron hyperconjugation) (XXXI ↔ XXXII) not involving metal participation (J. C. Ware and T. G. Traylor, Tetrahedron Letters, 1965, 1295; J. D. Fitzpatrick, L. Watts and R. Pettit, ibid., 1966, 1299):

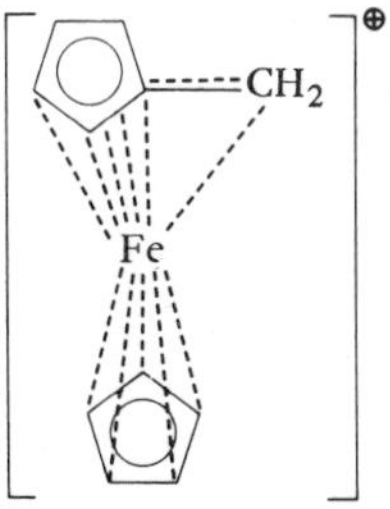

(XXIX) 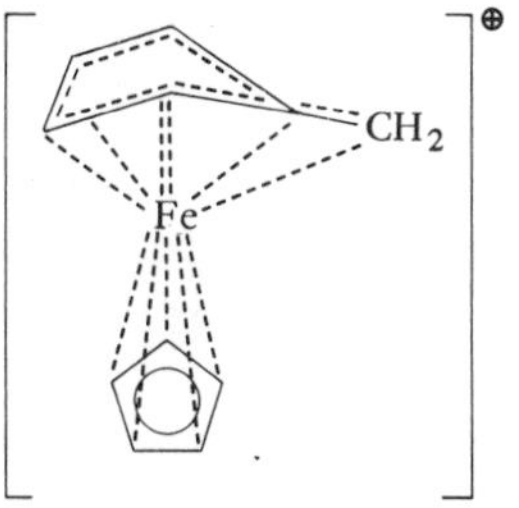 (XXX) 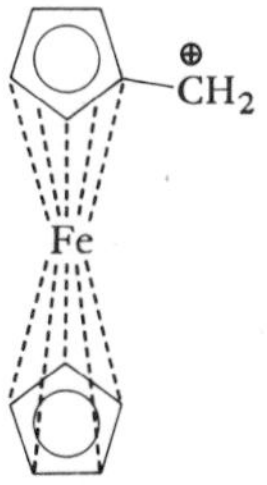 (XXXI) 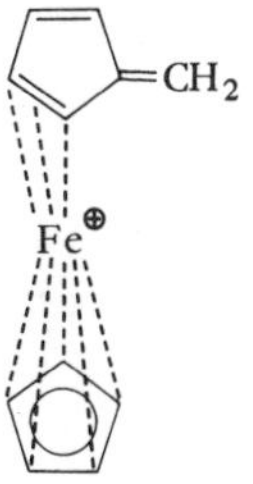 (XXXII)

N.m.r. results have been interpreted in favour of (Cais et al., ibid., 1966, 1695) and against (J. Feinberg and M. Rosenblum, J. Amer. chem. Soc., 1969, 91: 4324) metal participation, or as being consistent with either XXIX or XXXII (Traylor and Ware, ibid., 1967, 89: 2304). A study of the transmission of substituent effects, from one ring to the other, in the solvolysis of ferrocenylmethyl acetates has been interpreted as implying that the metal atom carries a considerable proportion of the charge in the intermediate carbonium ion (D. W. Hall, E. A. Hill and J. H. Richards, J. Amer. chem. Soc., 1968, 90: 4972). Although the Mössbauer effect has been interpreted in favour of metal participation (J. H. Dannenberg and Richards, Tetrahedron Letters, 1967, 4747) the situation must be regarded as un-resolved in the absence of X-ray crystallographic analysis. One application of the stability and ease of formation of α-ferrocenyl alkyl carbonium ions of great potential is in the use of optically active α-ferrocenyl alkylamines in stereoselective asymmetrically induced syntheses of peptides, by four-component condensations between an acid, amine, aldehyde, and isocyanide as shown below:

$$PhCO_2H + R^*NH_2 + i\text{-}C_3H_7CHO + CN \cdot Bu^t \longrightarrow$$

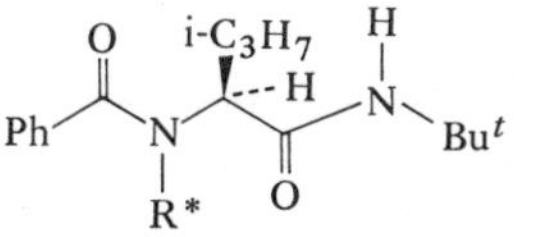

(XXXIII) 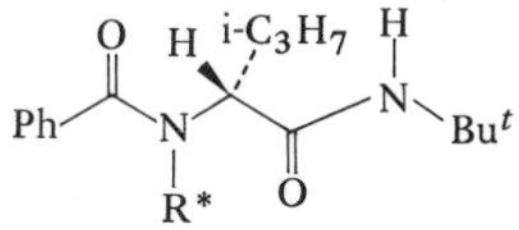 (XXXIV)

The group R* of the optically active amine must be replaceable by hydrogen [viz., XXXIII (R* = $C_5H_5FeC_5H_5CH_2\cdot$) → XXXIII (R* = H)] under mild conditions, so that cleavage or racemization of the peptide chain is prevented, and the asymmetric inducing power of the amine (i.e. the ratio of XXXIII:XXXIV or vice versa) must be high. α-Ferrocenyl alkyl amines are the most promising class of amines found for this purpose (I. Ugi, Record of Chem. Progress, 1969, 30: 289). Similarly dimethylaminomethylferrocene methiodide reacts with a variety of nucleophilic reagents, such as azides (D. E. Bublitz, J. organometal. Chem., 1970, 23: 225) and enamines (T. I. Bieber and M. T. Dorsett, J. org. Chem., 1964, 29: 2028) to give the corresponding ferrocenylmethyl substituted derivative formed by displacement of trimethylamine.

The powerful electron-releasing property of the ferrocene system is also reflected in its enhanced reactivity to electrophilic aromatic substitution (M. Rosenblum, Chemistry of the Iron Group Metallocenes, Part I, New York, 1965), its greater migratory aptitude (compared to phenyl) in pinacol rearrangements (S. I. Goldberg and W. G. Bailey, Chem. Comm., 1969, 1059), the greater thermodynamic stability of ferrocenylvinyl relative to styryl systems (M. J. A. Habib, J. Park and W. E. Watts, J. organometal. Chem., 1970, 21: P59), and the reduced stabilization of radical anions (C. Elschenbroich and M. Cais, ibid., 1969, 18: 135; cf. G. Marr and J. Ronayne, Chem. Comm., 1970, 351). (For reactions of α-ferrocenylcarbenes see A. Sonada et al., Tetrahedron, 1970, 26: 3075.)

(vi) Bridged ferrocenes, ferrocenophanes

Bridged ferrocenes are designated as [*m*]ferrocenophanes (where *m* = 2—10) when the two rings of one ferrocene molecule are bridged once, [*m*][*n*]ferrocenophanes when bridged twice, and [*m.n*]ferrocenophanes when two ferrocene molecules are joined by two chains (of *m* or *n* atoms in length). The general chemistry of ferrocenophane systems has been reviewed (W. E. Watts, Organometal. Chem. Reviews, 1967, 2: 231; for synthetic aspects see also D. E. Bublitz and K. L. Rinehart, Organic Reactions, Vol. 17; M. I. Bruce, Organometal. Chem. Reviews, 1969, 5B: 379; T. H. Barr and Watts, Tetrahedron, 1968, 24: 3219; 1969, 25: 861). Various heterobridged ferrocenes have also been described (Ger. P. 59,785/1966; U.S.P. 3,415,859/1965; R. A. Abramovitch, C. I. Azogu and R. G. Sutherland, Chem. Comm., 1969, 1439 and M. Kumada et al., ibid., 1969, 207) containing nitrogen, oxygen, sulphur and silicon hetero-atoms. A novel ferrocenophane system is found in the so-called *bis(π-azulene)iron*, formed from ferric chloride and azulene in the presence of a Grignard reagent (E. O. Fischer and J. Müller, J. organometal. Chem., 1964, 1: 464) which has been shown to consist of two azulene fragments dimerized by *ortho—para* coupling (XXXV) (M. R. Churchill and J. Wormald, Chem. Comm., 1968, 1033). *Ortho—ortho* coupling has been observed in diazulenetetrairon decacarbonyl (XXXVI) (Churchill and P. H. Bird, J. Amer. chem. Soc., 1968, 90: 3241). (See M. R. Churchill, Progress in Inorg. Chem., 1970, 11:

53, for a review of transition-metal complexes of azulene and related ligands.)

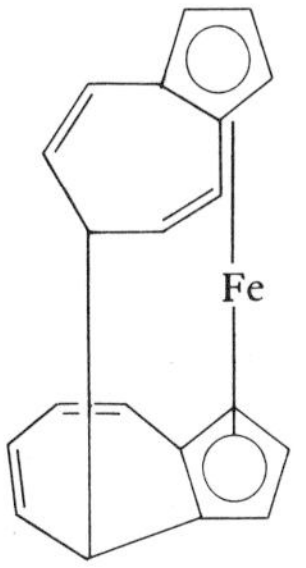

(XXXV) 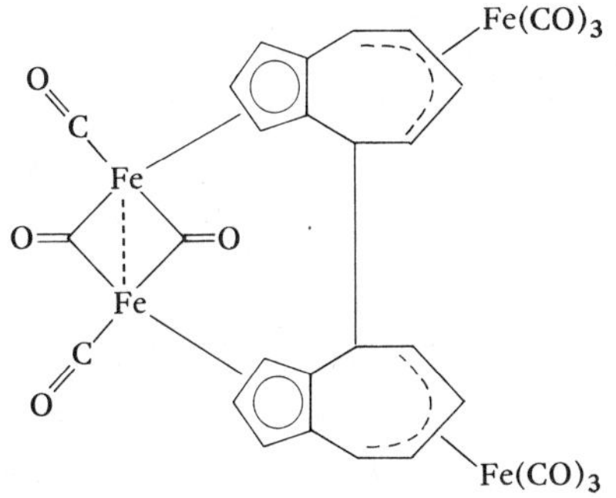   (XXXVI)

The simplest members of the $[m \cdot n]$ series, $[O \cdot O]$ (F. H. Hedberg and H. Rosenberg, J. Amer. chem. Soc., 1969, 91: 1258; M. D. Rausch, R. F. Kovar and C. S. Kraihanzel, ibid., p. 1259) and $[1 \cdot 1]$ ferrocenophane (T. H. Barr, H. L. Lentzner and W. E. Watts, Tetrahedron, 1969, 25: 6001) have been reported. An off-centre configuration of the iron atoms is proposed for the $[O \cdot O]$ derivative (XXXVII) in order to account for the p.m.r. spectrum.

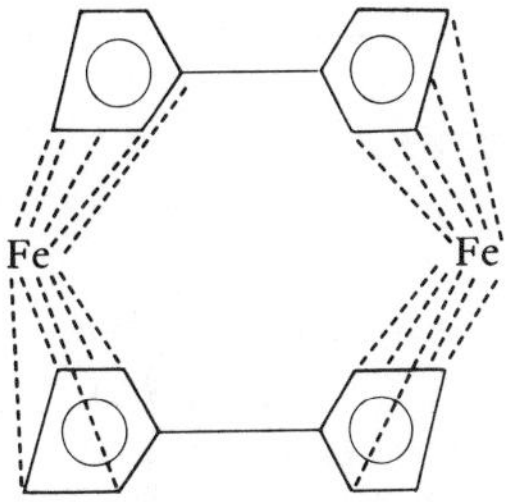

(XXXVII) 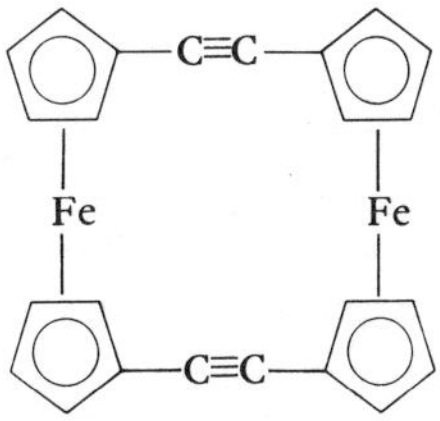   (XXXVIII) 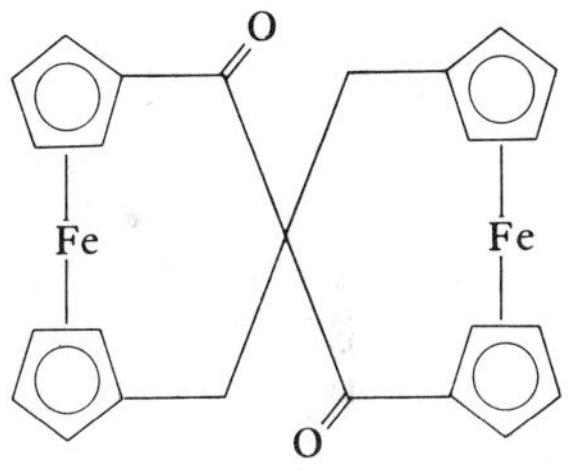   (XXXIX)

Coupling of the cuprous salt of 1-ethynyl-1′-iodoferrocene gives the $[2 \cdot 2]$ ferrocenophane-1,3-diyne (XXXVIII) (M. Rosenblum et al., J. organo-metal. Chem., 1970, 24: 469), and condensation of dimethylaminomethyl-ferrocene methiodide (2 mole) with malonic ester gives the spiroferroceno-phane (XXXIX) (A. Dormond and J. Décombe, Bull. Soc. Chim. Fr., 1968, 3673). Oligomeric ferrocenophanes (XL) (n = 1—4) and bridged titanocenes (XLI) have been isolated (T. J. Katz, N. Acton and G. Martin, J. Amer. chem. Soc., 1969, 91: 2804; Katz and Acton, Tetrahedron Letters, 1970,

2497) from the reaction of biscyclopentadienylmethane dianion with transition-metal halides.

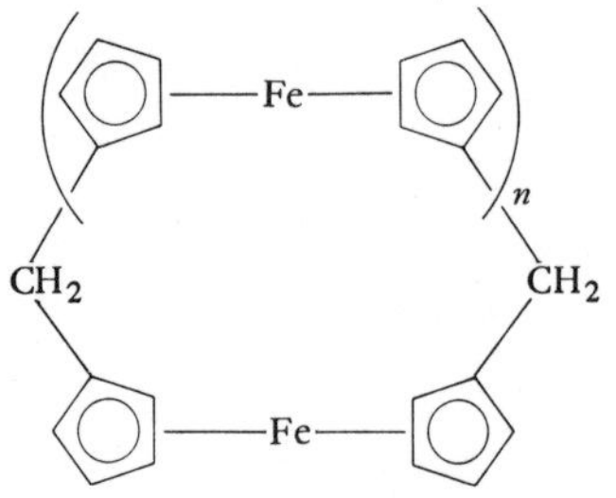

(XL) 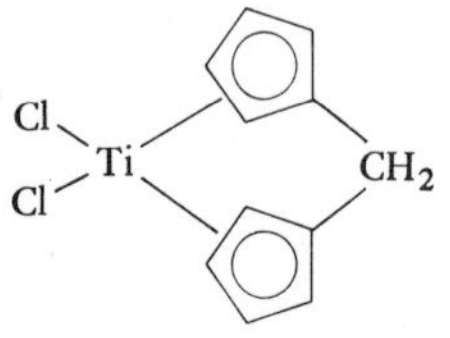   (XLI)

Cyclization of dimeric α-ferrocenylcarbonium ions gives the ferrocenophane
(XLII) and the homo-annular ferrococarbocyclic derivative (XLIII) (W. M.
Horspool, R. G. Sutherland and J. R. Sutton, Canad. J. Chem., 1970, 48:
3542):

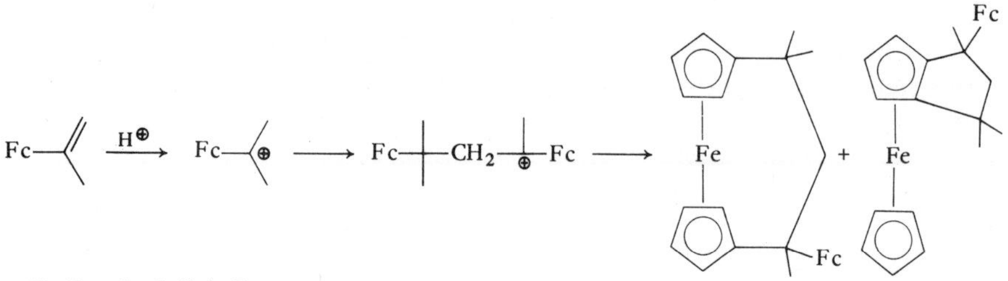

H–Fc = $(\pi\text{-}C_5H_5)_2Fe$

(XLII)   (XLIII)

The effect of ring-tilt deformations caused by interannular bridging on the
p.m.r. and u.v. spectral characteristics (T. H. Barr and W. E. Watts,
Tetrahedron, 1968, 24: 6111; J. organometal. Chem., 1968, 15: 177) and
Lewis basicity of ferrocenophanes (H. L. Lentzner and Watts, Chem. Comm.,
1970, 26) has been discussed.

(vii) Bi- and poly-ferrocenes

The acylation and formylation of biferrocenes has been investigated
(M. D. Rausch and T. M. Gund, J. organometal. Chem., 1970, 24: 463 and refs.
therein) and the intermediacy of a ferrocyne precursor proposed in their
synthesis (J. W. Huffman, L. H. Keith and R. L. Asbury, J. org. Chem.,
1965, 30: 1600). The formation of isomeric 2,7- and 2,10-bis(N,N-dimethyl-
aminomethyl)biferrocenyls (XLIV and XLV) (G. Marr, R. E. Moore and
P. W. Rockett, Tetrahedron, 1969, 25: 3477) and ferrocene oligomers (XLVI,

$n = 1-5$) (Rausch, P. V. Roling and A. Siegel, Chem. Comm., 1970, 502) has been reported.

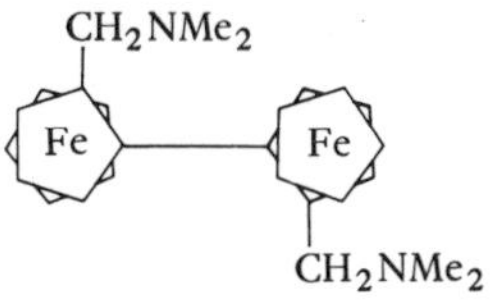

(XLIV) 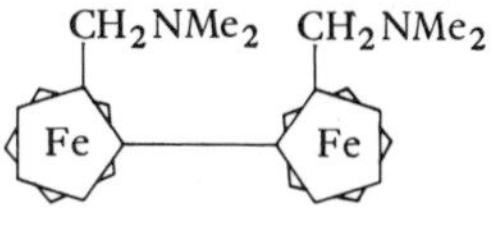   (XLV) 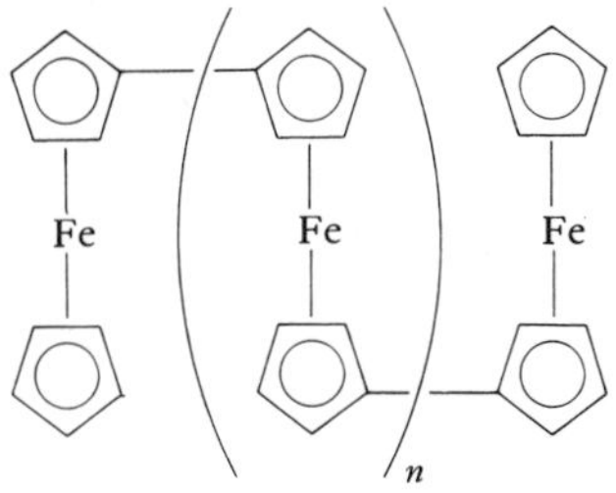   (XLVI)

1,2,4- and 1,3,5-*Triferrocenylbenzene* derivatives (XLVII and XLVIII) have been obtained by trimerization of ferrocenylacetylenes (at 180°) or acetyl-ferrocene (in the presence of ethyl orthoformate and hydrogen chloride) respectively (K. Schlögl and H. Soukup, Tetrahedron Letters, 1967, 1181).

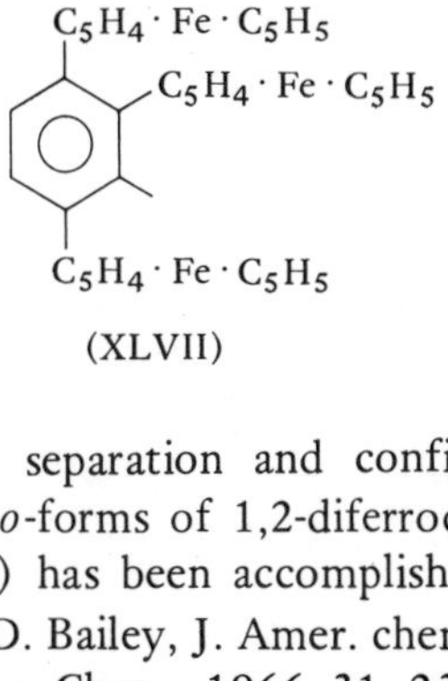

(XLVII) 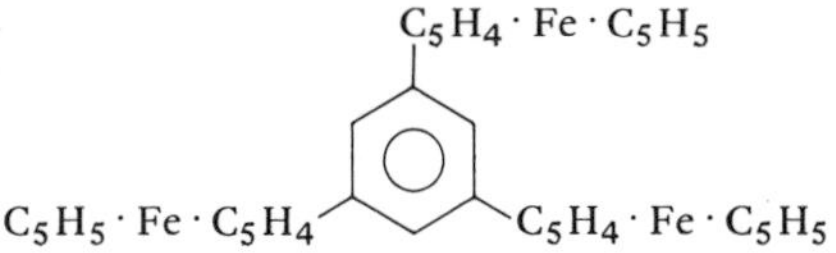   (XLVIII)

The separation and configurational assignment of the *erythro-* and the *threo*-forms of 1,2-diferrocenyl-1,2-diphenylethane (and the corresponding diol) has been accomplished by asymmetric selection (S. I. Goldberg and W. D. Bailey, J. Amer. chem. Soc., 1969, 91: 5685; Goldberg and F.-L. Lam, J. org. Chem., 1966, 31: 2336).

(viii) Di- and poly-nuclear complexes

Many dinuclear complexes containing different metal atoms are known, such as *tricarbonyliron-μ-carbonyl-μ-diphenylphosphido-π-cyclopentadienyl-nickel* (XLIX) (K. Yasufuku and H. Yamazaki, Bull. chem. Soc., Japan, 1970, 43: 1588, and refs. therein) and *tricarbonylcyclopentadienyl-chromium complexes* (viz., $(C_5H_5)_2M(C_5H_5(CO)_3Cr$, M = Co, Cr, V, Ti) (A. Miyake, H. Kondo and M. Aoyama, Angew. Chem., intern. Edn., 1969, 7: 520), and the preparation and properties of those with metal—metal bonds have been reviewed (N. S. Vyazankin, G. A. Razuvaev and O. A. Kruglaya, Organometal. Chem. Reviews, 1968, 3A: 323).

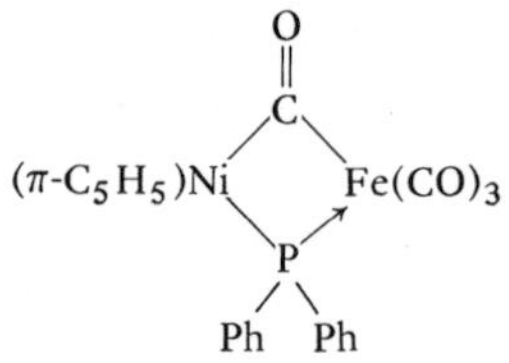

(XLIX)

Reaction of rhodium trichloride and cyclopentadienyl magnesium bromide gives $[(C_5H_5)_2Rh]^{\bullet}$ and L, in which the fourth cyclopentadienyl ring is parallel to the triangle of rhodium atoms (E. O. Fischer et al., Chem. Comm., 1967, 643). Photolysis of $[C_5H_5Rh(CO)_2]$ gives $[(C_5H_5)_2Rh_2(CO)_3]$, containing a single bridging carbonyl group (O. S. Mills and J. P. Nice, J. organometal. Chem., 1967, 10: 337) and two isomers (LI and LII) differing only in the terminal and bridging carbonyl groups (Mills and E. F. Paulus, ibid., 1967, 10: 331; Paulus et al., ibid., p. P3).

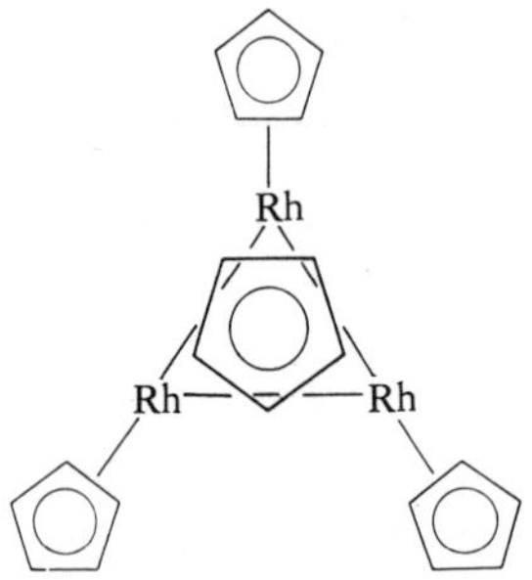

(L) 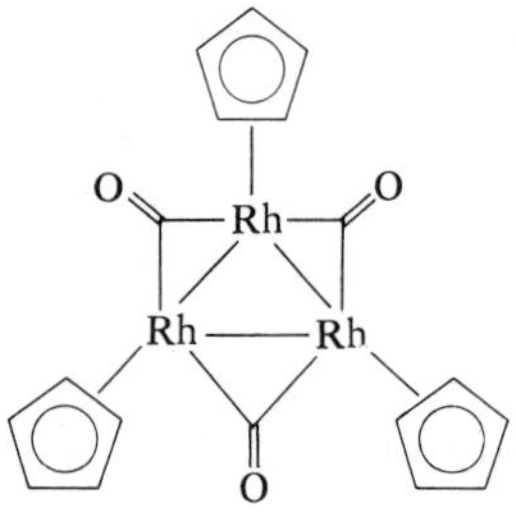   (LI) 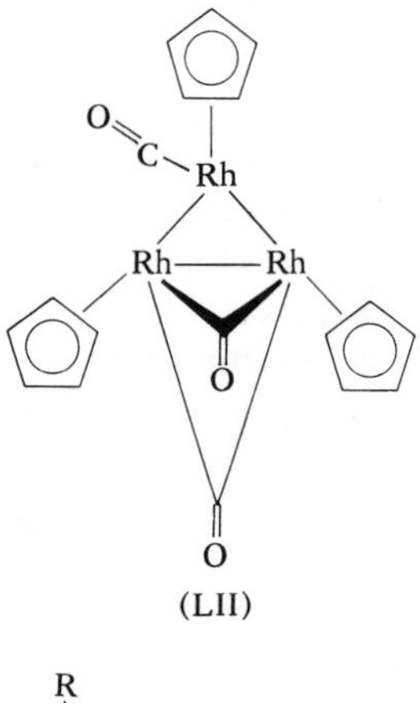   (LII)

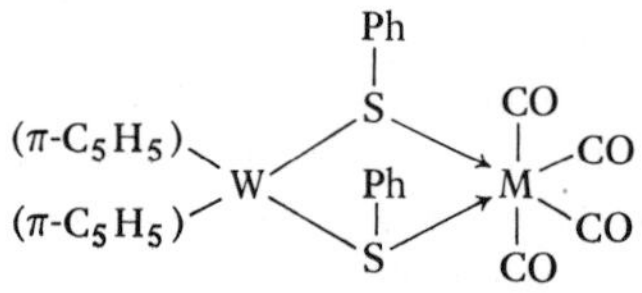

(LIII) 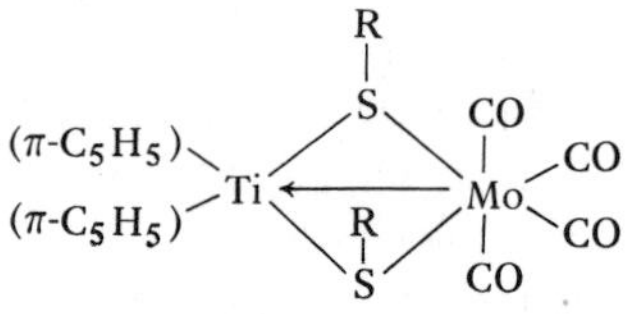   (LIV)

However, not all dinuclear complexes contain metal–metal bonds. For example X-ray analysis of $(\pi\text{-}C_5H_5)_2W(SPh)_2M(CO)_4$, (M = Cr, Mo and W) (LIII) and $(\pi\text{-}C_5H_5)_2Ti(SMe)_2Mo(CO)_4$ (LIV), together with infrared and p.m.r. data, indicate metal–metal bonding specific to the titanium–molybdenum complex (LIV) only (T. S. Cameron et al., Chem. Comm., 1971, 14).

X-ray analysis of the tetranuclear mixed metal complex $(\pi\text{-}C_5H_5)_2$-$Rh_2Fe_2(CO)_8$ shows a tetrahedral metal-atom cluster with one $\pi\text{-}C_5H_5$ ligand being associated with each of the rhodium atoms, which are bridged by one carbonyl group, and three carbonyl groups attached to one iron atom and four to the other (two of these latter carbonyl groups form asymmetric bridges between the iron atom and the two rhodium atoms) (M. R. Churchill and M. V. Veidis, Chem. Comm., 1970, 529). The stereochemistry of metal-cluster compounds, as derived from X-ray diffraction studies (B. R. Penfold, Perspectives in Structural Chemistry, 1968, II: 71) and metal–metal bonding in transition-metal compounds (M. C. Baird, Progress in Inorg. Chem., 1968, 9: 1; T. G. Spiro, ibid., 1970, 11: 1) have been reviewed.

(ix) Carboranes

Numerous π-bonded transition-metal carborane complexes analogous to metallocenes (see Vol. II A, p. 144) have recently been reported, such as $(\pi\text{-}C_5H_5)Fe[\pi\text{-}(3)\text{-}1,2\text{-}B_9C_2H_{11}]$ (M. F. Hawthorne et al., J. Amer. chem. Soc., 1968, 90: 879), $(\pi\text{-}C_5H_5)Co(\pi\text{-}(2)\text{-}1,6\text{-}B_7C_2H_9)$ (LV) and $(\pi\text{-}C_5H_5)Co(\pi\text{-}(2)\text{-}1,10\text{-}B_7C_2H_9)$ (LVI) (T. A. George and Hawthorne, ibid., 1969, 91: 5475), $(\pi\text{-}C_5H_5)Cr(\pi\text{-}(3)\text{-}1,7\text{-}B_9C_2H_{11})$ and $(\pi\text{-}C_5H_5)Cr(\pi\text{-}(3)\text{-}1,2\text{-}B_9C_2H_{10}C_6H_5$ (H. W. Ruhle and Hawthorne, Inorg. Chem., 1968, 7: 2279). Preliminary investigations into the electrophilic substitution reactions of these "π-sandwich" cobalt complexes indicate that the boron atoms of the carborane cage are attacked rather than carbon of the π-cyclopentadienyl ring (B. M. Graybill and Hawthorne, Inorg. Chem., 1969, 8: 1799):

$$(\pi\text{-}C_5H_5)Co(\pi\text{-}1,6\text{-}B_7C_2H_9) \rightarrow (\pi\text{-}C_5H_5)Co(\pi\text{-}1,6\text{-}B_7C_2H_8\text{-}8\text{-}COCH_3)$$

Metallocene analogues of copper, gold and palladium, derived from the (3)-1,2-dicarbollide ion have been reported (L. F. Warren and Hawthorne, J. Amer. chem. Soc., 1968, 90: 4823; R. M. Wing, ibid., p. 4828). For reviews on carboranes and transition metal–carborane complexes see L. J. Todd (Adv. organometal. Chem., 1970, 8: 87) and T. Onak (ibid., 1965,

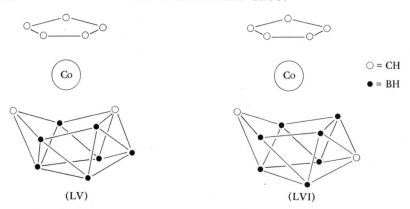

○ = CH
● = BH

(LV)   (LVI)

3: 263). σ-Bonded carborane systems, $[1\text{-}(\pi\text{-}C_5H_5)Fe(CO)_2\text{-}2\text{-}(CH_3)\text{-}(\sigma\text{-}1,2\text{-}B_{10}C_2H_{10})]$ and $1,10\text{-}[(\pi\text{-}C_5H_5)Fe(CO)_2]_2\text{-}1,10\text{-}(\sigma\text{-}B_8C_2H_8)$ have also been prepared (J. C. Smart, P. M. Garrett and Hawthorne, J. Amer. chem. Soc., 1969, 91: 1031).

(c) Physical properties and structure of cyclopentadienyl metal compounds

(i) Structure and fluxional behaviour

The possibility that *triscyclopentadienylnitrosylmolybdenum* has three different forms of C_5H_5-to-metal bonding (as LVII), viz. mono*hapto** (h'- or σ-), tri*hapto* (h^3- or π-allyl), and penta*hapto* (h^5 or π- or "sandwich") has been ruled out by X-ray analysis (F. A. Cotton, Discuss. Faraday Soc., 1969, 47: 79). The true structure (LVIII) shows that all five carbon atoms of two rings are participating in the bonding to varying degrees and that no simple electronic description is possible. R. B. King has suggested that this may be attributed to interligand resonance (LIXa $\leftrightarrow$ LIXb) (X = C_5H_5) (Inorg. Nucl. Chem. Let., 1969, 5: 901).

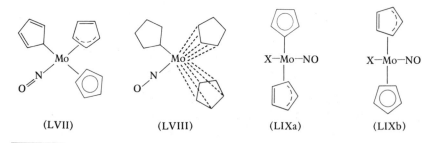

(LVII)   (LVIII)   (LIXa)   (LIXb)

---

* See footnote on p. 82 for nomenclature.

The structure of the first compound containing a benzyl group bonded to metal through an allylic sequence of three carbon atoms, 1,2,7-*tri*hapto-*benzylpenta*hapto*cyclopentadienyldicarbonylmolybdenum* (LX), has been confirmed by X-ray crystallographic analysis (F. A. Cotton and M. D. LaPrade, J. Amer. chem. Soc., 1968, 90: 5418). There is marked bond alternation in LX, indicating significant reduction in benzenoid resonance. The product derived from cyclopentadienylirondicarbonyl anion and tropylium tetrafluoroborate is assigned the tri*hapto* structure (LXII) rather than the mono*hapto* structure (LXI) (D. Ciappenelli and M. Rosenblum, ibid., 1969, 91: 3673, 6876). Similarly complexes of type $C_5H_5CuPR_3$, originally assumed to have a σ-bonded ring, are π-bonded "sandwich" structures (Cotton and T. J. Marks, ibid., 1969, 91: 7281; Cotton and J. Takats, ibid., 1970, 92: 2353).

(LX)   (LXI)   (LXII)

Fluxional behaviour of π- and σ-bonded organometallic systems, involving rapid interconversion of structurally equivalent molecules, is a well established phenomenon and has been reviewed (Cotton, Accounts Chem. Res., 1968, 1: 257). Rearrangement of the metal—carbon bonds results in the metal moving along the periphery of the ring in a series of 1,2-shifts. These have been regarded as proceeding through dipolar (Cotton et al., J. Amer. chem. Soc., 1966, 88: 4371; 1967, 89: 6136) or dissociated polar transition states (G. M. Whitesides and J. S. Fleming, ibid., 1967, 89: 2855) but are probably more correctly regarded as degenerate [1,5]sigmatropic rearrangements (Rosenblum et al., ibid., 1968, 90: 5293). This behaviour accounts for the apparent equivalence of the protons in the n.m.r. spectra of σ-bonded cyclopentadienyl complexes, such as the *dialkylaluminium cyclopentadienyls* (W. R. Kroll and W. Naegele, Chem. Comm., 1969, 246), *dicyclopentadienylmercury* (vide infra), or (π-C_5H_5)M(CO)$_2$(σ-C_5H_5) (M = Fe, Ru) (C. H. Campbell and M. L. H. Green, J. chem. Soc., A, 1970, 1318). Dicyclopentadienylmercury has been represented as fluxional mono-, tri-, or penta*hapto* structures (LXIII, LXIV, or LXV respectively) (T. S. Piper and G. Wilkinson, J. inorg. nucl. Chem., 1956, 2: 32; 1956, 3: 104; A. N. Nesmeyanov et al., Doklady Akad. Nauk, S.S.S.R., 1964, 159: 847; 1966,

166: 868; Chem. Comm., 1969, 105). Recent p.m.r., i.r., and Raman spectroscopic studies are interpreted in favour of the σ-bonded structure (LXIII) (P. West, M. C. Woodville and M. D. Rausch, J. Amer. chem. Soc.,

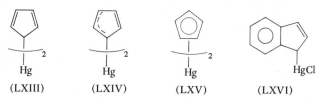

(LXIII)   (LXIV)   (LXV)   (LXVI)

1969, 91: 5649; E. Maslowsky and K. Nakamoto, Inorg. Chem., 1969, 8: 1108; F. A. Cotton and T. J. Marks, J. Amer. chem. Soc., 1969, 91: 7281). The mono*hapto* structure (LXVI) for *indenylmercuric chloride* has been confirmed (W. Kitching and B. F. Hegarty, J. organomet. Chem., 1969, 16: P39) and the p.m.r. spectrum of *bis(pentamethylcyclopentadienyl)mercury* is that of a static σ-bonded structure (three non-equivalent methyl groups); this is attributed to retardation of the fluxional behaviour by the substituents (B. Floris, G. Illuminati and G. Ortaggi, Chem. Comm., 1969, 492). *Dicyclopentadienylberyllium,* originally regarded as a σ-complex (E. O. Fischer and H. P. Hoffmann, Ber., 1959, 92: 482) has now been shown to have a sandwich structure with the beryllium atom unequally spaced from the two staggered, parallel cyclopentadiene rings (H. P. Fritz and D. Sellmann, J. organometal. Chem., 1966, 5: 501). The reported fluxional behaviour of trans,trans-1,4-*bis-(dicarbonyl-π-cyclopentadienyliron)buta-*1,3-*diene,* based on the apparent equivalence of the butadiene protons at 60 MHz, has not been substantiated by measurements at 100 and 250 Mhz (F. A. L. Anet and O. J. Abrams, Chem. Comm., 1970, 1611). Evidence for fluxional behaviour of a benzene ring, in *pentamethylpenta*hapto*cyclopenta-dienyltetra*hapto*benzene-rhodium* and -*iridium* complexes, has been observed (J. W. Kang, R. F. Childs and P. M. Maitlis, J. Amer. chem. Soc., 1970, 92: 720). An intramolecular interchange of the mono- and penta-*hapto* rings occurs in $Ti(C_5H_5)_4$ (viz., $h^1 \rightleftharpoons h^5$; activation energy 16.1 kcal · mole^{-1}). There is no such equilibrium in $Fe(C_5H_5)_2(CO)_2$, presumably due to a lack of empty metal orbitals whereby both rings can be bound in an equivalent fashion (F. A. Cotton et al., ibid., 1970, 92: 3801). The complex $(C_5H_5)(CO)_2Mo(C_7H_7)Fe(CO)_3$ (LXVII) contains both molybdenum and iron atoms bound to the cycloheptatrienyl ring (tri*hapto* and tetra*hapto* respectively). The molecule is fluxional and gives a p.m.r. spectrum consistent with this structure only below −50° (F. A. Cotton and C. R. Reich, J. Amer. chem. Soc., 1969, 91: 847). Both hydrogen and trimethylsilyl migration occur in 5-*trimethylsilylcyclopentadiene,* the latter

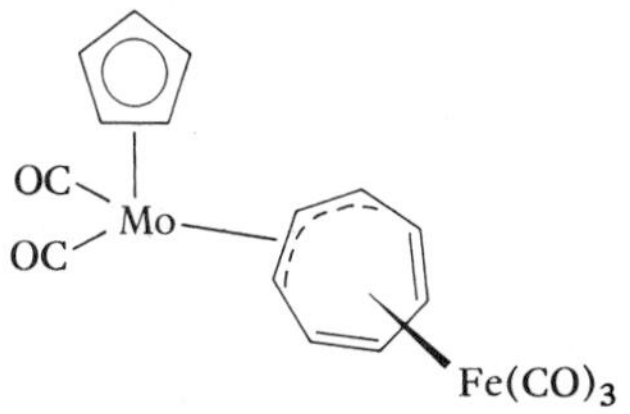

(LXVII)

process being the most rapid by a factor of 10^6 (A. J. Ashe, J. Amer. chem. Soc., 1970, 92: 1233). Similar stereochemical non-rigidity is evident in methylcyclopentadienyl silicon and tin systems (A. Davison and P. E. Rakita, ibid., 1968, 90: 4479).

Evidence for the formation of a "triple-decker" sandwich ion (LXVIII) in the mass fragmentation of tetranuclear cyclopentadienylmetal carbonyls has been cited (R. B. King, Chem. Comm., 1969, 436). Mass spectral evidence indicates that the metal—carbon bond to an unsubstituted cyclopentadiene ring is weaker than that to a ring having an unsaturated group at the α-position (M. Cais, M. S. Lupin and J. Sharvit, Israel J. Chem., 1969, 7: 73). An intramolecular rearrangement of *π-cyclopentadienyl-π-cycloheptatrienyl-vanadium* to bis(benzene)vanadium follows electron impact (M. F. Retting et al., J. Amer. chem. Soc., 1970, 92: 5100). Electron-diffraction studies of gaseous nickelocene indicates that the metal—carbon bonds are weaker than

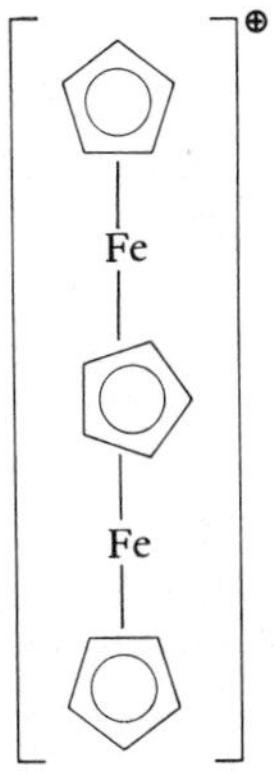

(LXVIII) 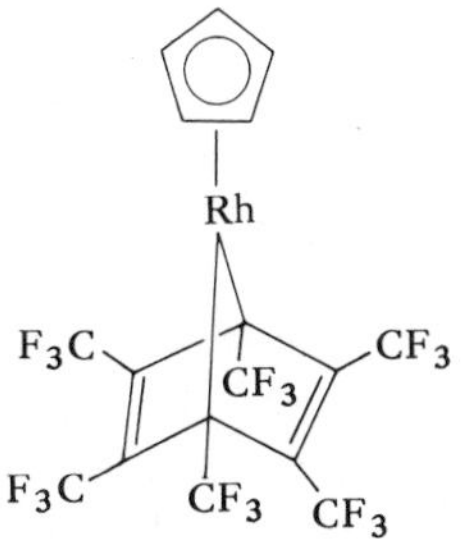   (LXIX) 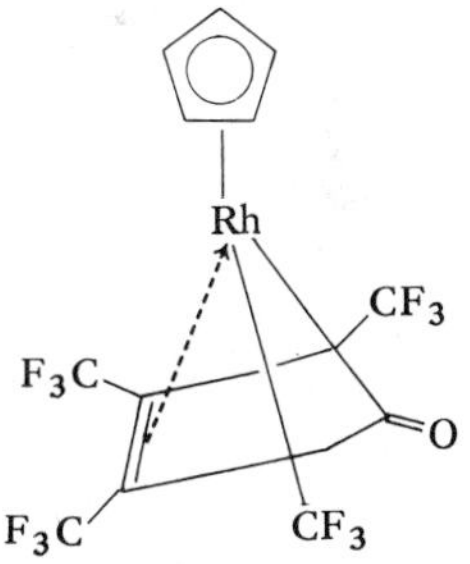   (LXX)

in ferrocene and that the rings consequently have much greater freedom of rotation (L. Hedberg and K. Hedberg, J. chem. Phys., 1970, 53: 1228).

The reaction of cyclopentadienylrhodiumdicarbonyl with hexafluoro-2-butyne gives a "Dewar benzene" (LXIX) containing two σ- and one π-bond (R. S. Dickson and G. Wilkinson, Chem. and Ind., 1963, 1432), the structure being confirmed by X-ray analysis (M. R. Churchill and R. Mason, Proc. chem. Soc., 1963, 365), together with LXX. [The formation of cyclopentadienone complexes from the reaction of alkynes with π-$C_5H_5M(CO)_2$ complexes (M = Co, Rh) is well documented (J. W. Kang, S. McVey and P. M. Maitlis, Canad. J. Chem., 1968, 46: 3189; F. L. Bowden and A. B. P. Lever, Organometallic Chem. Reviews, 1968, 3A: 227; R. S. Dickson and G. R. Tailby, Austral. J. Chem., 1970, 23: 1531, and references therein).]

(ii) Stereochemistry

σ-Benzyl-π-$C_5H_5Mo(CO)_2$ [$P(OC_6H_5)_3$] and analogues have been shown to exist in two isomeric forms in solution (LXXI and LXXIII), identified by the non-equivalent methylene proton signals of the benzyl group when attached to a chiral centre as in the cis-isomer (LXXI). The facile cis—trans interconversion presumably occurs via an intermediate such as LXXII (J. W. Faller and A. S. Anderson, J. Amer. chem. Soc., 1969, 91: 1550; Faller, Anderson and C.-C. Chen, J. organometal. Chem., 1969, 17: P7). The

(LXXI)   (LXXII)   (LXXIII)   (LXXIV)

introduction of bulky substituents into silylmethyl and silylacetyl derivatives of π-cyclopentadienyliron carbonyl phosphine complexes (LXXIV) (R = Me_3SiCH_2, Me_3SiCH_2CO; L = Ph_3P, Ph_2MeP, Ph_2EtP) results in restricted rotation about the metal—carbon bond. The presence of discrete rotamers is indicated by the temperature-dependent carbonyl absorptions and non-equivalent methylene protons (K. H. Pannell, Chem. Comm., 1969, 1346). The preparation, resolution, and the use of physical and chemical methods in determining the relative and absolute configurations of chiral metallocenes has been reviewed (K. Schlögl, "Topics in Stereochemistry", Vol. I, eds. N. L. Allinger and E. L. Eliel, Wiley, New York, 1967). More recent publications are concerned with chiral arylferrocenes (H. Falk, H. Lehner and K.

Schlögl, Monatsh., 1970, 101: 967; Lehner and Schlögl, ibid., 1970, 101: 895), racemization of optically active ferrocenes (K. Bauer et al., ibid., 1970, 101: 941), conformational analysis of [3]ferrocenophanes (Falk and O. Hofer, ibid., 1970, 101: 477), circular dichroism of ferrocenes (idem, ibid., 1969, 100: 1499 et seq.), biferrocenyl (Schlögl and M. Walser, ibid., 1969, 100: 1515), and resolution by countercurrent distribution (Bauer, Falk and Schlögl, ibid., 1968, 99: 2186).

(iii) Spectroscopic properties

Review articles have appeared on the structure elucidation of cyclopentadienyl metal compounds by mass spectrometry (I. J. Spilners and J. G. Larson, Org. Mass. Spectrom., 1970, 3: 915; Cais and Lupin, Adv. organometal. Chem., 1970, 8: 211; M. I. Bruce, ibid., 1968, 6: 273), nuclear magnetic resonance spectroscopy (M. L. Maddox, S. L. Stafford and H. D. Kaesz, ibid., 1965, 3: 1; see also M. D. Rausch and A. Siegel, J. organometal. Chem., 1969, 17: 117, and W. M. Horspool and R. G. Sutherland, Canad. J. Chem., 1968, 46: 3453; D. W. Slocum et al., J. chem. eng. Data, 1968, 13: 378), and X-ray diffraction of π-complexes (P. J. Wheatley, Perspectives in Structural Chemistry, 1967, I: 1) and σ-complexes (M. R. Churchill, ibid., 1970, III: 91). Molybdenum—carbon bond lengths are remarkably sensitive to changes in bond order; X-ray analysis of π-$C_5H_5Mo(CO)_2$- $P(C_6H_5)_3COCH_3$ provides direct evidence for $d\pi$–$p\pi$ metal-to-acyl back donation (viz., M—C=$O\cdot R$ ↔ $\overset{\oplus}{M}$=C—$O^{\ominus}\cdot R$) (M. R. Churchill and J. P. Fennessey, Inorg. Chem., 1968, 7: 953) (See also Spectroscopic Properties of Inorganic and Organometallic Compounds, The Chemical Society, London, 1968 et seq.).

(d) Orbital symmetry considerations

In the presence of certain metal ions or complexes symmetry forbidden pericyclic reactions may become allowed owing to formation of organometallic π-complexes. The extension of the Woodward—Hoffmann rules to organometallic systems and the role of the catalyst, in providing a ground-state reaction path by re-routing electron pairs in the reactants has been discussed (F. D. Mango, Adv. in Catal., 1969, 20: 291; R. Pettit et al., Discuss. of the Faraday Soc., 1969, 47: 71). For example J. L. von Rosenberg et al., have shown that the endo-alcohol (LXXV) is converted into the ketone (LXXVII) but the epimer (LXXVIII) is unaffected. A transition state such as LXXVI is envisaged, involving a metal-assisted suprafacial concerted 1,3-hydrogen shift (idem, ibid., 1969, 77):

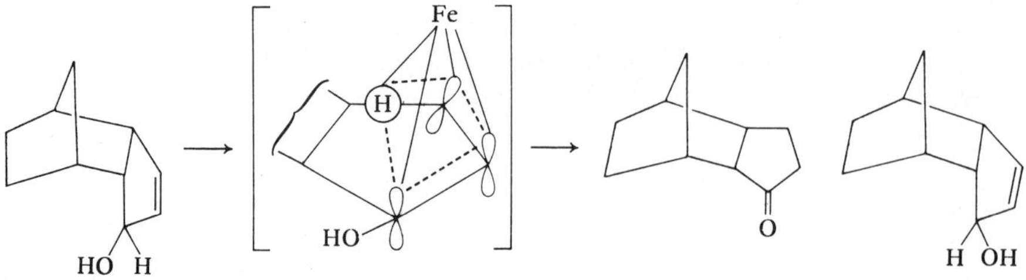

HO H
(LXXV)　　(LXXVI)　　(LXXVII)　　H OH
(LXXVIII)

These considerations are only justified if the reactions are concerted, otherwise a sequence of several discrete steps are possible while retaining stereospecificity owing to the intermediate stages being complexed with the metal atom (T. J. Katz and S. Cerefice, J. Amer. chem. Soc., 1969, 91: 2405). Dewar regards the stabilization of the transition state rather than symmetry correlations between reactants and products as being the decisive factor in pericyclic reactions, and has predicted that the catalysis of electrocyclic reactions by transition metals will be confined to processes which are thermally "forbidden" since these have anti-aromatic transition states. Anti-aromatic conjugated hydrocarbons form exceptionally stable π-complexes with transition metals owing to strong co-ordination to the metal by weakly bound electrons in the highest occupied molecular orbital of the hydrocarbon, and back donation by the metal d-electrons into a low-lying unoccupied molecular orbital of the hydrocarbon. By analogy, an anti-aromatic transition state would therefore be expected to be strongly complexed with the metal and thereby stabilized relative to aromatic or non-aromatic reactants (M. J. S. Dewar, Chemical Society Symposium on Asymmetry in Chemistry, Manchester, 1971).

4. Halogen derivatives

Controlled vapour-phase chlorination of cyclopentane at 250° gives chloro-cyclopentane (95%) (F. Gaymard, R. Guedj and J. Jullien, Bull. Soc. chim. Fr., 1970, 1864). Fluorination of cyclopentane with cobalt trifluoride gives a complex mixture of polyfluorocyclopentanes (J. C. Tatlow et al., Tetrahedron, 1966, 22: 43). The C—C bond joining two tertiary carbon atoms has been shown to be particularly weak in fully fluorinated molecules (G. B. Barlow and Tatlow, J. chem. Soc., 1952, 4695) and pyrolysis of perfluoro-bicyclohexyl has been shown to give perfluoro-1,1-dimethylcyclopentane by a process involving homolysis of the 1,1'-carbon—carbon bond, ring-contraction of the resulting cyclohexyl radical, and a radical disproportionation involving the transfer of a CF_3 group (D. R. Mackenzie, V. H. Wilson and E. W. Anderson, J. chem. Soc., B, 1968, 762):

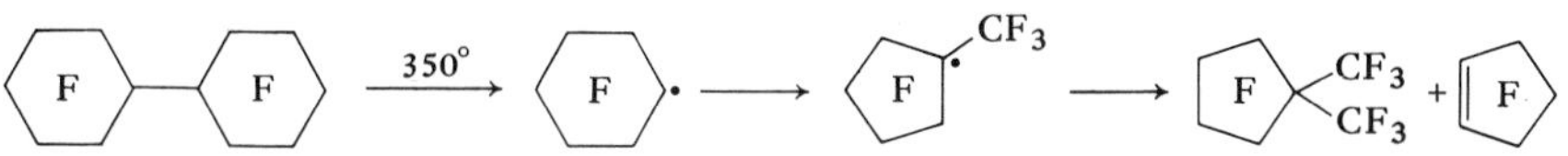

Halogen substituents may be replaced by hydrogen, by lithium aluminium hydride reduction (W. J. Feast, D. R. A. Perry and R. Stephens, Tetrahedron, 1966, 22: 433; Tatlow et al., J. chem. Soc., 1965, 2382), but violent explosions have sometimes resulted (Nachr. Chem. Tech., 1964, 12: 488; C.A., 1965, 62: 9025).

The base-catalysed ring-expansion of bromomethylenecyclobutane to 1-bromocyclopentene has been reported and the mechanism discussed (K. L. Erickson, B. E. Vanderwaart and J. Wolinsky, Chem. Comm., 1968, 1031). The use of sodium borohydride in diglyme at 0° for the selective reduction of vinylic halogen has been described; 1,2-dichlorohexafluorocyclopentene gives 1-hydro-2-chlorohexafluorocyclopentene (88%) and octafluorocyclopentene gives 1-hydro-heptafluorocyclopentene (73%) (D. J. Burton and R. L. Johnson, J. Amer. chem. Soc., 1964, 86: 5361). Similarly the reaction of methoxide ion with perfluorocyclopentene gives 1-fluoro-2-methoxy- and 1,2-dimethoxyhexafluoro-cyclopentene (R. F. Stockel et al., Canad. J. Chem., 1964, 42: 2880); under the influence of ultraviolet light, or peroxides, or ν-irradiation, alcohols undergo free-radical addition to the double bond (Stockel and M. T. Beachem, J. org. Chem., 1967, 32: 1658). Treatment of 1,2-dichlorohexafluorocyclopentene with cyanide results in ring-opening and isolation of the red fluoro-1,1,4,5,5-pentacyano-2-azapentadienide anion (λ max. 502 nm, ϵ 35,200). The mechanism probably involves cyanide addition to the activated monocyano derivative, followed by a series of elimination—addition reactions leading to intermediates which are progressively more reactive (W. R. Carpenter and G. J. Palenik, J. org. Chem., 1967, 32: 1219):

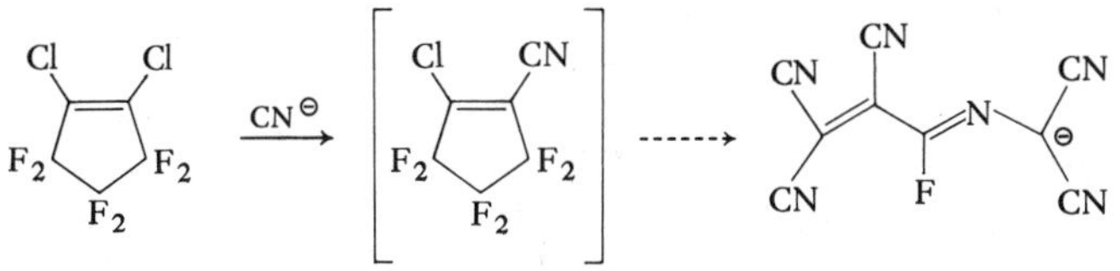

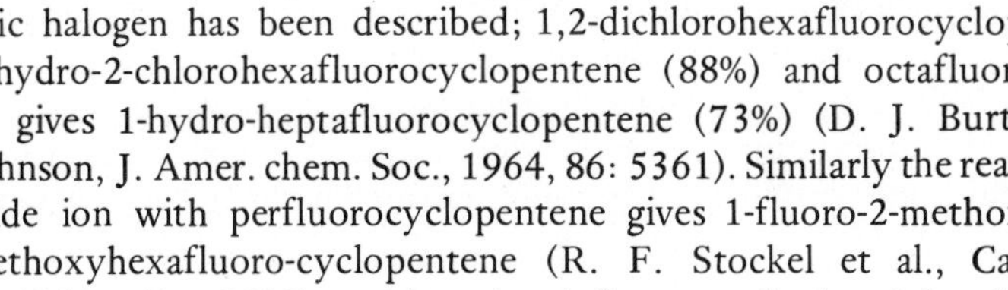

Antimony pentafluoride catalysed sulphur trioxide reaction of 1,2-dichlorohexafluorocyclopentene gives 2,3-dichlorotetrafluoro-2-cyclopenten-1-one and 2,3-dichlorotetrafluoro-2-cyclopentene-1,4-dione (R. F. Sweeney et al., J. org. Chem., 1966, 31: 3174):

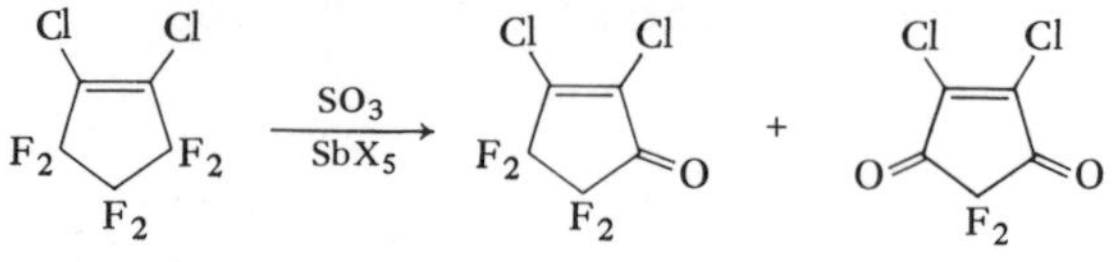

Hexachlorocyclopentadiene undergoes a novel reaction with nitrogen dioxide at 60° to give tetrachlorocyclopentene-1,2-dione (90%) (R. M. Scriber, J. org. Chem., 1965, 30: 3657):

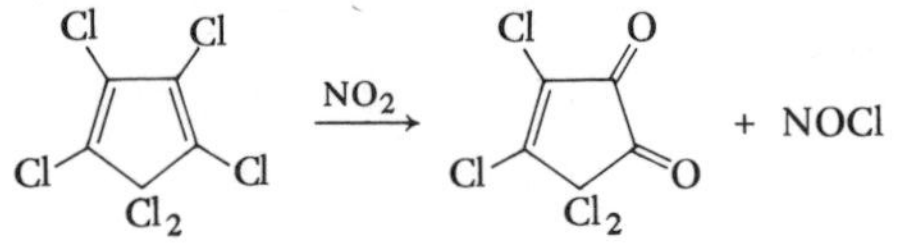

Dehydrofluorination of hexafluorocyclopentenes gives 1*H*- and 2*H*-pentafluorocyclopentadienes (A. Bergomi, J. Burdon and J. C. Tatlow, Tetrahedron, 1966, 22: 2551); similarly *hexafluorocyclopentadienes* are obtained from heptafluorocyclopentenes (Tatlow et al., J. chem. Soc., 1965, 808). The Diels—Alder reactions of perfluorocyclopentadiene have been described (R. E. Banks, M. Bridge and R. N. Haszeldine, J. chem. Soc., C, 1970, 48; Banks, L. E. Birks and Haszeldine, ibid., 201):

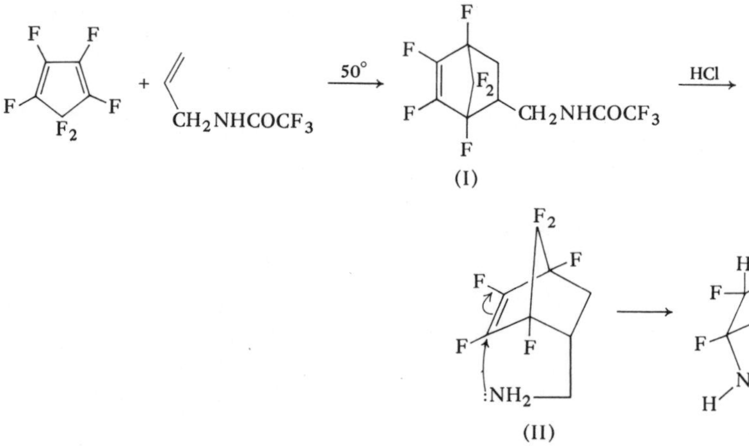

The spontaneous isomerization of the amine (II) to 1,2,3,7,8,8-hexafluoro-4-azatricyclo[4,2,1,0^{3,7}]nonane (III) showed the amide adduct (I) to be the *endo*-isomer. Hexafluorocyclopentadiene can act as both a diene and a dienophile with cyclopentadiene (Haszeldine et al., Chem. Comm., 1965, 41; 1966, 338):

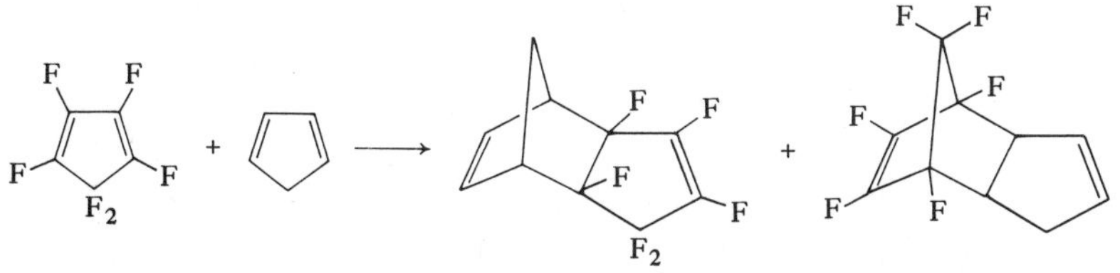

(16%)   (84%)

Hexabromocyclopentadiene undergoes reductive coupling to octabromoful-valene (R. West and P. T. Kwitowski, J. Amer. chem. Soc., 1968, 90: 4697) and dimerization to a perbromo analogue of "Prins dimer" (dodecabromo-pentacyclo[5.3. 0.0^{2,6}0^{3,9}0^{4,8}]decane) (U.S.P. 3,212,973/1965):

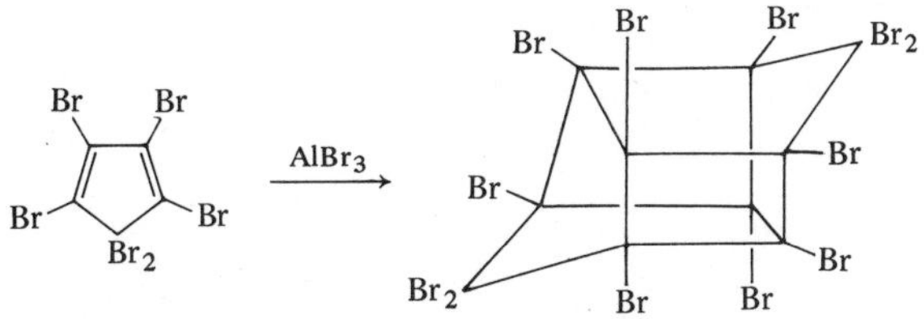

Nucleophilic substitution by methoxide ion gives 1,2,3,4-tetrabromo-5,5-dimethoxycyclopentadiene which acts as an electron-rich diene in the Diels—Alder reaction; the *p*-benzoquinone adduct (80% yield) undergoes photoisomerization to 1,8,9,11-tetrabromopentacyclo[6.3. 0.0^{2,7}0^{4,11} 0^{5,9}]undecane-3,6,10-trione-10,10-dimethyl acetal (R. G. Pews, C. W. Roberts and C. R. Hand, Tetrahedron, 1970, 26: 1711):

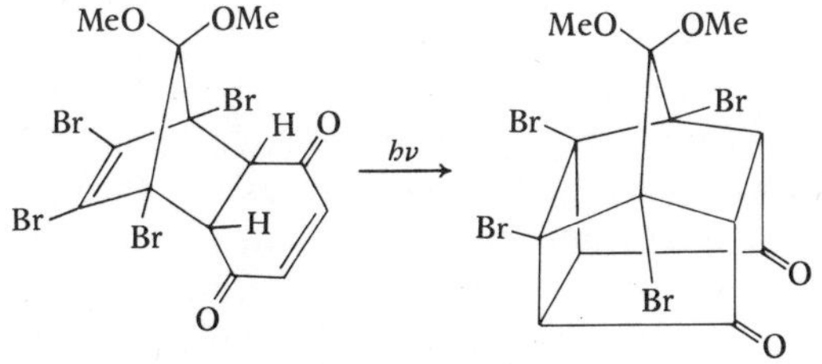

5. Nitro and amino derivatives

The diacetylnitrocyclopentadiene obtained by condensation of hexane-2,5-dione with nitromalondialdehyde (W. J. Hale, Ber., 1912, 45: 1596; J. Amer.

chem. Soc., 1912, **34**: 1580) has been shown to have the fulvenoid structure
(I) rather than the previously suggested unconjugated structure (II) or the
O-protonated *aci*-form (III) (A. N. Campbell-Crawford, A. M. Gorringe and
D. Lloyd, Chem. and Ind., 1966, 1961):

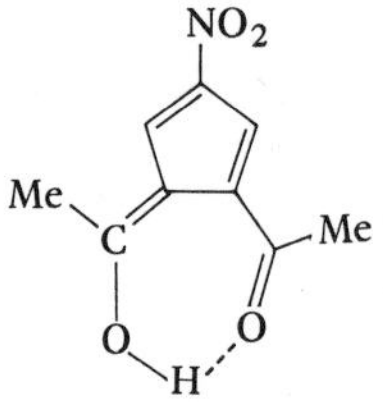

(I) 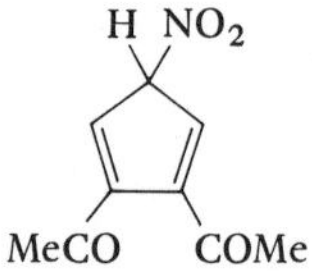   (II) 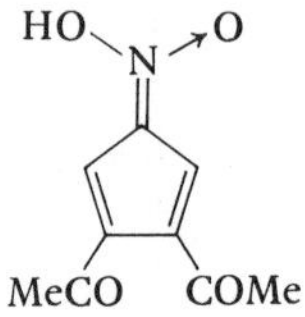   (III) 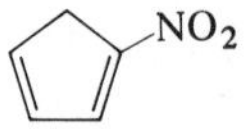   (IV)

Similarly protonation of sodium nitrocyclopentadienide has been shown to
give IV rather than the O-protonated *aci*-form (R. C. Kerber and M. J. Chick,
J. org. Chem., 1967, **32**: 1329).

1,3-Bis(*o*-chlorobenzylaminomethyl)camphocean (V) is of use in lowering
the blood-cholesterol level (U.S.P. 3,239,558/1963):

(V)

6. Alcohols of the cyclopentane and cyclopentene series

(*a*) *Secondary and tertiary cyclopentanols*

The synthesis and configurational assignments of optically active diastereoisomers of 1,2-dimethyl-2-phenylcyclopentanols have been described and
their base-catalysed epimerization, racemization, and cleavage reactions
investigated (T. D. Hoffmann and D. J. Cram, J. Amer. chem. Soc., 1969,
91: 1000, 1009). Marked stereoselectivity of hydrogenation of *exo*- and
endo-unsaturated cyclopentanols over nickel catalysts has been observed, the
hydrogen adding from the same side of the double bond as the hydroxyl
group (S. Mitsui, K. Hebiguchi and H. Saito, Chem. and Ind., 1967, 1746).
Addition of lithium alkyls and Grignard reagents to 2-methylcyclopentanone
gives 1-alkyl-2-methylcyclopentanols in which the alkyl groups are predominantly *trans*-orientated (J.-P. Battioni, M.-L. Capmau and W.

Chodkiewicz, Bull. Soc. chim. Fr., 1969, 976). The synthesis, configurational and conformational assignments of 3,3- and 3,4-diphenylcyclopentanols have been reported (A. Warshawsky and B. Fuchs, Tetrahedron, 1969, 25: 2633). *cis*-3-(2-Hydroxyethyl)cyclopentanol has been prepared by Baeyer—Villiger oxidation of norcamphor and reduction of the resulting lactone (J. L. Marshall, J. P. Brooks and G. W. Hatzenbuehler, J. org. Chem., 1969, 34: 4193). The primary hydroxyl group is the more reactive and can be selectively mono-tosylated, thus allowing selective oxidation of the secondary alcohol to give 3-(2-hydroxyethyl)cyclopentanone:

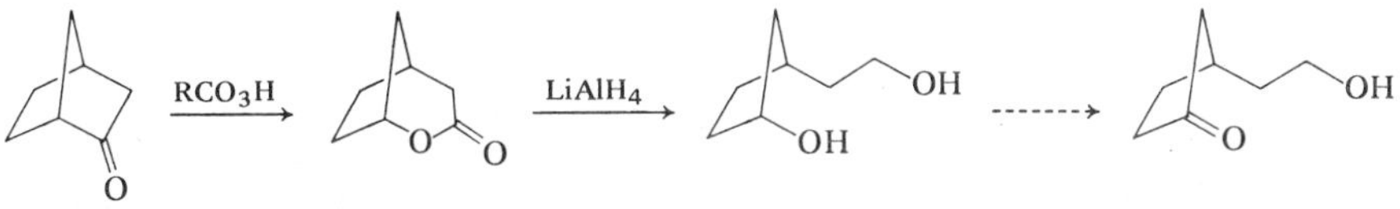

2-(2-Hydroxyethyl)cyclopentanone has been synthesized by the photolysis of 2-cyclopentylethyl nitrite (the Barton reaction) (H. Obara and H. Kimura, Bull. chem. Soc., Japan, 1969, 42: 2705).

(b) Cyclopentanediols

(+)-*Caldariomycin* (2,2-dichloro-*trans*-1,3-cyclopentanediol), m.p. 121°, racemic (±) m.p. 88.5—89.5° [bis-d-camphor sulphonate, m.p. 142—143°, $[\alpha]_D^{25}$ + 36° (acetone)] is a mold metabolite first isolated by Raistrick et al. (Biochem. J., 1940, 34: 663). (±) Caldariomycin has been synthesized by chlorination and reduction of 1,3-cyclopentanedione (J. R. Beckwith and L. P. Hager, J. org. Chem., 1961, 26: 5206) or by alkaline chlorination of phenol or 2,4,6-trichlorophenol (A. W. Burgstahler, T. B. Lewis and M. O. Abdel-Rahman, J. org. Chem., 1966, 31: 3516) presumably via a Favorsky-type ring contraction:

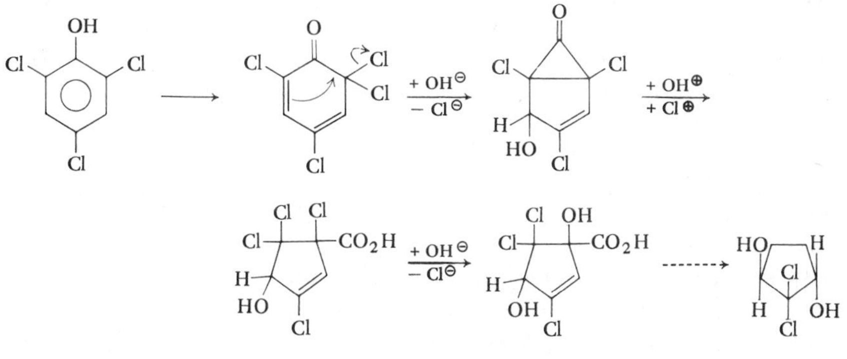

Each of the asymmetric centres has the S-configuration.

(c) Unsaturated cyclopentanols

Cyclopent-1-en-4-ols have been obtained by lithium aluminium hydride reduction of cyclopent-2-ene-1,4-diols (G. Rio and M. Charifi, Bull. Soc. chim. Fr., 1970, 3593 and refs. therein):

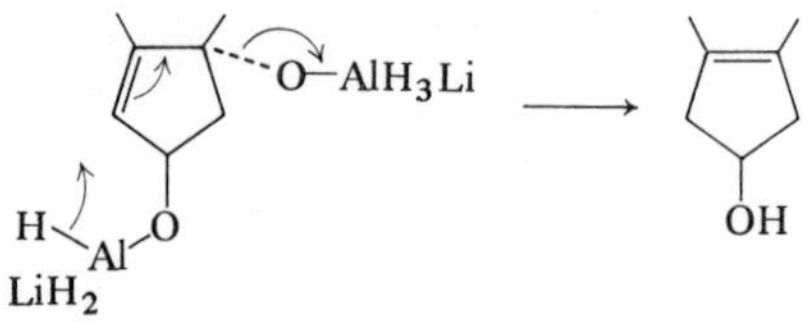

The synthesis of *loganin* (III), the key iridoid glucoside in indole alkaloid biogenesis, involves photocycloaddition of the enol (II) to the protected cyclopent-1-en-4-ol (I) (G. Büchi et al., J. Amer. chem. Soc., 1970, 92: 2165):

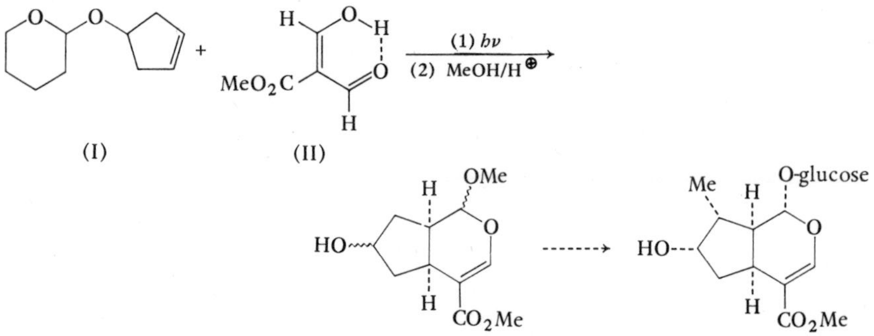

Selective reduction of the carbonyl group in cyclopentenones can be effected without concomitant saturation of the double bond, by inverse addition of aluminium hydride (H. C. Brown and H. M. Hess, J. org. Chem., 1969, 34: 2206).

2-n-*Butylcyclopenten-3-ol* is formed by lithium aluminium hydride reduction of 2-butylcyclopenta-1,3-dione or its methyl enol ether (K. Matoba et al., Yakugaku Zasshi, 1969, 89: 505; C.A., 1969, 71: 38404 p).

(d) Cyclopentanepentols

These can be considered as lower homologues of the biologically important inositols. There are four symmetric stereoisomers theoretically possible, having the following configurations: 1, 2, 3, 4, 5/0* (IV), 1, 2, 3, 4/5 (V), 1, 2, 3/4, 5 (VI), and 1, 2, 4/3, 5 (VII). Of these isomers V, VI, and VII have been prepared by hydroxylation of cyclopentene derivatives (Th. Posternak and G. Wolczunowicz, Naturwissenschaften, 1968, 55: 82; H. Z. Sable et al., Helv., 1963, 46: 1157). Nitro- and amino-cyclopentanetetrols, obtained by condensation of nitromethane with dialdehyde precursors (R. Ahluwalia, S. J. Angyal and B. M. Luttrell, Austral. J. Chem., 1970, 23: 1819) have also been converted, by deamination and epoxidation, into cyclopentanepentols (V, VI and VII) (Angyal and Luttrell, ibid., 1970, 23: 1831).

7. Aldehydes

Epoxides undergo a facile rearrangement to carbonyl compounds (aldehydes or ketones) on treatment with lithium bromide—tri-*n*-butylphosphine oxide complex (B. Rickborn and R. M. Gerkin, J. Amer. chem. Soc., 1968, 90: 4193). Thus the epoxide from 1-methylcyclohexene yields a mixture of 1-formyl-1-methylcyclopentane and 1-acetylcyclopentane. The suggested mechanism involves diaxial opening of the epoxide to give a bromohydrin salt followed by expulsion of an *equatorial* bromide anion:

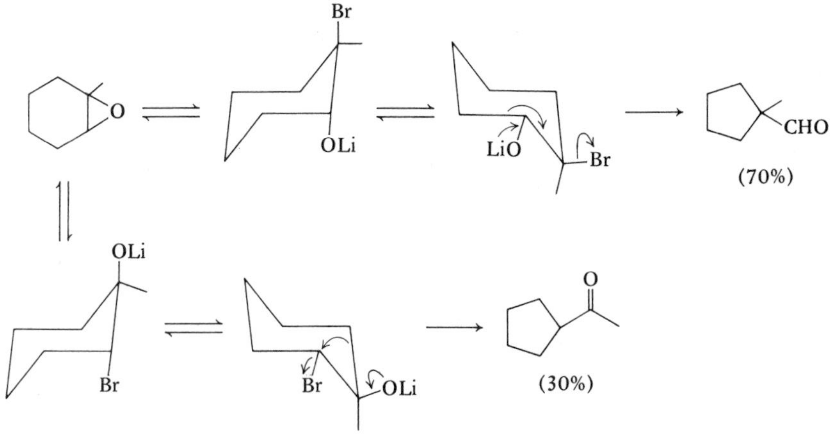

Thermocyclization of hept-6-enal at 320° gives cis and trans-2-*methylcyclo-pentanealdehyde* (J. M. Conia, Fr. Addn., 92,065/1967; C.A., 1969, 71: 90925y). 3- and 4-Cyclopentenylacetaldehydes have been isolated from the ceric ammonium nitrate oxidation of *exo-* or *endo*-2-norbornanol (W. S. Trahanovsky, P. J. Flash and L. M. Smith, J. Amer. chem. Soc., 1969, 91:

---

* Numerals on the same side of the / sign refer to substituents having a *cis*-relationship.

5068). Transvinylation of cyclopentenylpropenols with ethyl vinyl ether gives a mixture of dienal isomers resulting from a [3,3]sigmatropic rearrangement of the intermediate ether (S. Bancel and P. Cresson, Compt. rend., 1969, 268: 1535):

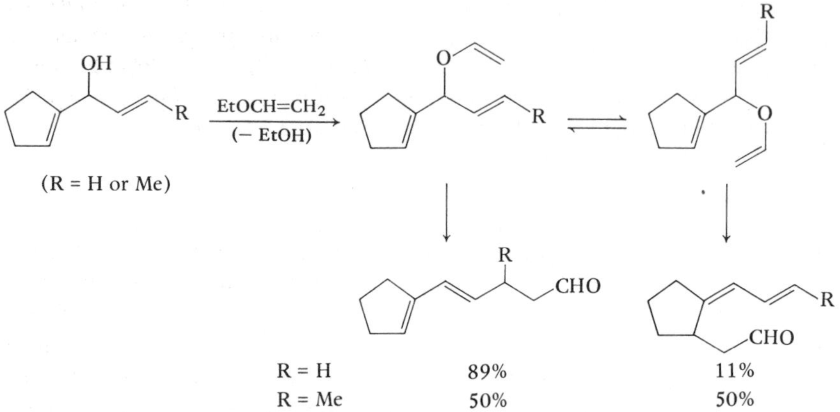

R = H	89%	11%
R = Me	50%	50%

The action of heat on *endo*-6-formylbicyclo[3,1,0]hex-2-ene (I) gives the dienal (II) (F. Bickelhaupt, W. L. de Graaf and G. W. Klumpp, Chem. Comm., 1968, 53):

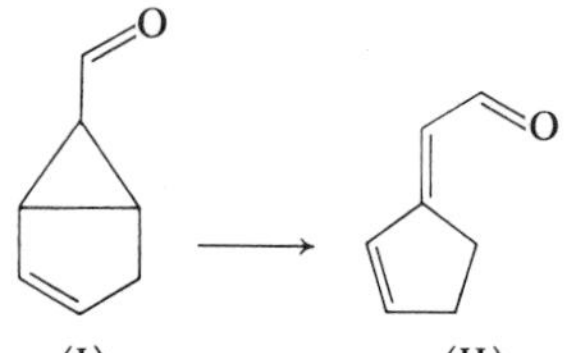

(I)   (II)

A stereospecific synthesis of an enantiomer of *iridodial* (III) has been effected by conversion of (+)-*trans*-pulegenic acid into 4,8-dimethylbicyclo-[3,3,0]oct-3-en-2-one followed by hydroxylation, reduction, and glycol cleavage (S. A. Achmad and G. W. K. Cavill, Austral. J. Chem., 1965, 18: 1989):

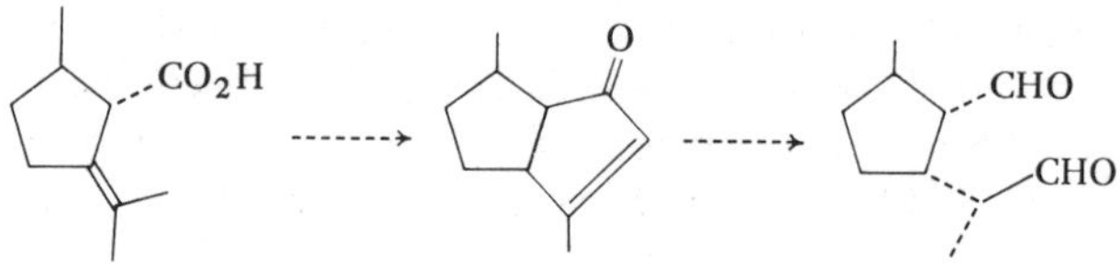

(III)

8. Ketones

(a) *Ring ketones*

(i) *Saturated monoketones*

A general approach to the synthesis of five-, six-, and seven-membered cyclic ketones involves the introduction and intramolecular alkylation of a 2-chloro-1-ene side-chain and solvolysis of the α-chlorocarbonium ion (P. T. Lansbury et al., J. Amer. chem. Soc., 1970, 92: 5649):

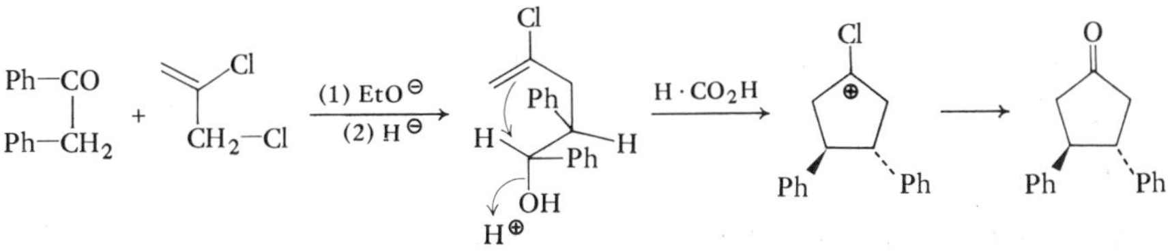

Diphenylcyclopropenethione reacts with enamines to give the corresponding cyclopentenethione (J. W. Lown and T. W. Maloney, Chem. and Ind., 1970, 870):

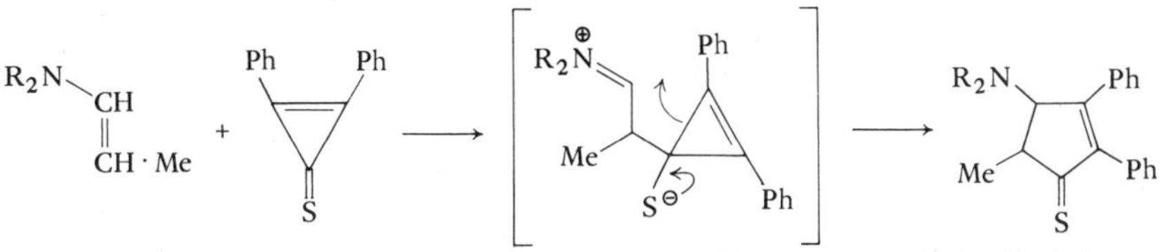

The corresponding reaction with diphenylcyclopropenone yields acyclic products (J. Ciabattoni and G. A. Berchtold, J. org. Chem., 1966, 31: 1336).

Cyclopentanone condenses with aniline in the presence of zinc chloride at 110° to give 2-anilino-1-cyclopentylidenecyclopentan-2-ene (J. Schoen and K. Bogdanowicz-Szwed, Rocz. Chem., 1968, 42: 1849). Tertiary enamines derived from cyclopentanone have been α,α'-annelated with ethyl α-bromo-methylacrylate to give ethyl bicyclo[3,2,1]octan-8-one-3-*endo*-carboxylate (R. P. Nelson, J. M. McEuen and R. G. Lawton, J. org. Chem., 1969, 34: 1225). Alternatively α,α-disubstitution occurs on alkylation of the enamine or the ketone with tetramethylene halides to give spiro[4,4]nonan-1-one and dispiro[4,1,4,2]tridecan-6-one (H. Krieger, H. Ruotsalainen and J. Montin, Ber., 1966, 99: 3715; J. Brugidou and H. Christol, Compt. rend., 1966, 262: 1595):

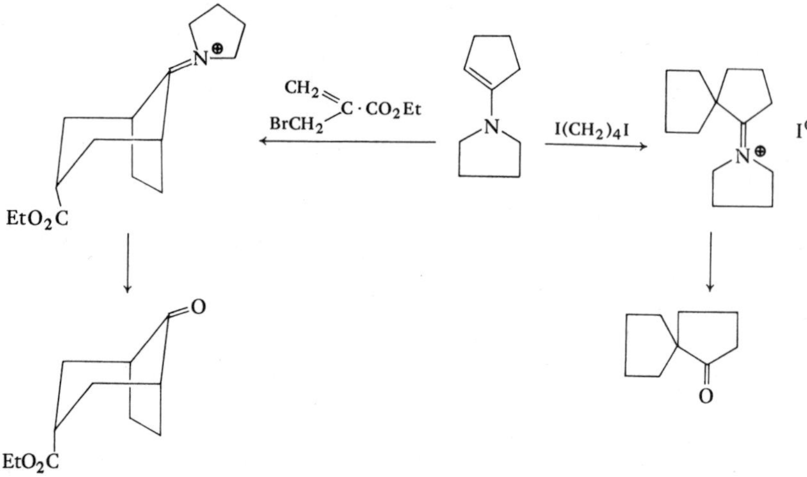

Two-carbon ring-expansion to give cycloheptene carboxylic acids can be effected by acid-catalysed transformation of β-(2-oxocyclopentyl)ethylaryl ketones (G. L. Buchanan et al., Tetrahedron, 1967, 23: 4729):

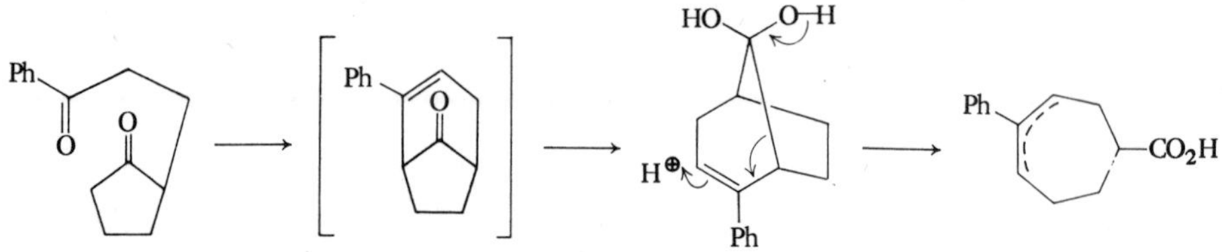

Anodic oxidation of cyclopentanone sodium bisulphite gives 4-hydroxy-pentanoic acid lactone (Jap. P. 6,804,261/1965), and treatment of cyclopentanone oxime with halogen and triphenylphosphine gives δ-valerolactam (M. Ohno and I. Sakai, Tetrahedron Letters, 1965, 4541).

(ii) Unsaturated monoketones

A novel rearrangement of α,α'-*dibromocyclopentanone* and α,α'-*dibromo-cyclohexanone* occurs on treatment with morpholine (K. Sato, Y. Kojima and H. Sato, J. org. Chem., 1970, 35: 2374):

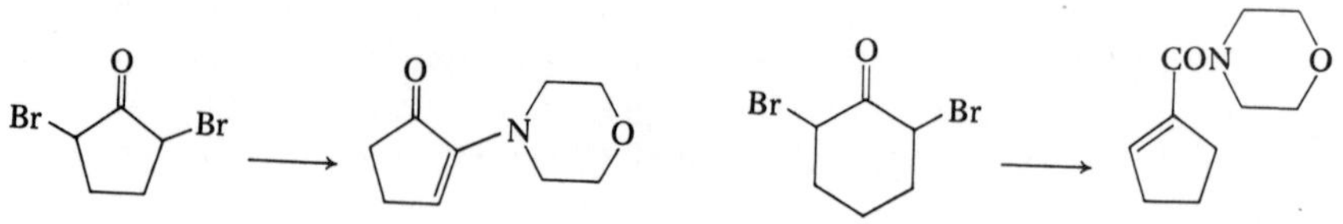

The reaction of furfuraldehyde and aniline is reported to yield 2,4-dianilino-cyclopent-2-en-1-one (K. G. Lewis and C. E. Malquiney, Chem. and Ind., 1968, 1249) and 2-hydroxy-3-methylcyclopent-2-en-1-one and aniline gives 2-anilino-5-methylcyclopent-2-en-1-one (I. McWatt, D. Phillips and G. R. Proctor, J. chem. Soc., C, 1970, 593).

Action of carbon tetrachloride on 1-piperidinocyclopentene gives 2-*di-chloromethylenecyclopentanone* on hydrolysis (J. Wolinsky and D. Chan, Chem. Comm., 1966, 567; Wolinsky and R. V. Kasubick, J. org. Chem., 1970, 35: 1211). Acylation of alkynes with $\alpha\beta$-unsaturated acid chlorides gives cyclopentenones (G. J. Martin, Cl. Rabiller and G. Mabon, Tetrahedron Letters, 1970, 3131; Martin and G. Daviaud, Bull. Soc. chim. Fr., 1970, 3098):

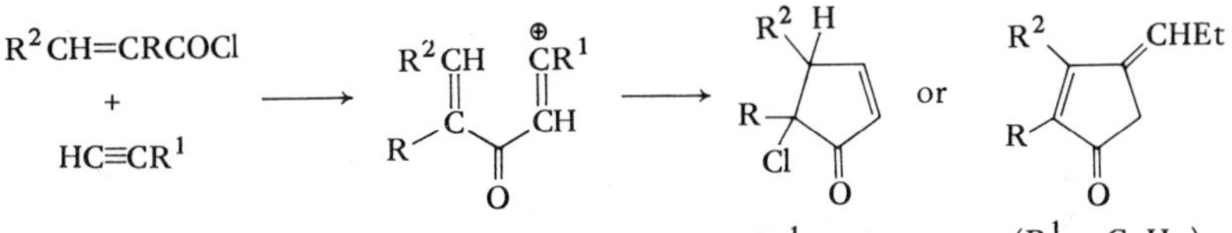

5-Substituted cyclopent-2-en-1-ones are obtained by treatment of allyl halides with alkynes and carbon monoxide in the presence of nickel complexes (Ital. P. 792,602/1967; C.A., 1969, 71: 3047x; G. P. Chiusoli, G. Bottaccio and C. Venturello, Tetrahedron Letters, 1965, 2875). Photolytic ring-contraction of 4-methyl-4-trichloromethylcyclohexa-2,5-dien-1-one gives the cyclopentenone (I) as the major product (D. L. Schuster and V. Y. Abraitys, Chem. Comm., 1969, 419, and refs. therein):

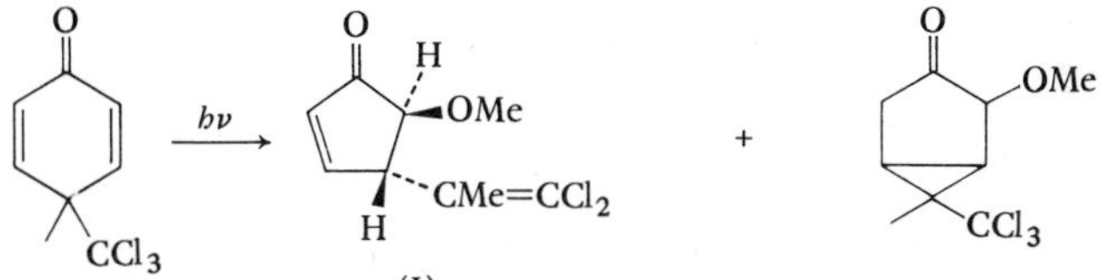

(I)

Although cyclopentenone undergoes photodimerization (Vol. II A, pp. 170—171) photolysis of 5,5-disubstituted cyclopent-2-en-1-ones in aqueous solution results in ring-contraction (W. C. Agosta et al., Tetrahedron Letters, 1969, 4517):

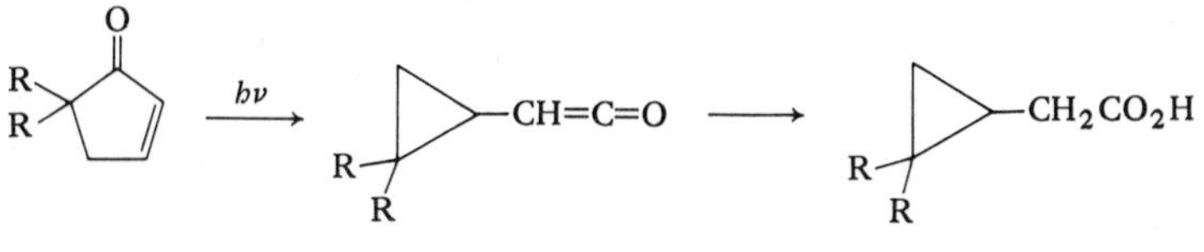

This is a reverse photochemical example of the thermal vinylcyclopropane—cyclopentene rearrangement (G. D. Gutsche and D. Redmore, Adv. alicyclic Chem., Suppl. 1, 1968, 163).

2-Alkylidenecyclopentanones rearrange to 2-alkylcyclopent-2-en-1-ones on treatment with polyphosphoric acid (J.-M. Conia and P. Amice, Bull. Soc. chim. Fr., 1968, 3327). ·

Cyclopentadienone, in contrast to fulvene, is unstable and dimerizes with such ease that until recently, all attempts to isolate the monomer have failed (M. A. Ogliaruso, M. G. Romanelli and E. I. Becker, Chem. Reviews, 1965, 65: 261). However pyrolysis of *o*-benzoquinone or *o*-phenylene sulphite at 550°, or *o*-phenylene carbonate at 660° in a furnace connected to a low temperature i.r. cell results in deposition of monomeric cyclopentadienone on the sodium chloride plate at −196°. The monomer (ν_{CO} 1709 cm^{-1}) is stable at −196° but dimerizes at −80° (O. L. Chapman and C. L. McIntosh, Chem. Comm., 1971, 770). The first authentic monomeric monosubstituted derivative to be reported, 3-tert-*butylcyclopentadienone* (II) is obtained by allylic bromination and dehydrobromination of 3-*tert*-butylcyclopentenone (E. W. Garbisch and R. F. Sprecher, J. Amer. chem. Soc., 1969, 91: 6785) and 2,4-*di-tert-butylcyclopentadienone* (III) was similarly obtained from the corresponding disubstituted cyclopentenone. The relative rates of dimerization of III and II to the *endo* adducts V and IV (1 and 5 x 10^6 respectively)

(II)   (III)   (IV)   (V)

is controlled by non-bonded steric effects. The unique reactivity of cyclopentadienone is attributed to the small difference in energies between the highest occupied and lowest unoccupied π-molecular orbitals and the unusually high-field chemical shifts of τ5.07 and 3.50 for the olefinic protons of III are attributed to increased ring π-electron density (idem, loc. cit.) or to a paramagnetic ring current (G. M. Pilling and F. Sondheimer, ibid., 1971, 93: 1977). The cycloaddition of cycloheptatriene and 2,5-*di-methyl-3,4-diphenylcyclopentadienone* gives a complex mixture of products (K. N. Houk and R. B. Woodward, ibid., 1970, 92: 4143 and refs. therein). However, the [6 + 4] cycloaddition process has been shown to proceed stereospecifically *exo*, to yield VI, whereas the competing [4 + 2] process goes stereospecifically *endo*, to give VII.

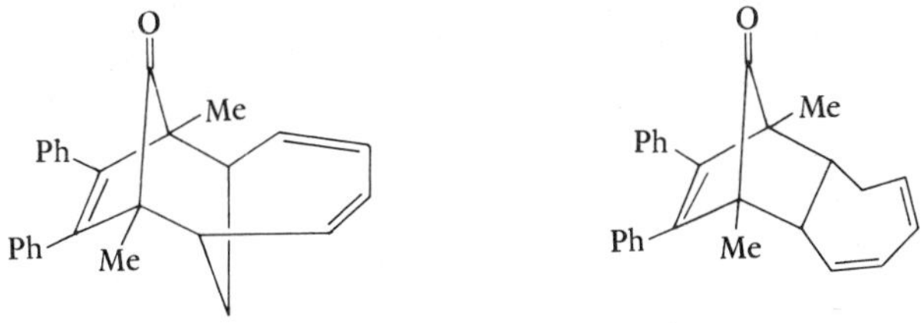

(VI)   (VII)

This result is attributed to the affect of secondary orbital interactions which destabilize an *endo*-transition state in a [6 + 4] addition, but stabilize an *endo*-transition state in a [4 + 2] addition (idem, loc. cit.). The cycloaddition of tropone to the same cyclone is equally complex and results in the formation of *exo* [6 + 4] and [8 + 2], and *endo* [4 + 2], cycloadducts (idem, ibid., 1970, 92: 4145). Cycloaddition of cumulenes to *tetraphenylcyclopentadienone* (*tetracyclone*) has been reported to give a [4 + 2] cycloadduct (W. Ried and R. Neidhardt, Ann., 1970, 739: 155):

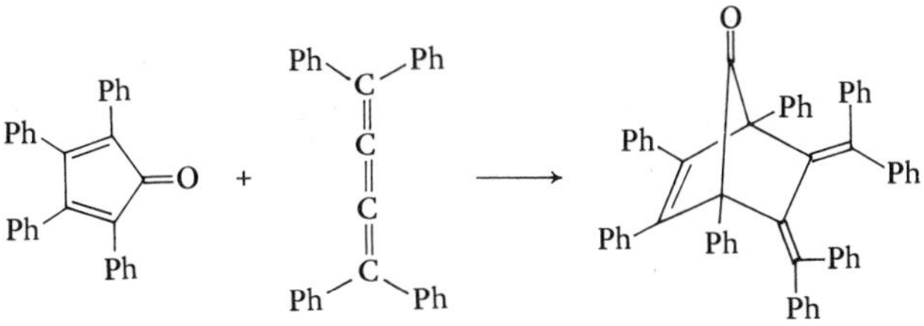

Cyclopent-2-en-1-one dimer (*cis,anti,cis*-tricyclo-[5,3,0,0^{2,6}]deca-4,9-diene-3,8-dione (VIII; X = H) undergoes an intramolecular rearrangement to *endo*-cyclopentadienone dimer (IX; X = H) on irradiation in benzene with light of $\lambda > 300$ nm. When the dibromo derivative is used (VIII; X = Br) the photoproduct (IX; X = Br) can be converted into 1,3-dimethoxy carbonyl-cubane (P. E. Eaton and T. W. Cole, Chem. Comm., 1970, 1493):

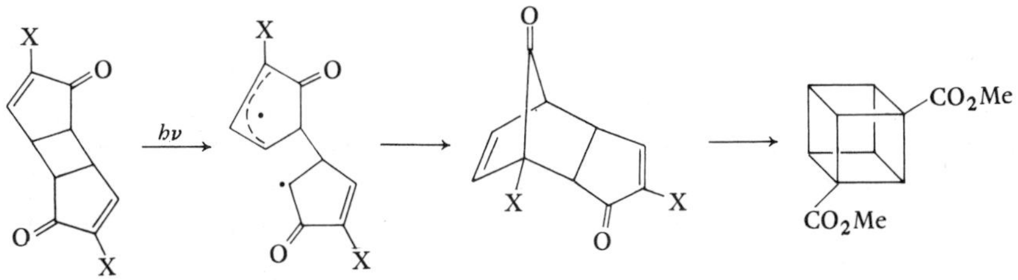

(VIII)   (IX)

cis,syn,cis-Tricyclo[5,3,0,0^{2,6}]deca-4,8-diene-3,10-dione (X) has been
synthesized and shown to rearrange photochemically to *exo*-dicyclo-
pentadienone (XI) (Eaton and S. A. Cerefice, ibid., 1970, 1494):

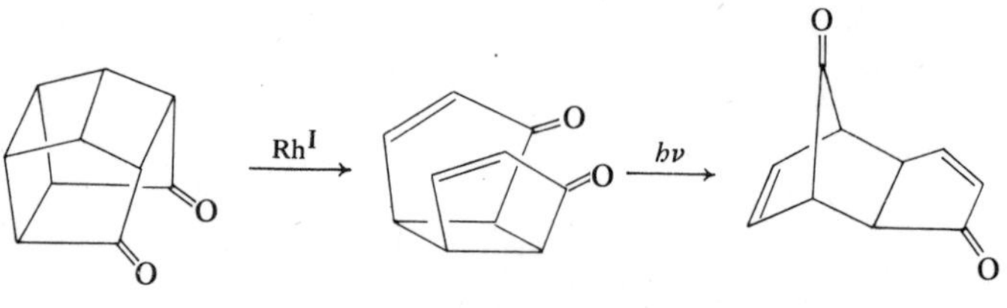

(X)   (XI)

The reaction of trimethyl phosphite with tetraphenylcyclopentadienone
proceeds by way of Arbuzov rearrangements of an intermediate zwitterion
formed by attack of phosphorus on the carbonyl oxygen atom (M. J.
Gallagher and I. D. Jenkins, J. chem. Soc., C, 1969, 2605):

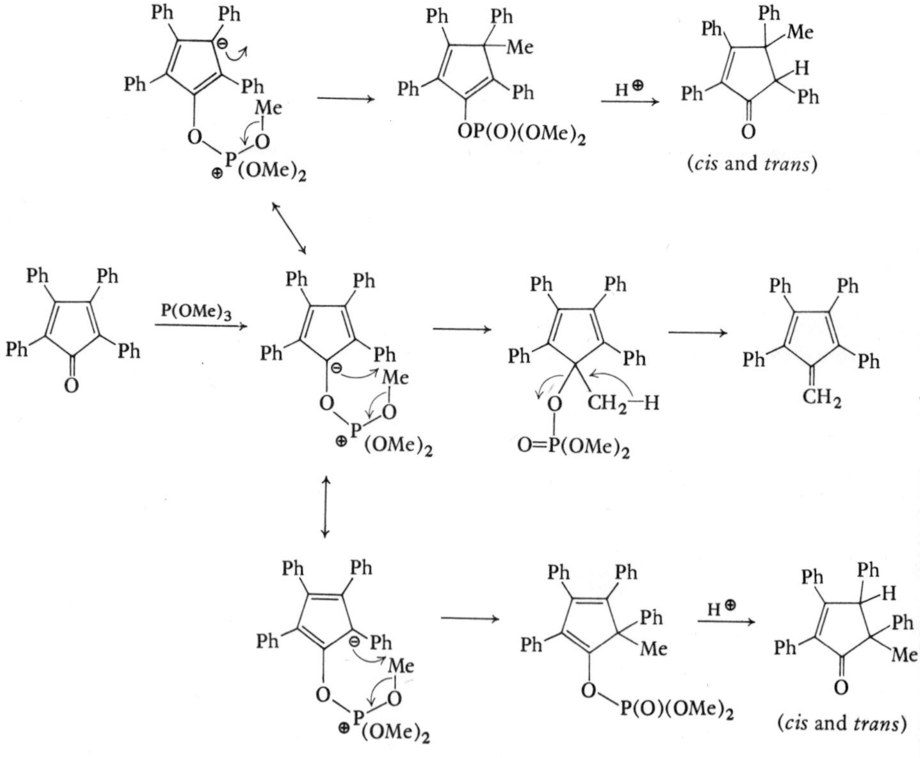

In the presence of water proton transfer results in the reduction of one double bond (A. J. Floyd et al., Tetrahedron Letters, 1970, 1735).

9. Hydroxyketones and polyketones of the cyclopentene series

(a) Hydroxycyclopentenones; cyclopentenolones

(i) Pyrethrins, rethrolones, and rethrones

The keto-alcohol *cinerolone* (Ia), *jasmolone* (Ib), and *pyrethrolone* (Ic), are known collectively as *rethrolones*. These are the alcohol components of the insecticidally active constituents of the flower-heads of Chrysanthemum cinerariaefolium (pyrethrum). Esterified with chrysanthemic acid (IIa) and pyrethric acid (IIb) they form the six active principles (the "pyrethrins"), *cinerin I* (IIIa), *cinerin II* (IIIb), *jasmolin I* (IIIc), *jasmolin II* (IIId), *pyrethrin I* (IIIe), and *pyrethrin II* (IIIf) (L. Crombie and M. Elliott, Fortschr. Chem. org. Naturstoffe, 1961, 19: 121).

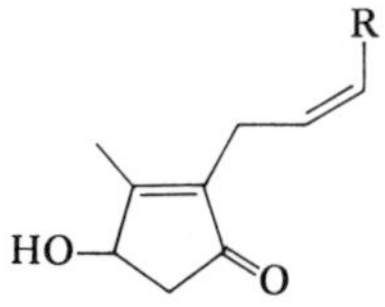

(I) 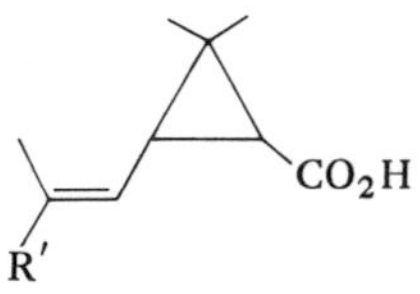   (II) 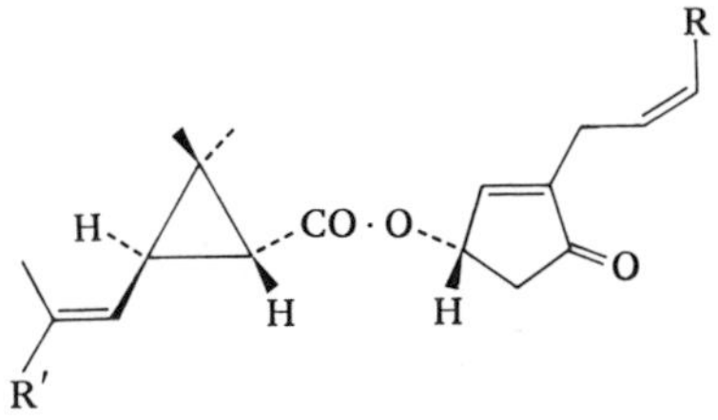   (III)

(a) R = Me   (a) R' = Me   (a) R = R' = Me
(b) R = Et   (b) R' = CO_2Me   (b) R = Me, R' = CO_2Me
(c) R = CH=CH$_2$ (c) R = Et, R' = Me
                                         (d) R = Et, R' = CO_2Me
                                         (e) R = CH=CH$_2$, R' = Me
                                         (f) R = CH=CH$_2$, R' = CO_2Me

Improved syntheses of the rethrolones (I) are reported in which the *cis* side-chains are introduced by means of the Wittig reaction (Crombie, P. Hemesley and G. Pattenden, J. chem. Soc., C, 1969, 1016):

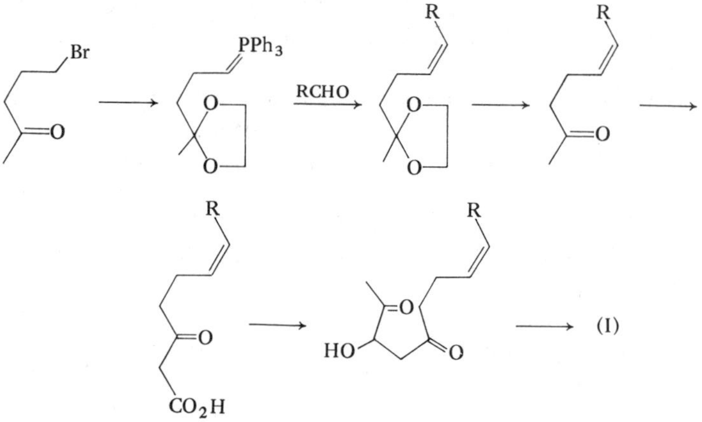

Cinerone (IVa), jasmone (IVb) and pyrethrone (IVc) (the "rethrones") are degradation products of the rethrolones and are generally more accessible synthetically (idem, ibid., p. 1024; G. Büchi and H. Wuest, J. org. Chem., 1966, 31: 977, and refs. therein; M. Fetizon and J. Schalbar, Fr. Ses Parfums, 1969, 12: 330; T. Yoshida, A. Yamaguchi and A. Komatsu, Agr. Biol. Chem., Tokyo, 1966, 30: 370; C.A., 1966, 65: 2136g).

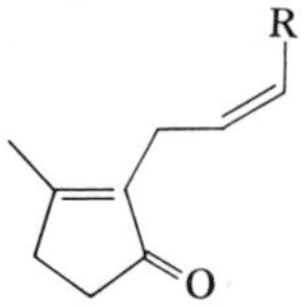

(a) R = Me
(b) R = Et
(c) R = CH=CH$_2$
(d) R = H

(IV)

cis-Jasmone, the important constituent of the essential oil of jasmine, can be obtained by a three-stage synthesis from 2-methylfuran in 40–45% yield:

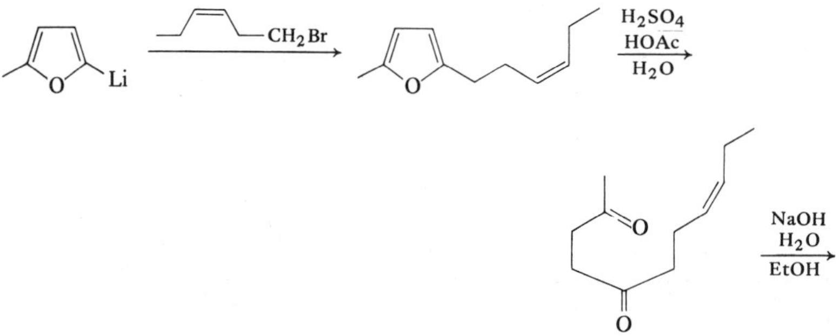

Methods are available for conversion of rethrones (IV) into rethrolones (I). Although allylic bromination and oxidation methods have failed, the Diels—Alder photo-addition of oxygen to the cyclopentadiene (V), formed by reduction and dehydration of cinerone, gave a mixture consisting mainly of cinerolone (R. A. LeMahieu, M. Carson and R. W. Kierstead, J. org. Chem., 1968, 33: 3660):

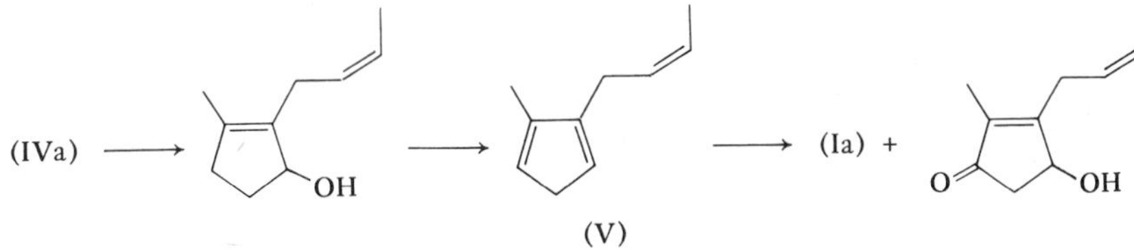

$$(IVa) \longrightarrow \quad \longrightarrow \quad \longrightarrow (Ia) +$$

(V)

Treatment of 2-N-pyrrolidino-5-methyl-2-cyclopenten-1-one with an alkyl-magnesium bromide, followed by dehydration and hydrolysis of the enamine function, gives allethrone (IVd). This has been converted to allethrolone (VI), as the major product, by microbiological hydroxylation with *Aspergillus niger* (NRRL 3228) (R. A. LeMahieu et al., ibid., 1970, 35: 1687):

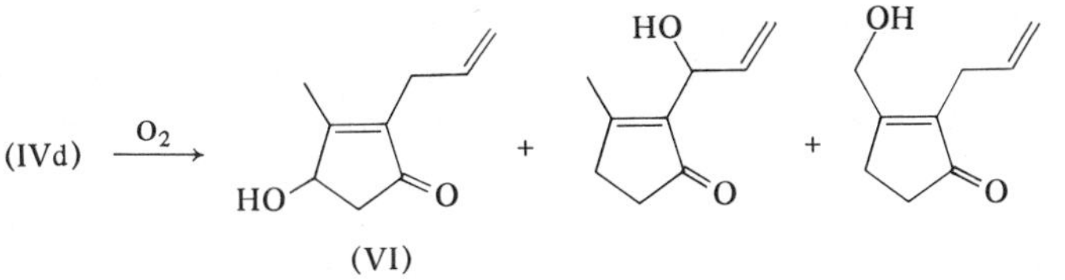

$$(IVd) \xrightarrow{O_2} \qquad + \qquad + \qquad$$

(VI)

Manganese dioxide oxidation of allethrolone results in oxidative coupling to give the tetraone (VIII) in addition to the dione (VII) (M. Elliott et al., Tetrahedron Letters, 1969, 373):

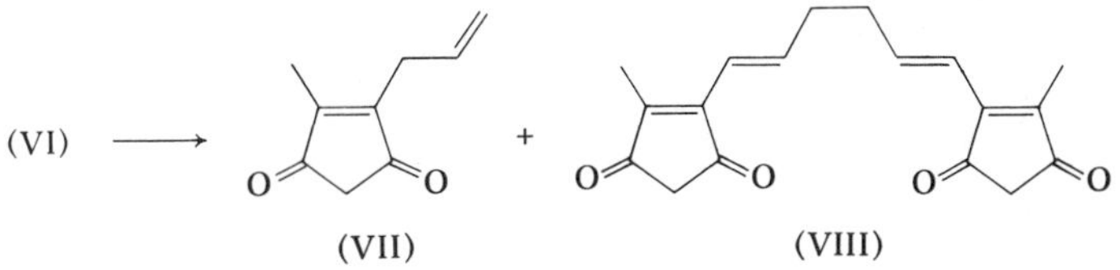

$$(VI) \longrightarrow \qquad + $$

(VII) (VIII)

(ii) "Anhydroacetonebenzils"

The condensation of benzil with many ketones yields 3,4-diphenylcyclo-pent-2-en-4-ol-1-ones which are known generically as "anhydroacetone-benzils". Herbicidal properties have been claimed for compounds in this category, such as (IX) or (X) derived from the condensation of diethyl ketone with benzil or phenanthraquinone respectively (B.P. 1,052,951/1962); C.A., 1967, 66: 94727e):

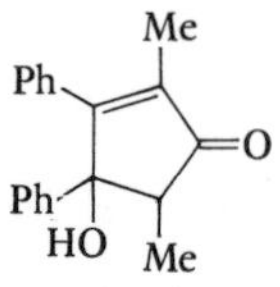

(IX) 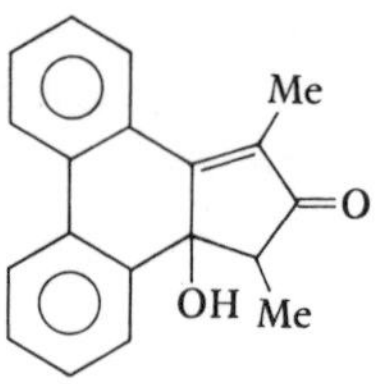   (X)

(b) Ring-polyketones

(i) 1,2-Diketones

These exist as the fully conjugated enolic structures (XI) (J. Bredenberg, Acta chim. Scand., 1959, 13: 1733; 1960, 14: 214). 3-*Methylcyclopent-2-en-2-ol-1-one* (*cyclotene*) (XI; R = H, R′ = Me), which is a constituent of coffee aroma and a precursor for dihydrojasmone, has been prepared by nitrosation or chlorination of 2-carbethoxy-2-methylcyclopentanone followed by hydrolysis and decarboxylation (K. Sato, S. Suzuki and Y. Kojima, J. org. Chem., 1967, 32: 339 and refs. therein; C. M. Leir, ibid., 1970, 35: 3203). 3-*Bromocyclopent-2-en-2-ol-1-one* (XI; R = H, R′ = Br) may be obtained by bromination of cyclopentanone and hydrolysis of the tribromo derivative (C. Rappe, Acta chem. Scand., 1965, 19: 270); bromination of XI (R = Br) gives 3,5-*dibromocyclopent-2-en-2-ol-1-one* (idem, ibid., p. 274). Allylic bromination of 2-acetoxycyclopent-2-en-1-one (XI; R = COCH$_3$, R′ = H) gives *cyclopent-3-en-1,2-dione* (XII) (C. Maignan, J. C. Grandguillot and F. Rowssac, Bull. Soc. Chim. Fr., 1970, 2019):

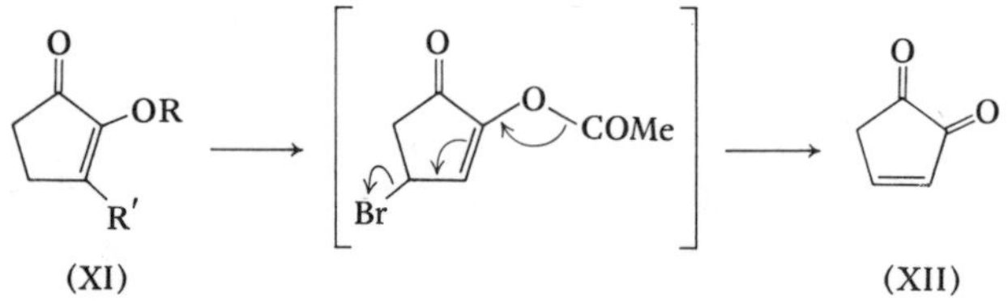

(XI)   (XII)

Di- and tetramethyl derivatives of reductic acid (XI; R = R′ = H) have been synthesized (G. Hesse and B. Wehling, Ann., 1964, 679: 100).

(ii) 1,3-Diketones

2-*Methylcyclopentane-1,3-dione* and homologues, key intermediates in the total synthesis of steroids (T. B. Windholz and M. Windholz, Angew. Chem., intern. Edn., 1964, 3: 353; S. Danishefsky and H. B. Migdalof, J. Amer. chem. Soc., 1969, 91: 2806) have been obtained in good yield by condensation of succinic anhydride with 2-buten-2-ol acetate (the major enol acetate of methyl ethyl ketone) or by intramolecular cyclization of γ-keto-carboxylic acids (V. J. Grenda et al., J. org. Chem., 1967, 32: 1236; H. Schick, G. Lehmann and G. Hilgetag, Angew. Chem., intern. Edn., 1967, 6: 371). 2-Methylcyclopentane-1,3-dione is completely enolized in polar solvents and there is a greater tendency for *O*-alkylation of the enolate anion than in the case of six-membered-ring ketones (D. Rosenthal and K. H. Davis, J. chem. Soc., C, 1966, 1973).

2-*Formyl*- and 2-*acyl-cyclopentane*-1,3-*diones* are also completely enolized, enolization of the ring-carbonyl group being predominant (S. Forsen, F. Merenyi and M. Nilsson, Acta chem. Scand., 1967, 21: 620). Di- and poly-alkylcyclopentadiones and cyclopentenediones have been obtained by ring-contraction of various six-membered ring derivatives (T. Balwe, W. Riedl and H. Simon, Ber., 1966, 99: 3277; H. Stetter and H.-J. Sandhagen, ibid., 1967, 100: 2837; M. L. Kaplan, J. org. Chem., 1967, 32: 2346; H. W. Moore, W. Weyler and H. R. Shelden, Tetrahedron Letters, 1969, 3947). Cycloaddition of cyclopropenones to ynamines gives substituted cyclopent-4-en-1,3-diones (M. Franck-Neumann, Tetrahedron Letters, 1966, 341):

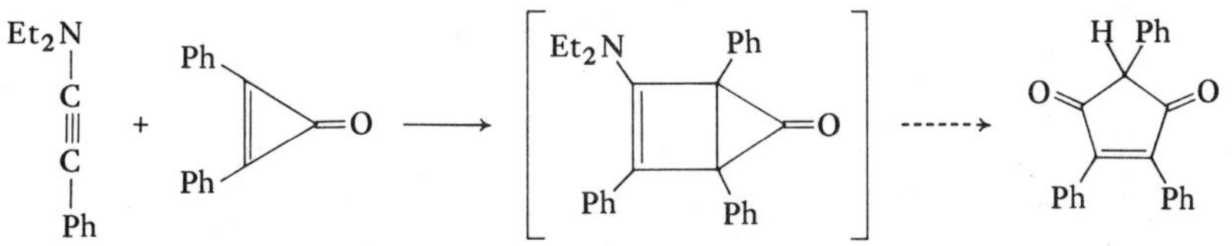

The poor reactivity of 2,2-*dimethylcyclopent-4-en*-1,3-*dione* as a dienophile in the Diels—Alder reaction has been ascribed to steric interactions between the geminal methyl groups and the approaching reagent. No such retardation is observed in the reaction with 1,3-dipolar reagents, suggesting a transition state like XIII rather than XIV (W. C. Agosta and A. B. Smith, Chem. Comm., 1970, 685).

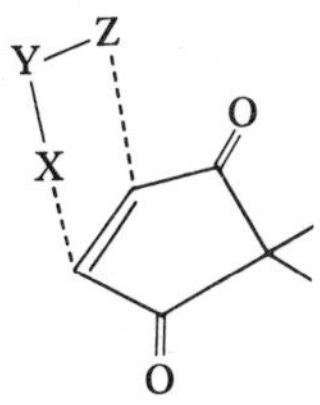

(XIII) 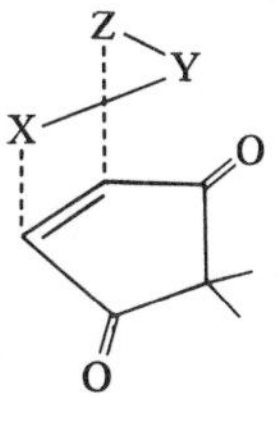   (XIV)

The constitution of *limocrocin*, a decrocetin derivative from *Streptomyces limosus*, has been reported (H. Brockmann et al., Ber., 1969, 102: 3217).

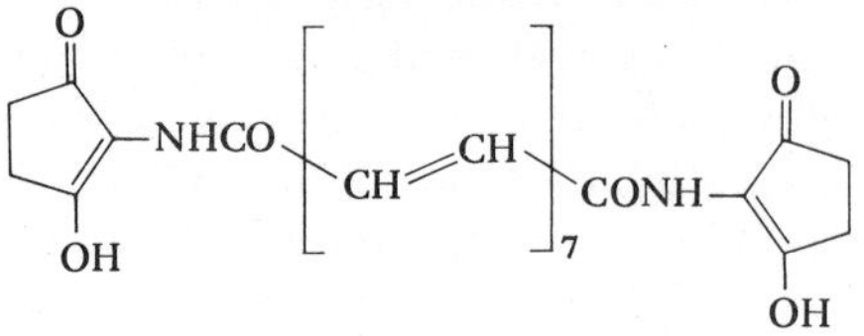

Limocrocin

(c) Poly- and exocyclic ketones

Spectroscopic investigations of cyclopentanetriones and acylcyclopentane-triones show that the former are monoenolic (i.e. XVa $\rightleftharpoons$ XVb) and the latter dienolic (i.e. XVIa $\rightleftharpoons$ XVIb $\rightleftharpoons$ XVIc) (J. A. Elvidge and R. Stevens, J. chem. Soc., 1965, 2251):

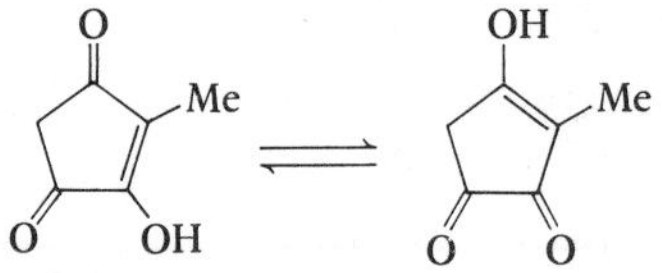

(a)   (b)

(XV)

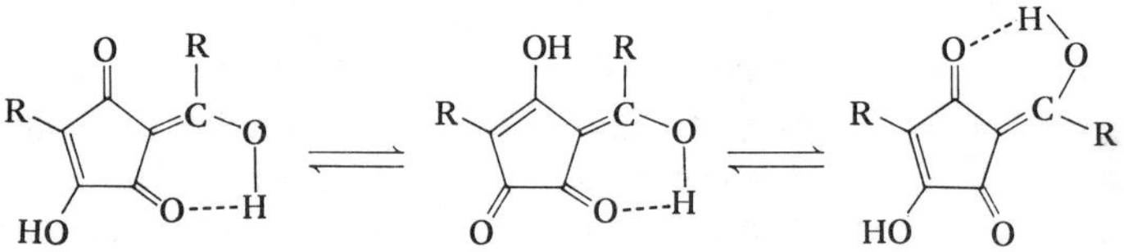

(a)   (b)   (c)

(XVI)

The hop resin* *humulone* (XVIII; R = i-Bu, R' = HO) undergoes rapid autoxidation to 3-hydroxy-3-(3,4-dihydroxy-4-methylpentanoyl)-5-iso-valerylcyclopentane-1,2,4-trione (XIX). The reaction involves ring-contrac-tion and loss of a side-chain, analogous to the conversion of lupulone (XVIII; R = i-Pr, R' = CH₂CH : CMe₂) to hulupone (XVII), and mild oxidation of

---

* For a review of the chemistry of hop constituents see R. Stevens, Chem. Review, 1967, 67: 19.

the other unsaturated side-chain (P. R. Ashurst and Elvidge, ibid., C, 1966, 675, and refs. therein):

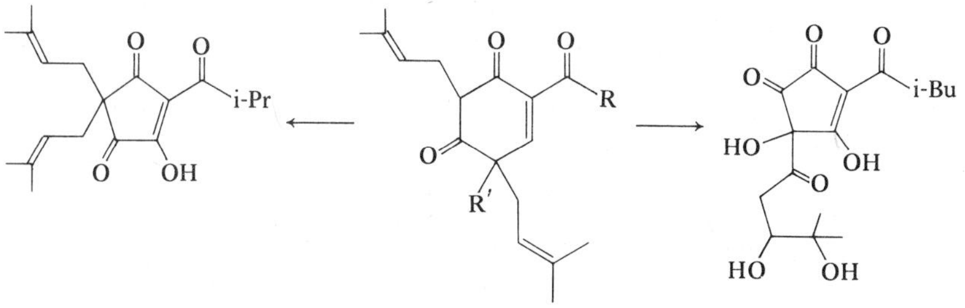

(XVII)   (XVIII)   (XIX)

These products are monoenolic owing to the presence of disubstituted ring-carbons. Magnesium ion catalysed isomerization of humulone gives a mixture of cis- and trans-*isohumulone* (XXa and XXb) the principal bittering materials in beer (H. Koeller, J. Inst. Brew., 1969, 75: 175).

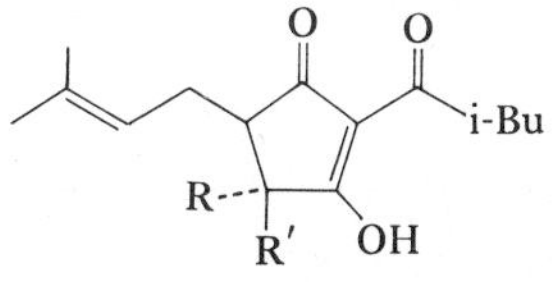

(a) R = OH, R' = $COCH_2CH=CMe_2$
(b) R = $COCH_2CH=CMe_2$, R' = OH

(XX)

A similar ring-contraction occurs on treatment of 2-acetyl-4,6-diethyl-1,3,4-trihydroxycyclohexa-2,6-dien-5-one with alkali to give cis- and trans-2-acetyl-5-ethyl-3,4-dihydroxy-4-propionylcyclopent-2-en-1-one; in the presence of acid an alternative rearrangement to 5-ethyl-3-hydroxy-3-propionyl-cyclopent-4-en-1-one occurs (S. J. Shaw and P. J. Smith, J. chem. Soc., C, 1968, 1882; Shaw, J. Inst. Brew., 1968, 74: 464). A remarkable ring-contraction occurs on treatment of the magnesium enolate (XXI; R = MgI) derived from 2-methylcyclohex-1-enyl acetate (XXI; R = COMe) with chloroacetone to give 1-(1-methylcyclopentyl)butan-1,3-dione (XXII) (C. J. R. Adderley, G. V. Baddeley and F. R. Hewgill, Tetrahedron, 1967, 23: 4143):

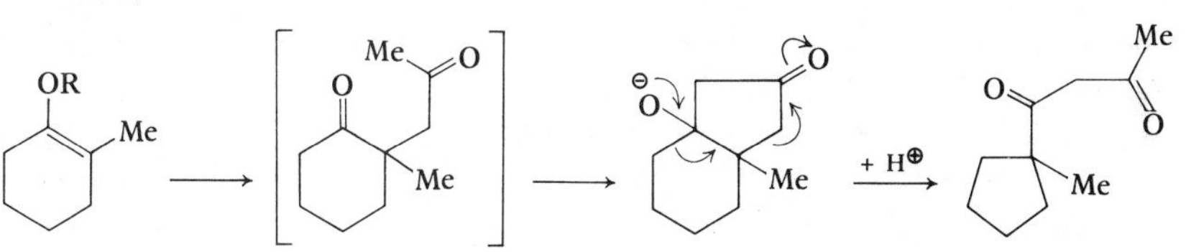

(XXI)   (XXII)

The synthesis of complicated organic molecules from relatively simple precursors in the presence of transition-metal complexes, whereby the reactants and intermediate stages are co-ordinated to the metal, is an area of great potential in synthetic organic chemistry. A remarkable example is the reaction of allyl chloride, phenylacetylene, water, and nickel carbonyl which gives XXXIII in high yield (G. P. Chiusoli, G. Bottaccio and C. Venturello, Tetrahedron Letters, 1965, 2875):

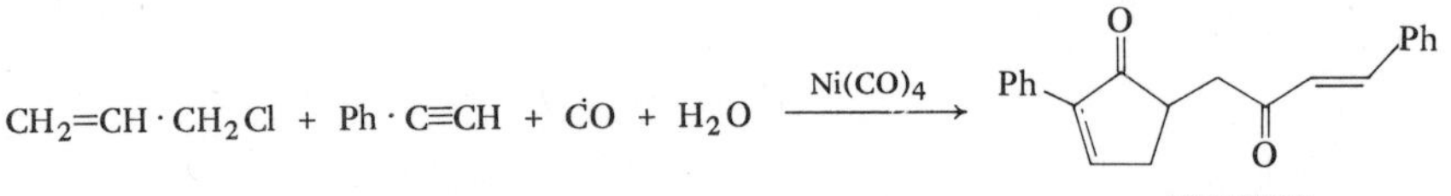

$$CH_2{=}CH{\cdot}CH_2Cl + Ph{\cdot}C{\equiv}CH + \dot{C}O + H_2O \xrightarrow{Ni(CO)_4}$$

(XXXIII)

The mechanism by which these reagents react, by way of five addition reactions to produce one product, has been discussed (R. F. Heck, Accounts chem. Res., 1969, 2: 10).

10. Carboxylic acids, amides and esters

(a) Saturated monocarboxylic acids

Selenium dioxide—hydrogen peroxide oxidation of alkylcyclohexanones gives the corresponding alkylcyclopentane carboxylic acid (R. Granger et al., Bull. Soc. chim. Fr., 1969, 2806). Methyl 4,4-dimethylcyclopent-1-ene-1-carboxylate and 4,4-dimethylcyclopent-2-ene-1-carboxylate are obtained in good yield from methyl 3,3-dimethylcyclobutene-1-carboxylate (T. H. Kinstle, R. L. Welch and R. W. Exley, J. Amer. chem. Soc., 1967, 89: 3660):

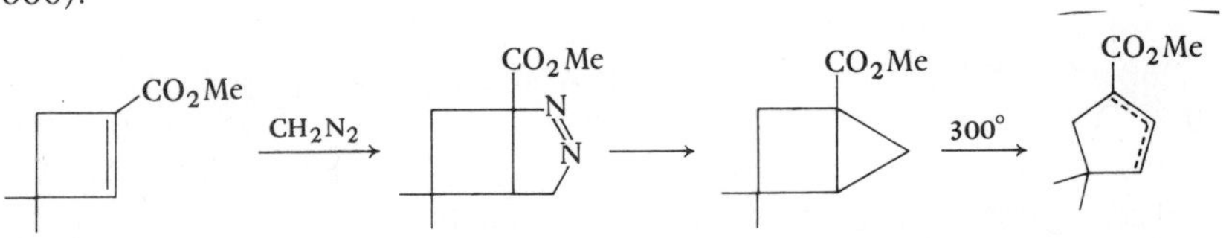

The separation of and assignment of stereochemistry to the cis- and the trans-1,2-dimethylcyclopentanecarboxylic acids has been achieved on the basis of the much slower rate of hydrolysis of the more hindered ester group of the trans-isomer (M. J. Jorgenson, A. J. Brattesani and A. F. Thacher, J. org. Chem., 1969, 34: 1103).

(b) Unsaturated monocarboxylic acids and derivatives

Ring-contraction of (+)pulegone (I) by means of a Favorski rearrangement (in dimethylamine) gives the endo- and the exo-unsaturated amides II and III which are reduced by lithium aluminium hydride to the amine IV (K. S. Schorno, G. R. Waller and E. J. Eisenbraun, J. org. Chem., 1968, 33: 4556). Similarly methyl 4-methyl-2-oxocyclohexane carboxylate gives *dimethyl 3-methylcyclopent-1-ene-1,2-dicarboxylate* (Schorno, G. H. Adolphen and Eisenbraun, ibid., 1969, 34: 2801).

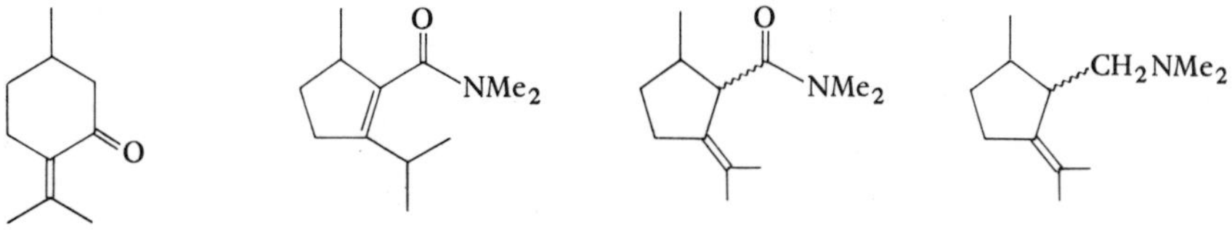

(I)   (II)   (III)   (IV)

The action of chlorine on an alkaline solution of phenol to give VI, a reaction reported by Hantzsch in 1887, is suggested to involve ring-contraction of an intermediate such as V (C. J. Moye and S. Sternhell, Tetrahedron Letters, 1964, 2411) or of a dienone intermediate (see p. 113). Similarly *m*-cresol gives VII, in low yield (G. M. Strunz, A. S. Court and J. Kombassy, ibid., 1969, 3613):

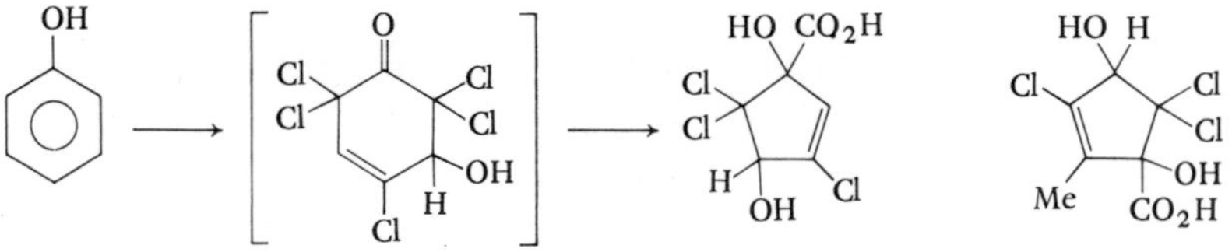

(V)   (VI)   (VII)

Cycloalkylation of malonic ester with *trans*-1,4-dichlorobut-2-ene gives diethyl 2-vinylcyclopropane-1,1-dicarboxylate (VIII). Pyrolysis of the latter then gives IX which is readily decarboxylated to *cyclopent-3-ene-1-carboxylic acid* (G. H. Schmid and A. W. Wolkoff, J. org. Chem., 1967, 32: 254):

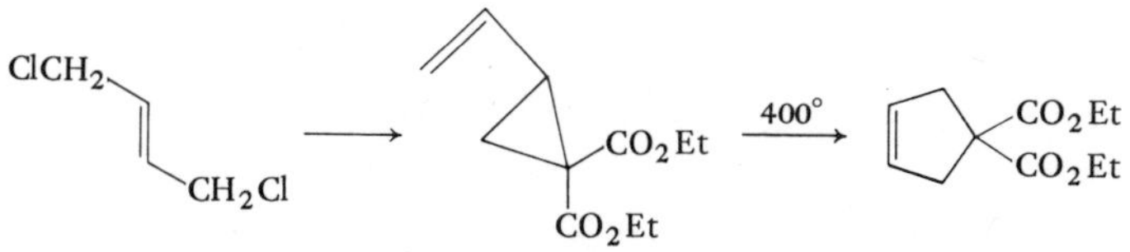

(VIII)   (IX)

An elegant, biogenetic type conversion of *dl-α(Δ³-cyclopentenyl)butyric acid* (X) into the indole alkaloid ajmaline (XI), which contains six rings and nine asymmetric centres, has been described (E. E. van Tamelen and L. K. Oliver, J. Amer. chem. Soc., 1970, 92: 2₁36):

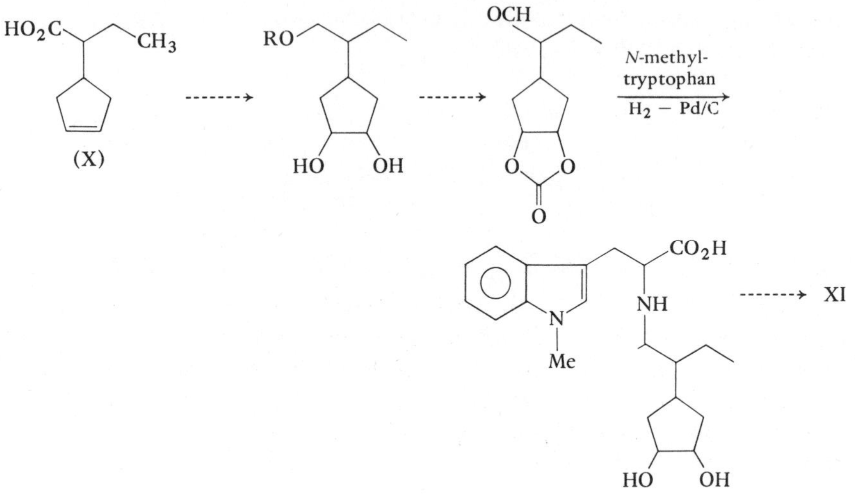

The structure of the antibiotic *borrelidin* has been determined by chemical and spectroscopic methods (W. Keller-Schierlein, Experientia, 1966, 22: 355; Helv., 1967, 60: 731).

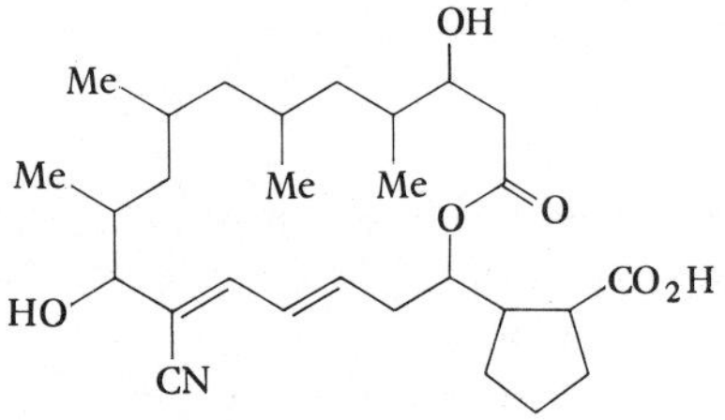

Borrelidin

(c) Polycarboxylic acids

The previously reported *cyclopentene*-1,5-*dicarboxylic acid* has been shown to be a mixture of *cyclopentene*-1,2- and -1,5-*dicarboxylic acids* (m.p. 178° and 182—183° respectively). *Cyclopentene*-1,3- and -1,4-*dicarboxylic acids* (m.p. 150.5° and 181—182° respectively) have also been isolated (H. C. Stevens and G. M. Trenta, Chem. Comm., 1970, 1407). The reaction of dimethyl malonate and dimethyl acetylenedicarboxylate in the presence of pyridine and acetic acid has been shown to give tautomeric octamethyl cycloheptadienoctacarboxylate (XII) which on treatment with potassium acetate gives potassium pentamethoxycarbonylcyclopentadienide (XIV), presumably via a transannular Michael reaction (XII → XIII) and elimination of trimethyl ethylenetricarboxylate (R. C. Cookson et al., J. chem. Soc., C, 1967, 1968, and refs. therein):

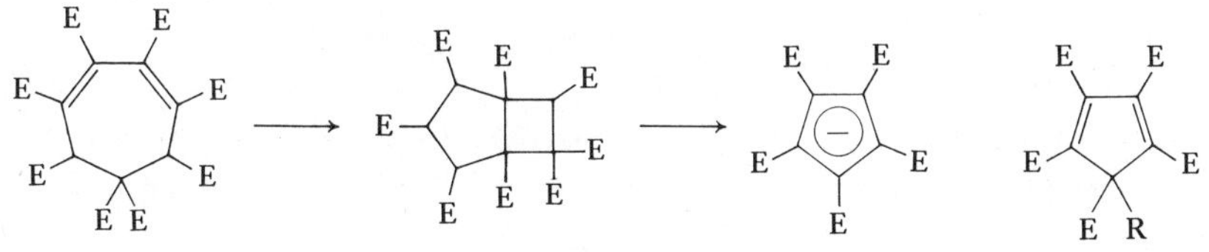

(XII)　　　　(XIII)　　　　(XIV)　　　　(XV)

E = CO₂Me

The derived 1,2,3,4,5-*pentamethoxycarbonylcyclopentadiene* (XV; R = H) is an extremely strong acid, aqueous solutions of which dissolve iron with evolution of hydrogen and formation of the ferrous salt. As with the tetracyano- and dicyanodiethoxycarbonyl analogues, the anion is too stable to rearrange to the ferrocene. Treatment of XV (R = H) with diazomethane gives the *C*-methyl derivative XV (R = Me) which is readily hydrolysed to 1,2,3,4-tetramethoxycarbonyl-5-methylcyclopentadiene (idem, loc. cit.).

(*d*) *Keto-acids*

Michael addition of malonic esters to 4-substituted cyclopent-2-en-1-ones is kinetically controlled to give the *trans*-adduct XVI (R = H, COMe), but under more vigorous conditions the *cis*-lactone (XVII) is formed (A. Ichihara et al., Tetrahedron, 1970, 26: 1331).

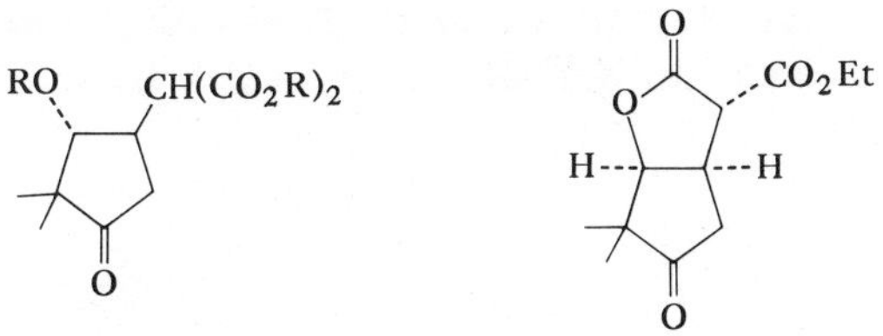

(XVI)　　　　(XVII)

Stereoselective syntheses of the *cis*- and the *trans*-2-phenyl-4-oxocyclopentane-1-carboxylic acid are reported (A. W. Frahm, Ann., 1969, 728: 21).

The reaction of 2-methoxybutadiene with 2,4-dicarboalkoxy-3-(2-cyano-ethyl)cyclopent-2-en-1-one (XVIII) gives XX by a [3,3]sigmatropic re-arrangement of the intermediate allylic enol ether XIX (L. J. Dolby et al., J. org. Chem., 1968, 33: 4508):

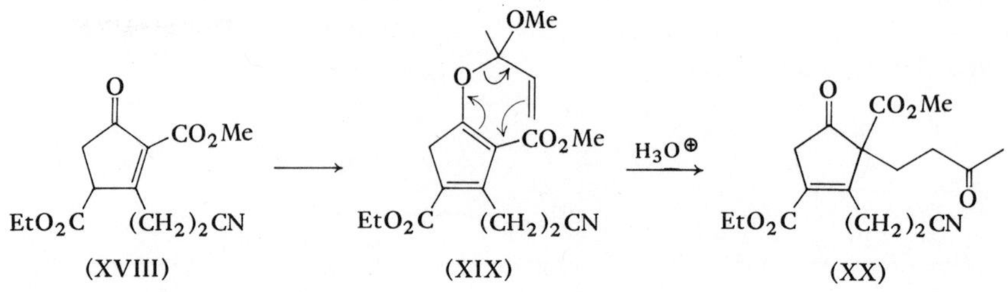

(XVIII)   (XIX)   (XX)

Chlorinated metabolites having the *terrein* skeleton (XXI) (D. H. R. Barton and E. Miller, J. chem. Soc., 1955, 1028) have been isolated from fermentations of *S. affinis*, and the major component shown, by X-ray analysis, to be a (1S,5S)*cyclopent-2-ene-1-carboxylic ester* (XXII) (W. J. McGahren, J. H. van den Hende and L. A. Mitscher, J. Amer. chem. Soc., 1969, 91: 157).

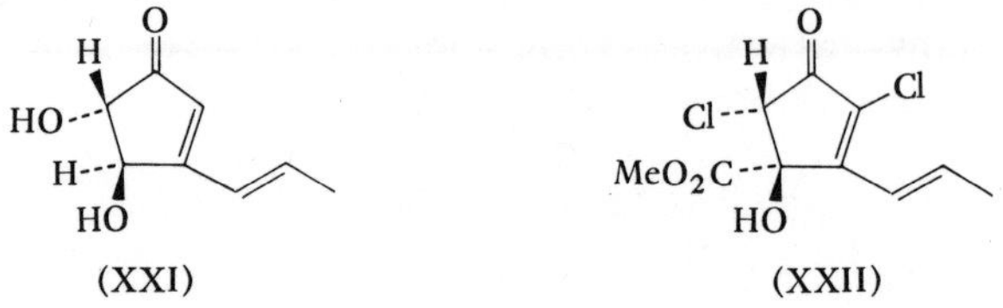

(XXI)   (XXII)

The absolute configuration of *methyl jasmonate* has been established as XXIII (R. K. Hill and A. G. Edwards, Tetrahedron, 1965, 21: 1501). The structure XXIV (R = OH, R' = Cl) previously proposed for the antifungal agent *cryptosporiopsin* (G. M. Strunz et al., Canad. J. Chem., 1969, 47: 2087) has been amended to XIV (R = Cl, R' = OH) (idem, ibid., 1969, 47: 3700).

(XXIII)   (XXIV)

Cycloaddition of *N,N*-diethylaminopropyne to cyclopentenone gives an aminobicycloheptenone (XXV) which can be converted stereoselectively to either XXVI or XXVII. Acid hydrolysis of XXV gives XXVI (R = OH) and basic hydrolysis gives XXVII (R = NEt$_2$) (J. Ficini and A. Krief, Tetrahedron Letters, 1970, 1397).

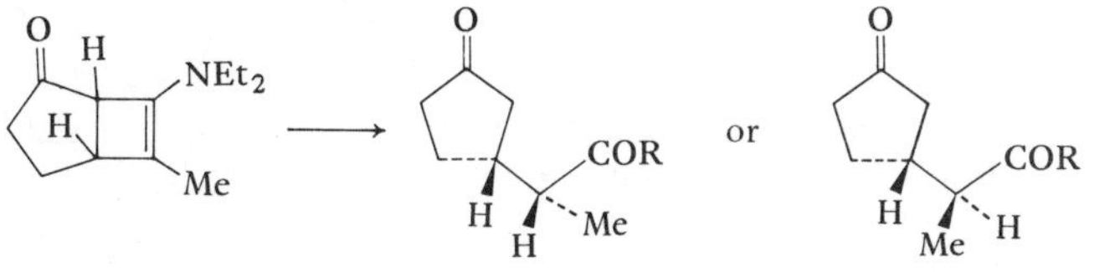

(XXV)   (XXVI)   (XXVII)

Palladium-catalysed carbonylation of hexa-1,5-diene gives the methyl ester of 5-*methylcyclopentan*-1-*one*-2-*acetic acid* (XXVIII) (S. Brewis and P. R. Hughes, Chem. Comm., 1965, 489):

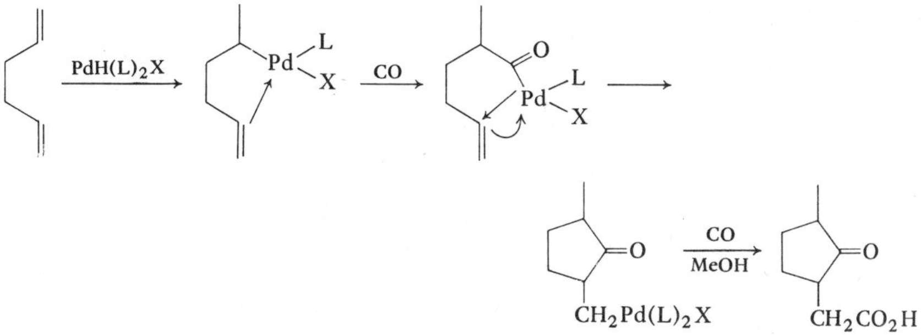

(XXVIII)

(e) Prostaglandins

Prostaglandins occur in many animal species and tissues. They are C$_{20}$-carboxylic acids containing one five-membered ring, one to three hydroxyl groups and carbon—carbon double bonds, and (except for the F-series) a keto group. They have great potential in birth control, as a consequence of the stimulating effect of the primary prostaglandins on smooth muscle, and possess a bewilderingly wide spectrum of pharmacological actions. This has stimulated very considerable activity in this field in recent years and numerous reviews are now available (S. Bergström, Science, 1967, 157: 382; U. Axen, Ann. Rep. Med. Chem., 1967, 290; E. W. Horton, Phys. Rev., 1969, 49: 122; E. J. Corey, Proceedings of the Robert A. Welch Foundation on Chemical Research XII; Organic Synthesis, ed. W. O. Milligan, 1969, p. 51; P. W. Ramwell and E. G. Daniels, Lipid Chromatographic Analysis, ed. G. W. Marinetti, Dekker, New York, 1969, 2: 313; E. W. Horton, Recent Advances in Pharmacology, 4th Edn., eds. R. S. Stacey and J. M. Robson, Churchill, London, 1968, p. 185; P. W. Ramwell et al., Progress in the Chemistry of Fats and Other Lipids, ed. R. T. Holman, Pergamon, Oxford, 1968 9(2): 233; S. Bergström, L. A. Carlson and J. R. Weeks, Pharmacol. Review, 1968, 20: 1; V. R. Pickles, Biol. Review, 1967, 42: 614; U.S. von Euler and R. Eliasson, Medicinal Chemistry Monograph, Academic Press, New York, 1967, p. 8; P. R. Ramwell and J. E. Shaw, Rec. Progr. in Hormone Res., 1970, 26: 139; "Prostaglandins", Nobel Symposium 2 (June 1966, Stockholm), eds. S. Bergström and B. Samuelson, Almqvist and Wiksell, Stockholm, 1967).

The structures and absolute configurations of the primary prostaglandins are shown in Fig. 1. With the exception of $PGF_{3\alpha}$, all these have been isolated from human seminal plasma, together with those shown in Fig. 2 and related compounds. The systematic method of nomenclature is illustrated in Table 1.

Table 1

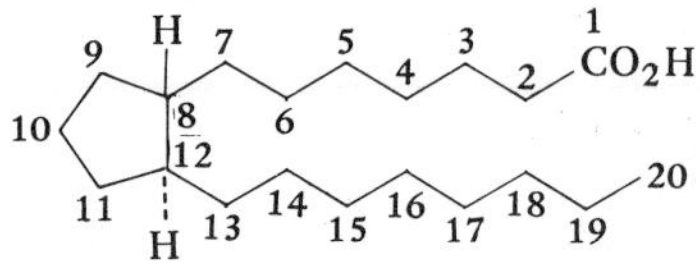

Prostanoic acid

PGE$_1$ = 11α,15(S)-dihydroxy-9-oxo-13-*trans*-prostenoic acid

PGE$_2$ = 11α,15(S)-dihydroxy-9-oxo-5-*cis*,13-*trans*-prostadienoic acid

PGF$_{1\alpha}$ = 9α,11α,15(S)-trihydroxy-13-*trans*-prostenoic acid

PGF$_{1\beta}$ = 9β,11α,15(S)-trihydroxy-13-*trans*-prostenoic acid

PGB$_1$ = 15(S)-hydroxy-9-oxo-8(12),13-*trans*-prostadienoic acid

PGE$_1$ is very sensitive to alkali, which rapidly destroys the biological activity to give the dienone PGB$_1$. With weak base, PGE$_1$ is converted into the enone PGA$_1$.

Most syntheses of prostaglandins rely on the modification of a five-membered ring. D. E. Orr and F. B. Johnson (Canad. J. Chem., 1969, 47: 47) approach the problem using a six-membered ring as starting material; the ring-contraction is triggered off by the action of base on a vicinal hydroxy-tosylate.

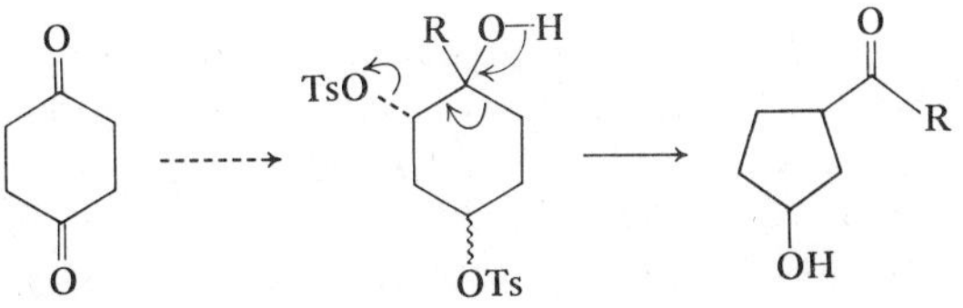

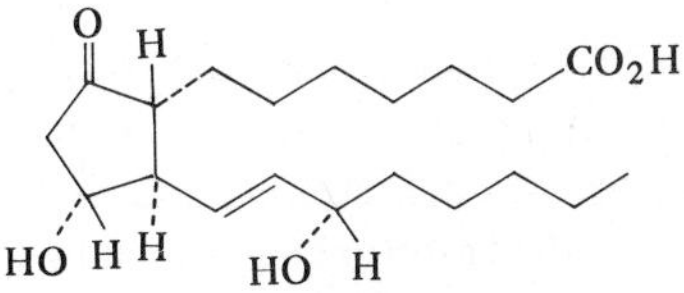
PGE$_1$

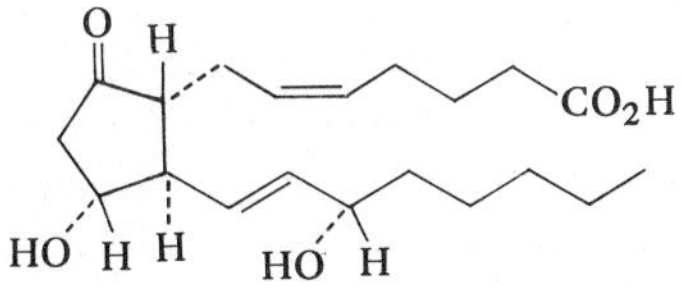
PGE$_2$

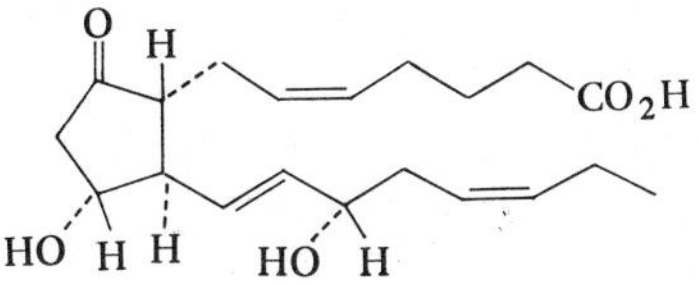
PGE$_3$

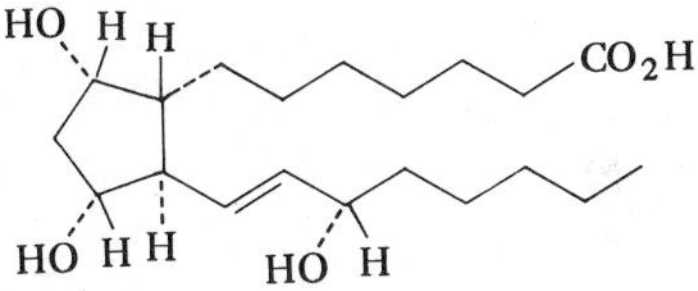
PGF$_{1\alpha}$

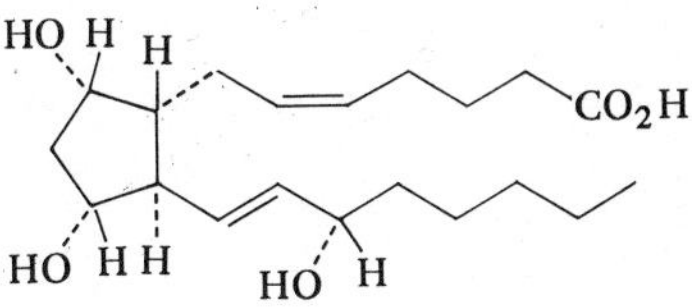
PGF$_{2\alpha}$

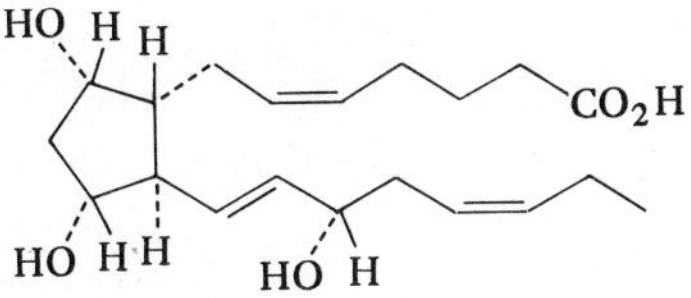
PGF$_{3\alpha}$

Fig. 1.

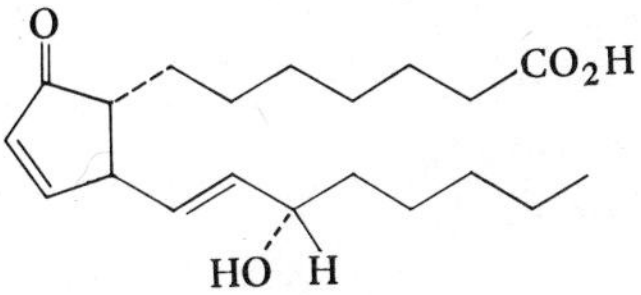
PGA$_1$

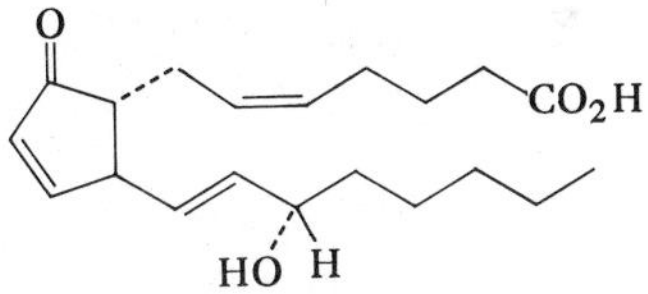
PGA$_2$

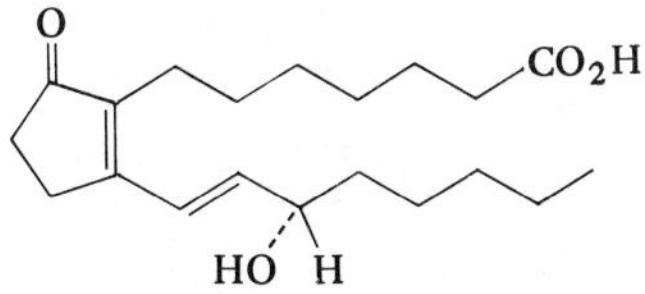
PGB$_1$

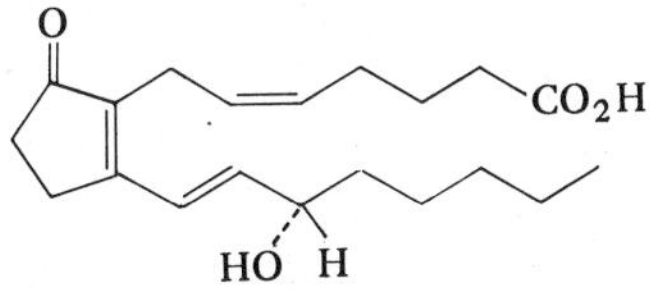
PGB$_2$

Fig. 2.

Attachment of side-chains to an appropriately substituted benzene ring and subsequent conversion to a prostaglandin has so far been unsuccessful (R. B. Morin et al., Tetrahedron Letters, 1968, 6023). E. J. Corey et al. have successfully synthesized the prostaglandins dl E_1, $F_{1\alpha}$, $F_{1\beta}$, A_1, and B_1 by synthesis and elaboration of a six-membered ring followed by ring-cleavage, cyclization to a five-membered ring and further elaboration as summarized in Scheme I (J. Amer. chem. Soc., 1968, 90: 3245, 3247).

The total synthesis of optically active prostaglandins E_1, E_2, $E_{1\alpha}$ and $F_{2\alpha}$ in their natural forms, from a common intermediate III (R = THP) (Scheme 2) has been described. Optically resolution was achieved at an early stage in the synthesis by hydrolysis of lactone I and resolution of the acid product

Scheme I

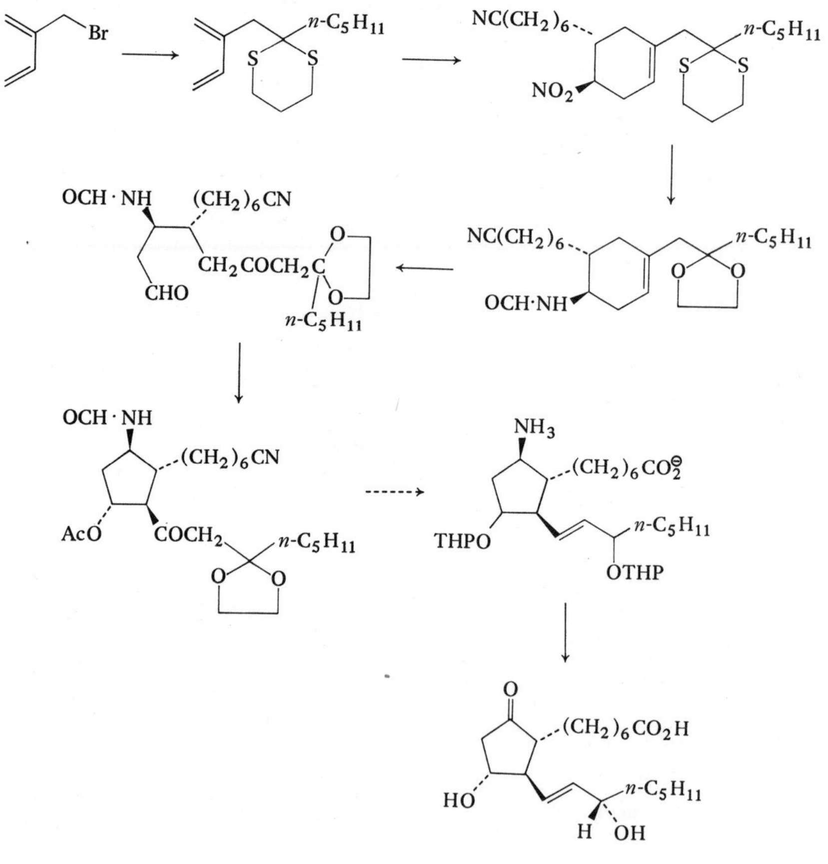

Scheme II

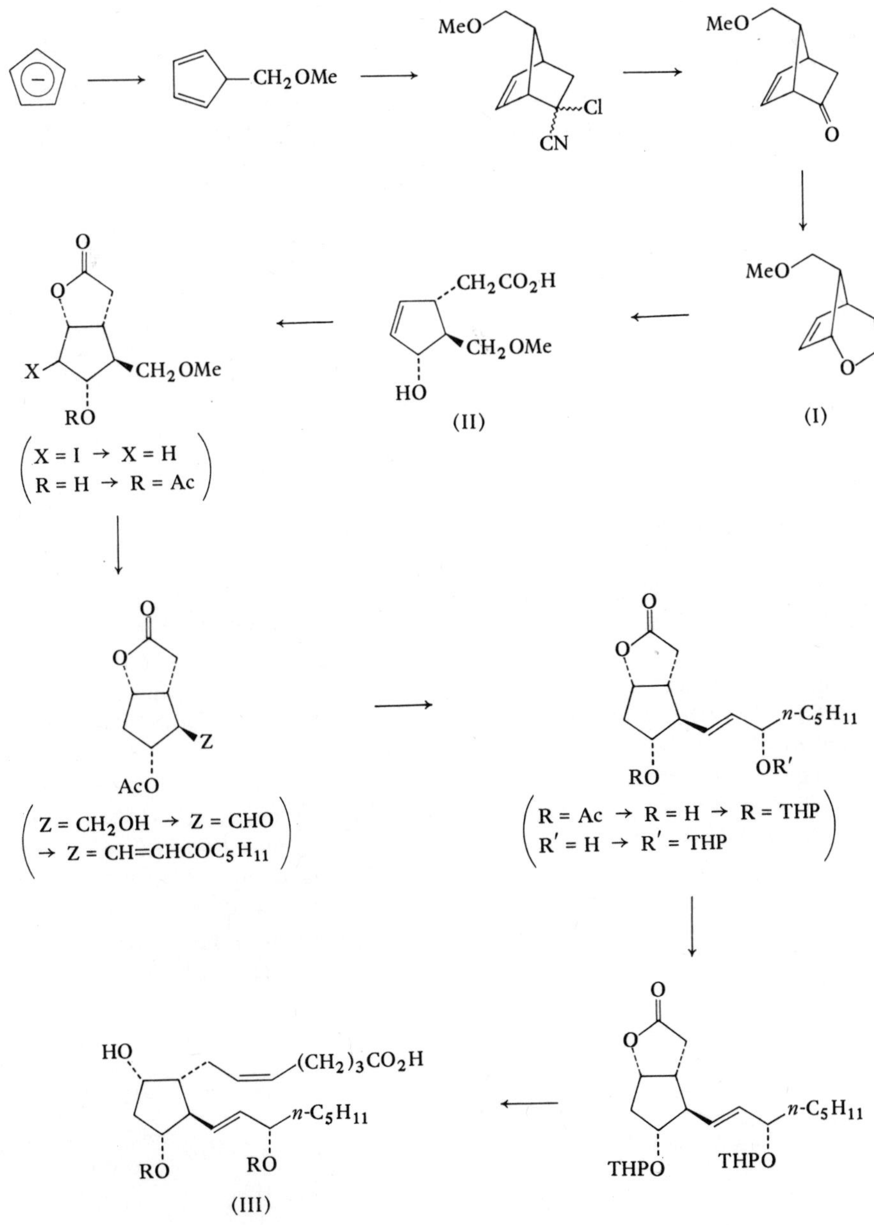

MeO
CH_2OMe
Cl
CN
O
CH_2CO_2H
X
CH_2OMe
RO
HO
(II)
(I)
X = I → X = H
R = H → R = Ac
Z
AcO
Z = CH_2OH → Z = CHO
→ Z = CH=CHCOC_5H_{11}
n-C_5H_{11}
OR′
R = Ac → R = H → R = THP
R′ = H → R′ = THP
HO
$(CH_2)_3CO_2H$
THPO
(III)

with (+)-ephedrine to give dextrorotatory acid II converted as shown to the dextrorotatory prostaglandin precursor III (R = THP). Hydrolysis of III (R = THP) gives $PGF_{2\alpha}$ (III; R = H); oxidation and removal of the tetrahydro-pyranyl-protecting groups gives PGE_2; selective hydrogenation of the *cis*-double bond and removal of the protecting groups gives $PGF_{1\alpha}$; selective hydrogenation, oxidation, and removal of the protecting groups gives PGE_1 (Corey et al., J. Amer. chem. Soc., 1969, 91: 5675; 1970, 92: 397, 2586).

Other syntheses have been reported of racemic prostaglandin E_1 (U. Axen, F. H. Lincoln and J. L. Thompson, Chem. Comm., 1969, 303; W. P. Schneider et al., J. Amer. chem. Soc., 1968, 90: 5895; 1969, 91: 5372; N. Finch and J. J. Fitt, Tetrahedron Letters, 1969, 4639), E_1 and $F_{1\alpha}$ (Corey, I. Vlattas and K. Harding, J. Amer. chem. Soc., 1969, 91: 535), $F_{1\alpha}$ (G. Just et al., ibid., 1969, 91: 5364), E_2 and $F_{2\alpha}$ (Corey, Z. Arnold and J. Hutton, Tetrahedron Letters, 1970, 307, 311), E_2, $F_{2\alpha}$, and $F_{2\beta}$ (Schneider, Chem. Comm., 1969, 304), E_3 (Axen, Thompson and J. E. Pike, ibid., 1970, 602), B_1 (J. Kalsube and M. Matsui, Agr. Biol. Chem., 1969, 33: 1078; C.A., 1969, 71: 80764; P. Collins, C. J. Jung and R. Pappo, Israel J. Chem., 1968, 6: 839; R. Klok, H. J. J. Pabon and D. A. van Dorp, Rec. Trav. chim., 1968, 87: 813; E. Hardegger, H. P. Schenk and E. Broger, Helv., 1967, 50: 2501), and various prostanoic derivatives such as dihydro-PGE_1 (P. F. Beal, J. C. Babcock and F. H. Lincoln, J. Amer. chem. Soc., 1966, 88: 3131; D. P. Strike and H. Smith, Tetrahedron Letters, 1970, 4393), 15-dehydro-PGE_1 (M. Miyano and C. R. Dorn, ibid., 1969, 1615), 11-deoxy-$PGF_{1\beta}$ (J. F. Bagli, T. Bogri, R. Deghenghi and K. Wiesner, Tetrahedron Letters, 1966, 465; Bagli and Bogri, ibid., 1967, 5), and 11-deoxydihydro-$PGF_{1\alpha}$ and $F_{1\beta}$ (idem, ibid., 1969, 1639).

The enzymatic and large-scale bioconversion of all-*cis*-8,11,14-eicosatri-enoic acid (IV) into prostaglandin E_1 has been described (D. H. Nugteren, R. K. Beerthuis and D. A. van Dorp, Rec. Trav. chim., 1966, 85: 405, and refs. therein; E. G. Daniels and J. E. Pike, Prostaglandin Symposium of the Worcester Foundation of Experimental Biology, eds. P. W. Ramwell and J. E. Shaw, Interscience, Wiley, New York, 1968) and a new prostaglandin isomer, 8-isoprostaglandin E_1 (V) isolated (Daniels et al., J. Amer. chem. Soc., 1968, 90: 5894).

(IV)   (V)

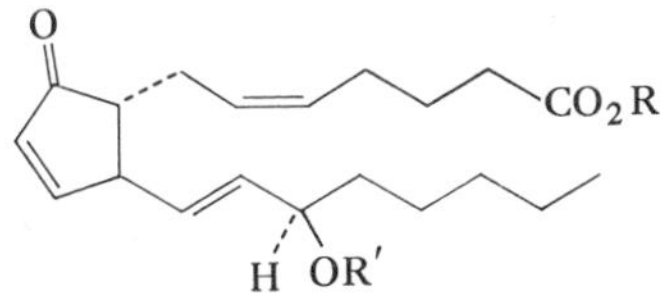

(VI) (a) R, R' = H
(b) R = Me, R' = Ac

Non-enzymatic autoxidation of IV gives prostaglandin E_1 only in very low yield (0.1%) (Nugteren, H. Vonkeman and Van Dorp, Rec. Trav. chim., 1967, 86: 1237). New prostaglandins, 15-epi-*prostaglandin* A_2 (VIa) and its diester (VIb), have also been isolated from coral (A. J. Weinheimer and R. L. Spraggins, Tetrahedron Letters, 1969, 5185).

Prostaglandin antagonists are substances structurally related to prostaglandins but which, although devoid of smooth muscle activity themselves, retain affinity for the receptor sites and so competitively inhibit the action of prostaglandins. Examples are 9,11-di- and 9,11,15-tri-deoxy analogues of 7-oxaprostaglandin $F_{1\alpha}$ (VII) and their ring homologues (VIII) (J. Fried et al., Nature, 1969, 223: 208). The prostaglandin antagonist 7-oxa-13-prostynoic acid has been shown to block the stimulatory effect of prostaglandins E_1 and E_2 on adenosine 3',5'-monophosphate formation and steroidogenesis in mouse ovaries (F. A. Kuehl et al., Science, 1970, 169: 883). The 7-oxa-prostaglandins (E_1 and $F_{1\alpha}$) have been shown to possess smooth muscle activity (J. Fried et al., Tetrahedron Letters, 1970, 2695; Chem. Comm., 1968, 634).

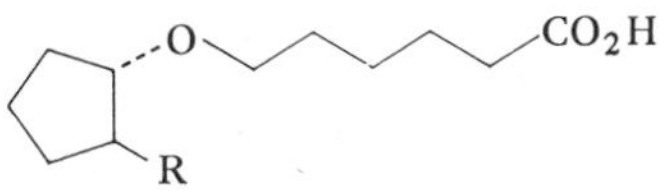

(VII) 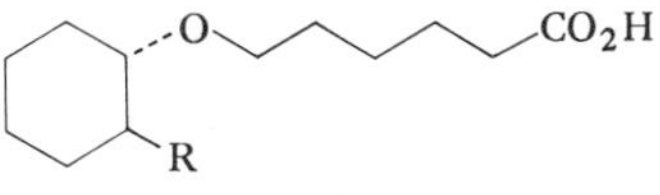   (VIII)

R = *n*-octyl, $C\equiv C \cdot C_6H_{13}$, $CH=CH \cdot CH(OH)C_5H_{11}$, $CH=CH \cdot C_6H_{13}$, etc.

Conformational analysis of the five-membered ring in prostaglandins has been reported. Optical rotatory dispersion and circular dichroism data have been interpreted in favour of a half-chair conformation for prostaglandin E_1 and E_2, with the three ring-substituents in (pseudo)equatorial positions (O. Korver, Rec. Trav. chim., 1969, 88: 1070), whereas X-ray analysis shows prostaglandin F_2 to have an envelope conformation (S. Abrahamsson, Acta Cryst., 1963, 16: 409).

Chapter 5

The Cyclohexane Group

A. J. BELLAMY

1. Formation and properties

(a) Methods of formation

(i) Reduction of benzenoid compounds

The electrochemical reduction of benzene and some benzene derivatives can be controlled to give either non-conjugated dihydro-derivatives or tetrahydro-derivatives by the absence or presence of a cell divider in the electrolytic cell (R. A. Benkeser, E. M. Kaiser and R. F. Lambert, J. Amer. chem. Soc., 1964, 86: 5272; H. W. Sternberg et al., ibid., 1969, 91: 4191). Thus, similar products to those produced by the Birch reduction (R. G. Harvey, Synthesis, 1970, 161) are obtained. The course of the further reduction of the initial products from Birch reductions is found to be dependent upon the alcohol present (A. J. Birch and G. S. R. Subba Rao, Austral. J. Chem., 1970, 23: 1641). Elaboration of non-conjugated dienes produced by the Birch reduction has led to a new synthesis of acyclic 1,5-dienes (J. A. Marshall and J. H. Babler, Tetrahedron Letters, 1970, 3861):

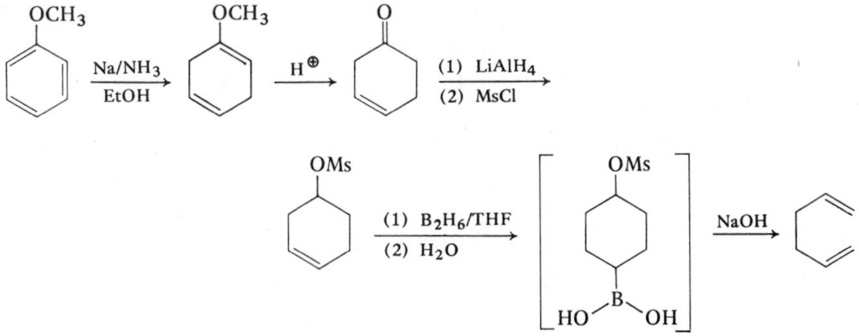

(ii) The Diels—Alder diene synthesis

Both preparative and mechanistic aspects of this method have been reviewed (J. Sauer, Angew. Chem., intern. Edn., 1966, 5: 211; 1967, 6: 16; S. Seltzer, Adv. Alicyclic Chem., 1968, 2: 1).

(iii) Other general synthetic methods

Intramolecular reductive cyclisation of diethyl deca-2,8-diene-1,10-dioate can be effected electrolytically in aqueous acetonitrile to give a high yield of diethyl cyclohexane-1,2-diacetate (*trans/cis* = 2.3) (J. D. Anderson, M. M. Baizer and J. P. Petrovich, J. org. Chem., 1966, 31: 3890).

Cyclisation to cyclohexyl derivatives induced by radical initiators has been much studied (M. Julia and M. Maumy, Bull. Soc. chim. Fr., 1969, 2415, 2427; M. Julia, Acc. Chem. Res., 1971, 4: 386, and refs. therein). For example, cyclisation of ethyl 2-cyanohept-6-enoate using dibenzoyl peroxide in cyclohexane gives a good yield of ethyl 1-cyanocyclohexanecarboxylate, together with some methylcyclopentyl derivative. Suitably substituted acetylenes may be cyclised by a similar process; in this way 7-bromo-1-phenylhept-1-yne gives benzylidenecyclohexane (75%) (J. K. Crandall and D. J. Keyton, Tetrahedron Letters, 1969, 1653). Cyclisation of non-conjugated alkenones to cyclohexanols may be effected electrochemically when the double bond and the carbonyl group are separated by four carbon atoms (T. Shono and M. Mitani, J. Amer. chem. Soc., 1971, 93: 5284).

Cyclohexanones may be prepared from 1,5-dibromopentanes by condensation with two moles of methylenetriphenylphosphorane, and treatment with oxygen (H. J. Bestmann and E. Kranz, Angew. Chem., intern. Edn., 1967, 6: 81):

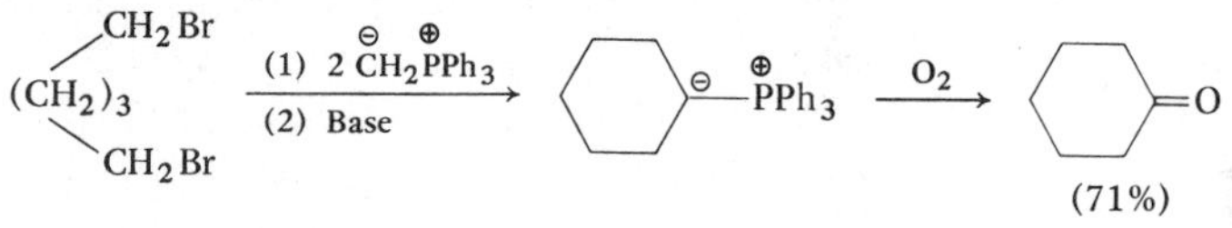

Functionalisation of cyclohexane occurs on treating cyclohexane (as the solvent), with substituted olefins in the presence of dibenzoylperoxide (J. I. G. Cadogan, D. H. Hey and S. H. Ong, J. chem. Soc., 1965, 1939):

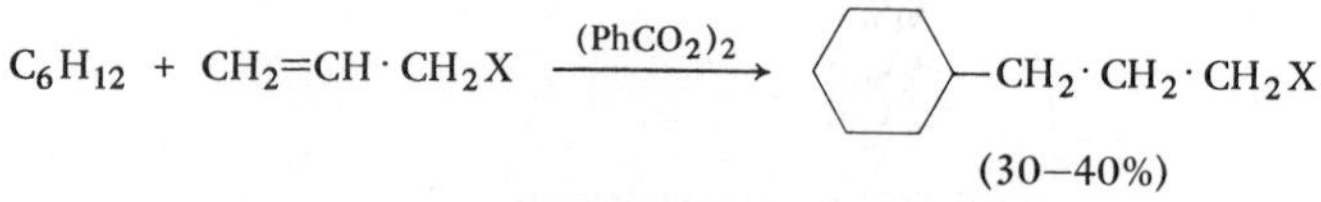

The methods of formation of cyclohexane derivatives by ring-expansion from 4- or 5-membered rings (C. D. Gutsche and D. Redmore, Adv. Alicyclic Chem., Suppl. 1, 1968), and by ring-contraction from 7- and larger membered rings (ibid., 1971, 3: 1), have been reviewed.

(b) Configuration of the cyclohexane ring

Electron-diffraction studies on gaseous **cyclohexane** have shown that the C—C—C angles are $111.05°$, and that the C—C—C—C dihedral angles are $55.9°$ (H. R. Buys and H. J. Geise, Tetrahedron Letters, 1970, 2991). These values are in better agreement with X-ray and n.m.r. data, and theoretically calculated values, than those reported earlier by M. Davis and O. Hassel (Acta chem. Scand., 1963, 17: 1181).

The methods available for conformational analysis of cyclohexane derivatives have been reviewed by E. L. Eliel (Angew. Chem., intern. Edn., 1965, 4: 761). The use of n.m.r. spectroscopy has been particularly important (J. E. Anderson, Quart. Reviews, 1965, 19: 426; N. C. Franklin and H. Feltkamp, Angew. Chem., intern. Edn., 1965, 4: 774). An extensive and critical review of conformational preferences in cyclohexane and cyclohexene derivatives, and methods for their study, has been published by F. R. Jensen and C. H. Bushweller (Adv. Alicyclic Chem., 1971, 3: 140).

There has been some disagreement concerning the accuracy of the chemical shift n.m.r. method for the determination of $-\Delta F$ values. F. R. Jensen and B. H. Beck (J. Amer. chem. Soc., 1968, 90: 3251) have shown that with some cyclohexyl derivatives, $-\Delta F$ values determined by this method differ by ~ 0.2 kcal/mole from values determined by other methods (see also: F. A. L. Anet and P. M. Henrichs, Tetrahedron Letters, 1969, 741; H. Booth, Tetrahedron, 1966, 22: 615). An n.m.r. study of specifically deuteriated *tert*-butylcyclohexane has shown that the *tert*-butyl group distorts the chair form of the ring, $H_{(1)}$ being bent towards the centre (J. D. Remijnse, H. van Bekkum and B. M. Wepster, Rec. Trav. chim., 1971, 90: 779, and refs. therein).

Although most of the n.m.r. studies have used ^1H resonance, the use of ^{19}F resonance in *gem*-difluorocyclohexanes is of great value since the chemical shift difference between an axial ^{19}F and an equatorial ^{19}F resonance is much larger than that for axial and equatorial ^1H (884 Hz at 56.4 MHz) (J. D. Roberts, Chem. Brit., 1966, 2: 529). The activation parameters for chair—chair interconversion in cyclohexane and 1,1-difluorocyclohexane are very similar ($\Delta H\ddagger = 10.9$ and 10.4 kcal/mole, and $\Delta S\ddagger = +2.9$ and $+3.0$ e.u. respectively) and a 1,3-diaxial methyl-fluorine interaction has a negligible effect on the chair—chair interconversion (C. W. Jefford, D. T. Hill and K. C. Ramey, Helv., 1970, 53: 1184). Roberts has shown that a *gem*-difluoro group can have a large effect upon the $-\Delta F$ value of a polar substituent at $C_{(4)}$. Thus $-\Delta F$ for cyclohexanol is 0.87 kcal/mole whereas $-\Delta F$ for

1,1-difluorocyclohexan-4-ol is 0.02 kcal/mole (Tetrahedron Letters, 1968, 5777).

Values for the free-energy difference between substituents in axial and equatorial positions have been determined for 21 substituents in carbon bisulphide solution at $-80°$, by n.m.r. (F. R. Jensen, C. H. Bushweller and B. H. Beck, J. Amer. chem. Soc., 1969, 91: 344); for the hydroxyl group in protic and aprotic solvents at $-83°$ by n.m.r. (Bushweller, J. org. Chem., 1970, 35: 2086); for MgBr and MgC_6H_{11} in ether at $-83°$ by n.m.r. (Jensen and K. L. Nakamaye, J. Amer. chem. Soc., 1968, 90: 3248); for the groups CHO, CH_2OH, $COCH_3$ by equilibration (E. L. Eliel, D. G. Neilson and E. C. Gilbert, Chem. Comm., 1968, 360); for the amino group in protic, aprotic and acidic solvents by n.m.r. (Eliel, E. W. Della and T. H. Williams, Tetrahedron Letters, 1963, 831); for free and protonated NH_2 and NMe_2 groups in 80% aqueous methylcellosolve by pKa (J. Sicher, J. Jonas and M. Tichy, ibid., 1963, 825). The axial conformer ($\sim$1%) of methylcyclohexane has been detected at $-110°$ by ^{13}C n.m.r. (F. A. L. Anet, C. H. Bradley and G. W. Buchanan, J. Amer. chem. Soc., 1971, 93: 258).

The enthalpies of 1,3-diaxial methyl—hydrogen and methyl—methyl interactions are 1.9 and 5.6 kcal/mole respectively (H. Werner et al., Tetrahedron Letters, 1970, 3563). Despite the presence of four 1,3-diaxial methyl—hydrogen interactions, 1,1,4,4-tetramethylcyclohexane exists in a chair conformation (W. Reusch and D. F. Anderson, Tetrahedron, 1966, 22: 583). The activation parameters for chair—chair interconversion for this molecule have been determined by R. W. Murray and M. L. Kaplan (ibid., 1967, 23: 1575), and for 1,1-, cis-1,2-, trans-1,3-, and cis-1,4-dimethylcyclohexane by D. K. Dalling, D. M. Grant and L. F. Johnson (J. Amer. chem. Soc., 1971, 93: 3678).

The preferred conformation of a hydroxyl group has been determined from the C—D infrared stretching vibration in 1-deuteriocyclohexanol (I. O. C. Ekejiuba and H. E. Hallam, J. chem. Soc., B, 1970, 1029), and from the ^{13}C n.m.r. chemical shift of the carbinol carbon (G. W. Buchanan, D. A. Ross and J. B. Stothers, J. Amer. chem. Soc., 1966, 88: 4301). Both methods give good agreement with previous values namely 73% equatorial OH at 20°. Eliel and Gilbert have studied the effect of solvent, concentration, and temperature on the conformational preference of the hydroxyl group using an equilibration method (ibid., 1969, 91: 5487), and R. D. Stolow, T. Groom and P. D. McMaster found that some 4-substituents have a marked effect on $-\Delta F$ for OH. Thus in 4-oxocyclohexanol there is only 44% of the equatorial-hydroxy conformer (Tetrahedron Letters, 1968, 5781).

When a neat sample of either chlorocyclohexane or trideuteriomethoxy-cyclohexane is cooled very slowly to $-150°$, the pure equatorial conformer crystallises out. This can be studied by n.m.r. in CD_2CDCl solution at $-150°$. On warming to $-100°$, equilibration with the axial conformer occurs (Jensen and Bushweller, J. Amer. chem. Soc., 1969, 91: 3223).

The factors which force a cyclohexane derivative to adopt a non-chair conformation (e.g. boat or twist-boat), and the physical methods available to detect these conformations have been reviewed by D. L. Robinson and D. W. Theobald (Quart. Review, 1967, 21: 314).

For cyclohexane derivatives, such as methylenecyclohexanes and cyclohexanones, with one sp^2-hybridised carbon atom in the ring, the chair form is the predominant conformation. The activation parameters for chair—chair interconversion in these systems have been determined by J. T. Gerig and R. A. Rimerman (J. Amer. chem. Soc., 1970, 92: 1219), and M. St-Jacques, M. Bernard and C. Vaziri (Canad. J. Chem., 1970, 48: 2386, 3039). Inversion occurs faster than for cyclohexane. From X-ray crystallographic examination, it is found that 4,4-diphenylcyclohexanone is severely flattened at the carbonyl end of the ring, and more puckered at the $C_{(4)}$ end, when compared with cyclohexane (J. B. Lambert, R. E. Carhart and P. W. R. Corfield, J. Amer. chem. Soc., 1969, 91: 3567). Due to the dipole—dipole interaction, 2-bromocyclohexanone exists preferentially in the chair conformation with bromine axial. The proportion of the latter increases as the dielectric constant of the solvent decreases (e.g., 91% axial bromine in CCl_4) (E. W. Garbisch Jr., ibid., 1964, 86: 1780). The conformational preference of 2-halogenocyclohexanones may be different in 'super acid' media to that obtaining in CCl_4. For example, *trans*-2-fluoro-6-methylcyclohexanone prefers the methyl-equatorial, fluorine-axial conformer in CCl_4, but the reverse in 'super acids'. This is caused by hydrogen bonding between fluorine and hydrogen attached to the carbonyl oxygen, which is most effective with fluorine equatorial (R. Jantzen and J. Cantacuzene, Tetrahedron Letters, 1971, 2925).

Electron diffraction studies on gaseous **cyclohexene** have confirmed the preference for the half-chair conformation (J. F. Chiang and S. H. Bauer, J. Amer. chem. Soc., 1969, 91: 1898; H. J. Geise and H. R. Buys, Rec. Trav. chim., 1970, 89: 1147). The activation parameters for the half-chair, half-chair interconversion via the boat conformation have been determined by n.m.r. using *cis*-3,3,4,5,6,6-hexadeuteriocyclohexene. At $25°$, the average life-time before inversion is 10^{-9} sec cf. $4 \cdot 10^{-4}$ sec for cyclohexane (F. A. L. Anet and M. Z. Haq, J. Amer. chem. Soc., 1965, 87: 3147). The preferred conformation for 4-substituted cyclohexenes has the substituent equatorial for CHO and CO_2Me (N. S. Zefirov et al., Doklady Akad. Nauk, S.S.S.R., 1970, 190: 345), and for the halogens except iodine (Jensen and Bushweller, J. Amer. chem. Soc., 1969, 91: 5774).

Allylic strain in 6-membered rings has been reviewed by F. Johnson (Chem. Review, 1968, 68: 375). Two types of allylic strain are recognised: $A^{(1,2)}$ and $A^{(1,3)}$. $A^{(1,2)}$ *strain* may occur with *endo*-cyclic double bonds:

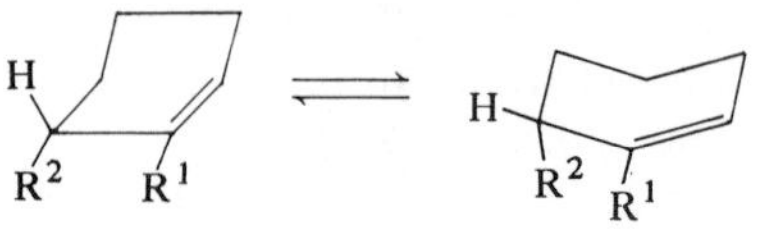

H
R^2 R^1
(E)

H—
R^2 R^1
(A)

The dihedral angle $R^1-C_{(1)}-C_{(6)}-R^2$ is larger in conformation A than in conformation E (85° vs. 35°). Thus, any steric interaction between R^1 and R^2 will be less in conformation A than in conformation E. Conformation A, with R^2 in a pseudo-axial position, may therefore be the preferred conformation. For example, equilibration of 4-*tert*-butyl-6-nitro-1-phenylcyclohexene gives mainly the isomer with the nitro-group in a pseudo-axial position (E. W. Garbisch Jr., J. org. Chem., 1962, 27: 4249):

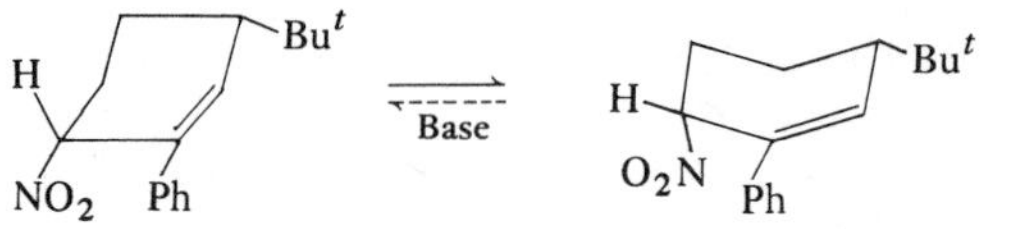

H
—Bu^t
NO_2 Ph

$\xrightarrow{\text{Base}}$

H—
—Bu^t
O_2N Ph

(85–95 %)

$A^{(1,2)}$ strain may also affect the stereochemical course of reactions, if, for example, the *endo*-cyclic double bond is part of an enamine or enolate ion (F. Johnson, loc. cit.).

$A^{(1,3)}$ *strain* may occur with *exo*-cyclic double bonds:

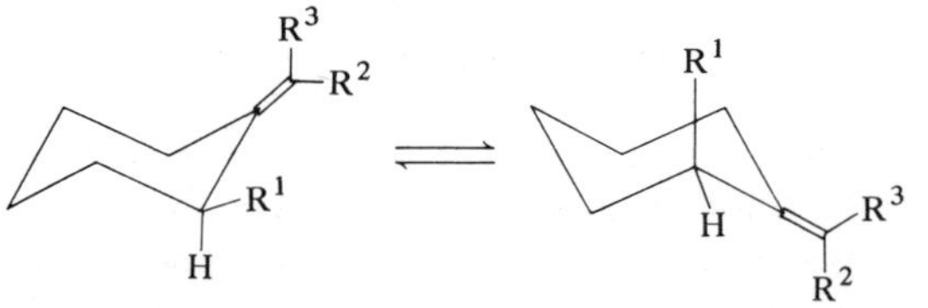

R^3
—R^2
—R^1
H
(E)

R^1
H
—R^3
R^2
(A)

If R^1 and R^2 are moderately large, the steric interaction between them may be sufficient to make conformation A, with R^1 in an axial position, the preferred conformation. This may affect the stereochemistry of reactions involving the formation of *exo*-cyclic double bonds, and also conformational and isomerisational equilibria. For example, C-protonation of the *aci* form of

1-nitro-2-phenylcyclohexane, which involves axial protonation from the least hindered side, gives the least stable product (F. Johnson, loc. cit.):

S. K. Malhotra and F. Johnson (Chem. Comm., 1968, 1149) have shown that the magnitude of $A^{(1,3)}$ methyl–hydrogen and methyl–methyl interactions are 1.4 and 5.9–6.2 kcal/mole respectively. These are comparable to 1,3-diaxial interactions involving the same groups.

The effects of stereochemistry and conformation on the elimination reactions of cyclohexane derivatives have been reviewed by N. A. LeBel (Adv. Alicyclic Chem., 1971, 3: 195).

2. Hydrocarbons

(a) Saturated hydrocarbons: Cyclohexanes

Although cyclohexane itself on treatment with acetyl chloride and aluminium trichloride in chloroform gives 1-acetyl-2-methylcyclopentene, methylcyclohexane gives a mixture of isomeric 1-acetyl-2-methylcyclohexenes (I. Tabushi, K. Fujita and R. Oda, Tetrahedron Letters, 1968, 4247, 5455). G. A. Olah et al. have shown that cyclohexyl cations readily rearrange to 1-methylcyclopentyl cations (J. Amer. chem. Soc., 1967, 89: 2692). The anodic oxidation of cyclohexane to cyclohexyl cations in fluorosulphonic acid/acetic acid also gives 1-acetyl-2-methylcyclopentene (J. Bertram, M. Fleischmann and D. Pletcher, Tetrahedron Letters, 1971, 349). Acylation of cyclohexanes also occurs via cyclohexyl radicals. Thus, acetylcyclohexane is formed in 70% yield by the reaction of cyclohexane with biacetyl in the presence of dibenzoyl peroxide (W. G. Bentrude and K. R. Darnall, J. Amer. chem. Soc., 1968, 90: 3588).

Bis(trifluoromethyl)carbene reacts with cyclohexane under suitable conditions with C–H insertion as the major reaction path (D. M. Gale, W. J. Middleton and C. G. Krespan, ibid., 1965, 87: 657), and cyanonitrene is found to undergo a similar reaction with tertiary C–H bonds. For example, cis- and trans-1,2-dimethylcyclohexane give the corresponding 1-cyanoamino-1,2-dimethylcyclohexane with retention of configuration. Reduction of the latter with lithium aluminium hydride then gives the 1-methylaminoderivative (A. G. Anastassiou and H. E. Simmons, ibid., 1967, 89: 3177).

Substitution of hydrogen in cyclohexane by NH_2 or NO can be accomplished photolytically using respectively ammonia (V. I. Stenberg and C.-H. Niu, Tetrahedron Letters, 1970, 4351), and *tert*-butyl nitrite (A. Mackor, J. U. Veenland and Th. J. de Boer, Rec. Trav. chim., 1969, 88: 1249).

(b) Unsaturated hydrocarbons

(i) Cyclohexenes

Preparation. Cyclohexenes can be prepared in high yield from substituted cyclohexanones by hydroboration of the corresponding pyrrolidine enamine (J. W. Lewis and A. A. Pearce, Tetrahedron Letters, 1964, 2039). Reduction of the enamine may also be accomplished using aluminium hydride (J. M. Coulter, J. W. Lewis and P. P. Lynch, Tetrahedron, 1968, 24: 4489):

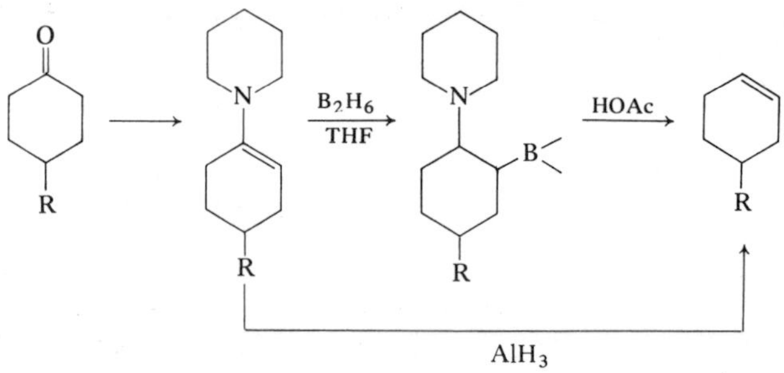

S. Wolfe and P. G. C. Campbell (Canad. J. Chem., 1965, 43: 1184) synthesised *3,3,6,6-tetradeuteriocyclohexene* by Diels–Alder addition of ethylene to 1,1,4,4-tetradeuteriobutadiene, and used it to study the stereochemistry of additions to cyclohexene.

Properties. Isomerisation of cyclohexenes may be accomplished in a variety of ways. Thermally induced as well as catalysed isomerisations have been reviewed by A. J. Hubert and H. Reimlinger (Synthesis, 1969, 97; 1970, 405). A method developed by H. C. Brown et al. (J. Amer. chem. Soc., 1967, 89: 567) involves hydroboration of the cyclohexene, thermal isomerisation to the least substituted organo-borane, followed by liberation of the rearranged olefin by treatment with dec-1-ene. Thus, 1-methylcyclohexene gives methylenecyclohexane (56%) and 1-ethylcyclohexene gives vinylcyclohexane (62%). Alternatively, the rearranged organo-borane may be oxidised to an alcohol e.g. 1-ethylcyclohexene gives mainly 2-cyclohexylethanol (idem, ibid., p. 561). 1-Methylcyclohexene can also be converted into methylenecyclohexane (60–75%) by successive treatment with hydrogen chloride, and potassium triethylmethoxide (S. P. Acharya and H. C. Brown, Chem. Comm., 1968, 305).

Examination of the relative stabilities of 1,3- and 1,5-disubstituted cyclohexenes by equilibration gives an estimate of the $A^{(1,2)}$ methyl—methyl interaction as 0.45 kcal/mole (Y.-H. Suen and H. B. Kagan, Bull. Soc. chim. Fr., 1970, 3552). Double-bond isomerisation in unsaturated esters and enol ethers has been analysed in terms of stabilising and destabilising interactions by S. J. Rhoads, J. K. Chattopadhyay and E. E. Waali (J. org. Chem., 1970, 35: 3352, 3358).

Hydroboration of cyclohexene to give tricyclohexylborane, followed by treatment with carbon monoxide under a variety of conditions gives high yields of cyclohexyl derivatives. For example, cyclohexanecarboxaldehyde (93%), cyclohexylmethanol (80%), and tricyclohexylmethanol (80%) can be obtained:

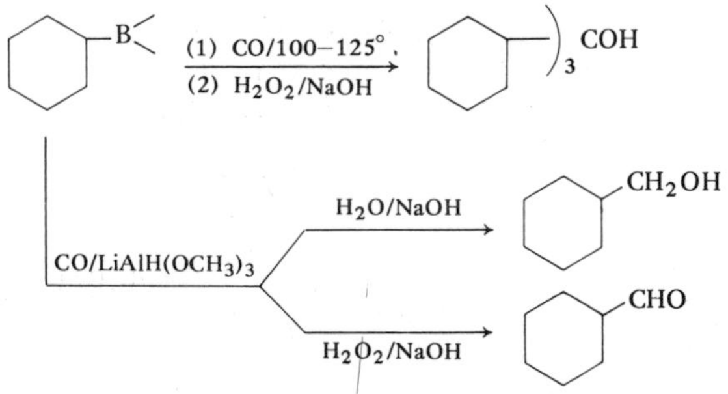

Similarly, preparation of dicyclohexylborane followed by reaction with another alkene e.g. but-1-ene or acrylonitrile, gives an alkyldicyclohexylborane, which on treatment with (a) aqueous carbon monoxide, and (b) alkaline hydrogen peroxide, produces a mixture of dicyclohexylketone and alkylcyclohexylketone. Thus butyldicyclohexylborane gives a 3.5:1 mixture of butylcyclohexyl ketone and dicyclohexyl ketone in 72% yield (see Brown, Acc. chem. Res., 1969, 2: 65, for refs.). Dicyclohexylborane is also utilised in the conversion of acetylenes into primary alcohols, carboxylic acids, cyclohexyl-substituted alkenes, and alkylcyclohexylketones, the product depending upon the substitution pattern of the alkyne and the reaction conditions (G. Zweifel, H. Arzoumanian and C. C. Whitney, J. Amer. chem. Soc., 1967, 89: 291, 3652, 5086).

The stereochemistry and direction of hydroboration of 3- and 4-chloro- and -methoxy-cyclohexenes (D. J. Pasto and J. Hickman, ibid., 1968, 90: 4445), and of 1,4-di-*tert*-butylcyclohexene (Pasto and F. M. Klein, Tetrahedron Letters, 1967, 963) have been studied. The latter compound gives *trans,trans*-2,5-di-*tert*-butylcyclohexanol, which exists in a twist-boat conformation.

The stereochemistry of oxymercuration of 4-*tert*-butylcyclohexene and 4-*tert*-butyl-1-methylcyclohexene is exclusively *trans*-diaxial. Reduction of the intermediate addition products gives approximately equal amounts of *cis*-4-*tert*-butylcyclohexanol and *trans*-3-*tert*-butylcyclohexanol from the former olefin, and *cis*-4-*tert*-butyl-1-methylcyclohexanol only from the latter. D. J. Pasto and J. A. Gontarz have concluded that oxymercuration of cyclohexenes proceeds via mercurinium ion intermediates which are formed in a fast, reversible step before the rate and product determining addition of a nucleophile (J. Amer. chem. Soc., 1971, 93: 6902). Cyclohexyl-mercury derivatives can be prepared electrochemically from cyclohexene using a mercury anode (N. L. Weinberg, Tetrahedron Letters, 1970, 4823).

The addition of dihalogenocarbene to 1-alkoxy-cyclohexene and -cyclohexa-1,4-diene has been utilised in the synthesis of cyclohepta-3,5-dien-1-one and -2,4,6-trien-1-one (W. E. Parham, R. W. Soeder and R. M. Dodson, J. Amer. chem. Soc., 1962, 84: 1755, and A. J. Birch, J. M. H. Graves and F. Stansfield, Proc. chem. Soc., 1962, 282 respectively):

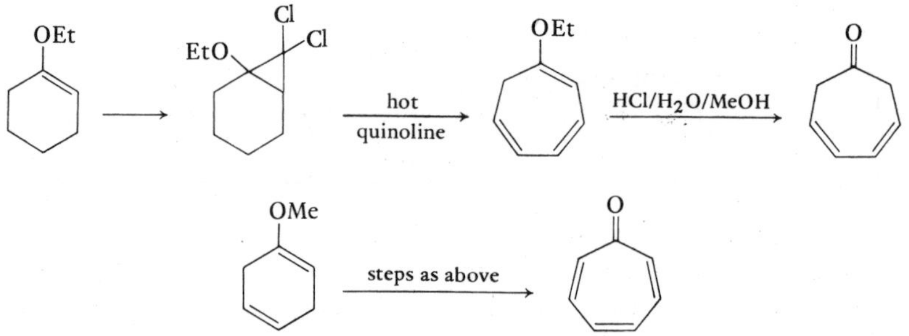

Free-radical additions to olefins, including cyclohexenes, have been reviewed by C. Walling and E. S. Huyser (Org. Reactions, 1963, 13: 91), F. W. Stacey and J. F. Harris Jr. (ibid., p. 150), and B. A. Bohm and P. I. Abell (Chem. Review, 1962, 62: 599). Photosensitised ionic additions to cyclohexenes has been reviewed by J. A. Marshall (Acc. chem. Res., 1969, 2: 33). For the photochemical reaction between cyclohexene and acetone in the liquid phase (313 nm), the quantum yields of seven products have been determined (P. Borrell and J. Sedlar, Trans. Farad. Soc., 1970, 66: 1670).

Electrochemical allylic oxidation of cyclohexene is found to give 3-substituted cyclohexenes. The substituent is derived from the solvent: OAc from acetic acid (55%); OMe from methanol (21%); and NH·CO·Me from acetonitrile (17%) (T. Shono and T. Kosaka, Tetrahedron Letters, 1968, 6207). Electrochemical oxidation of cyclohexene/chlorine in acetonitrile gives *N*-acetyl-2-chlorocyclohexylamine at low potentials, and 3-chlorocyclohexene at high potentials (G. Faita, M. Fleischmann and D. Pletcher, J. Electroanalyt. Interfacial Electrochem., 1970, 5: 455).

(ii) Alkylidene- and alkenyl-cyclohexanes
The use of extrusion reactions has facilitated the synthesis of sterically hindered cyclohexylidene derivatives. Thus, cyclohexanone may be converted into diphenylmethylenecyclohexane (80%) or cyclohexylidenecyclohexane (73%) in two and three steps respectively (D. H. R. Barton, E. H. Smith and B. J. Willis, Chem. Comm., 1970, 1225, 1226):

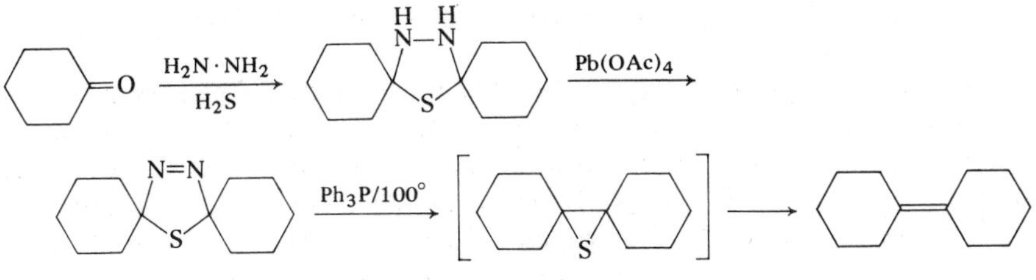

(not isolated)

1-Chloro-2-chloromethylprop-2-ene, on treatment with magnesium in tetrahydrofuran, gives 1,4-*dimethylenecyclohexane* (20%) together with other products (see F. Weiss, Quart. Reviews, 1970, 24: 299). 1,2,4,5-*Tetramethylenecyclohexane* is obtained in 8% yield on pyrolysis of the tetra-acetate of 1,2,4,5-tetrahydroxymethylcyclohexane. It reacts with two moles of maleic anhydride, giving a derivative which can be readily converted into anthracene (W. J. Bailey, E. J. Feller and J. Economy, J. org. Chem., 1962, 27: 3479).
cis-1,3-*Divinylcyclohexane* has been prepared by a Wittig reaction on cis-cyclohexane-1,3-dicarboxaldehyde (G. C. Corfield et al., Chem. Comm., 1966, 238). Alkylidenecyclohexanes have been utilised in a one-carbon ring-expansion reaction to give cycloheptanones. Ethylidenecyclohexane on treatment with cyano-azine gives an intermediate, which on hydrolysis produces 2-methylcycloheptanone (80%) (J. E. McMurry, J. Amer. chem. Soc., 1969, 91: 3676). The stereochemistry of hydroboration (J. Klein and D. Lichtenberg, J. org. Chem., 1970, 35: 2654), and of hydrogenation (heterogeneous and homogeneous catalysts) (T. R. B. Mitchell, J. chem. Soc., B, 1970, 823) of methylenecyclohexanes has been studied. The kinetics of acid-catalysed equilibration of *exo*- and *endo*-cyclic olefins has been examined by D. B. Bigley and R. W. May (ibid., p. 1761).
Photochemical cyclisation of 1,1'-bicyclohexenyl gives *cis*-tricyclo-[6,4,0,0^{2,7}]dodec-1-ene (W. G. Dauben et al., J. Amer. chem. Soc., 1966, 88: 2742).

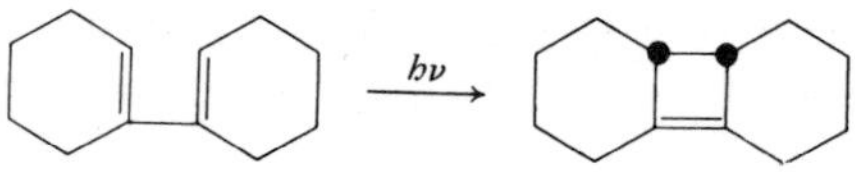

(iii) Cyclohexadienes

Cyclohexa-1,3-dienes may be prepared from cyclohex-2-enones by treatment of the tosylhydrazone with methyl-lithium. In this way 5,5-dimethylcyclohexa-1,3-diene (85%) is obtained from 4,4-dimethylcyclohex-2-enone (W. G. Dauben et al., J. Amer. chem. Soc., 1968, 90: 4762). The reaction between butadiene and acetylene over a copper–zeolite catalyst can give either cyclohexa-1,4-diene or 4-vinylcyclohexene, depending on the reaction conditions (H. Reimlinger, U. Kruerke and E. de Rinter, Ber., 1970, 103: 2317).

2-Methoxycyclohexa-1,3-diene has been prepared by treatment of 3-bromo-1-methoxycyclohexene with potassium tert-butoxide. The diene is stable at −30°, but in chloroform at 20° readily disproportionates into anisole and other products. It is not detected in the Birch reduction of anisole (E. R. de Ward, J. Kattenberg and H. O. Huisman, Tetrahedron Letters, 1970, 4427).

The photolysis of solutions of triethylamine, diethyl ether, and tetramethylethylene, in benzene is found to give 3-substituted cyclohexa-1,4-dienes by 1,4-addition of the solute (D. Bryce-Smith and coworkers, Chem. Comm., 1967, 263, 862; 1971, 794, 915, 916):

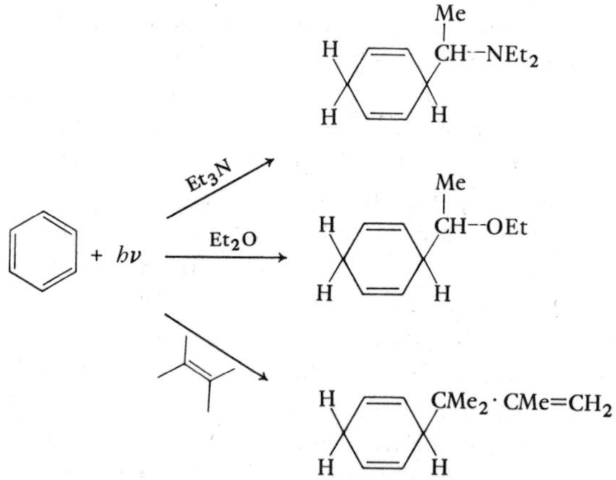

The base-catalysed interconversion of cyclohexadienes, via the cyclohexadienyl anion, has been studied by R. B. Bates, R. H. Carnighan and C. E. Staples. Protonation of the anion at $C_{(3)}$ occurs eight times faster than at $C_{(1)}$ (J. Amer. chem. Soc., 1963, 85: 3032). The disproportionation of cyclohexa-1,3-diene into benzene and cyclohexene is also base-catalysed, and involves hydride transfer in the rate-determining step (J. E. Hofmann, P. A. Argabright and A. Schriesheim, Tetrahedron Letters, 1964, 1005).

The predictions made by R. B. Woodward and R. Hoffmann (Angew. Chem., intern. Edn., 1969, 8: 781) based on the concept of conservation of orbital symmetry have been verified for the thermal transformation of *trans,cis,trans*-octa-2,4,6-triene into *cis*-5,6-dimethylcyclohexa-1,3-diene (E. N. Marvell, G. Caple and B. Schatz, Tetrahedron Letters, 1965, 385); for the photochemical interconversion of the octatriene and *trans*-5,6-dimethylcyclohexa-1,3-diene (G. J. Fonken, ibid., 1962, 549), and for the concerted, thermal elimination of hydrogen from cyclohexa-1,4-diene, but not from cyclohexa-1,3-diene (I. Fleming and E. Wildsmith, Chem. Comm., 1970, 223).

The mercury-sensitised photorearrangement of 1,1,4,4-tetramethylcyclohexa-1,4-diene gives 2,2,6,6-tetramethylbicyclo[3,1,0] hex-3-ene (W. Reusch and D. W. Frey, Tetrahedron Letters, 1967, 5193). 4,4-Dimethylcyclohexa-2,5-dienylidene has been generated photolytically from 1-diazo-4,4-dimethylcyclohexa-2,5-diene, and its addition to alkenes, alkynes, and benzene studied (M. Jones Jr., A. M. Harrison and K. R. Rettig, J. Amer. chem. Soc., 1969, 91: 7462):

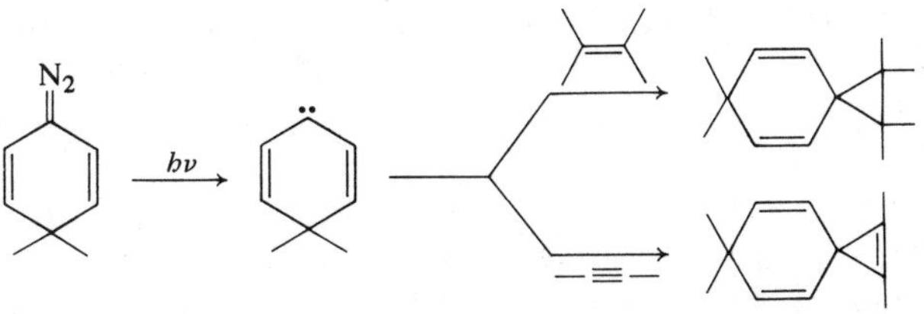

Cyclohexa-1,2-*diene* appears to be the reactive intermediate formed from 6,6-dibromobicyclo[3,1,0] hexane on treatment with methyl-lithium. Unless trapped by styrene, it readily forms a dimer and two tetramers (W. R. Moore and W. R. Moser, J. org. Chem., 1970, 35: 908):

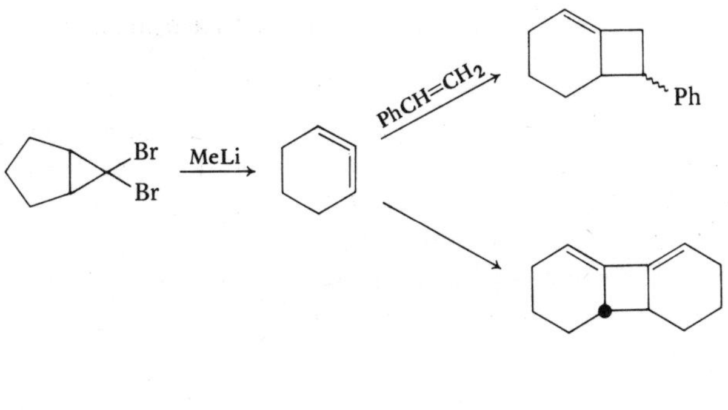

The factors which favour the formation of cyclohexa-1,2-diene rather than cyclohexyne in the elimination of hydrogen halide from 1-halogenocyclo-hexene on treatment with potassium *tert*-butoxide have been analysed by A. T. Bottini et al. (Tetrahedron Letters, 1970, 4753, 4757).

(iv) Alkylidene- and alkenyl-cyclohexenes

3-Methylenecyclohexa-1,4-*diene* has been synthesised by heating the quaternary ammonium salt of 3-aminomethylcyclohexa-1,4-diene with alkali (H. Plieninger and W. Maier-Borst, Angew. Chem., intern. Edn., 1964, 3: 62). The structures of the dimers formed from triphenylmethyl radicals and some diphenylmethyl radicals have been revised. They are now shown to be 3-alkylidenecyclohexa-1,4-dienes, and not substituted ethanes (H. Lankamp, W. Th. Nauta and C. MacLean, Tetrahedron Letters, 1968, 249):

cis-1,2-Divinylcyclohexane and *cis*-1,2-dimethyl-4,5-divinylcyclohexene have been prepared by thermal rearrangement of *cis,trans*-cyclodeca-1,5-diene and *cis,cis,trans*-1,2-dimethylcyclodeca-1,4,8-triene respectively. The cyclo-hexene, on treatment with strong base, is converted into 1,2-diethyl-4,5-dimethylbenzene (P. Heimbach et al., Ann. 1969, 727: 194; Angew. Chem., intern. Edn., 1968, 7: 727). 1-Vinylcyclohexa-1,3-diene has been prepared by elimination of acetic acid from 1,2-diacetoxycyclo-oct-5-ene (W. Ziegenbein, ibid., 1965, 4: 70), and cyclodeca-1,2,6,7-tetraene, on heating at 300°, rearranges to 2,3-divinylcyclohexa-1,3-diene. The latter compound forms adducts with two and three moles of maleic anhydride (L. Skattebol and S. Solomon, J. Amer. chem. Soc., 1965, 87: 4506).

The photorearrangements of substituted 3-methylene-cyclohexenes and -cyclohexa-1,4-dienes have been studied (H. E. Zimmerman and coworkers, ibid., 1967, 89: 5971, 5973; 1971, 93: 3653; W. G. Dauben and W. A. Spilzer, ibid., 1968, 90: 802; H. Hart et al., Chem. Comm., 1968, 1650).

3. Halogeno derivatives of the hydrocarbons

The preferred conformations of *cis*- and *trans*-1,2-dichlorocyclohexane, 1,1,2-trichlorocyclohexane, and α-, β-, γ-, and δ-hexachlorocyclohexane in solution have been determined from dipole moments and molar Kerr constants by K. E. Calderbank, R. J. W. Le Fèvre and R. K. Pieren (J. chem. Soc., B, 1970, 1608). The stereochemistry of the ionic addition of hydrogen bromide, and of the free-radical addition of hydrogen bromide and methane-thiol to 4-*tert*-butylcyclohexene has been studied. Predominantly diaxial addition occurs in all cases (P. D. Readio and P. S. Skell, J. Amer. chem. Soc., 1966, 31: 753, 759). The configuration of the products obtained by the addition of bromine to *trans*-3,4-disubstituted cyclohexenes, under both kinetic and thermodynamic product control, has also been examined (P. L. Barili et al., Chem. Comm., 1970, 1437).

Treatment of cyclohexanones with sodium chlorodifluoroacetate in the presence of tributylphosphine and *N*-methylpyrrolidine, gives a single-step synthesis of difluoromethylenecyclohexanes (46%) (S. A. Fuqua, W. G. Duncan and R. M. Silverstein, Tetrahedron Letters, 1965, 521).

The stereochemistry of some substitution reactions of *trans*-4-*tert*-butylcyclohexyl-lithium has been studied. Carbonation, protonation, and coupling give predominantly retention of configuration, while halogenation gives some inversion (W. H. Glaze et al., J. org. Chem., 1969, 34: 641).

Cyclohex-3-enyl-magnesium bromide when heated in solution above 80° equilibrates with cyclopent-2-enylmethyl-magnesium bromide, the latter predominating (K = 8–9; concentration and solvent dependent). The bicyclo[3,1,0]hex-2-yl derivative is presumed to be an intermediate (A. Maercker and R. Geuss, Angew. Chem., intern. Edn., 1970, 9: 909):

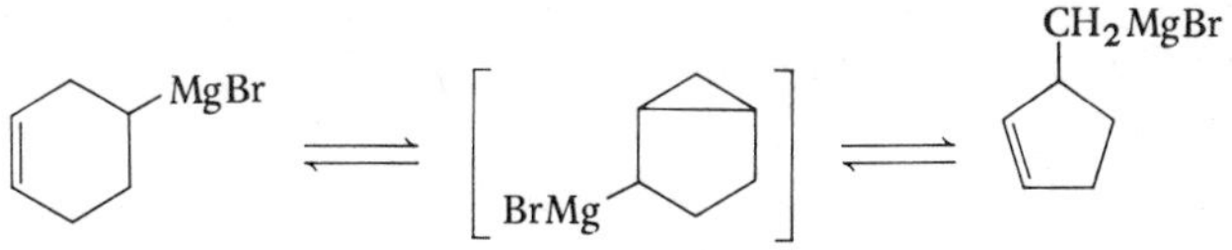

Electrochemical oxidation of iodocyclohexane in aqueous acetonitrile produces *N*-acetylcyclohexylamine (85%) and *N*-acetylcyclohex-2-enylamine via cyclohexyl cations (A. Laurent and R. Tardivel, Compt. rend., 1970, 271C: 324).

4. Alcohols and their derivatives

(a) Alcohols

(i) Monohydric alcohols

The conversion of cyclohexane into cyclohexanol may be effected by the photolysis of pyridazine *N*-oxide in cyclohexane (H. Igeta et al., Chem. and Pharm. Bull., Japan, 1968, 16: 767). Formate esters of cyclohexanols are obtained by cyclising *trans,trans*- and *cis,cis*-octa-2,6-diene in a mixture of formic acid and sulphuric acid. These cyclisations are models for terpene biosynthesis (H. E. Ulery and J. H. Richards, J. Amer. chem. Soc., 1964, 86: 3113).

All four isomers of 2,5-di-*tert*-butylcyclohexanol have been synthesised. Both the *cis,trans*- and the *trans,cis*-isomers exist in the chair conformation, the *cis,cis*-isomer exists in the twist-boat conformation, and the *trans,trans*-isomer exists 34% in the chair and 66% in the twist-boat conformations. These systems have been used to study the effect of conformation on various reactions of the hydroxyl group (D. J. Pasto and D. R. Rao, ibid., 1970, 92: 5151). M. Tichy has reviewed intramolecular hydrogen bonding between OH and other groups e.g. OH, NH_2, S, including cyclohexane and cyclohexene derivatives in several conformations (Adv. org. Chem., 1965, 5: 115). Conformational analysis of 1-methylcyclohexanol shows the hydroxyl group to be in an axial position in the predominant chair conformation ($-\Delta F$ = 0.24 kcal/mole in aqueous dioxane at 75°) (J. J. Uebel and H. W. Goodwin, J. org. Chem., 1968, 33: 3317; N. L. Allinger and C. D. Liang, ibid., 3319). The rate of oxidation of an axial secondary hydroxyl group by bromine is only 20–30% faster than the rate for an equatorial secondary hydroxyl group cf. 250–300% for chromic acid oxidation (I. R. L. Barker, W. G. Overend and C. W. Rees, J. chem. Soc., 1964, 3263).

The direction of elimination from the toluene-*p*-sulphonate esters of cyclohexanols, and from cyclohexylamines has been reviewed by W. Huckel and M. Hanack (Angew. Chem., intern. Edn., 1967, 6: 534). The crystal structure of cyclohexyl toluene-*p*-sulphonate shows a slight flattening of the chair conformer (OTs equatorial) (V. J. James and J. F. McConnell, Tetrahedron, 1971, 27: 5475).

(ii) Dihydric and trihydric alcohols

cis-Hydroxylation of cyclohexene may be accomplished by treatment with boron subchloride, followed by alkaline hydrogen peroxide (M. Zeldin, A. R. Gatti and T. Wartik, J. Amer. chem. Soc., 1967, 89: 4217). Oxymercuration of cyclohex-2-enol gives *trans*-cyclohexane-1,3-diol (50%) (S. Moon and B. H. Waxman, Chem. Comm., 1967, 1283).

The reaction between epichlorohydrin and cyanide ion in aqueous solution provides a novel synthesis of a substituted cyclohexa-1,3-diene (F. Johnson and J. P. Heeschen, J. org. Chem., 1964, 29: 3252):

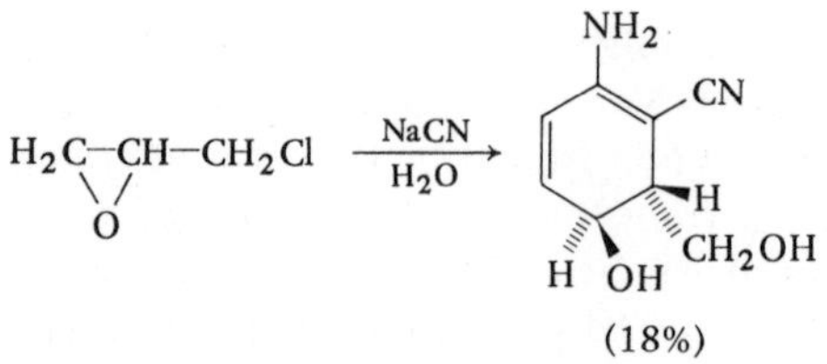

The acid-catalysed pinacol rearrangement of both *cis*- and *trans*-1,2-dimethylcyclohexane-1,2-diol gives 1-acetyl-1-methylcyclopentane (93 and

97%), and only a small amount of 2,2-dimethylcyclohexanone (7 and 3% respectively) (C. A. Bunton and M. D. Carr, J. chem. Soc., 1963, 5854).

The first synthesis of the four cyclohexane-1,2,4-triols, starting from cyclohex-3-enol, has been accomplished by G. E. McCasland, M. O. Naumann and L. J. Durham (J. org. Chem., 1966, 31: 3079; see also refs. therein) who have also confirmed the configurations of all nine cyclo-hexanetriols (Table 1) by n.m.r. studies.

Table 1

CYCLOHEXANETRIOLS[1]

Compound	m.p.($^\circ$C)	Tribenzoate m.p.($^\circ$C)
1,2,3-Cyclohexanetriol	148	142
1,2/3-Cyclohexanetriol	125	181
1,3/2-Cyclohexanetriol	108	142
1,2,4-Cyclohexanetriol	138	
1,2/4-Cyclohexanetriol	161	129
1,4/2-Cyclohexanetriol	138	116
2,4/1-Cyclohexanetriol	150	154
1,3,5-Cyclohexanetriol	185	
1,3/5-Cyclohexanetriol	145	

Reference

1   G. E. McCasland, M. O. Naumann and L. J. Durham, J. org. Chem., 1966, 31: 3079.

(iii) Polyhydric alcohols (tetrols, pentols, and hexols)

Racemic conduritol-B (1,3/2,4-cyclohex-5-enetetrol) has been synthesised in three steps (overall yield 24%) from 1,4,5,6-tetra-O-acetyl-*myo*-inositol, using the Corey—Winter method for converting 1,2-diols into olefins (T. L. Nagabhushan, Canad. J. Chem., 1970, 48: 383):

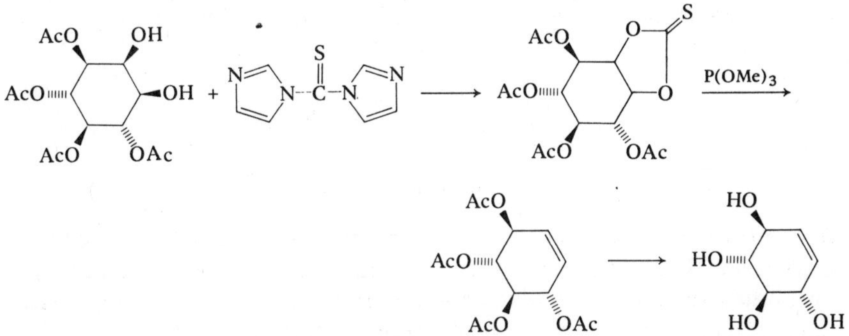

The crystal structures of *epi*-inositol (G. A. Jeffrey and H. S. Kim, Carbohydrate Res., 1970, 15: 310), and *myo*-inositol (T. R. Lomer, A. Miller and C. A. Beevers, Acta Cryst., 1963, 16: 264) have been reported.

On treatment with potassium *tert*-butoxide in dimethylsulphoxide, tetra-O-alkyl-1,2-O-isopropylidene-*myo*-inositols are converted into 1,2,4-tri-alkoxybenzenes, via 3,4,5,6-tetra-alkoxycyclohex-2-enols (P. A. Gent and R. Gigg, J. chem. Soc., C, 1970, 2253).

The biosynthesis of cyclitols has been reviewed by H. Kindl, R. Scholda and O. Hoffmann-Ostenhof (Angew. Chem., intern. Edn., 1966, 5: 165). The biosynthesis of spectinomycin, a broad spectrum antibiotic, from D-glucose has been studied using ^{14}C and ^{3}H labelling. *myo*-Inositol was shown to be an intermediate (L. A. Mitscher et al., Chem. Comm., 1971, 1541).

(b) Epoxycyclohexanes

Treatment of *trans*-cyclohexane-1,2-diol with the dimethyl acetal of di-methylformamide gives 1,2-epoxycyclohexane (88%) in one step (N. Neumann, Chimia, Switz., 1969, 23: 267). The epoxide is converted into pure *cis*-1,2-dichlorocyclohexane (70%) on treatment with sulphuryl chloride in chloroform, in the presence of pyridine (J. R. Campbell, J. K. N. Jones and S. Wolfe, Canad. J. Chem., 1966, 44: 2339).

The reaction of methyl organometallic compounds with 3,4- and 4,5-epoxycyclohex-1-ene has been studied (J. Staroscik and B. Rickborn, J. Amer. chem. Soc., 1971, 93: 3046; D. M. Wieland and C. R. Johnson, ibid., p. 3047). Lithium dimethyl cuprate adds to 3,4-epoxycyclohex-1-ene to give both *trans*-2-methylcyclohex-3-en-1-ol and *trans*-4-methylcyclohex-2-en-1-ol in roughly equal amounts. Methyl-lithium reacts with both 3,4- and 4,5-epoxycyclohex-1-ene to give cyclohexa-2,4-dienol ('*benzene hydrate*', b.p. 36°/2 mm) as the major product (63 and 75% respectively). The product may be kept for several weeks at −15°. It gives Diels—Alder adducts.

'Benzene hydrate'

Determination of the crystal structure of crotepoxide iodohydrin (P. Coggan, A. T. McPhail and G. A. Sim, J. chem. Soc., B, 1969, 534) has established the constitution and absolute stereochemistry of *crotepoxide*, a tumour-inhibiting compound isolated from *Croton macrostachys*:

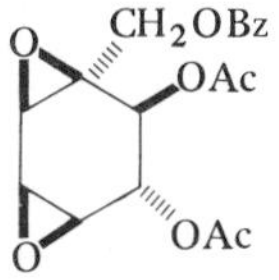

Crotepoxide

(c) Peroxides

Cyclohexa-1,3-diene endoperoxide, **norascaridole**, rearranges both thermally and photochemically to give mixtures of 1,2:3,4-bisepoxycyclohexane and 3,4-epoxycyclohexanone. The latter rearranges further to 4-hydroxycyclo-hex-2-enone. *Ascaridole* on photolysis in cyclohexane gives the corresponding bisepoxide (iso-ascaridole; 85%) (K. K. Maheshwari, P. de Mayo and D. Wiegand, Canad. J. Chem., 1970, 48: 3265).

The mode of formation of di- and tri-cyclohexylidene peroxides from cyclohexanone has been further studied (P. R. Storey et al., J. org. Chem., 1970, 35: 3059).

5. Nitro- and amino-cyclohexanes

(a) Nitro compounds

Methods of synthesis of aliphatic and alicyclic nitro compounds have been reviewed by N. Kornblum (Org. Reactions, 1962, 12: 101). Cyclisation of glutaraldehyde with nitromethane gives *trans,trans*-2,6-dihydroxy-1-nitrocyclohexane (50%). On catalytic hydrogenation, this is converted into *trans,trans*-2,6-dihydroxycyclohexylamine (F. W. Lichtenthaler, Angew. Chem., intern. Edn., 1964, 3: 211).

(b) Amines

The formation of cyclohexylamines from cyclohexenes or cyclohexanols by the Ritter reaction (H_2SO_4/HCN) has been reviewed (L. I. Krimen and D. J. Cota, Org. Reactions, 1969, 17: 213).

cis-1,2-*Diaminocyclohexane* (di-HCl, m.p. 312°) has been synthesised from 1,2-epoxycyclohexane via *trans*-2-azidocyclohexanol and *cis*-1,2-di-azidocyclohexane (G. Swift and D. Swern, J. org. Chem., 1967, 32: 511). Both cis,cis- and cis,trans-1,3,5-*triaminocyclohexane* are obtained on reduction of cyclohexane-1,3,5-trione trioxime with sodium in liquid ammonia. Only the *cis,cis*-isomer forms a *bis*-complex with Co[III] and Rh[III], in which the ligand has the all—axial conformation (R. A. D. Wentworth and J. J. Felton, J. Amer. chem. Soc., 1968, 90: 621).

The syntheses of the *cis-* and the *trans-*isomers of 2-hydroxymethylcyclohexylamine and 2-aminomethylcyclohexanol (G. Bernath, K. Kovacs and K. L. Lang, Acta Chim., Budapest, 1970, 64: 183), and of 4-hydroxymethylcyclohexylamine (W. Schneider and K. Lehmann, Tetrahedron Letters, 1970, 4285) have been reported.

The pKa values of *cis-* and *trans-*4-*tert-*butylcyclohexylamine (9.24 and 9.50 respectively, in 80% aqueous methylcellosolve at 20°) indicate that an equatorial amino-group is more basic than an axial amino-group. Cyclohexylamine has pKa 9.51 under the same conditions (M. Tichy, J. Jonas and J. Sicher, Coll. Czech. Chem. Comm., 1959, 24: 3434). The pKa values for the three isomers of 2,6-dimethylcyclohexylamine, and their *N,N*-dimethyl derivatives, in aqueous 2-methoxyethanol have also been determined (G. Bellucci, F. Macchia and M. Poggianti, Gazz., 1969, 99: 1217).

Thermodynamic parameters for the dissociation of *cis-* and *trans-*1,2-diaminocyclohexane, and for complex formation with several metal cations, in aqueous solution between 10 and 40° have been determined (C. R. Bertsch, W. C. Fernelius and B. P. Block, J. phys. Chem., 1958, 62: 444). The pKa values at 20° for the *cis-*isomer are 9.93 and 6.13, and for the *trans-*isomer are 9.94 and 6.47.

Elimination from quaternary ammonium compounds and amine *N*-oxides in cyclohexane systems is included in a review by A. C. Cope and E. R. Trumbull (Org. Reactions, 1960, 11: 317). Hofmann elimination in cyclohexyl derivatives has been shown to occur by 4% *syn-* and 96% *anti-*elimination (M. P. Cooke Jr. and J. L. Coke, J. Amer. chem. Soc., 1968, 90: 5556).

Dicyclohexylcarbodiimide, $C_6H_{11} \cdot N=C=N \cdot C_6H_{11}$, is widely used as a mild dehydrating agent, and to effect condensation reactions e.g. peptide and nucleotide synthesis, and cyclisation reactions, in which water is eliminated. During these reactions, the diimide is converted into *N,N'-*dicyclohexylurea. The diimide (m.p. 34°) is prepared by either the action of toluene-*p*-sulphonyl chloride and pyridine on *N,N'-*dicyclohexylurea (82%), or of mercuric oxide on the corresponding thiourea (86%) (F. Kurzer and K. Douraghi-Zadeh, Chem. Reviews, 1967, 67: 107).

6. Alicyclic aldehydes

Formylcyclohexane is obtained in high yield on treating cycloheptene with thallic nitrate in methanol (A. McKillop et al., Tetrahedron Letters, 1970, 5275). Cyclohexanone can be converted into 1-allylcyclohexanecarboxaldehyde (65%) by (i) conversion into cyclohexylidenemethyl allyl sulphide by reaction with $CH_2=CH\cdot CH_2\cdot S\cdot CH^-\cdot PO(OEt)_2$, and (ii) thio-Claisen rearrangement at 190° in the presence of mercuric oxide (E. J. Corey and J. I. Shulman, J. Amer. chem. Soc., 1970, 92: 5522):

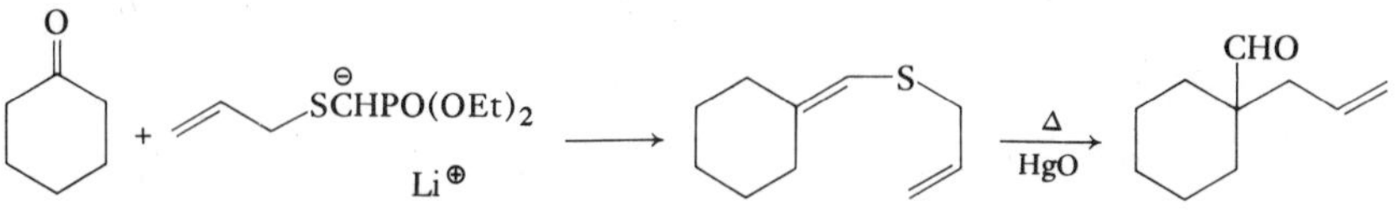

The stereochemistry of the intramolecular Aldol condensation shown below, has been investigated (A. T. Nielsen, J. org. Chem., 1965, 30: 3650):

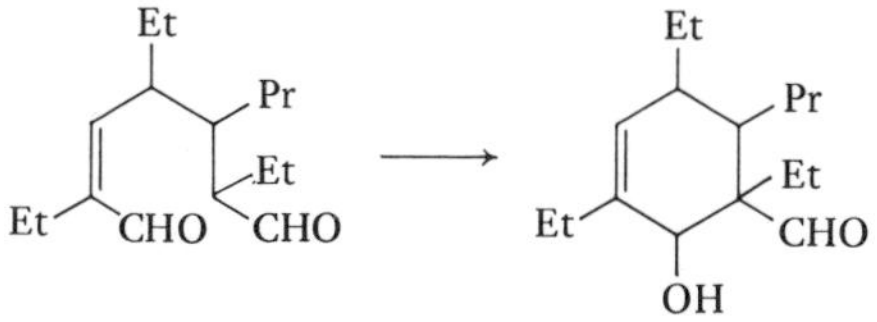

7. Alicyclic ketones

(a) Saturated ring ketones

(i) Cyclohexanones

A synthesis of 4-substituted 4-methylcyclohexanones in five steps from p-cresol (overall yield > 50%) involves electrophilic substitution by bromine (which may be replaced later by a methoxyl, acetoxyl or hydroxyl group) assisted by a neighbouring carboxylate ion (E. J. Corey, S. Barcza and G. Klotmann, J. Amer. chem. Soc., 1969, 91: 4782):

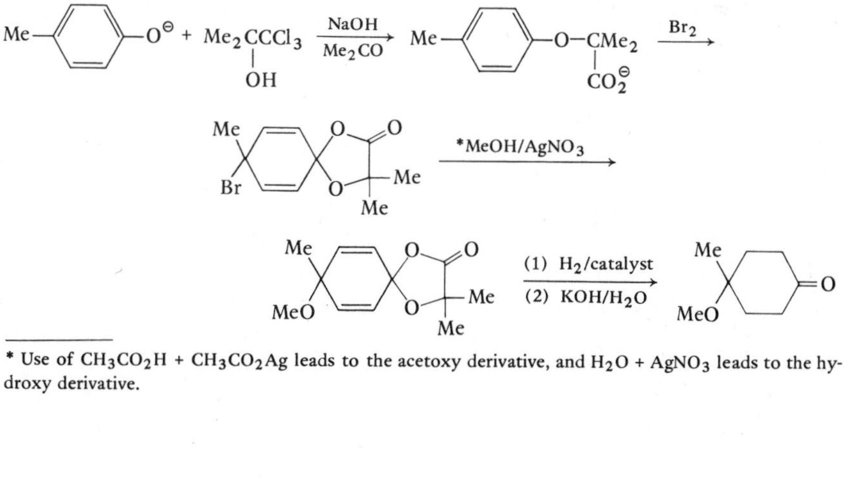

---

* Use of $CH_3CO_2H + CH_3CO_2Ag$ leads to the acetoxy derivative, and $H_2O + AgNO_3$ leads to the hydroxy derivative.

Ring expansion of alkylidenecyclopentanes, via the bromohydrin and the corresponding alkoxymagnesium bromide, gives 2-alkylcyclohexanones in good yield (A. J. Sisti, J. org. Chem., 1970, 35: 2670):

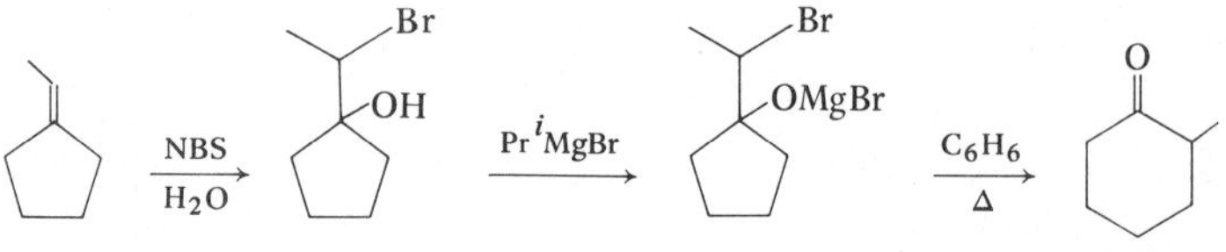

(65%)

The propane-1,3-dithiol ketals of cyclohexanone and cyclohexane-1,3-dione can be prepared by di-alkylation of 1,3-dithiane (D. Seebach, N. R. Jones and Corey, ibid., 1968, 33: 300):

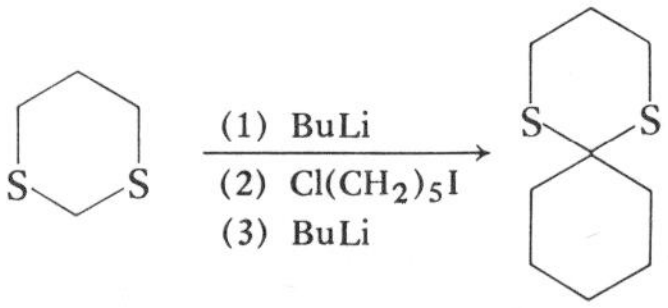

A conversion of *cis*-2,6-dialkylcyclohexanones into the less stable *trans*-isomer utilises the $A^{(1,3)}$ strain present in the semicarbazone of the *cis*-isomer (F. Johnson and L. G. Duquette, J. chem. Soc., D, 1969, 1448).

Methods for preparing 2-deuterio- and 2,2,6,6-tetradeuterio-cyclohexanone from the morpholine enamine of cyclohexanone (J. P. Schaefer and D. S. Weinberg, Tetrahedron Letters, 1965, 1801), 2,2-dideuterio-cyclohexanone from cyclohexan-1-one-2-carboxylic acid (J. Deutsch and A. Mandelbaum, ibid., 1969, 1351), and 3-deuterio-alkylcyclohexanones from alkylcyclohex-2-enones (M. Fetizon and J. Gore, ibid., 1966, 471) have been described.

2-Alkylcyclohexanones can be prepared in high yield by the alkylation of the quaternary derivative of 2-aminomethylcyclohexanone using a trialkyl-boron (H. C. Brown et al., J. Amer. chem. Soc., 1968, 90: 4166). Alkylation of cyclohexanone enamines gives predominantly axial 2-alkylation (S. Karady, M. Lenfant and R. E. Wolff, Bull. Soc. chim. Fr., 1965, 2472). The stereochemistry of the alkylation of the enolate ion of 4-*tert*-butylcyclohexanone has been investigated by H. O. House, B. A. Tefertiller and H. D. Olmstead (J. org. Chem., 1968, 33: 935, and refs. therein). Polyalkylation in the alkylation of the enolate ions of cyclohexanones can be largely eliminated by the addition of tributyltin chloride or triethylaluminium to the reaction mixture (P. A. Tardella, Tetrahedron Letters, 1969, 1117). Alternatively, the use of lithium enolate ions, generated by the reaction of enol acetates with methyllithium, gives good yields of mono-alkylated products (H. O. House and B. M. Trost, J. org. Chem., 1965, 30: 2502). The relative rates of alkylation of the enolate ions of 2,2- and 2,6-dimethyl-cyclo-hexanone have been studied (D. Caine and B. J. L. Huff, Tetrahedron Letters, 1967, 3399). Both G. Stork and P. F. Hudrlik (J. Amer. chem. Soc., 1968, 90: 4462, 4464), and H. O. House et al. (J. org. Chem., 1969, 34: 2324) have trapped the enolate ions of cyclohexanones as their trimethyl-silyl ethers in order to investigate the ratio of isomeric ions formed under both kinetically and thermodynamically controlled deprotonation. A useful method of controlling the position of alkylation involves alkylation of the dianion of a 1,3-dicarbonyl compound. Thus 2-formyl-6-methylcyclo-hexanone gives 2,2-dimethylcyclohexanone (60%) after alkylation and hydrolysis. This method has been reviewed by T. M. Harris and C. M. Harris (Org. Reactions, 1969, 17: 155):

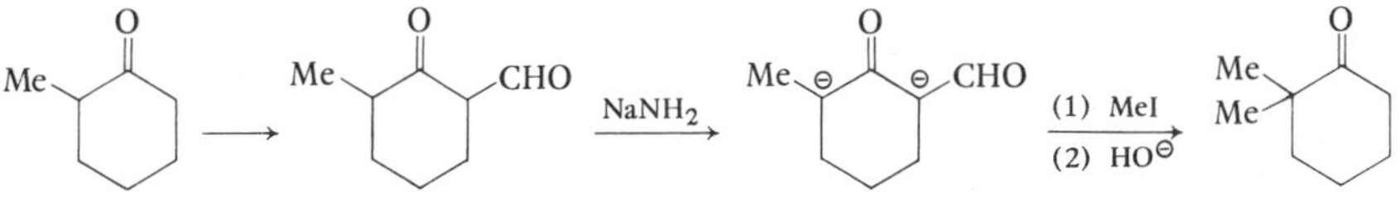

Procedures for controlling the stereochemistry of the Robinson annelation reaction, e.g. 2-methylcyclohexanone + *trans*-pent-3-en-2-one → $\Delta^{1,9}$-4,10-dimethyldecal-2-one, have been described (C. J. V. Scanio and R. M. Starrett, J. Amer. chem. Soc., 1971, 93: 1539). The usual conditions for this reaction are basic, but acid conditions may also be used (C. H. Heathcock et al., Tetrahedron Letters, 1971, 4995).

The mechanism of the acid-catalysed hydrolysis of 1-methoxycyclo-hexene, and the enolisation of cyclohexanone, have been studied (G. E. Lienhard and T.-C. Wang, J. Amer. chem. Soc., 1969, 91: 1146).

The concepts of 'steric approach control' and 'product development control' put forward by W. G. Dauben, G. J. Fonken and D. S. Noyce (J. Amer. chem. Soc., 1956, 78: 2579) to explain the stereochemistry of carbonyl addition, have been critically examined with respect to the addition to cyclohexanones. The former, but not the latter, concept appears relevant (E. L. Eliel and Y. Senda, Tetrahedron, 1970, 26: 2411 and refs. therein; B. Rickborn and M. T. Wuesthoff, J. Amer. chem. Soc., 1970, 92: 6894, and refs. therein; P. R. Jones, E. J. Goller and W. J. Kaufmann, J. org. Chem., 1969, 34: 3566, and refs. therein; M. Cherest and H. Felkin, Tetrahedron Letters, 1968, 2205). The ratio of isomeric products formed by the addition of 3-*tert*-butylallyl-magnesium bromide to ketones has been used to measure the amount of steric hindrance in the region of the carbonyl group in the transition state for addition. The products formed from 4-*tert*-butylcyclo-hexanone indicate that steric strain is the important factor impeding axial attack, and that torsional strain is the important factor impeding equatorial attack. The net stereochemistry of hydride and Grignard additions to simple cyclohexanones is suggested to result from a competition between these two effects (Cherest, Felkin and C. Frajerman, ibid., 1971, 379, 383):

$$Bu^t \cdot CH{=}CH \cdot CH_2 MgBr + R_2 C{=}O \rightarrow$$

$$R_2 C(OH)CH_2 \cdot CH{=}CH \cdot Bu^t + R_2 C(OH)CHBu^t \cdot CH{=}CH_2$$

Alkylation of 4-*tert*-butylcyclohexanone with $Me_3 Al$ in benzene using a reactant ratio of less than two gives mainly the *trans*-alcohol (~90%), in contrast to alkylation with $Me_2 Zn$, $Me_2 Cd$, $Me_2 Mg$, or MeMgX, where the major product is the *cis*-alcohol (~75%) (E. C. Ashby and S. Yu, Chem. Comm., 1971, 351).

The conversion of 4-*tert*-butylcyclohexanone into 4-*tert*-butylmethylene-cyclohexane oxide occurs with equatorial methylene transfer if dimethyl-oxosulphonium methylide is used, and predominantly axial methylene transfer if dimethylsulphonium methylide is used. The oxides are reduced to the corresponding 4-*tert*-butyl-1-methylcyclohexanols with lithium aluminium hydride (E. J. Corey and M. Chaykovsky, J. Amer. chem. Soc., 1965, 87: 1353).

The direction of elimination of water from the *cis*- and *trans*-isomers of the cyanohydrin of 2-methylcyclohexanone by (a) treatment with phosphorus oxychloride/pyridine, and (b) pyrolysis of the corresponding acetate, has been studied. 2-Methylcyclohexene- and 6-methylcyclohexene-nitrile were obtained (T. Holm, Acta chem. Scand., 1964, 18: 1577).

Oxidation of cyclohexanone with oxygen in alkaline hexamethylphosphoramide gives adipic acid in high yield (T. J. Wallace, H. Pobiner and A. Schriescheim, J. org. Chem., 1965, 30: 3768), while oxidation of cyclohexanones with lead tetra-acetate gives 2,6-diacetoxy-derivatives, which after reduction with lithium aluminium hydride and sodium periodate oxidation, produces substituted glutaraldehydes (G. W. K. Cavill and D. H. Solomon, Austral. J. Chem., 1960, 13: 121).

(ii) Cyclohexanediones

δ-Ketocarboxylic acids are converted into cyclohexane-1,3-diones on treatment with an acid chloride and aluminium trichloride in nitromethane e.g. 5-oxoheptanoic acid gives 2-methylcyclohexane-1,3-dione (35%) (H. Schick, G. Lehmann and G. Hilgetag, Ber., 1967, 100: 2973).

Photocycloaddition of cyclohexene with dimedone gives 4,4-dimethylbicyclo[6,4,0]dodecane-2,6-dione, after a retro-Aldol reaction of the cyclobutane intermediate (B. D. Challand et al., J. org. Chem., 1969, 34: 794):

2-Methylcyclohexane-1,3-dione (and 2-ethoxycarbonylcyclohexanone) can be annelated with 1,4-dichlorobutan-2-one. The product is a 1,9-epoxydecal-2-one (S. Danishefsky and G. A. Koppel, Chem. Comm., 1971, 367):

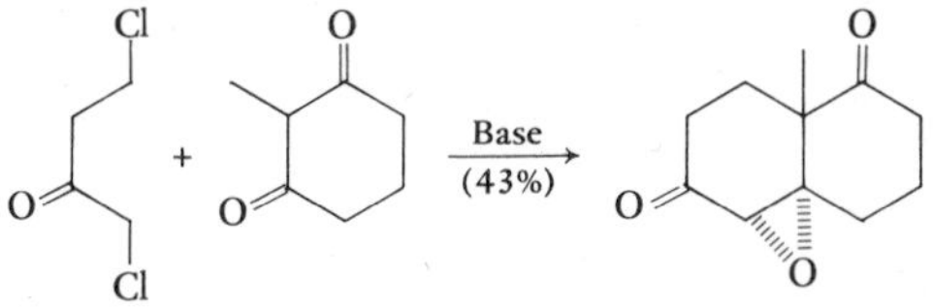

The ring-contraction via cyclopropanols, which occurs during the Clemmensen reduction of cyclohexane-1,3-diones and cyclohex-2-enones, has been reviewed (J. G. St. C. Buchanan and P. D. Woodgate, Quart. Review, 1969, 23: 522):

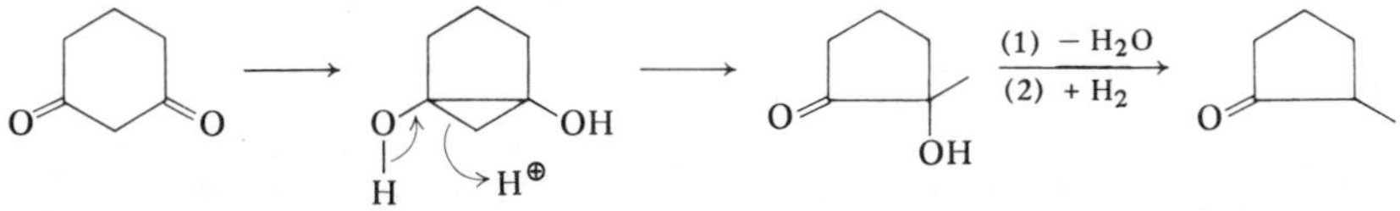

Chlorination of 2-alkylcyclohexane-1,3-diones, followed by treatment with base gives Favorskii type intermediates which decarbonylate to 2-alkylcyclopent-2-enones (G. Buchi and B. Egger, J. org. Chem., 1971, 36: 2021):

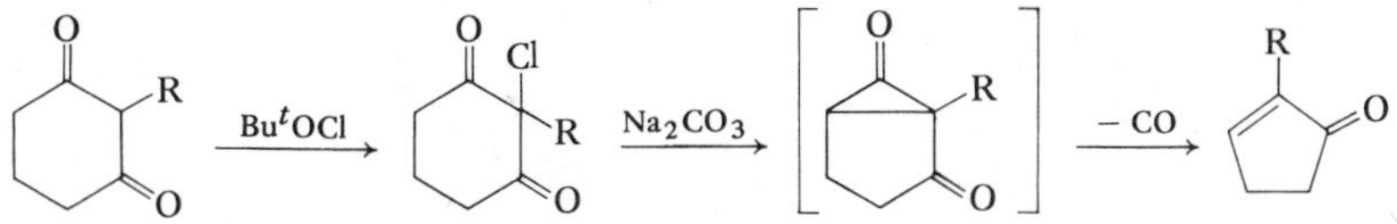

(b) Unsaturated ring ketones

Michael addition of tetrahydropyran-2-yloxycyclohexadienes with αβ-unsaturated ketones leads to the synthesis of 2-(3-oxo-alkyl)-cyclohex-2-enones in high yield (A. J. Birch, J. Diekman and P. L. MacDonald, Chem. Comm., 1970, 52). The corresponding 2-substituted cyclohexanones are obtained by a similar route:

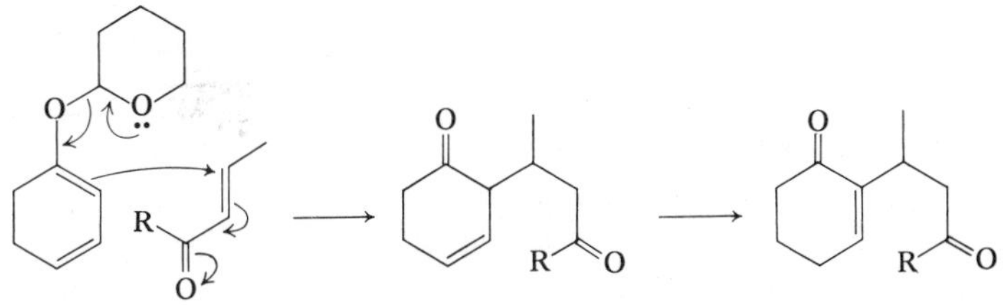

4-(3-Oxobutyl)cyclohex-2-enones (60%) may be synthesised by the Diels–Alder addition of 1-methoxycyclohexa-1,3-dienes with but-3-en-2-ones, followed by acid cleavage of the bicyclic intermediate (Birch and J. S. Hill, J. chem. Soc., C, 1966, 419):

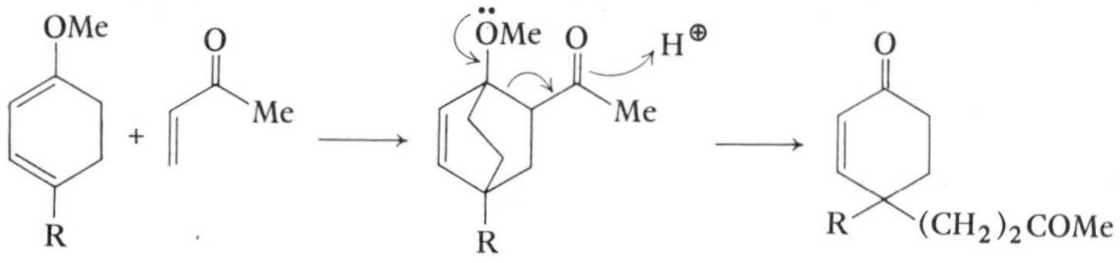

The photo-induced reaction between vinylcyclopropanes and $Fe(CO)_5$ or $Fe_2(CO)_9$ gives 3-substituted cyclohex-2-enones (R. Victor, R. Ben-Shoshan and S. Sarel, Tetrahedron Letters, 1970, 4253).

4-Ethylcyclohex-2- and -3-enone have been synthesised via Birch reduction of 4-ethylanisole. The equilibrated mixture of the two isomers contains 30% of the non-conjugated species (K. G. Lewis and G. J. Williams, Austral. J. Chem., 1970, 23: 807) cf. equilibrated cyclohexenone which contains only 1% of the non-conjugated isomer (N. Heap and G. H. Whitham, J. chem. Soc., B, 1966, 164). With larger alkyl groups at $C_{(4)}$, the proportion of the non-conjugated isomer increases further being 50% for 4-tert-butylcyclohexenone (Lewis and Williams, Tetrahedron Letters, 1965, 4573).

The extent and stereochemistry of the conjugate addition of organometallic reagents to cyclohex-2-enones has been studied (H. O. House, R. A. Latham and C. D. Slater, J. org. Chem., 1966, 31: 2667; House and W. F. Fischer Jr., ibid., 1968, 33: 949). House et al. have also studied the stereochemistry of dissolving metal reductions of 4-tert-butylcyclohex-2-enones and 4-tert-butylcyclohexylideneacetone (J. Amer. chem. Soc., 1970, 92: 2800). Hydroboration of cyclohex-2-enones gives trans-diequatorial cyclohexane-1,2-diols (J. Klein and E. Dunkelblum, Tetrahedron Letters, 1966, 6047).

In the base-catalysed deuterium exchange of 6,6-dimethylcyclohex-2-enone, the rate of exchange of the $C_{(2)}$ vinylic proton is much faster than that of the $C_{(4)}$ allylic protons when $DO^{\ominus}/D_2O$ is used, but the two rates are equal when $CD_3O^{\ominus}/CD_3OD$ is used. Exchange of the $C_{(2)}$ vinylic proton occurs via Michael addition of the base (J. Warkentin and L. K. M. Lim, Canad. J. Chem., 1964, 42: 1676).

Three further basic structures for the products of base-catalysed dimerisation of cyclohex-2-enones have been established. Cyclohex-2-enone itself, on treatment with 10% aqueous sodium hydroxide, gives 2-(3-oxocyclohexyl)-cyclohex-2-enone, but treatment with pyrrolidinium perchlorate followed by aqueous sodium hydroxide gives a tricyclic saturated diketone (N. J. Leonard and W. J. Musliner, J. org. Chem., 1966, 31: 639; see also N.

Schamp, E. de Bundel and E. Delarue, Bull. Soc. chim. Belg., 1966, 75: 230). 3,5,5-Trimethylcyclohex-2-enone on treatment with 60% aqueous sodium hydroxide at 150° gives a tricyclic unsaturated hydroxyketone (G. Kabas and H. C. Rutz, Tetrahedron, 1966, 22: 1219):

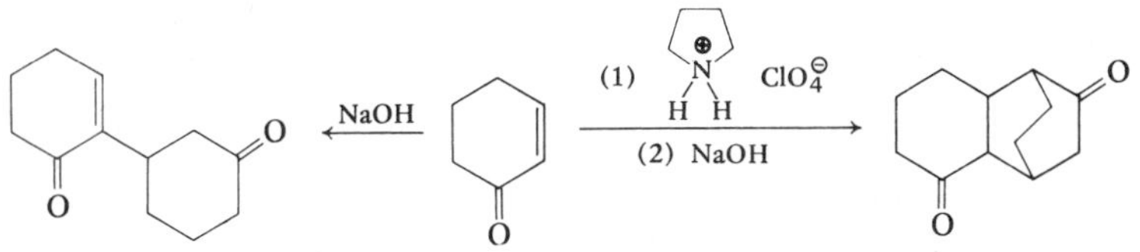

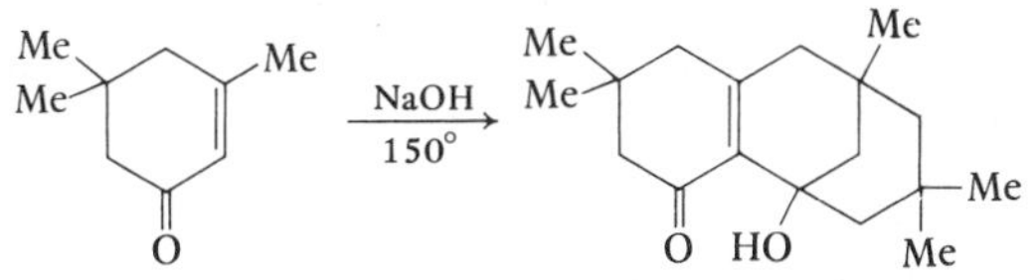

In the solvent system HF—SbF$_5$, cyclohex-2-enone is completely transformed into the O-protonated species. Further protonation at the double bond results in rearrangement of the di-cation to the di-cation of 3-methylcyclopent-2-enone; the free cyclopentenone can be isolated (75%) by pouring into ice (H. Hogeveen, Rec. Trav. chim., 1968, 87: 1295; see also T. S. Sorensen, J. Amer. chem. Soc., 1969, 91: 6398).

The photo-dimerisation, -cycloaddition, and -rearrangement of cyclohexenones has been reviewed by J. S. Swenton (J. chem. Educ., 1969, 46: 217), P. E. Eaton (Acc. chem. Res., 1968, 1: 50), P. de Mayo (ibid., 1971, 4: 41) and P. G. Bauslaugh (Synthesis, 1970, 287).

Reviews dealing with the complete chemistry of cyclohexa-2,4- and -2,5-dienones (A. J. Waring, Adv. alicyclic Chem., 1966, 1: 129), and the chemistry of *ortho*- and *para*-quinone methides (6-methylenecyclohexa-2,4-dienones and 4-methylenecyclohexa-2,5-dienones respectively) (A. B. Turner, Quart. Review, 1964, 18: 347) have been published. A one-step synthesis of substituted quinone methides involving the reaction of secondary amines with *ortho*- or *para*-hydroxybenzaldehydes is reported (L. P. Olekhonovich, L. E. Nivorochkin and V. I. Minkin, Zhur. org. Khim., 1968, 4: 1615).

Cyclohex-2-ene-1,4-dione (m.p. 54°) can be prepared by acid-catalysed hydrolysis of its mono-ketal. It is stable in the crystalline state at 0°, but in solution it tautomerises to hydroquinone (E. W. Garbisch Jr., J. Amer. chem. Soc., 1965, 87: 4971).

(c) Exocyclic unsaturated ring ketones

2-Alkylidenecyclohexanones may be synthesised in high yield by treatment of the boron difluoride complex of 2-formylcyclohexanone with an alkyl-lithium (R. A. J. Smith and T. A. Spencer, J. org. Chem., 1970, 35: 3220). 4-Methylenecyclohexanone has been obtained by photorearrangement of spiro[2,4]heptan-5-one (A. Sonoda et al., Tetrahedron Letters, 1969, 3187).

The isomerisation of 2-alkylidenecyclohexanones to 2-alkylcyclohex-2-enones using polyphosphoric acid has been studied by J. M. Conia and P. Amice (Bull. Soc. chim. Fr., 1970, 2972).

(d) Exocyclic ketones

Cyclohexyl ketones are formed by the thermal, intramolecular ene-reaction of enols. For example, a 1:1 mixture of cis- and trans-1-acetyl-2-methyl-cyclohexane is obtained in 80% yield by heating non-8-en-2-one at 360°. These reactions have been reviewed by J. M. Conia (Bull. Soc. chim. Fr., 1968, 3057):

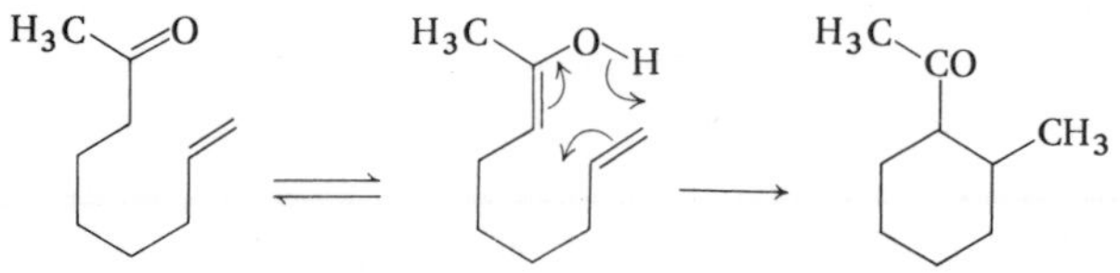

Another general method for preparing cyclohexyl ketones (70%) involves the Claisen rearrangement of 3,4-dihydro-2H-pyran-2-ylethylenes (G. Buchi and J. E. Powell Jr., J. Amer. chem. Soc., 1970, 92: 3126):

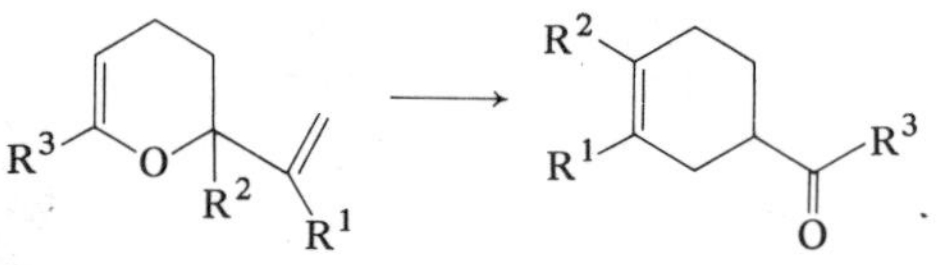

Treatment of cyclohex-2-enone with 2-lithio-2-methyl-1,3-dithiane gives initially 2-(1-hydroxycyclohex-2-enyl)-2-methyl-1,3-dithiane, which with acid rearranges to 2-(3-hydroxycyclohex-1-enyl)-2-methyl-1,3-dithiane. The latter may be hydrolysed to give 3-acetylcyclohex-2-en-1-ol (overall yield 53%), or oxidised and hydrolysed to give 3-acetylcyclohex-2-enone (overall yield 55%) (E. J. Corey and D. Crouse, J. org. Chem., 1968, 33: 298).

2-Acylcyclohexanones, on treatment with slightly acidic hydrogen perox-ide in tert-butyl alcohol, undergo ring-contraction to cyclopentanecarboxylic acids (80%) (G. B. Payne, ibid., 1961, 26: 4793):

8. Carboxylic acids

(i) Cyclohexanecarboxylic acids

All seven isomeric *tert*-butylcyclohexanecarboxylic acids have been synthesised, the *cis*- and the *trans*-isomers of 2-, 3-, and 4-*tert*-butylcyclohexanecarboxylic acid by the catalytic hydrogenation of the corresponding *tert*-butylbenzoic acids, and 1-*tert*-butylcyclohexanecarboxylic acid by treatment of 4-*tert*-butylcyclohexanol with formic acid/sulphuric acid in carbon tetrachloride (H. van Bekkum et al., Rec. Trav. chim., 1970, 89: 521).

meso-trans-1,3-Dimethylcyclohexane-2-acetic-1,3-dicarboxylic acid has been synthesised from 1,5-dimethylbicyclo[3,3,1]non-2-en-9-one, and shown to be identical with the tricarboxylic acid obtained by drastic oxidation of most diterpenoid resin acids (J. Martin, W. Parker and R. A. Raphael, Chem. Comm., 1965, 633).

cis- and *trans*-4-*tert*-Butylcyclohexanecarboxylic acid have been converted into seven other derivatives i.e. $CONH_2$, NH_2, CN, NH·CO·Me, CH_2OH, CHO, CO·Me (P. J. Beeby and S. Sternhell, Austral. J. Chem., 1970, 23: 1005).

The dissociation constant for an axial carboxyl group is lower than that for an equatorial carboxyl group. In 80% aqueous methylcellosolve at 20°, the pK values for *cis*- and *trans*-4-*tert*-butylcyclohexanecarboxylic acid are 7.91 and 7.43 respectively; the pK value for cyclohexanecarboxylic acid is 7.43 (M. Tichy, J. Jonas and J. Sicher, Coll. Czech. chem. Comm., 1959, 24: 3434). The difference is attributed to steric hindrance to solvation of the anion of the axial carboxyl group by 1,3-diaxial interactions (W. Simon, Angew. Chem., intern. Edn., 1964, 3: 661).

The thermal decarboxylation of *cis*- and *trans*-4-*tert*-butylperoxocyclohexanecarboxylic acid has been studied. The reaction of the *cis*-isomer is four times faster than that of the *trans*-isomer, but the ratio of 4-*tert*-butylcyclohexanols formed is identical for both isomers, the *cis*-cyclohexanol predominating (M. Gruselle, J. Fossey and D. Lefort, Tetrahedron Letters, 1970, 2069).

The X-ray crystal structure of *trans*-cyclohexane-1,2-dicarboxylic acid shows the molecules to be in the di-equatorial chair conformation (E. Benedetti, P. Corradini and C. Pedone, J. Amer. chem. Soc., 1969, 91: 4075).

(*ii*) *Unsaturated carboxylic acids*

The reaction of dimethyl *cis*-1,2-dichloro-3-vinylcyclobutane-1,2-dicarboxylate and of dimethyl *cis*-1,2-dichloro-3-methyl-3-isopropenylcyclobutane-1,2-dicarboxylate with $Ni(CO)_4$ in benzene–dimethylformamide gives dimethyl cyclohexa-1,4-diene-1,2-dicarboxylate and dimethyl 4,5-dimethylcyclohexa-1,4-diene-1,2-dicarboxylate (70%) respectively (H.-D. Scharf and F. Korte, Ber., 1966, 99: 3925).

The ionic addition of hydrogen bromide to cyclohex-1-enecarboxylic acid gives *cis*-2-bromocyclohexanecarboxylic acid as the kinetic product, and the *trans*-isomer as the thermodynamic product (W. R. Vaughan and R. Caple, J. Amer. chem. Soc., 1964, 86: 4928).

(*iii*) *Exocyclic carboxylic acids*

The reaction of 4-vinylcyclohexene with carbon monoxide in methanol at 300–700 atm. pressure, in the presence of $(Ph_3P)_2PdCl_2$ at 60°, gives methyl 2-(cyclohex-3-enyl)propionate (85%), whereas at 120° the isomeric methyl 2-(carbomethoxycyclohexyl)propionate (80%) is formed (K. Bittler et al., Angew. Chem., intern. Edn., 1968, 7: 329).

Carbethoxycarbene, generated photolytically from mercury bis(ethyldiazoacetate), gives with cyclohexene mainly ethyl cyclohex-2-enylacetate, and with cyclohexane a mixture of ethyl cyclohexylacetate and diethyl 2,3-dicyclohexylsuccinate (O. P. Strausz, T. DoMinh and J. Font, J. Amer. chem. Soc., 1968, 90: 1930).

cis- and *trans*-Cyclohexane-1,3-diacetic acid and -dipropionic acid have been synthesised by consecutive Arndt–Eistert reactions from *cis*- and *trans*-cyclohexane-1,3-dicarboxylic acid respectively (T. L. Westman, R. Paredes and W. S. Brey Jr., J. org. Chem., 1963, 28: 3512). The synthesis of *trans*-1-methylcyclohexane-1-carboxylic-2-propionic acid (52% overall) by Diels–Alder addition of dimethyl 2,5-dihydrothiophene-2-acetate-4-carboxylate with butadiene, followed by reduction and hydrolysis, was designed for application in the synthesis of steroids (G. Stork and P. L. Stotter, J. Amer. chem. Soc., 1969, 91: 7780).

(*iv*) *Hydroxycarboxylic acids*

The isomerisation of *quinic acid* in acetic acid/sulphuric acid gives, after hydrolysis, two diastereoisomers of quinic acid viz. (−)(3,4/1,5)-cyclohexanetetrolcarboxylic acid (*epi*-quinic acid), which readily forms a γ-lactone, and (4/1,3,5)-cyclohexanetetrolcarboxylic acid (*scyllo*-quinic acid), which forms a δ-lactone (J. Corse and R. E. Lundin, J. org. Chem., 1970, 35: 1904).

Chorismic acid and *prephenic acid* are both intermediates in the common biosynthetic route from shikimic acid to phenylalanine and tyrosine (B. A. Bohm, Chem. Rev., 1965, 65: 435); a review of the properties and attempted syntheses of prephenic acid has been published by H. Plieninger (Angew. Chem., intern. Edn., 1962, 1: 367). Some organisms (e.g. *Neurospora crassa*) use another biosynthetic route in which shikimic acid is converted via 3-dehydroshikimic acid to protocatechuic acid. The last step has been shown by ^{3}H-labelling to involve a *syn*-elimination of H_2O (K. H. Scharf et al., Chem. Comm., 1971, 765):

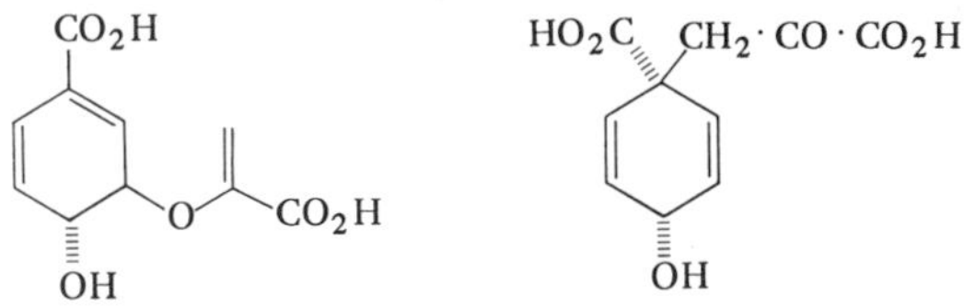

Chorismic acid   Prephenic acid

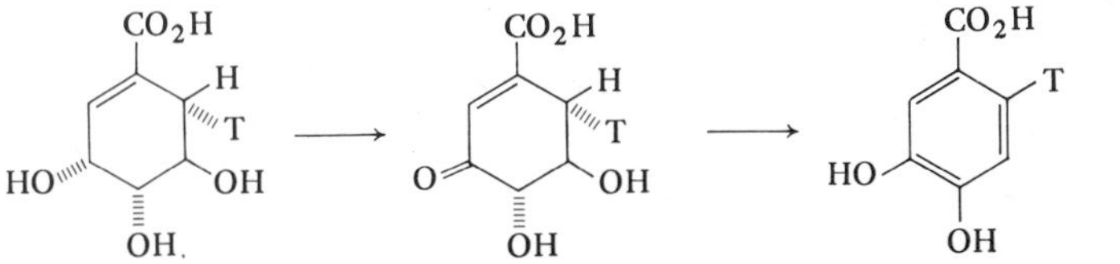

Shikimic acid   3-Dehydroshikimic acid   Protocatechuic acid

The electrolytic oxidation of the anion of 1-hydroxycyclohexylacetic acid in acetonitrile gives cycloheptanone (50%) (E. J. Corey et al., J. Amer. chem. Soc., 1960, 82: 2645):

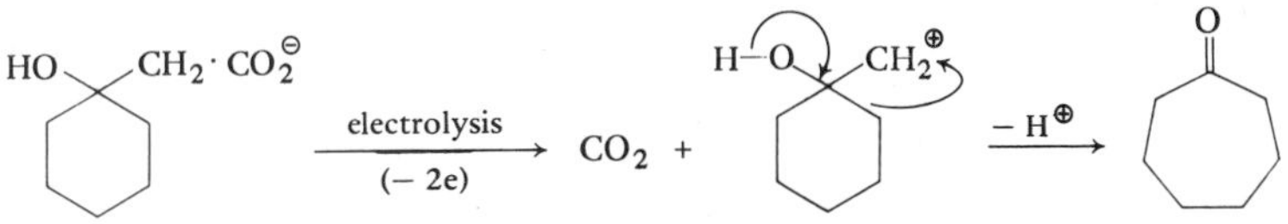

(v) Aminocarboxylic acids

The resolution of *trans*-2-aminocyclohexanecarboxylic acid, its ethyl ester, and the corresponding primary alcohol, has been performed. Nitrous acid deamination of these compounds gives the corresponding alcohols with retention of configuration (N. Hiroyuki, E. Kenji and M. Akira, Bull. chem. Soc. Japan, 1970, 43: 2230). In the crystalline state, *trans*-4-aminomethyl-cyclohexanecarboxylic acid exists as a zwitterion (P. Groth, Acta chem. Scand., 1968, 22: 143).

Chapter 6

The Acyclic and Monocyclic Monoterpenoids

S. H. HARPER

1. Introduction

The biosynthetic route to monoterpenoids through the alkylation of iso-pentenyl pyrophosphate by dimethylallyl pyrophosphate, themselves derived from mevalonic acid, is now well established (see e.g. T. A. Geissman and D. H. G. Crout, "Organic Chemistry of Secondary Plant Metabolism", Chap. 8, Freeman, Cooper & Co., San Francisco, 1969; D. Arigoni and E. L. Eliel, "Topics in Stereochemistry", Vol. 4, p. 209, Wiley/Interscience, New York, 1969; J. R. Hanson, Perfumery and Essential Oil Record, 1967, 58: 787; D. V. Banthorpe et al., J. chem. Soc., C, 1969, 541; J. chem. Soc., Perkin I, 1972, 1532, 1764, 1769).

2. Acyclic monoterpenoids

Presumably of mixed biogenetic origin are certain phenolic constituents of essential oils, in which a C_6-C_2 unit is coupled to a C_{10} unit (S. B. Challen and M. Kucera, Pharm. Weekblad, 1967, 102: 719). **Bakuchiol** from the seeds of *Psoralea corylifolia* has such a structure, elucidated by spectroscopic considerations and degradation (S. Dev et al., Tetrahedron Letters, 1966, 4561; 1967, 2897). Syntheses of (±)-bakuchiol have been effected from geraniol through the Claisen rearrangement of a vinyl ether (J. Carnduff and J. A. Miller, Chem. Comm., 1967, 606) (Scheme 1).

As models for the biogenesis of monoterpenes, it has been shown that the acid hydrolysis of geranyl, neryl, and linaloyl phosphates and pyro-phosphates leads to several naturally occurring acyclic and monocyclic systems (F. Cramer and W. Rittersdorf, Tetrahedron, 1967, 23: 3015, 3023). Geranyl and neryl diphenylphosphates decompose in inert solvents, such as boiling ether, to give mixtures of acyclic and monocyclic hydrocarbons (J. A. Miller et al., J. chem. Soc., C, 1968, 1837; 1969, 264).

[175]

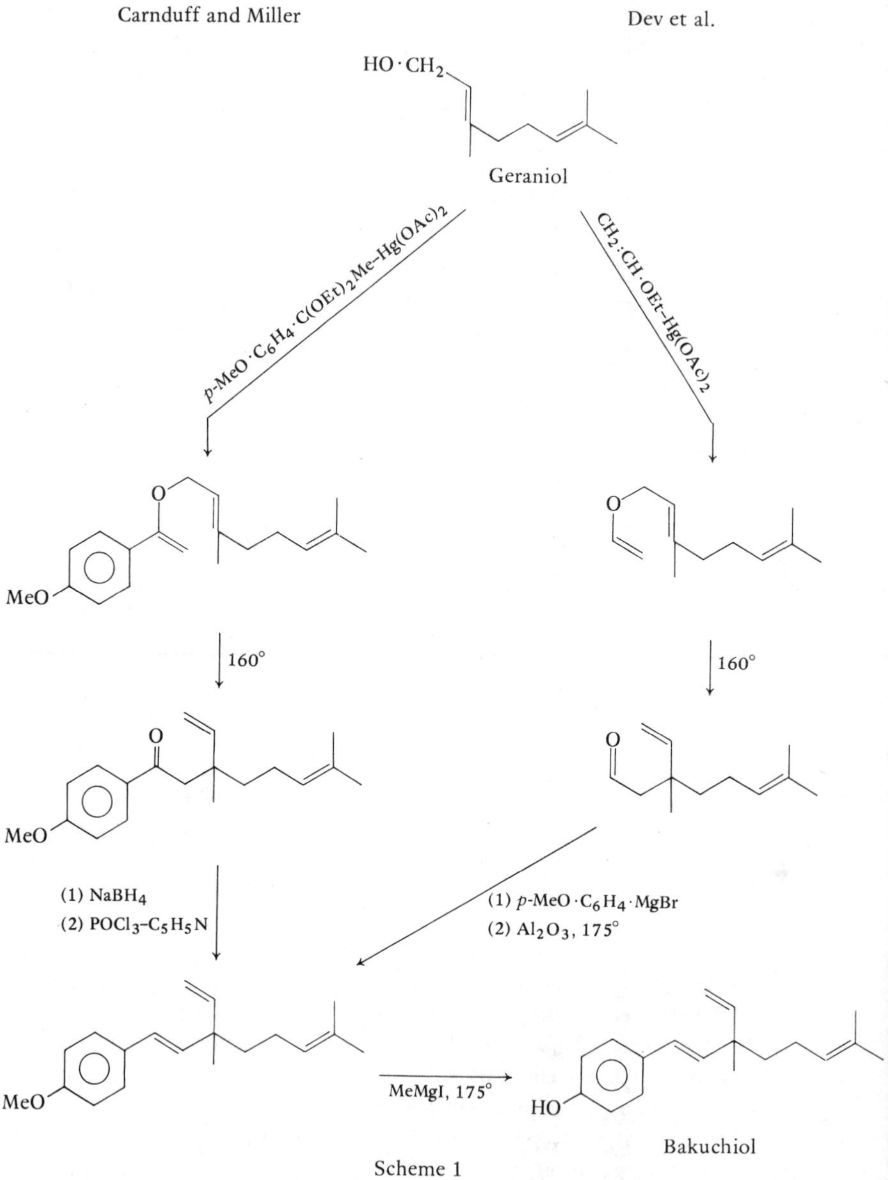
Carnduff and Miller
Dev et al.
HO·CH2
Geraniol
p-MeO·C6H4·C(OEt)2Me–Hg(OAc)2
CH2:CH·OEt–Hg(OAc)2
O
MeO
160°
(1) NaBH4
(2) POCl3–C5H5N
(1) p-MeO·C6H4·MgBr
(2) Al2O3, 175°
MeMgI, 175°
HO
Bakuchiol

Scheme 1

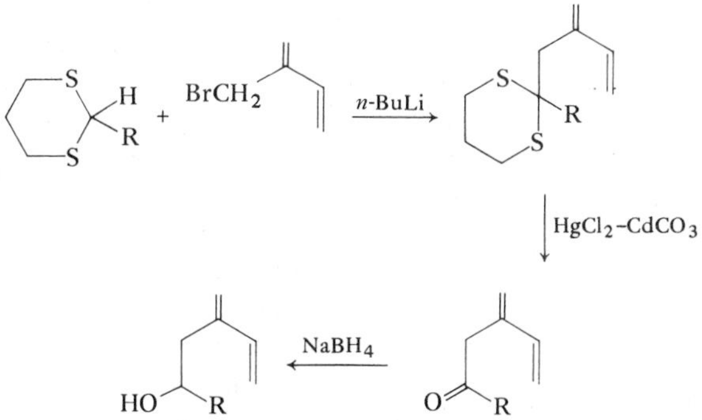

R = ·CH:CMe$_2$
= ·CH$_2$·CHMe$_2$

Scheme 2

The triene-alcohol (I), the diene-alcohol (II), and (±)-*cis*-verbenol are the principal components of the sex attractant produced by the male bark beatle, *Ips confusus*, and have been synthesised (R. M. Silverstein et al., Science, 1966, 154: 509; Tetrahedron, 1968, 24: 4249) (Scheme 2).

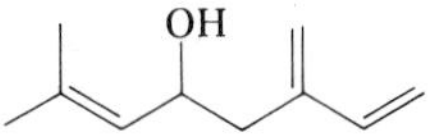

(I) 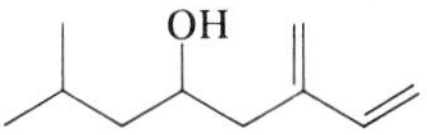   (II) 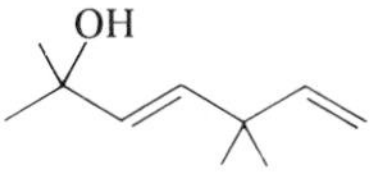   (III) 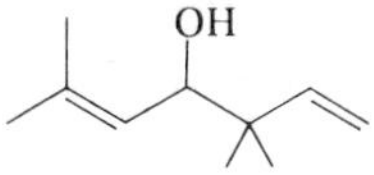   (IV)

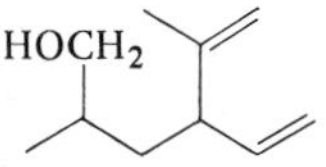

(V) 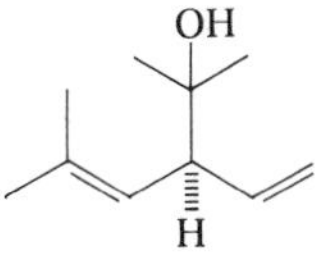   (VI) 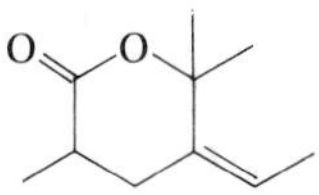   (VII)

Several monoterpenoid alcohols have been isolated belonging to the artemisyl, lavandulyl and santolinyl groups, where the head-to-tail linkage of isoprene units is not followed. All three skeletons are derivable in principle by opening of the cyclopropane ring of the chrysanthemic acid skeleton, which has led to speculation that this is the biosynthetic route to these groups of terpenes; although labelled chrysanthemic acid is not significantly incorporated into artemisia ketone. Representatives of these three types have been obtained by acid-catalysed ring-opening of chrysanthemic acid derivatives (L. Crombie et al., Tetrahedron Letters, 1967, 4553; J. chem. Soc.,

Perkin I, 1972, 642; T. Sasaki and M. Ohno, Chem. Letters, 1972, 503) (Scheme 3), whilst the reconversion of the artemisyl into the chrysanthemyl skeleton has been achieved (B. M. Frost et al., Chem. Comm., 1971, 1639).

Yomogi alcohol (III), isolated from the oil of *Artemisia feddei* (S. Hayashi et al., Tetrahedron Letters, 1968, 6241), is now known to be the allylic isomer of artemisia alcohol (IV) (K. Yano et al., Experientia, 1970, 26: 8). The latter has been synthesised by sigmatropic rearrangement of di-isopentenyl ether (V. Rautenstrauch, Chem. Comm., 1970, 4; J. E. Baldwin et al.,

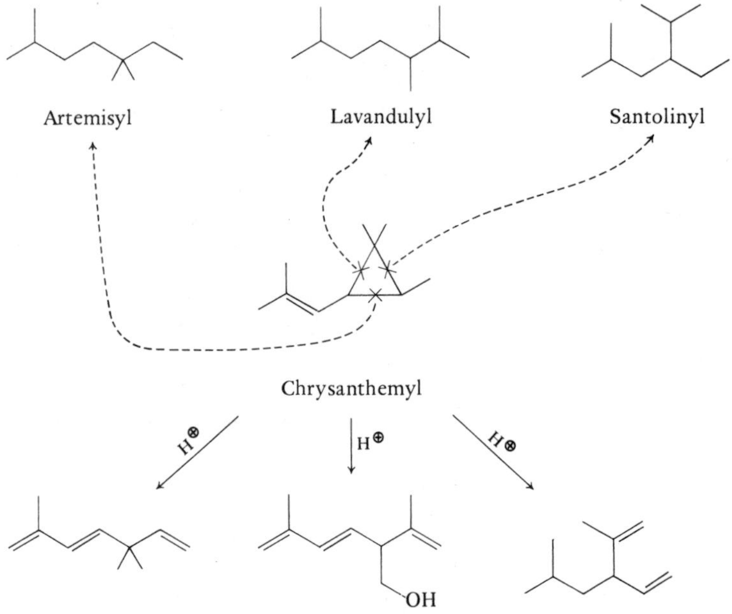

Scheme 3

Tetrahedron Letters, 1970, 353) (Scheme 4). Yomogi alcohol has been synthesised (W. Sucrow et al., Tetrahedron Letters, 1970, 1431; Ber., 1970, 103: 3771) (Scheme 4). The epoxide of yomogi alcohol is converted by acid-catalysed ring-opening, with a 1,2-vinyl shift, into compounds of the santolinyl type (A. F. Thomas et al., Chem. Comm., 1969, 1380; 1970, 1054; Helv., 1971, 54: 1822) (Scheme 4). The alcohol, lyratol (V), having the santolinyl skeleton has been isolated from the oil of *Cyathocline lyrata* (O. N. Devgan et al., Tetrahedron, 1969, 25: 3217; F. Bohlmann and M. Grenz, Tetrahedron Letters, 1969, 2413). Santolina alcohol (VI) has the same absolute configuration at $C_{(3)}$ as natural *trans*-chrysanthemic acid (C. D. Poulter et al., Tetrahedron Letters, 1972, 71). The lactone VII, having the santolinyl skeleton, has been isolated from the oil of *Chrysanthemum flosculosum* (Devgan, loc. cit.; Bohlmann and Grenz, loc. cit.).

The *Senecio* alkaloids are commonly composed of a dihydroxypyrrolizidine esterified with a ten-carbon di-basic '*necic*' *acid* to form a cyclic diester. The chemistry of these 'necic' acids has been well summarised by

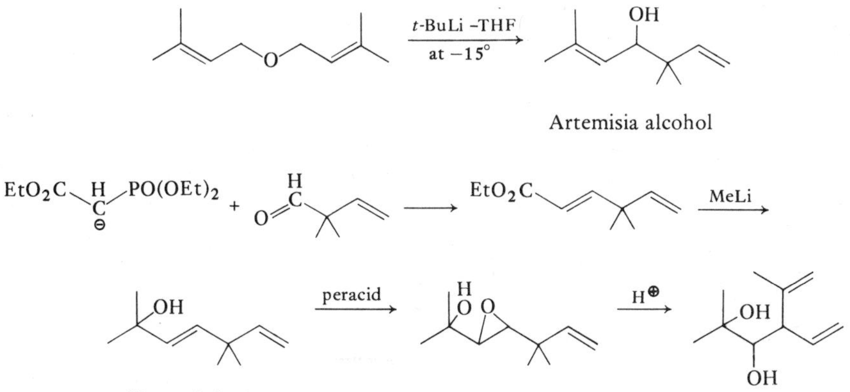

Artemisia alcohol

Yomogi alcohol

Scheme 4

F. L. Warren (Fortschr. Chem. org. Naturstoffe, 1955, 12: 225; 1966, 24: 346). The structures of these acids can be dissected into two isoprene units, which led G. Barger in 1936 to predict their terpenoid nature. However, evidence accumulated in the past five years shows clearly that the biosynthesis of these acids does not occur via the acetate–mevalonate pathway, but they are derived from common branched-chain amino acids (e.g. D. H. G. Crout et al., Phytochemistry, 1966, 5: 1; J. chem. Soc., C, 1966, 1968; 1968, 1233; 1969, 1386; Chem. Comm., 1968, 429). Hence they will not further be considered here.

3. Monocyclic monoterpenoids

(a) Tetramethylcyclohexane derivatives

Geranyl acetate cyclises to the 6-*hydroxy*-γ-*cyclogeraniol* (I) (isolated as an ester), either photochemically or in the presence of benzoyl peroxide, cupric benzoate and cuprous chloride (R. Breslow et al., Tetrahedron Letters, 1966, 4717). The separated *cis*-diol (I) cyclodehydrates to karahana ether (II) (R. M. Coates and L. S. Melvin, J. org. Chem., 1970, 35: 8657), which has been

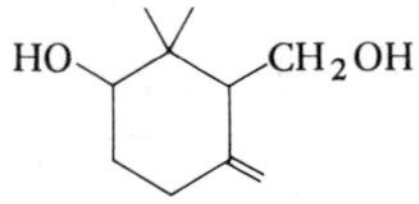
HO
CH_2OH

(I)

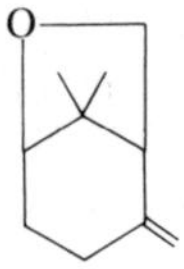
O

(II)

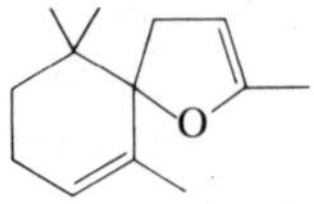
O

(III)

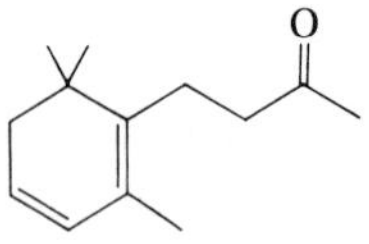
O

(IV)

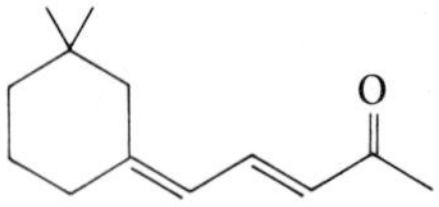
O

(V)

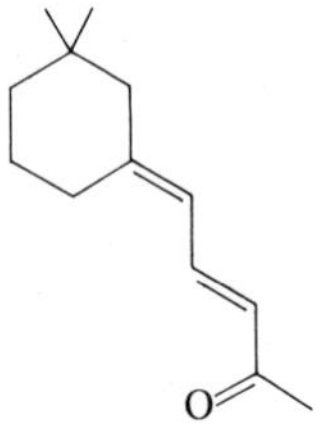
O

(VI)

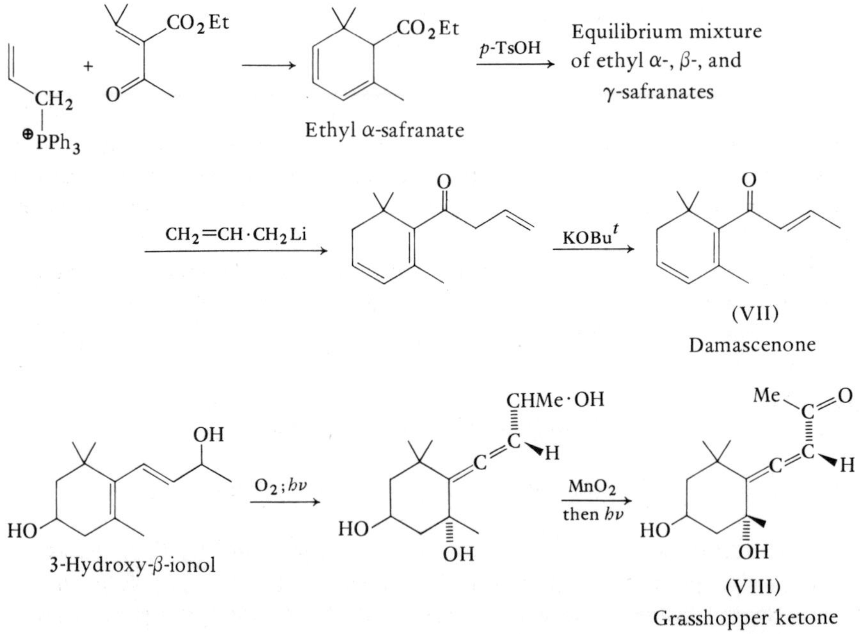
CH_2
$^{\oplus}PPh_3$
+
CO_2Et
O
CO_2Et
Ethyl α-safranate
p-TsOH
Equilibrium mixture of ethyl α-, β-, and γ-safranates
$CH_2{=}CH \cdot CH_2Li$
O
$KOBu^t$
O
(VII)
Damascenone
OH
HO
3-Hydroxy-β-ionol
O_2; hν
CHMe·OH
C
C
H
HO
OH
MnO_2
then hν
Me
C
O
C
C
H
HO
OH
(VIII)
Grasshopper ketone

Scheme 5

isolated from Japanese hop oil (Y. Maya and M. Kotake, Tetrahedron Letters, 1968, 1645). Further minor products formed in the acid-catalysed cyclisation of ψ-ionone to β-ionone, the spiro compound III and the ketones IV—VI, have been isolated by gas chromatography (C. Kruk et al., Rec. Trav. chim., 1968, 87: 641).

Further ketones having a trimethylcyclohexadienyl ring are damascenone (VII), a trace constituent of Bulgarian rose oil which has an exceptionally powerful, as well as pleasing odour, and grasshopper ketone (VIII). Several syntheses of damascenone have been accomplished (E. Demole et al., Helv., 1970, 53: 541; G. Büchi and H. Wüest, ibid., 1971, 54: 1767; K. H. Schulte-Elte et al., ibid., 1971, 54: 1805, 1899) and one is shown in Scheme 5. A synthesis of the allenic ketone (VIII) is also shown in Scheme 5 (J. Meinwald and L. Hendry, Tetrahedron Letters, 1969, 1657; B. C. L. Weedon et al., Chem. Comm., 1969, 85: 754; S. Isoe et al., Tetrahedron Letters, 1971, 1089).

A series of eleven-carbon lactones, having a cyclohomogeranic acid skeleton, are regarded as terpenoids.

The structure and configuration of **loliolide** (IX) has been confirmed by several syntheses, one of which is outlined in Scheme 6 (Z. Horii et al.,

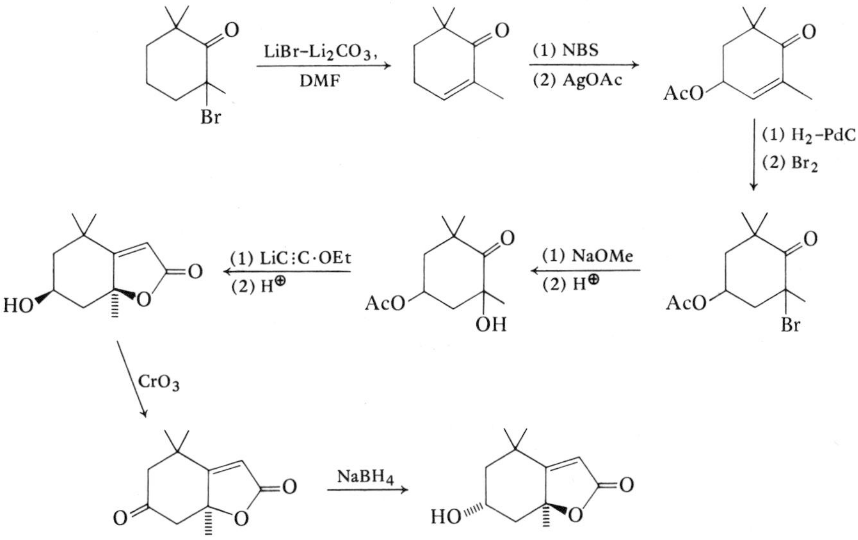

Scheme 6

Chem. Comm., 1966, 634; E. Demole and P. Enggist, Helv., 1968, 51: 481; J. N. Marx and F. Sondheimer, Tetrahedron, 1966, Suppl. 8, 1; S. Koe et al., Tetrahedron Letters, 1971, 1089).

Actinidiolide (X) and *dihydroactinidiolide* (XI) have been isolated from the leaves of *Actinidia polygama* and their structure confirmed by synthesis (S. Isoe et al., Tetrahedron Letters, 1967, 1623; 1968, 5561). Dihydroactinidiolide, m.p. 42–43°, is a major component of tea and tobacco aroma. (+)-*Tetrahydro-cis-actinidiolide* (XII) has been synthesised by homologation of α-cyclogeranic acid (M. Ridi and C. H. Eugster, Helv., 1969, 52: 1732).

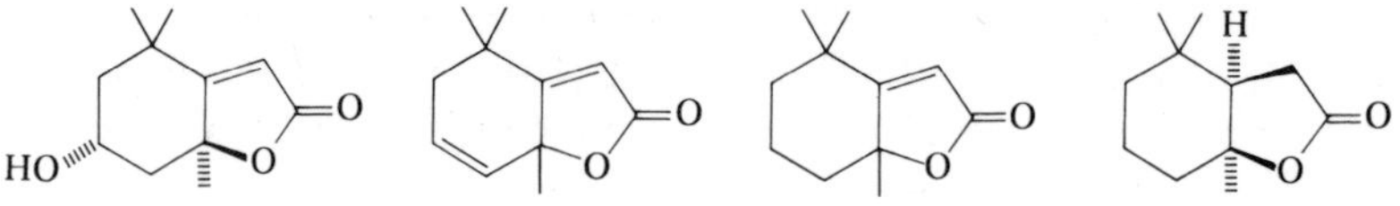

(IX)   (X)   (XI)   (XII)

(b) p-*Menthane derivatives*

p-*Mentha*-1,3,8-*triene* (XIII), a constituent of Yugoslavian parsley (R. L. Kenney and G. S. Fisher, J. Gas Chromat., 1963, 1: 19; J. Garnero et al., Bull. Soc. chim. Fr., 1967, 4679), has been synthesised (A. J. Birch and G.

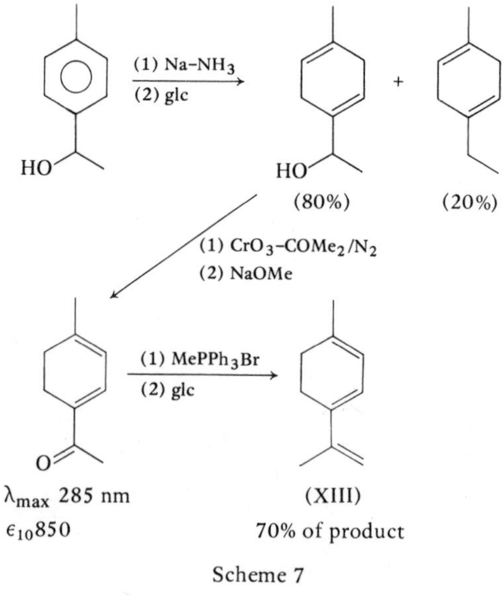

Scheme 7

SubbaRao, Austral. J. Chem., 1969, 22: 1037: A. F. Thomas and W. Bucher, Helv., 1970, 53: 770) (Scheme 7).

Uroterpene, *p*-menth-1-ene-8,9-diol (XIV), has been isolated as the β-D-glucuronide from human urine, probably derived from dietary limonene (F. M. Dean et al., J. chem. Soc., C, 1967, 1893). *p*-Menth-1-en-9-al (XV), isolated from Bulgarian rose oil (G. Ohloff et al., Helv., 1969, 52: 1531), is a mixture of diastereomers and has been synthesised.

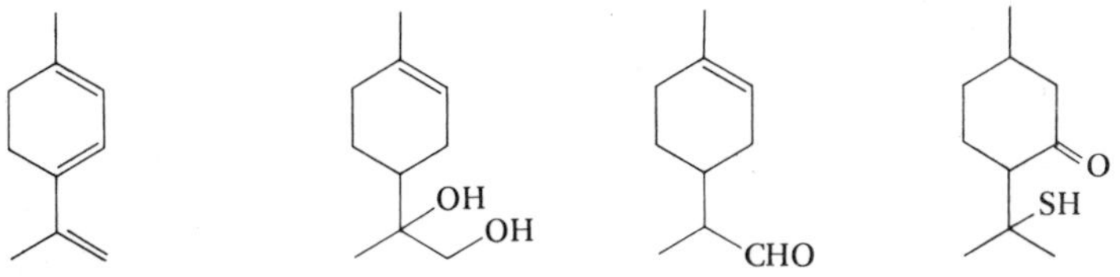

(XIII)   (XIV)   (XV)   (XVI)

The cassia flavour of buchu oil is due to the presence of diastereomeric 8-mercaptopulegones (XVI), which have been synthesised from pulegone (E. Sundt et al., Helv., 1971, 54: 1801).

(c) o-Menthane derivatives

Carquejol shows only end-absorption at 205 nm, so that its structure has been revised to the non-conjugated triene XVII, with the (2*R*,3*S*) configuration shown (G. S. Snatzke, A. F. Thomas and G. Ohloff, Helv., 1969, 52: 1253). The diol esters XVIII, also having the *o*-menthane structure, occur in the roots of *Baccharius timera* (F. Bohlmann and C. Zdero, Tetrahedron Letters, 1969, 2419). It will be of interest to establish if the *o*-menthane skeleton is derived biosynthetically from ring-opening of a pinene intermediate.

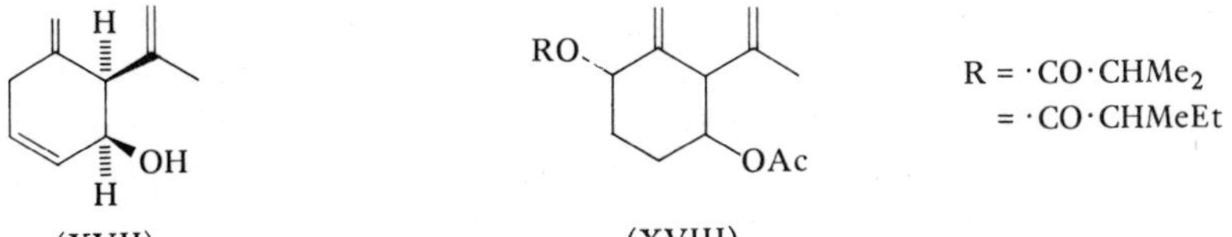

(XVII)   (XVIII)

$R = \cdot CO \cdot CHMe_2$
$= \cdot CO \cdot CHMeEt$

Y.-R. Naves et al. (Helv., 1970, 53: 201, 551, 1339) have reduced carquejol catalytically over platinum, and with lithium aluminium hydride, and separated the isomeric saturated *carquejanols* formed by gas chromatography. The carquejanols were then oxidised with chromic acid to the corresponding *carquejanones*. Configurations for these alcohols and ketones were deduced by analogy with the carvomenthols and carvomenthones, and by use of the von Auwers–Skita rule (Scheme 8, Table 1).

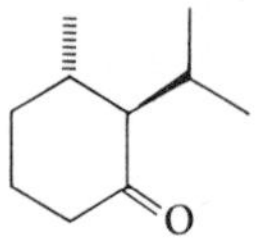

Carquejanone 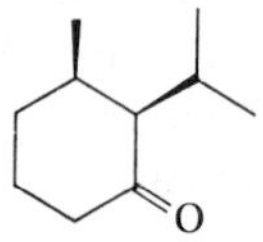   Isocarquejanone

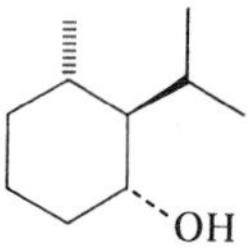

Carquejanol 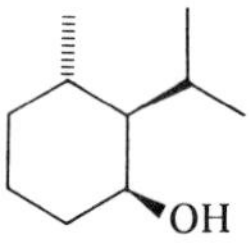   Neocarquejanol 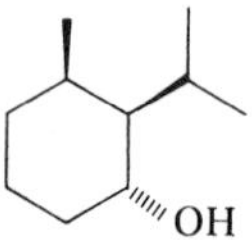   Isocarquejanol 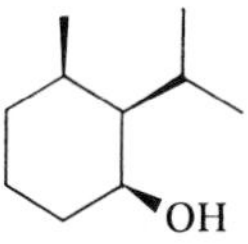   Neoisocarquejanol

Scheme 8

Table 1

CARQUEJANOLS

	Carquejanol	Neocarquejanol	Isocarquejanol	Neoisocarquejanol
m.p.	liquid	45–46°	liquid	27–28° and 38°
$[\alpha]_D^{20}$	+7.9°	−49.6°	−43.5°	−5.0°
3,5-Dinitro-benzoate (m.p.)	150–151°	119–120°	118–119°	113–114°

(d) Cyclopentane derivatives

Cyclopentane monoterpenoids having the iridane skeleton (XIX), or a cyclic variant (XX), the iridoids, are attracting increased attention as evidence accumulates that certain members are involved in the biosynthesis of indole and isoquinoline alkaloids. Other members show physiological activities in

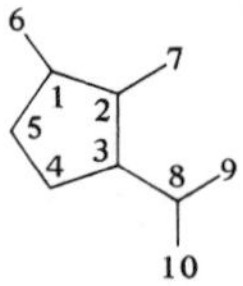

(XIX) 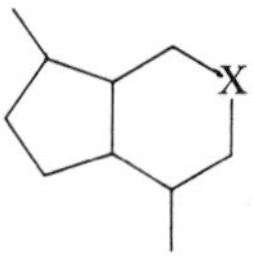   (XX) 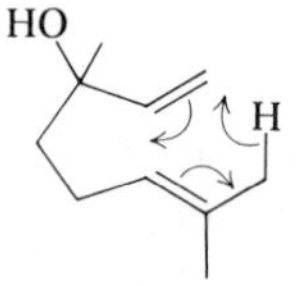   Linalool 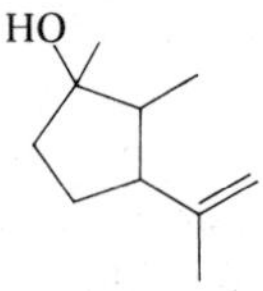   (XXI)

arthropod defence, as insect repellants and feline attractants (J. M. Bobbitt and K. P. Segebarth, in "Cyclopentanoid Terpene Derivatives". eds. W. I. Taylor and A. R. Battersby, Dekker, New York, 1969).

Brief heating of linalool at 650° with pyridine gives, by an intramolecular Alder-ene synthesis, four of the diastereomeric *plinols* (XXI). Their configurations have been elucidated by oxidative degradation and from optical considerations (H. Strickler, G. Ohloff and E. Kovats, Helv., 1967, 50: 759) (Scheme 9).

$[\alpha]_D^{20}$ −39.5° neat	−0.2° (CHCl$_3$)	−12.3° (CHCl$_3$)	+7.9° (CHCl$_3$)
m.p.(°C) liquid	45	37	92

Scheme 9

Loganin has achieved prominence with the discovery that it plays a central role in the biosynthesis of certain groups of indole and isoquinoline alkaloids (T. A. Geissman and D. H. G. Crout, "Organic Chemistry of Secondary Plant Metabolism", Chaps. 8 and 19, Freeman, Cooper & Co., San Francisco, 1969). In consequence the full stereostructure of loganin (XXII) has been determined, and the biosynthetic origin from geraniol established (A. R. Battersby et al., Chem. Comm., 1968, 131; S. Brechbühler-Bader et al., ibid., 1968, 136). The related *loganic acid* (XXIII), another monoterpene glucoside, occurs in *Swertia carolinensis* and has a common mevalonate—geraniol origin (C. J. Coscia and R. Guarnaccia, J. Amer. chem. Soc., 1967, 89: 1280; Chem. Comm., 1968, 138). Loganin has been synthesised (G. Büchi et al., J. Amer. chem. Soc., 1970, 92: 2165) (Scheme 10) as has the related genipin (XXIV) (idem, ibid., 1967, 89: 2776) (Scheme 11).

(XXII)   (XXIII)   (XXIV)

(±)-O-Methyldehydro-
loganin aglucone

(XXII)
(−)-Loganin

Scheme 10

New iridoids are *theveside* and its methyl ester from *Thevetia peruviana* (O. Sticher, Tetrahedron Letters, 1970, 3195; Helv., 1969, 52: 478); *galiridoside* from *Galeopsis tetrahit*, which is a $C_{(9)}$ epoxy iridoid; *valeroside* from *Valeriana* sp., which is an isovaleric ester (P. W. Thiess, Tetrahedron Letters, 1970, 2725); *gardenoside*, geniposide, the glucoside of genipin, the methyl ester of deacetyl-asperulosidic acid, *shanzhiside*, and the gentiobioside of genipin from *Gardenia jasminoides* (Scheme 12).

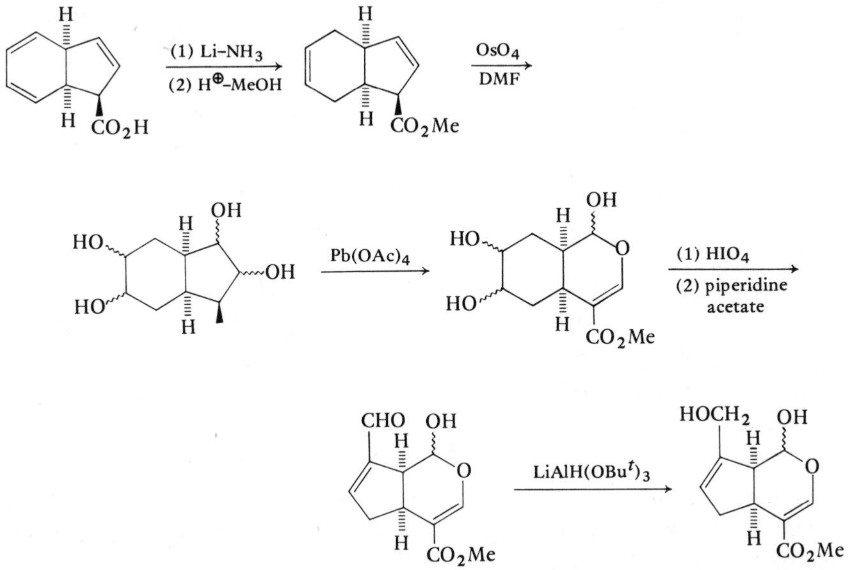

Scheme 11

New **secoiridoids** from the leaves of *Lonicera morrowii* are *morroniside* (XXV), *kingiside* (XXVI), *sweroside* (XXVII), and *loniceroside* (XXVIII) (I. Souzu and H. Mitsuhashi, Tetrahedron Letters, 1969, 2725; 1970, 191). Loniceroside is also a biosynthetic intermediate between loganin and the indole alkaloids. Cleavage of morroniside with emulsin gives the unstable aglycone, which spontaneously cyclises to the internal acetal (XXIX).

Campholenic aldehyde (XXX) containing the 1-ethyl-2,2,3-trimethyl-cyclopentane skeleton, and hitherto known only as a photochemical

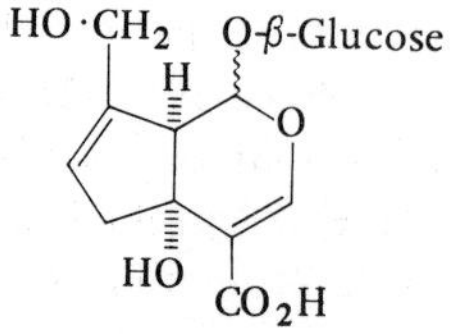

Theveside 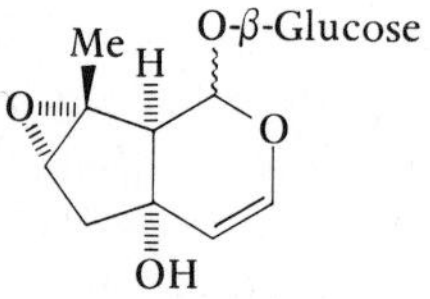   Galiridoside 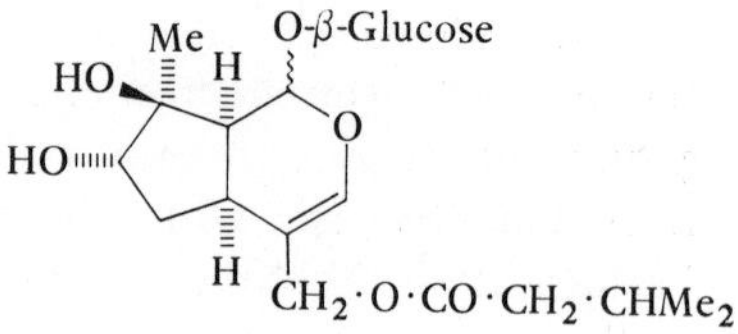   Valeroside

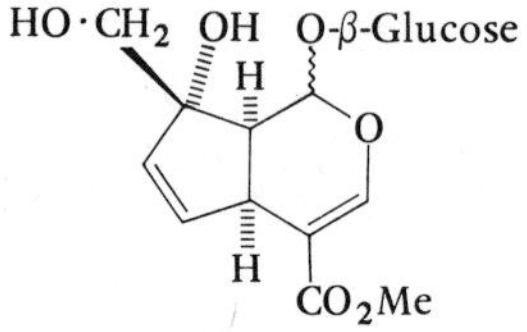

Gardenoside 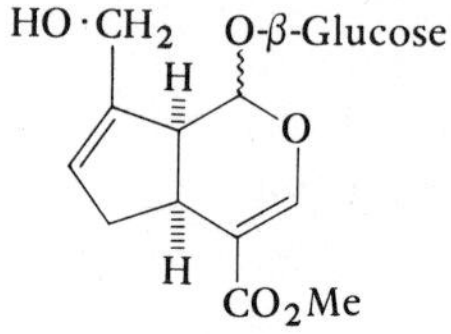   Geniposide

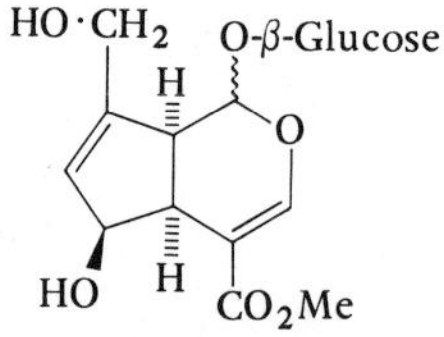

Methyl ester of
deacetylasperulosidic
acid 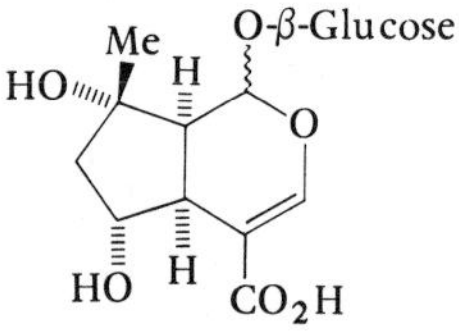   Shanzhiside

Scheme 12

rearrangement product of camphor, has been isolated from the oil of *Juniperus communis* together with its epoxide and the corresponding alcohol and its acetate (H. F. Thomas, *Helv.*, 1972, 55: 815).

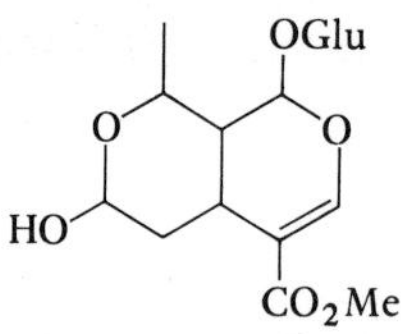

(XXV) 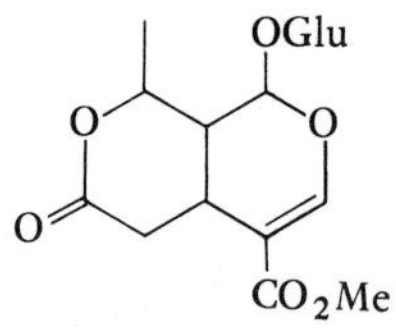   (XXVI) 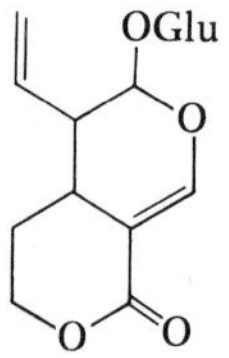   (XXVII)

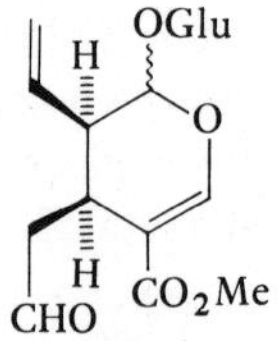

(XXVIII) 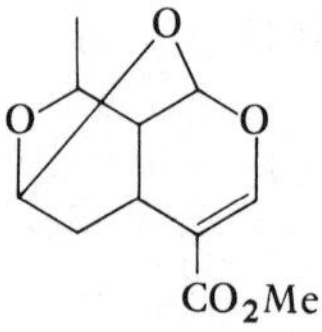   (XXIX)

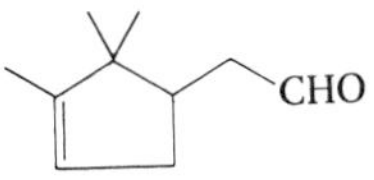

(XXX)

(e) Other monocyclic monoterpenoids

The sex pheromones produced by the male cotton boll weevil have structures derivable from isoprene units; the unusual cyclobutane (XXXI) and the *gem*-dimethylcyclohexane derivatives XXXII–XXXIV. The structure

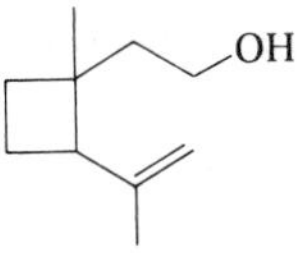

(XXXI) 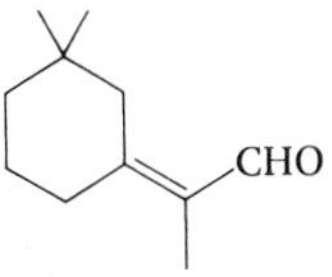　　　(XXXII) 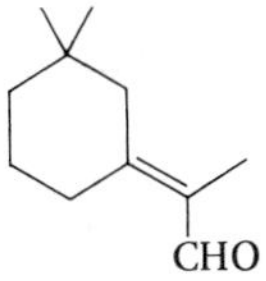　　　(XXXIII) 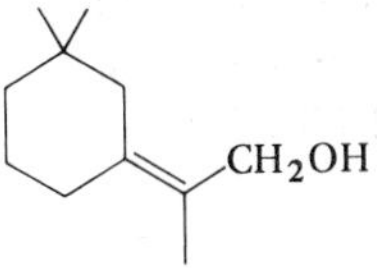　　　(XXXIV)

of XXXI has been confirmed by synthesis (J. H. Tumlinson et al., Science, 1969, 166: 1010; J. org. Chem., 1971, 36: 2616; R. Zurflüh et al., J. Amer. chem. Soc., 1970, 92: 425; J. H. Babler et al., Tetrahedron Letters, 1972, 669) (Scheme 13).

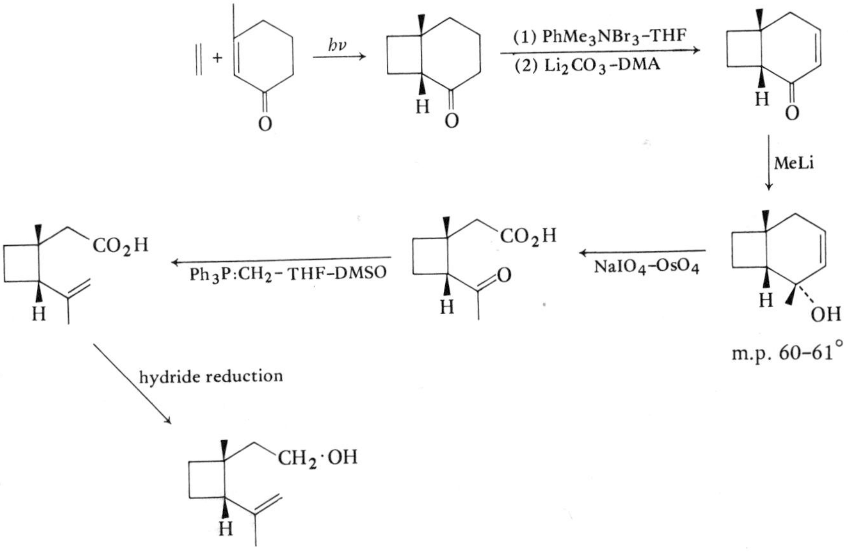

Scheme 13

Another example of hydroxyl group participation in terpene brominations is provided by the synthesis of the hop-oil cycloheptenone, *karahanaenone* (XXXV) by bromination of linalool, followed by dehydrobromination (E. Demole and P. Enggist, Chem. Comm., 1969, 264).

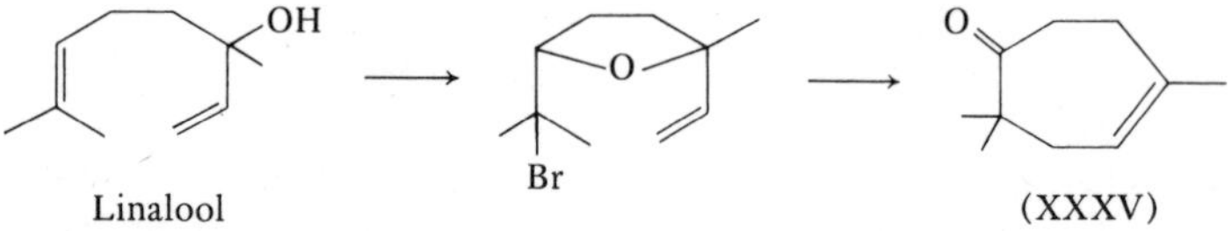

Linalool   Br   (XXXV)

The photoisomerisation of eucarvone has been further investigated and resulted in the isolation of two new bicyclic ketones, one of which further rearranges into dehydrocamphor (T. Takino and H. Hart, Chem. Comm., 1970, 450).

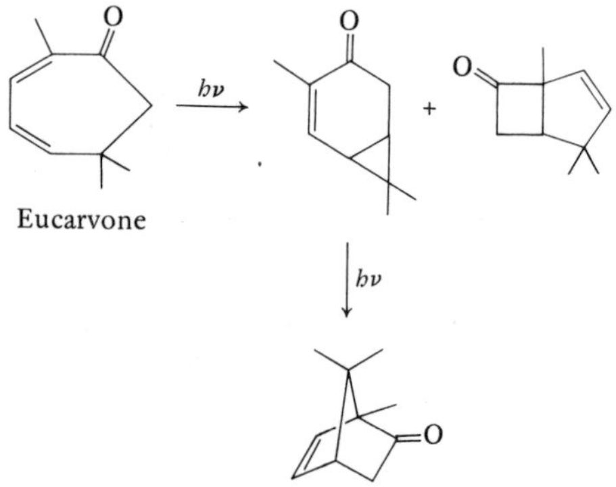

Eucarvone

Dehydrocamphor

Chapter 7

The Carotenoid Group

J. B. DAVIS

There has been a very rapid expansion of knowledge in this field since the manuscript for this chapter of the main work (2nd Edn., Vol. II B, pp. 231—346) was completed. In the area of naturally-occurring carotenoids, this expansion has resulted from the discovery of:

(i) new permutations of the well established types of structure;

(ii) entirely new families of compounds, such as the C_{50} (and C_{45}) compounds (p. 309), the *nor*-(C_{38})carotenoids (p. 327), and the acetylenic carotenoids (p. 300; the first examples being announced in time to be noted only briefly before);

(iii) new modifications of the traditional type of carotenoid structure, such as those with methyl groups oxidised to the hydroxymethyl (p. 320) or aldehyde (p. 268) level or terminal isopropylidene groups (cf. I) reduced to the 1,2-dihydro ($-CH_2 \cdot CHMe_2$) state (p. 222); carotenoids carrying sugar units (e.g. pp. 259, 263), thereby explaining the very high polarity of certain carotenoids; carotenoids having $-OH$ at $C_{(2)}$ (cf. Sections 7b, 7c) or on an aryl ring (hence phenolic: Section 7k); and those with a ring end-group in which the double bond is *exo* (at $C_{(5)}$) (p. 224).

For consistency, these new discoveries, which have largely resulted from the widespread application of physical methods (particularly p.m.r. and mass spectrometry), have been described using the same format as before, each compound being covered individually. However, for the novel types of carotenoid, general information on the discovery, methods of structure elucidation, and special properties associated with the novel structural feature, has been given at the beginning of the section, often in association with the first compound mentioned, the remaining examples usually being covered relatively briefly. Similarly, the overall plan of this chapter is as in the 2nd edition except that the aryl carotenoids have now been grouped together in one section (7k).

In addition to the above advances, one of the finer aspects of carotenoid

[191]

structure, the configuration at each asymmetric centre, has received detailed attention for the first time. Also there has been a continuing interest in the mode of biosynthesis of the carotenoids (Section 8), and in the nature of the visual pigments which are of basic importance to the sense of sight (Section 6c).

In addition to a number of reviews dealing with certain aspects of carotenoid research, and mentioned where appropriate, the proceedings of two international symposia devoted entirely to this topic, have been published (Pure Appl. Chem., 1967, 14: 215–278; 1969, 20: 365–553), and a new book has appeared ("Carotenoids", ed. O. Isler, Birkhäuser, Basel, 1971). Certain carotenoids are manufactured on an industrial scale for use as provitamins A and as food colourants (in place of the physiologically less desirable artificial dyestuffs), and the first of the above publications includes (p. 259) an illustrated survey, by O. Isler and co-workers, of the use of carotenoids as colourants.

The material in this chapter should be read in conjunction with Chapter 7 in Volume II B of the 2nd edition of "Rodd's Chemistry of Carbon Compounds". References back to that chapter will be in the style 2nd Edn., p. 231.

1. Basic constitution, classification, and nomenclature

Despite the above-mentioned proliferation in structural types, a carotenoid is still easily recognised by its methyl(or modified methyl)-substituted polyene chain. The C_{50} and C_{45} carotenoids, which barely fall within the classical

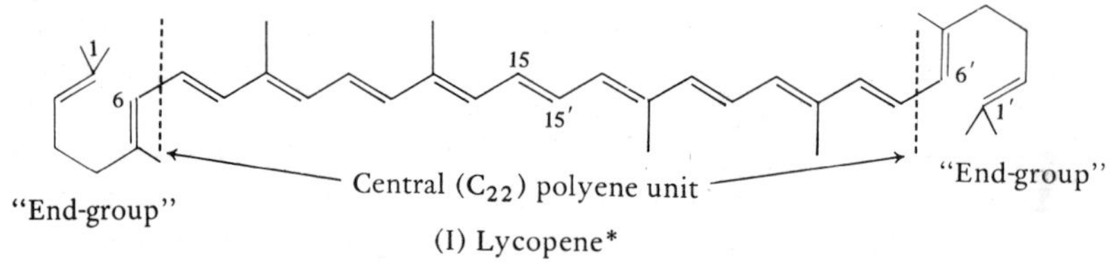

(I) Lycopene*

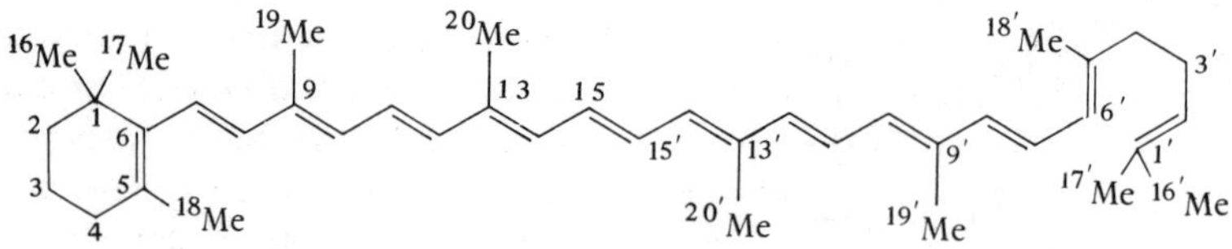

(II) γ-Carotene*

* Concerning the quasi-cyclic manner in which the acyclic end-groups of these and other carotenoids have been drawn throughout this chapter, see the footnote on p. 217.

definition of carotenoids, are considered as straightforward di- and mono-alkylated (at $C_{(2)}$ and/or $C_{(2')}$) derivatives of the more usual C_{40} skeleton. The alkylating agent is a C_5 (isoprene) unit and hence the generic term *"isoprenylated carotenoids"* is sometimes used to describe these compounds.

Nomenclature. Although the full structures of a large number of the many natural carotenoids now known have been elucidated, most of these compounds are still known by trivial, and chemically meaningless, names assigned to them at the time of their discovery. Hence, attempts have recently been made to devise systems of nomenclature which allow the structure of such compounds to be gleaned directly from the name*. The compound is named either as a derivative of one of the better known major carotenoids (a system which tends to be of restricted value) or, (and better) by using a semi-systematic method† in which the name of the basic skeleton of the molecule is first constructed by attaching symbols assigned to the common end-groups to the word "carotene". The name is then modified as appropriate to take account of substituents. The names suggested for some of the common end-groups are as follows:

ψ-

(as in ψ-ionone)

β-

(as in β-ionone
and β-carotene‡)

ε-

(as in ε-carotene§)

Thus the names given to lycopene (I), γ-carotene (II), β-carotene (p. 217), canthaxanthin (p. 275), and astaxanthin (p. 276) would be, respectively, ψ,ψ-carotene; β,ψ-carotene; β,β-carotene; β,β-carotene-4,4'-dione; and 3,3'-dihydroxy-β,β-carotene-4,4'-dione. Similarly, neurosporene (p. 339) and monadoxanthin (7,8-acetylenic: p. 305) would be 7,8-dihydro-ψ,ψ-carotene and 7,8-didehydro-β,ε-carotene-3,3'-diol. Absolute configuration, where known, can be incorporated in the usual way [e.g. zeaxanthin (p. 251) would be (3R,3'R)-β,β-carotene-3,3'-diol]. Using this system, the characteristic (aryl) end-groups of isorenieratene and renierapurpurin (2nd Edn., p. 277) are denoted by the letters Φ (phi) and χ (chi), and the trimethylcyclopentyl

---

* The systematic IUPAC name is generally too long and unwieldy for normal use.
† For further details, see the Appendix in "Carotenoids", ed. O. Isler, Birkhäuser, Basel, 1971, or Biochem. J., 1972, 127: 741.
‡ And so sometimes known as a "β-end-group".
§ And in α-ionone: hence sometimes known as an "α-end-group".

end-groups of capsorubin and related compounds (2nd Edn., p. 323) by κ (kappa) — so that capsorubin itself would be (3S,5R,3'S,5'R)-3,3'-dihydroxy-κ,κ-carotene-6,6'-dione.

Conformation. J. C. J. Bart and C. H. MacGillavry have summarised the results of the various earlier (cf. 2nd Edn., p. 236 and C. Sterling, Acta Cryst., 1964, 17: 1224) and their own more recent (ibid., 1968 [B], 24: 1569, 1587) X-ray diffraction studies of certain carotenoids and related compounds (Acta Cryst., 1968 [B], 24: 1587, 1600). In the five compounds studied (the all-*trans* forms of β-carotene and canthaxanthin, their central-acetylenic analogues, and retinoic acid), it appears — at least in the *crystalline state* — that:

(*a*) the polyene chain is virtually planar, and single/double bond alternation persists to a large degree throughout;

(*b*) the chain is not, however, a perfectly straight zig-zag, the two halves being distinctly bowed (probably by a 1,3-interaction of the side-chain methyls with allylic olefinic H's) as shown below for canthaxanthin:

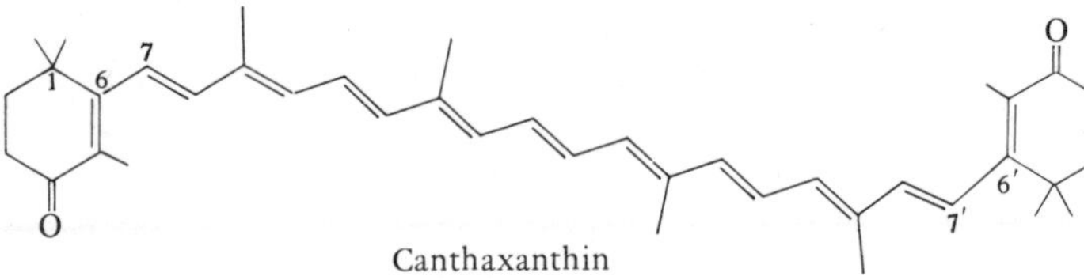

Canthaxanthin

(*c*) the 6,7 (and 6',7') bond approximates far more closely to the *s-cis* conformation (as drawn for canthaxanthin above) than the *s-trans* (as in III), the cyclohexene ring being tilted out of the plane of the chain double-bonds by ~30–40° in the above five compounds (this would be 0° in *pure s-cis*, but such is prevented by steric interaction between ring-methyls and polyene chain).

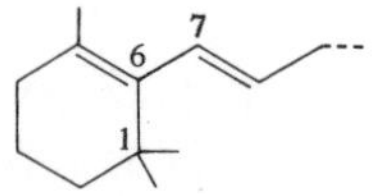

(III)

The retinal–iron carbonyl adduct on p. 243 also approximates to the 6,7-*s-cis* conformation in the solid state, as do the 11-*cis* and the all-*trans* isomers of retinal itself (R. Gilardi et al., Nature, 1971, 232: 187). Theoretical (free-energy) calculations on the isolated retinal molecule (B. Pullman et al., J. Mol. Structure, 1970, 6: 139) and spectral data (2nd Edn., p. 236; B. Honig, M. Karplus et al., Proc. Nat. Acad. Sci., U.S., 1971, 68:

1289) indicate that this also applies to carotenoids in free motion, as when in solution. For this reason, formulae are written throughout in the *s-cis* rather than the *s-trans* form, as before (and as in the majority of original papers).

Concerning the *configuration* of those carotenoids containing asymmetric centres [e.g. at $C_{(3)}$ in many of the xanthophylls (for definition, cf. footnote, p. 245) and at $C_{(6')}$ in α-carotene derivatives], see p. 210 and under the individual compounds (Sections 5 and 7).

2. Occurrence, function and isolation

Occurrence. Little definite can be added to the remarks made in the main work. Some carotenoids are of widespread distribution in nature, others appear to be restricted in occurrence to a very few closely related species. Around 250 individual carotenoid structures have now been identified in nature. As in other branches of natural product chemistry, the identity and/or pattern of compounds isolated from a given natural source can be of *chemotaxonomic* interest and value. Thus, in those cases where it has proved difficult to assign an organism to a particular biological family entirely by the traditional (morphological) methods, classification can sometimes be aided by comparing the carotenoids therein with those in various allied organisms (cf. p. 324 and T. R. Ricketts, Phytochem., 1970, 9: 1835). Similarly, a consideration of the differing patterns of carotenoids found in various classes of algae has been used as a basis for inferring the order in which the latter appeared during evolution [T. W. Goodwin in "Comparative Phytochemistry", ed. T. Swain, Academic Press, London, 1966, p. 123 (the actual structures of several of the carotenoids mentioned have since been modified); M. S. Aihara and H. Y. Yamamoto, Phytochem., 1968, 7: 497].

F. C. Czygan and others, extending the earlier discoveries of G. Dersch et al., have shown that when certain algae are grown under stressful conditions [e.g. on a nitrogen-deficient culture medium (see refs. below) or under unusually bright light (H. Kleinig, Ber. deutsch. bot. Ges., 1966, 79: 126)], they synthesise a group of carotenoids not normally present − the so-called "*secondary carotenoids*" (often keto- or hydroxyketo-compounds such as astaxanthin and adonixanthin). At the same time their normal (or "primary") carotenoids are formed in rather less quantity than usual, and chlorophyll synthesis drops dramatically [Czygan, E. Kessler, Kleinig et al., Z. Naturforsch., 1967, 22b: 1085; 1969, 24b: 927, 977; Arch. Mikrobiol., 1967, 55: 320; 1968, 61: 81 (short review)]. This phenomenon is of biogenetic and chemotaxonomic interest, and has sometimes led to a new source of an otherwise rare carotenoid.

There has been increased interest in the nature of the fatty acids esterifying the hydroxyl groups in hydroxy-carotenoids. In the few examples so far studied, saturated straight-chain $C_{10}-C_{14}$ acids have been the commonest (Kleinig et al., Phytochem., 1967, 6: 437; 1968, 7: 1171; S. Hertzberg et al., Acta chem. Scand., 1969, 23: 3290) but linoleic acid (C_{18}; diene) has also been found (N. Arpin et al., Phytochem., 1969, 8: 897).

Evidence that carotenoids were present on Earth 5×10^7 years ago has been presented by G. Eglinton et al. (Science, 1967, 157: 1040; cf. B. S. Cooper, J. Chromatog., 1970, 46: 112) who found a hydrocarbon tentatively identified (spectra; comparison with authentic material) as perhydro-β-carotene in Green River shale of that age.

Concerning the naturally-occurring carotenoid—protein complexes, or "chromoproteins", mentioned before (2nd Edn., p. 239), many such complexes have now been described. However, only relatively few have been obtained in a reasonable state of purity or investigated chemically. Two which have been investigated are the blue lobster-shell pigment, α-crustacyanin (2nd Edn., p. 321), and one of the visual pigments, rhodopsin (p. 229). Several reviews and papers describing methods of isolation, purification, and general properties have appeared (D. F. Cheeseman, P. F. Zagalsky et al., Biol. Rev. Camb. Phil. Soc., 1967, 42: 131 (review; Compar. Biochem. Physiol., 1967, 22: 851; D. Thirkell and M. I. S. Hunter, J. gen Microbiol., 1969, 58: 289; 1970, 62: 125; C. Subbarayan and H. R. Cama, Ind. J. Biochem., 1966, 3: 225). Chlorophyll—carotenoid—protein complexes, which may be implicated in light absorption during the photosynthetic process, have also been studied (J. P. Thornber et al., Biochemistry, 1967, 6: 2006; 1970, 9: 2688; L. P. Vernon et al., Biochim. biophys. Acta, 1967, 143: 144).

Function. Recent work dealing with the ability of certain carotenoids to act as vitamin A precursors and with their rôle in the visual process is covered on pp. 225 and 229. Considerable evidence in support of the suggestions that carotenoids also act as absorbers of light energy for subsequent "transfer" to chlorophyll and as protectors of air/light-sensitive systems, has been assembled by G. Mackinney (in "Metabolic Pathways", 3rd Edn., Vol. II, ed. D. M. Greenberg, Academic Press, New York, 1968, pp. 234, 268) and by N. I. Krinsky (in "Photophysiology", Vol. 3, ed. A. C. Giese, Academic Press, New York, 1968, pp. 123, 188; cf. also G. Tomita and T. Oku, Experientia, 1971, 27: 1406). The mechanism by which carotenoids exert their protective action has been discussed by C. S. Foote and co-workers (J. Amer. chem. Soc., 1970, 92: 5216, and refs. there cited) who suggest that it is due to their ability, readily observable *in vitro*, to act as an effective quencher of singlet oxygen (formed from the effect of *hv* on atmospheric oxygen in the presence of natural sensitising dyes). Under normal mild in vivo conditions, this involves no nett consumption of carotenoid (Foote et al., loc. cit.; J. Amer. chem. Soc., 1968, 90: 6233) but under more vigorous in vitro conditions, irreversible reactions of the type referred to on p. 299 occur. Krinsky (loc. cit.) also lists some of the several other proposed, but less certain, functions ascribed to carotenoids by various authors.

More recently suggested rôles are as precursors of: xanthoxin and allied compounds mentioned on p. 253 (from violaxanthin); "grasshopper ketone" (from neoxanthin?, p. 294); and the biologically important "fungal hormones" known as the trisporic acids (from β-carotene, by central cleavage to retinal then oxidative degradation?; cf. J. D. Bu'Lock et al., Experientia, 1970, 26: 348):

Trisporic acid "B" (and "C")

Isolation from natural sources; chromatography. Attempts have been made to isolate and identify the enzyme(s) responsible for the oxidative degradation of carotenoids which occurs on storing harvested plant material. The most active fraction from alfalfa was that obtained by extracting it with an aqueous detergent followed by chromatography on cellulose (S. Grossmann et al., Phytochem., 1969, 8: 2287; cf. J. W. Dicks, ibid., 1970, 9: 1433 for related work).

R. Ikan (in "Natural Products — a laboratory guide", Academic Press, New York, 1969, p. 89) outlines two simple laboratory procedures for isolating carotenoids from nature illustrated by capsanthin from paprika, and lycopene from tomato paste. B. H. Davies and E. I. Mercer (J. Chromatog., 1970, 46: 161) have described a semi-automatic column chromatographic separation for screening the radioactive mixture [of squalene, carotenoid precursors (phytoene, etc.), carotenoids, steroids. etc.] obtained from tracer experiments on biological systems, as used in work directed towards elucidating the mode of biosynthesis of carotenoids (cf. Section 8). Gradient elution (petroleum containing an increasing percentage of ether) was used and the eluate run from the column through consecutively: (*a*) a continuous-flow type of cell placed in an u.v./visible spectrophotometer (to determine the nature and amount of each light-absorbing component); (*b*) a continuous-flow scintillation counter (for ^{3}H and ^{14}C assays); (*c*) a hot-wire liquid chromatograph (to obtain the total mass of material being eluted); (*d*) a fraction collector. The readings from each instrument were dotted on a multi-pen chart recorder simultaneously with the fraction change to give a continuous record of the progress of the separation. In order to reduce decomposition of carotenoids during column chromatography, I. Stewart and T. A. Wheaton (ibid., 1971, 55: 325) have recommended incorporating an antioxidant in the solvent and the operation of the columns under increased pressure to reduce residence time. Repeated use of a column was claimed to have little adverse effect on resolution. Carbon tetrachloride and, to some degree, chloroform have a deleterious effect on carotenoids (p. 202) and so should be avoided during isolation/purification procedures. R. F. Bayfield et al. (J. Chromatog., 1968, 36: 54) recommend two-dimensional paper chromatography as a simple and mild method of surveying the carotenoids in crude extracts of plant material. The usual prior saponification procedure and other manipulations can be omitted, thereby minimising the opportunities for rearrangement and/or degradation. The first dimension is run using petroleum—acetone mixtures and then, after being briefly dried in the air, the paper is dipped in paraffin solution (to give a "reversed-phase" effect, 2nd Edn., p. 242) and eluted again at right angles to the first run.

For thin-layer (t.l.c.) separations involving carotenoid 5,6-epoxides (which are prone to rearrangement to furanoid oxides on alumina or silica) K. Egger recommends basic magnesium carbonate as adsorbent (Planta, 1968, 80: 65). Good separation of *cis/trans* isomers is also claimed with this adsorbent (see also H. Nitsche and Egger, Phytochem., 1969, 8: 1577 in this respect).

Information on the thin-layer, paper, and column chromatographic properties of many carotenoids and their derivatives has been collected together in tabular form by F. H. Foppen (Chromat. Reviews, 1971, 14: 133). For the t.l.c. separation of complex mixtures of hydroxy- and dihydroxy-carotenoids and their epoxides, R. E. Knowles and A. L. Livingston (J. Chromatog., 1971, 61: 133) recommend spotting the mixture on to the corner of a two-adsorbent plate (i.e. a strip of a second adsorbent along one side of an only partly covered plate) the mixture first being developed up to the narrow strip and then, after drying, developed at right angles up into the other adsorbent. A good separation of compounds with closely similar polarities, such as zeaxanthin/lutein, is obtained. L. R. G. Valadon and R. S. Mummery (Phytochem., 1972, 11: 413) have reported R_f values for a series of carotenoids on silica- and alumina-impregnated paper using circular chromatography. S. Shimizu (J. Chromatog., 1971, 59: 440) has used a column of Sephadex — which separates by molecular size rather than by adsorption — to separate carotenoids and chlorophylls.

H. Thommen (Intern. J. Vitamin Res., 1967, 37: 176) has described a new sensitive spray reagent for detecting polyene aldehydes and ketones.

3. Structure and synthesis

Determination of structure. Although the small-scale, and other, chemical tests are still used to some extent, the structure of a new compound is now mainly elucidated by the wholesale application of physical methods, coupled with an indication of polarity gleaned from partition and chromatographic tests carried out during isolation of the compound. The u.v./visible absorption spectrum (usually recorded in petroleum or hexane; but also in ethanol if conjugated-carbonyl is suspected; cf. 2nd Edn., p. 254) gives an indication of the chromophore and of the presence, or otherwise, of *cis*-bonds. Additional, and easily obtainable, information can be acquired at this stage by observing the effect of adding iodine or acid to the solution in the cell (cf. 2nd Edn., p. 245). Measurement of the i.r., and more particularly the p.m.r. and mass spectra (cf. p. 204) of the unknown, and comparison of the data obtained with those from compounds of known structure, often leads to the full structure being deduced.

A recent development has been the increased effort devoted to determining the absolute configuration at each chiral centre of the optically active carotenoids (cf. p. 210).

The detection of *tert*-hydroxyl groups in the presence of *primary/secondary* —OH(s) is often difficult by standard i.r., chemical reactivity, etc. methods, tending to rest on negative evidence. The discovery (A. McCormick and S. L. Jensen, Acta chem. Scand., 1966, 20: 1989) that although *tert*-OH's do not give acetates even under vigorous conditions, they do readily give trimethylsilyl (t.m.s.) ethers (on treatment with trimethylchlorosilane/hexamethyldisilazane/pyridine mixture), has recently been frequently* used to detect, and estimate the number of, *tert*-hydroxyl groups. Thus if the carotenoid itself, its fully acetylated derivative, and the t.m.s. ether of the latter are all subjected to mass spectrometry, the number of *primary/secondary* and of *tertiary* hydroxyls can be obtained from the masses of the respective molecular ions. In addition, the m.s. fragmentation patterns of mono-, di-, etc. t.m.s. ethers show characteristic ions at M-90, M-2 x 90, etc. due to loss of Me_3SiOH and, from the most common type of *tert*-OH (viz. $-CMe_2OH$; cf. Sections 7c, 7h), a strong ion at m/e 131:

---

* E.g. in the structure elucidation of the C_{50} carotenoids, such as bacterioruberin. The marked stepwise increase in Rf value on t.l.c. as successive *tert*-OH's are silylated also indicates the number present.

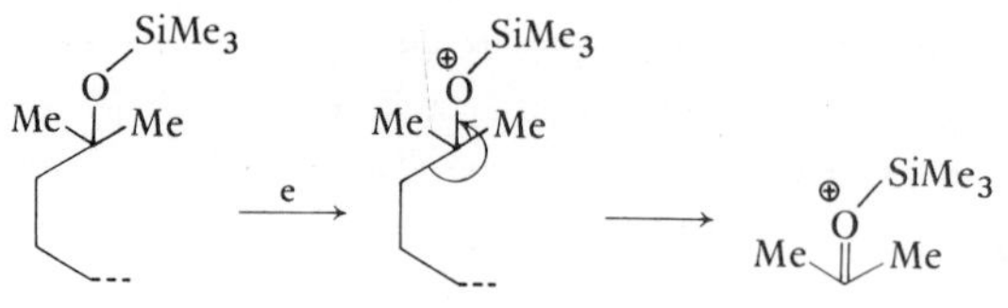

(*m/e* 131)

A further test for allylic hydroxyl has been described using the marked, and characteristic, lability of the corresponding trichloroacetate (details p. 215).

Synthesis. Advances in this area have been reviewed by B. C. L. Weedon (Pure Appl. Chem., 1967, 14: 265; 1969, 20: 531) and O. Isler et al. (ibid., 1967, 14: 245). Many carotenoid syntheses (cf. Isler et al., pp. 251, 253) start from one of two basic building blocks, the β-C_{19}-aldehyde (XX of 2nd Edn., p. 246) or crocetindial (C_{20}; XXIV of 2nd Edn., p. 248). New syntheses are given under the appropriate compound in Sections 5 to 7.

A new method of converting the configuration about a double bond from *cis* into *trans* has been described by C. S. Foote et al., as is given on p. 203.

The full papers describing the method of introducing *cis* double-bonds using the lactol (IV) have been published (G. Pattenden and Weedon, J. chem. Soc., C, 1968, 1984; C. F. Garbers et al., ibid., 1968, 1982). The method involves employing (IV), which contains the double-bond locked in the *cis*-configuration, in a Wittig reaction, under which conditions the lactol undergoes equilibration as shown, before coupling in the usual way:

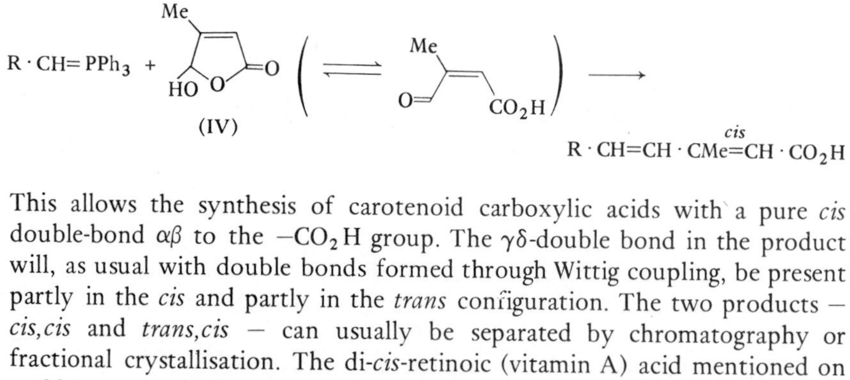

$$R \cdot CH=CH \cdot CMe=CH \cdot CO_2H \quad (cis)$$

This allows the synthesis of carotenoid carboxylic acids with a pure *cis* double-bond $\alpha\beta$ to the $-CO_2H$ group. The $\gamma\delta$-double bond in the product will, as usual with double bonds formed through Wittig coupling, be present partly in the *cis* and partly in the *trans* configuration. The two products — *cis,cis* and *trans,cis* — can usually be separated by chromatography or fractional crystallisation. The di-*cis*-retinoic (vitamin A) acid mentioned on p. 227 was synthesised using this technique. Similarly, carotenoid carboxylic acids with the $\gamma\delta$-double bond *cis* can be synthesised by carrying out a Wittig coupling with (V) (prepared as below, cf. Weedon et al., ibid., 1970, 235) as in the total synthesis of the dimethyl ester of natural (16-*cis*) bixin (p. 286):

$$(IV) \left(\rightleftharpoons \underset{HO_2C}{\overset{Me}{\diagup}}{=}O \right) + H_2C \overset{\overset{O}{\parallel}}{\underset{CO_2Me}{\diagup}}{P(OEt)_2} \xrightarrow{1.}$$

$$\underset{HO_2C}{\overset{cis}{\diagup}}\overset{trans}{\diagdown}_{CO_2Me} \xrightarrow{2,3.} O{=}\overset{c}{\diagup}\overset{t}{\diagdown}_{CO_2Me}$$

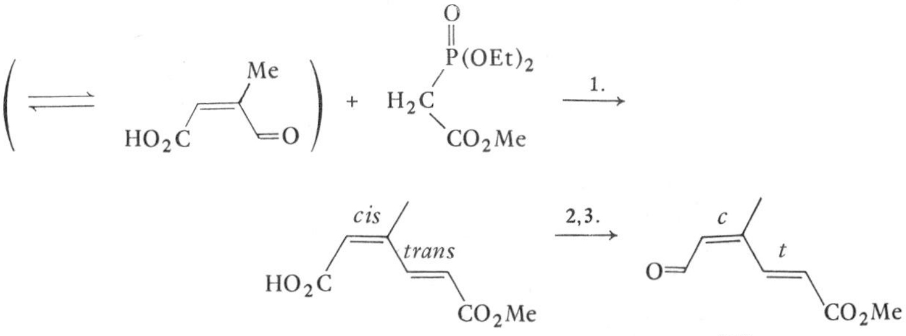

$$(V)$$

1. Horner reaction; 2. $SOCl_2$; 3. $LiAlH(OBu^t)_3$ reduction at $-78°$.

Pattenden and Weedon (1968, loc. cit.) also outline the possibilities of carrying out such, "*cis*-olefination", reactions using the isomer of IV (methyl group α to carbonyl function) or the unmethylated analogue, and review previous — and less stereospecific — methods of introducing *cis* double bonds into carotenoids and related compounds.

J. C. Leffingwell (Tetrahedron Letters, 1970, 1653) has shown that compounds of the following, potentially useful, type can be prepared in high yield from the readily available bulk chemicals isophorone and, respectively, methyl vinyl ketone and acrylonitrile:

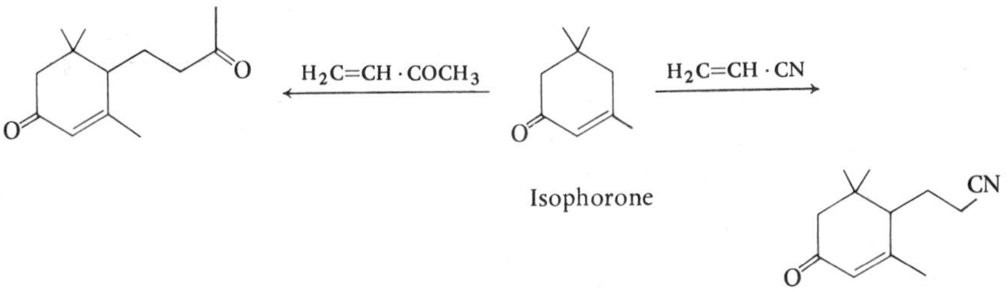

$$\xleftarrow{H_2C{=}CH \cdot COCH_3} \quad \text{Isophorone} \quad \xrightarrow{H_2C{=}CH \cdot CN}$$

4. Properties of carotenoids

(a) General properties; cis–trans isomerism

The oxidative degradation undergone by β-carotene and other carotenoids under various conditions has been further studied. A. H. El'Tinay and C. O. Chichester (J. org. Chem., 1970, 35: 2290) have confirmed that on treatment of β-carotene with oxygen in hot toluene, the 5,6- and 5,6:5',6'-epoxides and their furanoid derivatives (2nd Edn., p. 293, for definition) are the initial oxidation products. C. R. Seward et al. (Lipids, 1969, 4: 629) have found that β-carotene is destroyed particularly rapidly when dissolved in carbon tetrachloride—ethanol mixtures, even in the dark, the carbon tetrachloride being the destructive component; chloroform also showed evidence of activity in this respect. In neat carbon tetrachloride, nominally isolated from the atmosphere, the initial products are again the 5,6-epoxides, apparently produced from traces of oxygen, or water, in the solvent (J. B. Davis, unpubl.). The hv/O_2 bleaching of violaxanthin (which see), which is already epoxidised at the 5,6 and 5',6' positions, causes chain cleavage, at $C_{(11)}/C_{(12)}$, instead. The dye-sensitised photo-oxidation (hv/O_2/sensitiser — equivalent to singlet oxygen) of carotenoids leads to the formation of different products, at least partly derived from the 1,4-addition of O_2 across the $C_{(5)}-C_{(8)}$ diene unit (cf. p. 299).

Blatz and Pippert (J. Amer. chem. Soc., 1968, 90: 1296) have reiterated their point that to get meaningful and reproducible λ_{max} values for the intense blue colours obtained on treating carotenoid solutions with antimony trichloride and other acidic reagents, the absorption spectrum of the solution must be monitored from the moment of mixing, since decomposition products with different λ_{max} values soon start to appear. Much of the confusion, and lack of agreement, in the older literature concerning the precise position of such absorption bands can probably be ascribed to technique-factors of this kind. Whilst most carotenoids give *a* band in the 550 to 700 nm (mμ)* region with the Carr–Price reagent ($SbCl_3$/$CHCl_3$), this band is only the strongest one in the spectrum for those compounds, such as vitamin A and its derivatives, with a relatively *short* chromophore. For compounds such as β-carotene with a longer chromophore, the *major* peak occurs at correspondingly longer wavelengths ($\sim$1000 nm: cf. Wasserman, 1954, referred to before; also F. D. Collins, Nature, 1950, 165: 817), though subsidiary maxima (possibly due to decomposition products) are seen in the 550–700 nm region as well. The retinol (vitamin A) + acid reaction is mentioned on p. 242.

Cis–trans *isomerism*. There are several methods of converting a carotenoid with all its double bonds *trans* into an equilibrium mixture of *cis/trans* isomers (the number of isomers formed, and their relative proportions, depending on the particular compound). The most commonly used method involves briefly exposing a solution of the compound containing a trace ($\leqslant$

---

* Since this chapter in the main work was written, the unit used for wavelength in the ultraviolet/visible region has been changed by international agreement from millimicrons (mμ) to nanometres (nm). However, since 1 mμ = 1 nm, the λ_{max} figures given throughout this present article are unchanged from those given before.

1% w/w of carotenoid) of iodine to light. It is often necessary to use this procedure in reverse in synthetic work, to convert the final mono-*cis* carotenoid (often 15,15'-*cis*) into its all-*trans* form, but the usual methods (hv/I_2 as above or refluxing in petroleum) both lead to the compound's standard equilibrium mixture of *cis/trans* isomers from which the all-*trans* has to be isolated by crystallisation or chromatography:

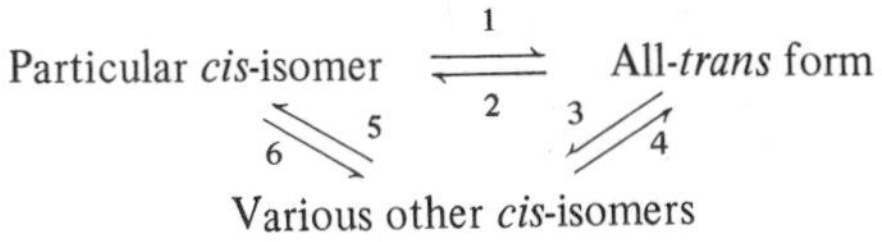

$$\text{Particular }cis\text{-isomer} \underset{2}{\overset{1}{\rightleftharpoons}} \text{All-}trans\text{ form}$$

Various other *cis*-isomers

Recently, C. S. Foote and co-workers (J. Amer. chem. Soc., 1970, 92: 5218) have discovered what promises to be a specific, *non-reversible* (i.e. catalyses only reactions 1 and 4 in the above scheme), and rapid method of converting a *cis*-carotenoid into its all-*trans* form in high yield. The *cis*-isomer is dissolved in a solution in which singlet oxygen is being generated, either chemically (from $H_2O_2/NaOCl$) or, by passing oxygen and irradiating the solution in the presence of a sensitising dye such as methylene blue.

(b) Physical properties

(i) Visible and ultraviolet absorption spectra

Concerning the change in wavelength units from $m\mu$ to nm, see the footnote on p. 202.

As mentioned under Section 3 above, determining the visible/ultra-violet (or "*electronic*") absorption spectrum is still usually one of the first measurements to be carried out on a new carotenoid, since it provides valuable information on the compound's chromophore, and on certain functional groups, for a minimal outlay of substance.

The manner in which the absorption peaks respond in position and intensity to changes in structure was outlined in the 2nd Edn., and refs. were given to reviews containing typical curves. In the latter respect, the reviews by S. L. Jensen and A. Jensen (Prog. Chem. Fats Lipids, 1965, 8: 141) and U. Schwieter et al. (Pure Appl. Chem., 1969, 20: 367) give further curves and information.

The steady λ_{max} shift towards longer wavelengths, seen in the visible absorption spectrum of every carotenoid as the solvent is changed from hexane through ether etc. to carbon disulphide (2nd Edn., p. 254) has long been realised to bear, with minor exceptions, a direct relationship to the refractive index of the solvent (ca. 1.37 for hexane to ca. 1.62 for carbon disulphide). Further theories rationalising this effect, in terms of the varying interactions to be expected between dissolved carotenoid molecules and the surrounding solvent, have been outlined by F. Feichtmayr et al. and others (cf. Tetrahedron, 1969, 25: 5383). Feichtmayr et al. also give λ_{max} values (in cm^{-1}; 500 nm = 20,000 cm^{-1}), extinction coefficients, and integrated absorption intensities (more meaningful, in theoretical terms, than extinction coefficients, which depend to some extent on the broadness of the band concerned) for a selection of carotenoid hydrocarbons. The spectra of a further selection of carotenoids have been recorded at low temperature (−196°), this time some conjugated-carbonyl compounds being included (B. Ke et al., Biochim. biophys. Acta, 1970, 210: 139). Each spectrum shows an increase in fine structure and in individual band intensity on lowering the temperature, fine structure becoming quite marked even for compounds such as canthaxanthin which show negligible fine structure at normal temperatures. This effect, which is essentially caused by a general *narrowing* of the absorption bands with decreasing temperature, has been ascribed to a tendency for the spread of molecules with slightly different conformations (and hence energies) in ground and excited states at normal temperatures, to freeze out into the two most favoured conformations as the temperature is lowered, thereby reducing the spread of ΔE values on π to π^* excitation (cf. Ke et al., loc. cit.; also B. Honig and M. Karplus, Nature, 1971, 229: 558).

(ii) Infrared spectra

There has been little development in this area, and the importance of this technique in structure-elucidation work has tended to diminish with the increasing use of p.m.r. and mass spectroscopy. However, it is still used as a convenient way of establishing the presence or absence of carbonyl, allene, acetylene (sometimes), and hydroxyl groups (as noted under the new compounds described in Section 7), and requires only ~0.1 mg of sample (potassium bromide disc; although OH groups do not always show up very clearly under these conditions, they can always be confirmed, if there is doubt, by mass spectroscopy and/or small-scale chemical tests).

C. Bodea et al. (Ann., 1966, 697: 201) have confirmed that *retro*-type (2nd Edn., p. 233 for definition) carotenoids give *two* (close) bands in the *trans* CH=CH region. Jensen and Jensen (1965, op. cit., p. 147) list pre-1965 references to published i.r. data on carotenoids.

(iii) Nuclear magnetic resonance

(a) Proton magnetic resonance (p.m.r., [1]H-n.m.r.)†. Further extensive use has been made of this technique in elucidating the structures of new

---

† Peak positions are given in τ (tau). To convert data given in δ values to τ, deduct from 10 (and vice versa). For representative curves see the Schwieter, Englert et al., and the Weedon refs.

compounds (cf. Sections 5 and 7), and it has often provided a vital clue in the discovery of a new structural type (e.g. lycoxanthin and lycophyll; decaprenoxanthin). It has also been used to determine the stereochemistry of each of the central three double bonds of the phytoene obtained from various sources (p. 339). This work has mostly been carried out on standard 60 or 100 MHz instruments. However, a selection of compounds has recently been run at 220 MHz, in an attempt to unravel the complex olefinic-proton region (see below).

The effect of adding a europium shift-reagent (commonly used nowadays to stretch out overlapping absorptions — in those molecules containing a polar group(s) for the Eu to "attach" itself to) to a series of carotenoids, has been examined by H. Kjosen and S. L. Jensen (Acta chem. Scand., 1972, 26: 2185). The result was similar to that found with other organic compounds (cf., e.g., J. K. M. Sanders and D. H. Williams, Nature, 1972, 240: 385) and led to information normally hidden within the complex olefinic-H and methylene-H multiplets being revealed. This technique was also used in the work on peridinin (p. 335); and it could be of value in determining the shape of (cis) carotenoids.

Methyl groups. The methyl group ("b") on the penultimate double bond in part-structure VI (e.g. in oscillaxanthin, plectaniaxanthin and myxoxantho-phyll) appears to occur at slightly but distinguishably higher field (i.e. higher τ value) than those methyl groups ("a") nearer the centre of the polyene chain. The terminal $Me_2C(OH)$-methyl groups of the last-mentioned compounds, being adjacent to an asymmetric carbon, are non-equivalent and

(VI)

(VII)

therefore (for reason, cf. e.g. E. I. Snyder, J. Amer. chem. Soc., 1963, 85: 2624) can, and usually do, appear as two separate 3H singlets. The effect of the 7,8-triple bond on the position of the end-group methyl groups in VII as compared with the analogous polyene is noted under alloxanthin (p. 302; see also p. 207). The effect of having an allene function at $C_{(6)}$–$C_{(8)}$ is noted on p. 288. A possible method of distinguishing between —OH at $C_{(3)}$ and at $C_{(2)}$ on a β-end-group (p. 193 for definition), which has always been difficult (the hydroxyl in a new carotenoid often having been assigned to

$C_{(3)}$ in the past only by analogy with other carotenoids), has been noted by Arpin and Jensen (Phytochem., 1969, 8: 185) and applied by them to this problem in the case of rubixanthin (p. 256). This depends on the assumption that the methyls at $C_{(1)}$ would be non-equivalent, and so give two separate peaks, if there were an —OH as close as $C_{(2)}$. However, the $C_{(1)}$-methyl groups in rubixanthin give a 6H *singlet*, as in 3-hydroxycarotenoids. Also acetylation of zeaxanthin, a typical 3-hydroxycarotenoid, causes a characteristic shift in the $C_{(1)}$-methyl group signals and makes them non-equivalent (due to the differing influence of the bulky, and anisotropic, —OAc group) and acetylation of rubixanthin had the same effect.

B. C. L. Weedon has listed, with references, the positions of the methyl-group signals (and —OMe, —OAc, —CHO, and certain other groups where present) for a large number of carotenoids and their derivatives (Fortschr. Chem. org. Naturst., 1969, 27: 98). For similar data on a range of retinol (vitamin A) derivatives, see p. 388 of the Schwieter, Englert et al. ref. given below.

Olefinic-protons. In contrast to spectra run at 60 or 100 MHz, in those run at 220 MHz (U. Schwieter, G. Englert et al., Pure Appl. Chem., 1969, 20: 383; cf. also Patel, p. 228) the complex olefinic-proton patterns are sufficiently spread out for the individual doublets, triplets, etc. to be picked out, and to be assigned to individual protons by comparison with model and partially deuterated compounds and by decoupling experiments, etc. (cf. Schwieter, Englert et al., p. 386). In this way, the latter authors (p. 400) have shown that in lycopene the lowest field protons are the $C_{(11)(11')}$*- and $C_{(15)}$-H's at ~τ 3.4; the $C_{(7)}$, $C_{(12)}$, $C_{(8/14)}$, and $C_{(10)}$-H's occur at progressively higher field (up to ca. τ 3.8), the $C_{(6)}$ (end-of-chain) -H at ca. 4.05, and the $C_{(2)}$ (unconjugated) -H at ca. 4.9. The peaks in the spectra of

Lycopene (= I)　　　　　　　　　　　　　　β-Carotene

the other symmetrical carotenes studied (β- and ϵ-) were assigned similarly, and those in unsymmetrical carotenes (α-, γ-, and δ-) by using the values found for the corresponding symmetrical compounds. The major olefinic-region difference between lycopene and its cyclised analogue β-carotene

* The $C_{(11)}$- and $C_{(11')}$-H's are of course equivalent in the symmetrical compounds; similarly the $C_{(15)}$ and $C_{(15')}$, the $C_{(7)}$ and $C_{(7')}$, etc.

(apart from the loss of the $C_{(2)}$- and $C_{(6)}$-H signals), is in the position of the $C_{(7)}$- and $C_{(8)}$-proton signals which both occur at rather higher field in the latter (positive shielding by the ring double bond) and, fortuitously, appear at virtually the same position as one another and so give a *characteristic 4H* (7,8- +7′,8′-protons) *singlet* at $\sim\tau$ 3.85. This is of value in detecting 7,8-triple bonds (cf. VII) when present (see p. 301).

Similar data have been obtained (Schwieter, Englert et al.; Patel; loc. cit.) for zeaxanthin (same positions as β-carotene), canthaxanthin (7 (and 7′), 8, 10, and 12 protons shifted to lower field by the conjugated carbonyl group), a series of compounds of type VIII (commonly used as synthetic intermediates), and a wide selection of retinol (vitamin A) derivatives with particular reference to the effect of having *cis*-bonds at various positions.

(VIII)

(*b*) ^{13}C-*n.m.r.* Of the carbon in the skeleton of an organic molecule 1.1% is ^{13}C-carbon which, like ^{1}H, is n.m.r.-active. Unfortunately, the ^{13}C nucleus gives an inherently weak magnetic resonance response, and the low natural concentration and the extensive splitting of the signals obtained, by coupling with neighbouring protons, normally leads to a very weak spectrum. However, developments such as the use of signal-accumulating techniques and the removal of the splitting effect by noise (broad-band) decoupling at the proton resonance frequency (and, more recently, of Fourier-transform techniques) has enabled useful natural-abundance ^{13}C-n.m.r. spectra to be obtained on a routine basis (cf. refs. given below and E. W. Randall, Chem. Brit., 1971, 7: 371).

In the carotenoid field, the spectra of retinyl (vitamin A) palmitate and acetate, and of β-carotene and its central (15,15′)-acetylenic and central-*cis* analogues have been recorded by E. Wenkert et al. and J. D. Roberts and co-workers (J. Amer. chem. Soc., 1969, 91: 6879; Proc. Nat. Acad. Sci., U.S., 1970, 65: 288), and most of the signals in each spectrum unambiguously assigned, partly instrumentally (by "off-resonance decoupling" to detect, selectively, the fully substituted carbon atoms) and partly by comparison with the spectra of other compounds and of model compounds such as the ionones and simple olefinic molecules. The olefinic carbon atoms (carbons* 5 to 14 and 5′ to 14′; plus 15 and 15′ in the β-carotenes) occurred mainly in the range 54 to 68 p.p.m. upfield from the $^{13}CS_2$ reference peak, well separated from the remaining, saturated carbons (mainly at 150 to 180

---

* Using standard carotenoid numbering (a slightly different system was used by Roberts).

p.p.m.). The acetylenic carbon atoms in the 15,15′-dehydro-β-carotene were readily detected, being well separated from all other peaks (at 93.5 p.p.m.). The triple bond's presence could also be inferred from the effect it had on the olefinic carbon α (shifted upfield from the normal olefinic-C position) and β (shifted downfield) to it. These effects will probably be of value in detecting new acetylenic carotenoid structures in nature (cf. Section 7g).

(iv) Mass spectrometry

As can be seen from the methods used for structure elucidation given under the paragraphs describing the individual new carotenoids in Sections 5 and 7, mass spectrometry has now become of major importance in this field, both as a method of determining the molecular weight and the molecular formula [by precise mass measurement of the molecular ion ($M^\oplus$), of particular value in detection of acetylenic, C_{50}, and glycosidic carotenoids, Sections 7g, h, b/c] ; and in elucidating part or all of the gross structure from a study of the fragmentation pattern. In this last area, basic research with a wide variety of known compounds is gradually leading to an understanding of the fragmentation pathways of many different types of carotenoid structure so that the appearance of an ion, or group of ions, at certain mass number(s) can often be used, with some degree of certainty, as evidence for the presence of a particular structural unit. This aspect of mass spectrometry was extensively reviewed in early 1969 by Weedon (1969, loc. cit., p. 110), Schwieter, Englert et al. (Pure Appl. Chem., 1969, 20: 365, p. 402), C. R. Enzell (ibid., p. 497), and Enzell et al. (Acta chem. Scand., 1969, 23: 727), but the field is still in a state of rapid development and much original work continues to appear (H. Budzikiewicz et al., Monatsh., 1970, 101: 579; G. W. Francis, Acta chem. Scand., 1969, 23: 2916; Enzell and co-workers, ibid., 1971, 25: 85, 271). Some of the characteristic fragmentations mentioned in Sections 5 and 7 are reviewed below.

(1) In *acyclic hydrocarbons* such as lycopene and the carotenoid precursors (phytoene, etc., p. 335), cleavage of the doubly (or "*bis*") allylic single bonds is, as expected, prominent. Thus the end-groups of such compounds undergo 3,4-cleavage giving $m/e = 69$ and M-69 ions:

$$Me_2C{=}CH \cdot CH_2^\oplus$$
$$(m/e = 69)$$

(uncharged residue)

and also —

$$Me_2C{=}CH \cdot CH_2$$
(uncharged residue)

$$H_2\overset{\oplus}{C}$$

(M-69)

The M-69 ion is relatively weak due to subsequent further fragmentations, including the loss of toluene (92) or xylene (106) units from the polyene chain. This is a characteristic of all carotenoid mass spectra due to coiling of the chain followed by extrusion.

In the precursors (or "hydrolycopenes"), where there are several doubly allylic bonds, those *adjacent to the chromophore* are cleaved preferentially (Weedon, 1969, loc. cit., p. 118; Weeks, Weedon et al., Nature, 1969, 224: 879). For example, ζ-carotene (p. 221) gives a weak M-69 ion from cleavage at the 3,4 position but a stronger ion from cleavage at 7,8 (M-137). Hence not only can the hydrogenation level of the compound be determined (from the molecular weight), but also the *position of the chromophore.* This method was used for elucidating the structures of "unsymmetrical ζ-carotene", nonaprenoxanthin, and the hydrospheroidenes (pp. 220, 317, 265).

In the $C_{(1)}$-*methoxylated* compounds of the spheroidene, spirilloxanthin, etc. type (Section 7c), the 69 and M-69$_\oplus$ ions mentioned above are replaced by ions at $m/e = 73$ [strong; $Me_2C=O$—Me, confirmed by precise mass measurement (Goodwin et al., Phytochem., 1969, 8: 1047)] and M-73, formed by cleavage of the 1,2-bond (allylic, and α to -O-) (Goodwin et al., loc. cit.; Davies et al., J. chem. Soc., C, 1969, 1266). The characteristic fragment ions produced by various other acyclic end-groups are listed by Enzell (1969, loc. cit., p. 512).

(2) *Cyclic end-groups.* α-End-groups (definition, p. 193) undergo a characteristic retro-Diels—Alder cleavage resulting in the production of an M-56 ion for an unsubstituted α-type ring, or of an ion showing correspondingly larger mass-loss for a 2-substituted ring (e.g. M-140 for decaprenoxanthin; p. 310 for formula and mechanism). This is of considerable use in the detection of α-rings in new compounds. β-End-groups are less easily detected, though an ion at M-137 (cleavage at 7,8) is usually visible (Schwieter, Englert et al., p. 409; Enzell, p. 510). Furanoid oxide (p. 246) derivatives of the latter give characteristic fragments at m/e 165 and 205 (or 181 and 221 for 3-hydroxy derivatives) probably due to the formation of ions of type (IX) and (X) (cf. Weedon, 1969, loc. cit., p. 114):

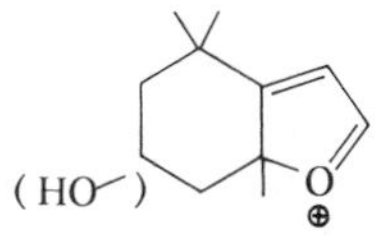

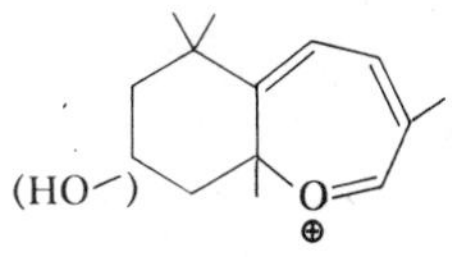

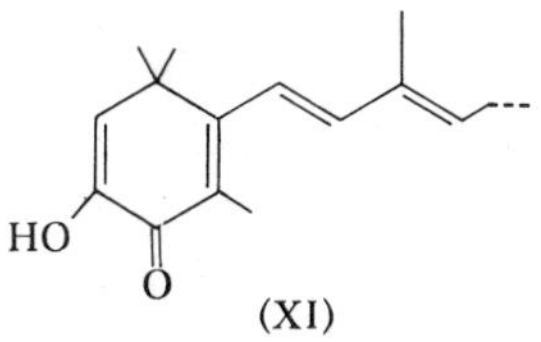

The corresponding 5,6-epoxides also usually exhibit such ions in their spectra, possibly due to spontaneous rearrangement to the furanoid form occurring on the heated probe (cf. Weedon, 1969, loc. cit.). The diosphenol end-group (XI; as in astacene, and the "dehydro" derivatives of adonirubin, etc. noted on p. 277) gives a strong, and characteristic, ion at m/e 203, as does the capsorubin (p. 279) type end-group, at m/e 109. Both are of diagnostic value (J. Baldas et al., Chem. Comm., 1969, 415). The characteristic m.s. peaks given by several other cyclic end-groups are listed by Enzell (loc. cit., p. 513), and some probable mechanisms of formation of same given by Weedon (loc. cit., p. 115).

The *technique* used for obtaining satisfactory mass spectra on these inherently unstable compounds is mentioned by Schwieter, Englert et al. (loc. cit., p. 405), D. Thirkell et al. (J. gen. Microbiol., 1967, 49: 157), and Enzell and co-workers (1971, loc. cit.). Direct insertion of the sample ($\sim$0.01 mg), placed on the tip of a probe, into the spectrometer's ion chamber is generally used. To minimise pyrolysis of the compound, as low a temperature as possible ($\sim$200°) is used to volatilise the sample, but even then some decomposition usually occurs, resulting in spectra being rather unreproducible as regards peak intensities (rhodopinal and allied compounds are particularly sensitive in this respect; cf. p. 268). The molecular ion, sometimes accompanied by a quite prominent M-2 peak (Enzell et al., 1969, loc. cit.), is reasonably intense in unsubstituted carotenoids but can become weak or even non-existent in the more heavily hydroxylated examples (when the highest-mass ion would probably be M-H$_2$O). The glycosides (Sections 7b,c,e,h) are usually examined as their per-acetate derivatives. In this way, even the octa-hydroxy compound oscillaxanthin (which see for ref.) gave a detectable molecular ion (at m/e 1144). For the mass spectrometric detection, and differentiation, of tertiary and other hydroxy-groups, see Section 3. In all cases, those spectra run at 70 eV contain a preponderance of relatively uninformative low-mass ions, due to subsequent break-down of the fragments formed initially, but this effect can be reduced, and the relative intensity of the molecular and other high-mass ions increased, by using a lower (12 eV) ionising voltage (Schwieter, Englert et al., loc. cit.; Budzikiewicz et al., 1970, loc. cit.). Metastable ions, although sometimes rather weak, have been reported with increasing frequency in recent papers, and used in the usual way to elucidate fragmentation pathways.

Mass spectrometry has also been used to determine, directly, the nature of the fatty-acid esterifying the carotenoid in a natural carotenoid ester by the appearance of R·CO$^{\oplus}$ etc. ions, from the fatty-acid residue, superimposed on the usual carotenoid spectrum (Arpin et al., Phytochem., 1969, 8: 897; cf. p. 259).

(v) Optical rotatory dispersion (o.r.d.) and circular dichroism (c.d.)

Many of the mono- and bi-cyclic carotenoids contain one or more chiral (asymmetric) centres. These are usually at $C_{(3)}$ (-OH substituent; very common; cf. Sections 7b, 7d, 7g, etc.), at $C_{(6)}$ (α-type end-group, as in α- and ϵ-carotenes), at $C_{(5)}$ ($+C_{(3)}$; in capsorubin-type compounds), at $C_{(5)}$ and $C_{(6)}$ (5,6-epoxy-carotenoids, p. 245, and azafrin), or associated with the allene group (Section 7f). The first large-scale attempt to obtain representative o.r.d. curves of a wide selection of carotenoids, and subsequently to infer the absolute configurations of chiral centres from such measurements, was made by P. M. Scopes, Weedon and co-workers in 1969 (Scopes et al., J. chem. Soc., C, 1969, 2527). The curves were recorded in the u.v. rather than the visible region of the spectrum to avoid the technical difficulties* associated with operating in regions of intense light absorption. Comparison of the various curves obtained suggested that [where A, A′ and B are respectively an optically active (chiral) end-group, a different optically active end-group, and an inactive end-group] : (a) the curve of a molecule A—A′ (a "hetero di-chiral" compound) would be a composite of the curves produced by the A half of the molecule and that produced by the A′ half; (b) the curve given by A—A (a "homo di-chiral" compound) would be similar in shape, including peak/trough positions, to that of a compound A—B but of twice the amplitude. It was assumed that converting an asymmetric centre into its enantiomer would invert the resulting curve. In this way it was shown that the two chiral centres (3,3′) in zeaxanthin have the same configuration as each other (hence an A—A molecule) and in turn the same as at the 3-position of β-kryptoxanthin (an A—B compound). Similarly the configurations at the 3- and the 6′-positions in α-kryptoxanthin, one half of which has the gross structure of zeaxanthin and one half that of α-carotene, could be related, through the respective o.r.d. curves, to the configurations of these compounds at these positions. However, the configurations were only known relative to one another and to deduce the absolute configurations of the various compounds studied required that the configuration of

Zeaxanthin

β-Kryptoxanthin

α-Kryptoxanthin

---

* Overcome to a large extent in more recently developed instruments; cf. M. Buchwald and W. P. Jencks, Biochemistry, 1968, 7: 834.

one or more key compounds, as determined by an independent and absolute method, be known. At the time of the Scopes et al. paper, this had been done only for the capsorubin molecule and other deductions had to be based on biogenetic reasoning. Since then, however, additional information has become available, so that the absolute configuration, R or S^*, at the various centres of several carotenoids is now known with some certainty. Details are given under the individual compound descriptions in Section 7.

In all 3-substituted carotenoids studied so far,† the configuration at the 3-position is R (that is, the -OH substituent is of the "β" type using steroid nomenclature), and in all carotenoids containing an α-type end-group that at the 6-position is also R (that is, the -H is "α").

C.d. measurements have been reported, on zeaxanthin (Scopes et al., loc. cit.), astaxanthin (Buchwald and Jencks, loc. cit.), rubixanthin/gazania-xanthin (Arpin and Jensen, Phytochem., 1969, 8: 185), and α- and δ-carotenes (Buchecker and Eugster, Helv., 1971, 54: 327).

(vi) Resonance-enhanced Raman (r.e.r.) spectroscopy

The ordinary Raman spectrum of β-carotene gives peaks at 1158 and 1527 cm^{-1} (C–C and C=C stretch of polyene chain) and is of limited value (cf. refs. given by L. Rimai et al., J. Amer. chem. Soc., 1970, 92: 3824). Recently however, Rimai et al. (loc. cit.) have measured carotenoid Raman spectra using as the excitation source an argon laser, whose main emission is in the region of maximal light absorption by the carotenoid (500–450 nm), and which can be tuned until maximum response is obtained from the carotenoid (near the latter's λ_{max} position). Under these conditions, Raman scattering from the carotenoid is greatly enhanced, so much so that even the very small concentrations present in for example tomato skin (Rimai et al., Nature, 1970, 227: 743) yield a clear spectrum in situ. The relatively massive amounts of other constituents present do not interfere because the system is tuned to detect only those compounds with strong absorption (of visible light) close to the λ_{max} value of the carotenoid being studied. In this way, carotenoids can be detected in living tissue without disrupting the system and without altering the compound by solvent extraction etc. Thus this technique has also been used (Rimai et al., Biochem. biophys. Res. Comm., 1970, 41: 492; J. Amer. chem. Soc., 1971, 93: 6776) to examine, directly, one of the many pigments (lumirhodopsin?) in the retina involved in the

---

* Concerning the R (and S for the enantiomer) method of denoting configuration, see Rodd, 2nd Edn., Vol. I A, p. 248; R. S. Cahn et al., Angew. Chem., intern. Edn., 1966, 5: 385.

† In fact in the 5-ring carotenoids, like capsorubin, and in α-end groups, as in lutein, the $C_{(3)}$ is said to be S because of the way in which the Sequence Rules operate (but the -OH is still of the "β" type).

visual process (cf. p. 231). Comparison of the spectrum (ν_{max} 1555 cm^{-1}) obtained from frozen ($-70°$) whole retina irradiated at $\sim$460 nm with that from all-*trans* retinal, neutral and acidified, and from a Schiff base of all-*trans* retinal and a primary amine, again both neutral and acidified, showed the latter to approximate most closely in ν_{max} (1560 cm^{-1}), indicating that the retinal is linked to the opsin (p. 230) at this stage by a protonated Schiff-base link ($-CH=\overset{\oplus}{N}H-$) (cf. pp. 236, 237).

For the use of this technique in distinguishing retinal *cis/trans* isomers, cf. p. 229.

(c) Chemical properties

(i) Oxidative cleavage at the central double bond of β-carotene (and ring-substituted β-carotenes) to give retinal (and substituted retinals)

R. K. Barua and A. B. Barua (Ind. J. Chem., 1969, 7: 528) have reported an improved (more reproducible yield; reaction easier to control) method of carrying out this, selective-oxidation, reaction. The carotene, in ether, is left to stand over a mixture of vanadium pentoxide and manganese dioxide at room temperature and the reaction followed by monitoring the absorption spectrum:

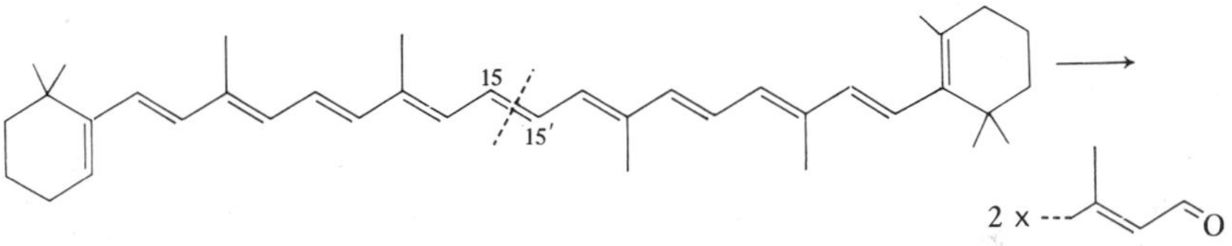

(Concerning the oxidation of retinal to the corresponding acid, see p. 227).

(ii) Reaction with N-bromosuccinimide (NBS)

A novel NBS-dehydrogenation reaction, involving the conversion of a $\cdots =CH \cdot CH=CH \cdot CH= \cdots$ unit into $\cdots =CH \cdot C\equiv C \cdot CH= \cdots$, is described under canthaxanthin (probably specific to conjugated-carbonyl compounds).

(iii) Reaction with titanium tetrachloride

V. Tamas and C. Bodea have studied the reaction of this Lewis acid with a selection of hydrocarbons (Rev. Roum. Chim., 1970, 15: 819; 1967, 12: 1517; Pure Appl. Chem., 1969, 20: 517), hydroxycarotenoids (Rev. Roum. Chim., 1969, 14: 405; 1967, 12: 1517; Pure Appl. Chem., 1969, 20: 517), and ketocarotenoids (Rev. Roum. Chim., 1969, 14: 141). In each case there was obtained, on mixing the components in benzene at 20°, a dark blue insoluble complex of carotenoid and reagent (usually in the ratio 1:1 but a 1:1/1:2 mixture with certain compounds; from $Cl^{\ominus}$ analyses) in which the reagent had attached itself (λ_{max} values, study of reaction products formed on acetone treatment) to the 1,2 and 5,6-double bonds (for acyclic hydrocarbon end-groups), to the 5,6 only (cyclic β-type end-groups), and to the oxygen (carbonyl compounds):

Lycopene
(= I, p. 217) $\longrightarrow$

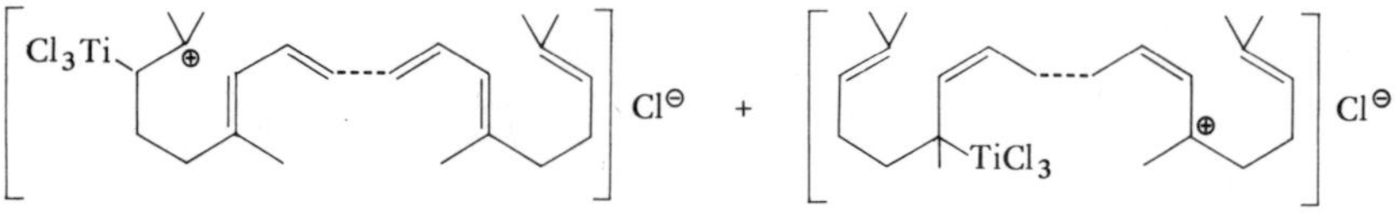

On treating the carotenoid-reagent mixture, soon after mixing, with acetone the reactions recorded below occur in good yield.

(a) Lycopene gives a mixture of cyclised (δ- and ϵ-carotenes) and dehydrogenated (3,4-dehydrolycopene) products:

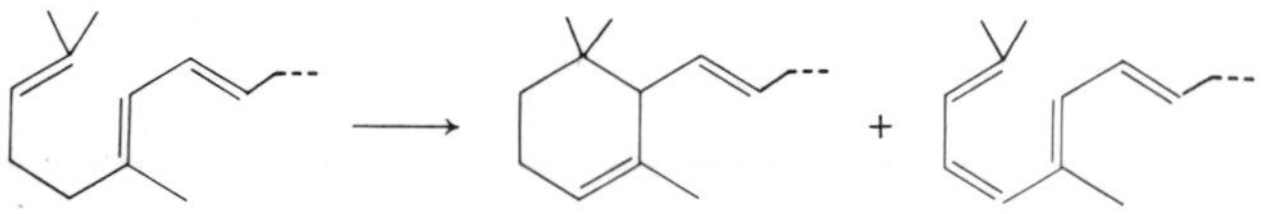

Similarly δ-carotene $\to$ ϵ-carotene, and γ-carotene $\to$ α-carotene. Cyclisation reactions of this type are difficult or impossible to achieve by straightforward treatment with mineral acids, thus making the above reagent particularly useful in this respect, but do apparently occur in nature (p. 354).

(b) β-Carotene yields retro-dehydro-β-carotene (also formed with boron trifluoride: 2nd Edn., Vol. II B, p. 264).

(c) β-Kryptoxanthin reacts as shown below:

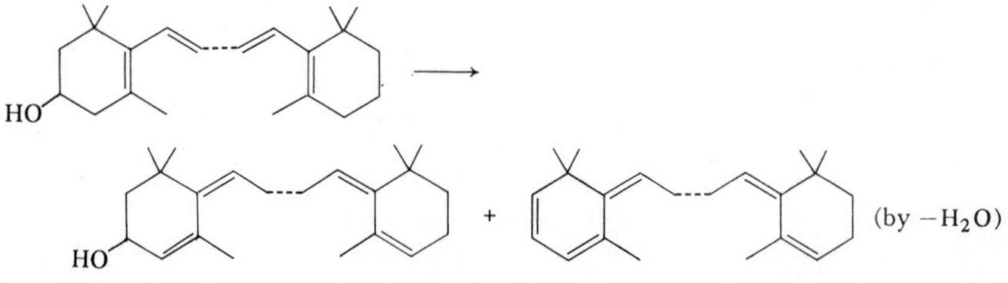

Zeaxanthin reacts in a similar manner, giving (80%) anhydroeschscholtzxanthin (a derivative of the natural carotenoid eschscholtzxanthin):

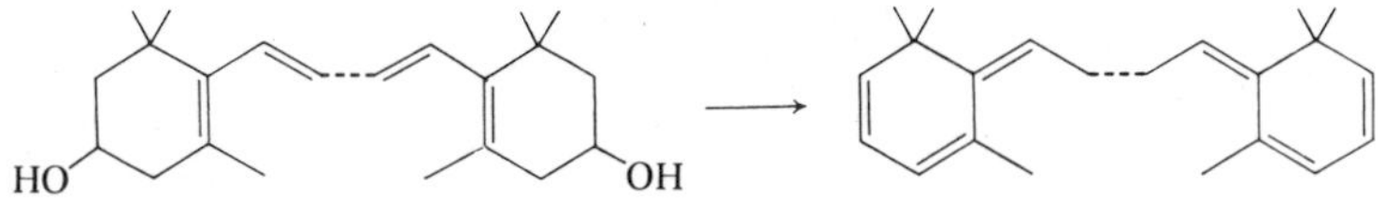

(*d*) The keto-carotenoids give back the starting materials. For the reaction with furanoid oxides, see Section 7a.

(*iv*) *Reactions of carotenoids containing allylic hydroxyl (or alkoxyl) groups*
    (1) The usual reaction with acid (allylic dehydration) does not occur in the special case of the allylic-OH, present as a side-chain -CH_2OH group, in rhodopin-20-ol (see p. 321; p. 2188 of the Aasen and Jensen, 1967, ref.).
    (2) Selective removal of allylic-OH groups from carotenoids containing both allylic and non-allylic-OH: D. Goodfellow, G. P. Moss and Weedon (Chem. Comm., 1970, 1578) have shown that on treatment of lutein with the pyridine—sulphur trioxide complex, only the allylic-OH reacts, giving the sulphate ester. Hydride ion ($LiAlH_4$) displaces the sulphate group (as with tosylates) giving α-k(c)ryptoxanthin:

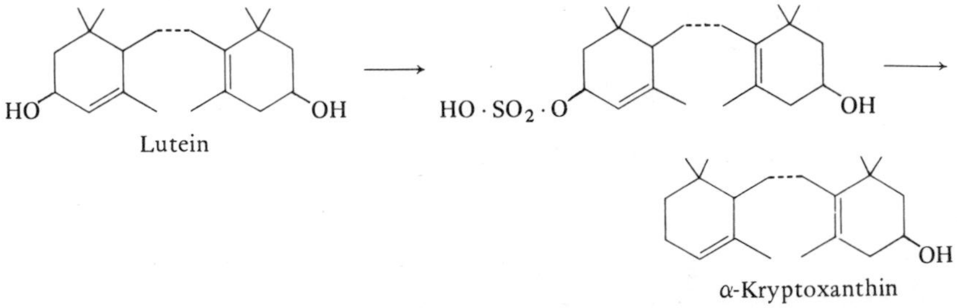

Lutein

α-Kryptoxanthin

    (3) A characteristic reaction of allylic-OH: J. Szabolcs and A. Rónai (Acta Chim. Acad. Sci. Hung., 1969, 61: 301) have shown that allylic trichloroacetates are solvolysed by methanol to give an allylic methyl ether (detectable by p.m.r., i.r.), whereas the corresponding non-allylic compounds are unaffected (used in the structural work carried out on α-kryptoxanthin, p. 254):

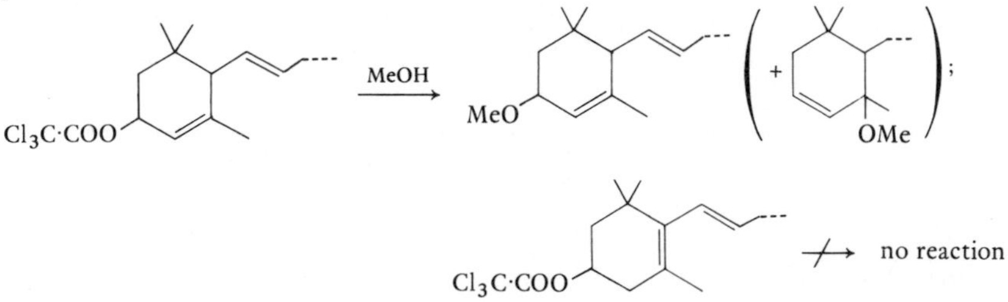

This reaction also allows the selective alkylation of allylic-OH groups in carotenoids containing both allylic and non-allylic hydroxyl groups (also possible with MeOH/H$^\oplus$: 2nd Edn., p. 265; Jensen and Hertzberg below). Thus the di-trichloroacetate of lutein (XII), prepared by treating lutein in benzene with trichloroacetyl chloride—pyridine mixture at 20°, undergoes methanolysis only at the allylic position to give XIII and its allylic rearrangement product (XIV). Alkaline hydrolysis then yields the corresponding 3-hydroxy compounds:

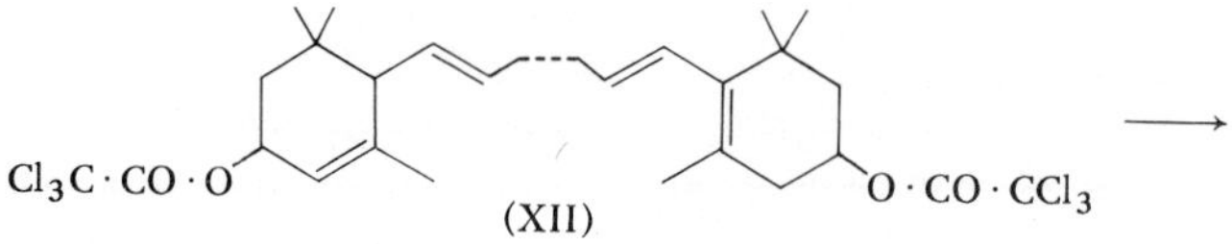

Cl$_3$C·CO·O

(XII)

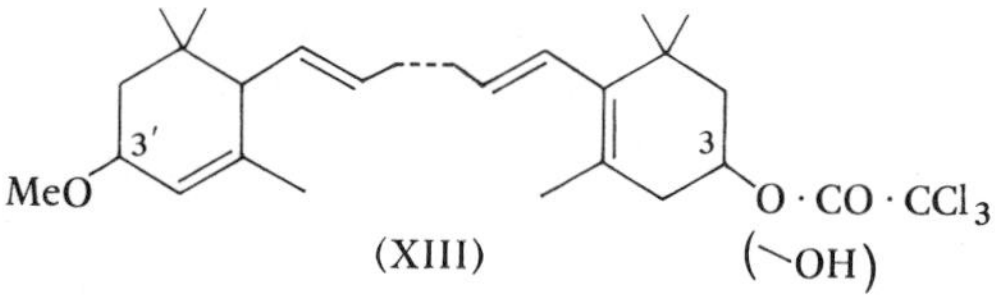

O·CO·CCl$_3$

MeO   3'

3

(XIII)   O·CO·CCl$_3$

(⁓OH)

+

OMe

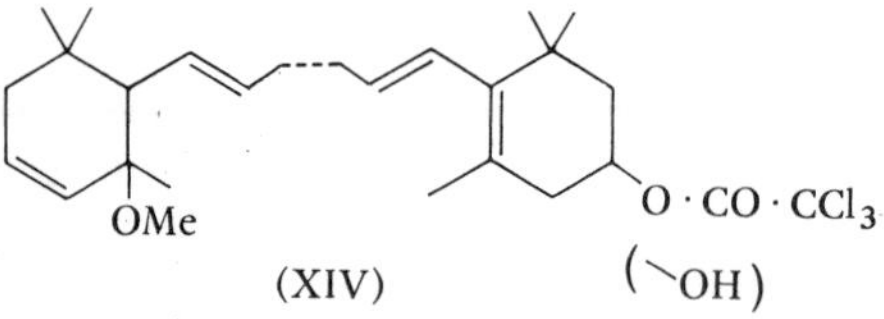

(XIV)   O·CO·CCl$_3$

(⁓OH)

(4) Oxidation to the corresponding ketone:

$$-CH=CH\cdot CHOH\cdot\ -\ \rightarrow\ -CH=CH\cdot CO\cdot\ -$$

Over recent years, treatment of the carotenoid, in benzene or ether, with a suspension of nickel peroxide (NiO$_2$) has been used increasingly often to effect this conversion. Unlike certain of the methods mentioned earlier, this method is effective even if the OH group is allylic to only a single double bond rather than a polyene chain (Jensen and Hertzberg, Acta chem. Scand., 1966, 20: 1703). Non-allylic hydroxyls are unaffected:

CH$_2$OH

CHO

As occurs with rhodopinol (p. 321) and pyrenoxanthin (p. 326).

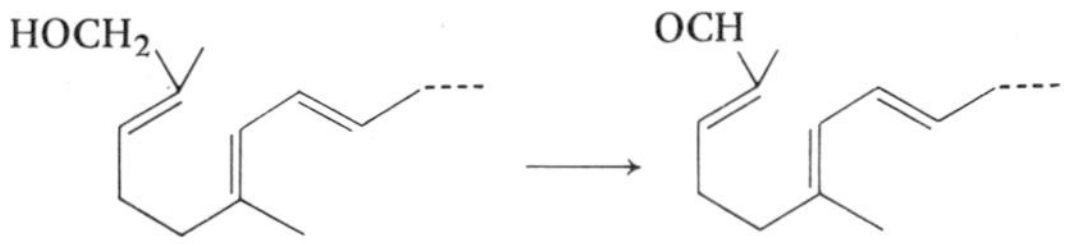

As occurs with lycoxanthin (p. 320).

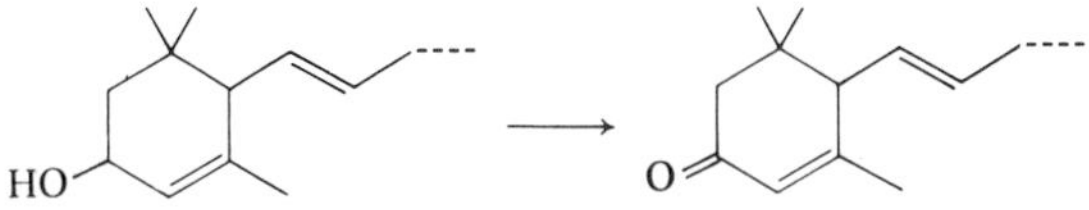

As occurs with lutein (Jensen and Hertzberg, loc. cit.).

5. Carotenoid hydrocarbons

Lycopene reacts with $Fe_3(CO)_{12}$ rather as for β-carotene below (which see for ref.), but gives a mixture of tetra- and penta-kis(iron-tricarbonyl)-lycopenes. Concerning its reaction with titanium tetrachloride, see p. 214.

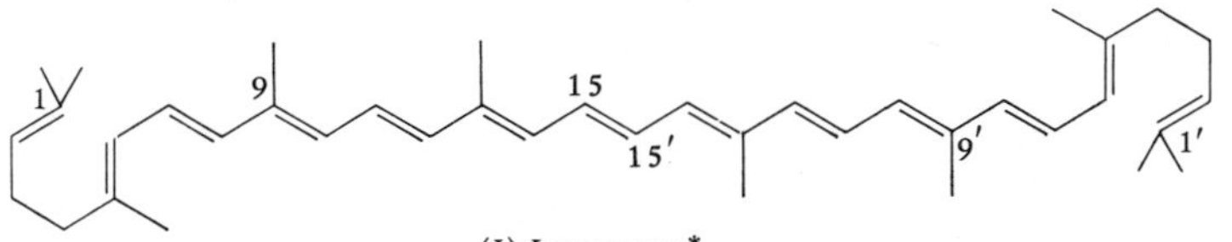

(I) Lycopene*

For further work on biosynthetic sequences involving lycopene see Section 8, and for additional (olefinic-H) p.m.r. data, p. 206.

β-**Carotene.** Further work on the production of β-carotene microbiologically from *Blakeslea trispora* has been reported. For high yields it is

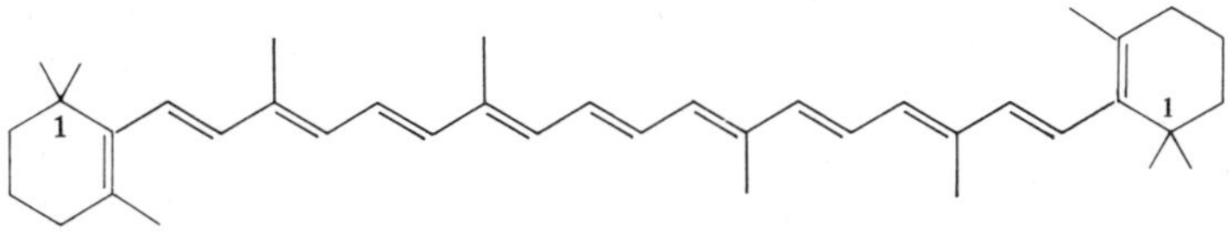

β-Carotene

---

* Carotenoids with acyclic end-groups have generally been written as here to emphasise the structural relationship they bear to those carotenoids with cyclic end-groups, and to aid comprehension. They are of course more accurately represented by the linear form:

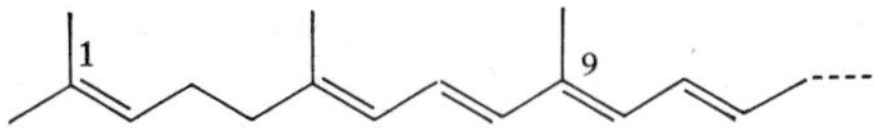

important to grow the organism in the dark (R. P. Sutter, J. gen. Microbiol., 1970, 64: 215). A. E. Purcell and W. M. Walter (J. Labelled Cmpds., 1968, 4: 94) have described a method for obtaining β-carotene uniformly labelled with ^{14}C by using the same organism and adding $^{14}CH_3 \cdot ^{14}CO_2 Na$ to the culture medium. β-Carotene is the major (86% of total) carotenoid in a variety of sweet potato examined by Purcell and Walter (J. agric. Food Chem., 1968, 16: 769).

Concerning n.m.r. data, in addition to that (methyl-group p.m.r.) referred to before, cf. p. 206 (olefinic-H p.m.r. and ^{13}C-n.m.r.).

Synthesis. By linking (NaOMe/pyridine: Michaelis—Arbuzov reaction) the following two compounds both of which are available from the industrial scale synthesis of vitamin A outlined in the 2nd Edn. (p. 285), the first being obtained by treating compound (LXXVIII; 2nd Edn., p. 285) with phosphorus tribromide (allylic rearrangement at $C_{(10)}/C_{(8)}$; bromination of both —OH groups) then triethyl phosphite (formation of phosphonate: loss of HBr), the second by manganese dioxide oxidation of vitamin A (retinol) itself. Selective reduction of the triple bond in the product so formed, followed by stereoisomerisation yields all-*trans*-β-carotene (Surmatis and Thommen, J. org. Chem., 1969, 34: 559):

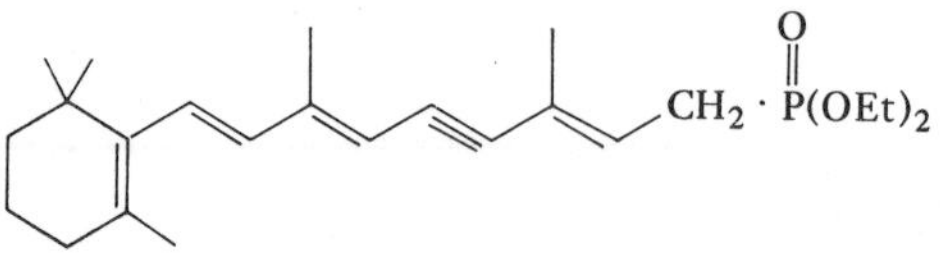

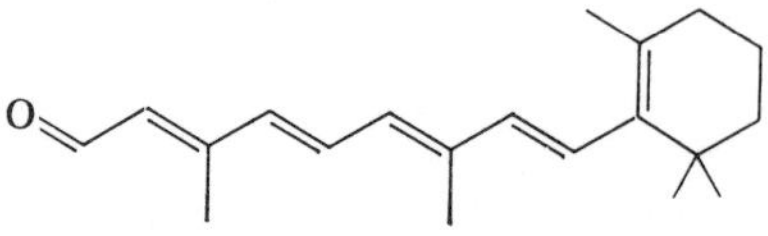

The advantage of this synthesis over that previously described from vitamin A (2nd Edn., p. 271) is that the above sequence avoids the use of (expensive) triphenylphosphine.

For the synthesis of a series of compounds based on the β-carotene skeleton but with —CO_2H groups attached to the chain at various positions, see under Section 7e.

β-Carotene reacts with dodecacarbonyl tri-iron, $Fe_3(CO)_{12}$, on prolonged refluxing in benzene to give a stable complex containing (microanalysis) four molecules of $Fe(CO)_3$ to each molecule of β-carotene [tetrakis(iron-tri-carbonyl)-β-carotene]. The $Fe(CO)_3$ units are probably attached to *cis*-diene units, formed by *s-trans* → *s-cis* isomerisation of the chain single bonds, at $C_{(7)}-C_{(10)}$, $C_{(11)}-C_{(14)}$, $C_{(15)}-C_{(13')}$, and $C_{(10')}-C_{(7')}$; i.r. and p.m.r.

data are given (M. Ichikawa et al., Z. Naturforsch., 1967, 22b: 376). Retinal
(p. 243) and lycopene (above) behave similarly.

For the oxidative fission of the β-carotene molecule to give retinal, see p.
213; and concerning its biogenesis, see p. 350.

α-Carotene. *Absolute configuration.* Although natural α-carotene has
long been known to be optically active, the absolute configuration at the
chiral centre, $C_{(6')}$, remained unknown. This has now been elucidated (C. H.
Eugster et al., Helv., 1969, 52: 1729). (±)-α-Ionone was resolved (menthyl-
hydrazone method) into the (+) and (−) enantiomers, and the latter shown
to have the S configuration at $C_{(6)}$ by demonstrating that it could be
converted into lactone (XV) [of known configuration; obtained on
degrading (+)-manool of known stereochemistry]. Hence (+)-α-ionone is R at
$C_{(6)}$. The (+) and the (−) enantiomers were then subjected to the reaction
sequence devised by Karrer to convert α-ionone into α-carotene (2nd Edn.,
p. 271). The (+) enantiomer gave an α-carotene with a positive $[\alpha]_{Cd}$ value
as for natural α-carotene (viz. + 385°), whereas the (−) enantiomer gave an
α-carotene with a negative rotation of similar magnitude, thereby proving
that natural α-carotene has the same configuration at the chiral centre as
(+)-α-ionone; that is R*

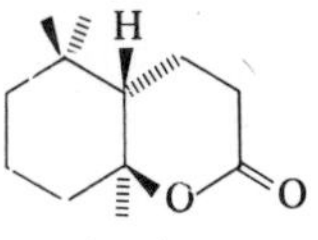

(XV)

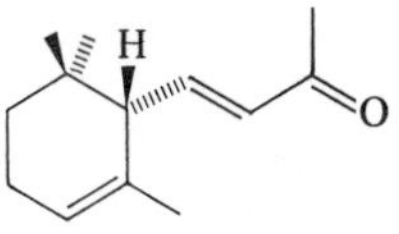

(−)-α-ionone
(6S)

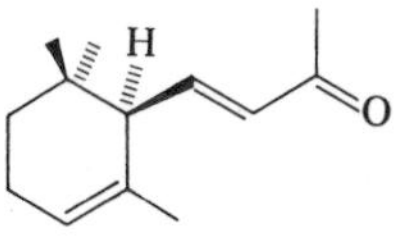

(+)-α-ionone
(6R)

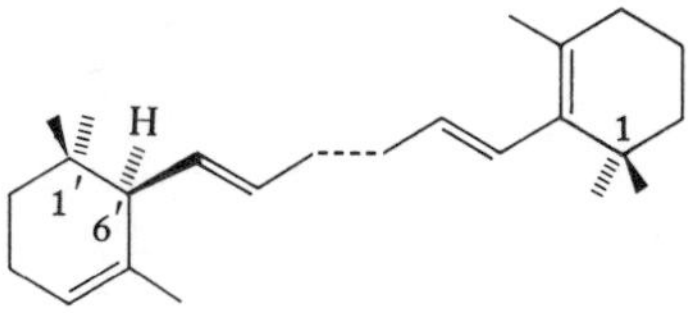

Natural [(+)-] α-carotene
(6'R)

The o.r.d. curve of natural α-carotene has been recorded (200—400 nm) by
Scopes, Weedon et al. (J. chem. Soc., C, 1969, 2527). They have also
described a higher-yield route to (+)-α-carotene than that used above (similar
in type to method (iii) described in the 2nd Edn., p. 272 but applied to (+)-,
rather than (±)-α-ionone). For c.d. measurements, see under δ-carotene
below.

---

* α-Zeacarotene (2nd Edn., p. 277) is also R at this position (Eugster, Helv., 1973, 56: 1124).

Synthesis of natural α-carotene: cf. above. For the biogenesis of α-carotene, see p. 354, and for p.m.r. data, see Schwieter, Englert et al. (Pure Appl. Chem., 1969, 20: 365, 400). For an improved (t.l.c.) isolation from carrot carotene, cf. Keefer et al., J. Chromatog., 1972, 69: 215.

δ-Carotene. *Absolute configuration.* The optical activity associated with the α-ionone end-group of this pigment has long been recognised but the absolute configuration of the asymmetric carbon responsible, unknown. R. Buchecker and Eugster (Helv., 1971, 54: 327) have recently measured the c.d. curves (200–400 nm) of δ-carotene (from "high-δ" tomatoes) and of natural α-carotene. The similarity of the two curves in general shape shows that the configuration of the asymmetric carbon of the α-end-group of δ-carotene is the same as in α-carotene, that is *R*.

For δ-carotene's (possible) formation from lycopene and conversion into ε-carotene, through the agency of titanium tetrachloride, see p. 214, and for p.m.r. data, see Schwieter, Englert et al., 1969, loc. cit.

ε-Carotene. This rather rare carotenoid has been tentatively identified by Fox and Hopkins in flamingoes (along with phoenicoxanthin, which see for refs.) and in a tomato mutant (Goodwin et al., Biochem. J., 1967, 105: 99). For p.m.r. data, see Schwieter, Englert et al., loc. cit.

Synthesis. Eugster et al. (1969, loc. cit.) have converted (+)- and (−)-α-ionones, above, into the corresponding ε-carotenes using Karrer's original sequence (2nd Edn., p. 274) and have obtained a (+)-ε-carotene ($[\alpha]_D$ + 630° in benzene) and a (−)-ε-carotene (−612°). Which of these corresponds to natural ε-carotene will not be known until the $[\alpha]_D$ value of the latter is measured.

"Unsymmetrical ζ-carotene" (*7,8,11,12-tetrahydrolycopene*). ζ-Carotene itself (heptaene chromophore placed *symmetrically*, see below) was first discovered in carrots and it was on material from that source that the structure was elucidated. Over the years, what appeared to be the same compound was discovered in many other sources. However, the material obtained from two of these sources (the bacteria *Rhodospirillum rubrum* and *Rhodopseudomonas spheroides*: cf. Jensen et al., Biochim. biophys. Acta, 1958, 29: 477; T. O. M. Nakayama, Arch. Biochem. Biophys., 1958, 75: 356) appeared to have absorption maxima at rather shorter wavelength than carrot ζ-carotene. B. H. Davies et al. (J. chem. Soc., C, 1969, 1266; cf. also Davies, Biochem. J., 1970, 116: 93) have now shown that this difference is not due to the presence of, for example, *cis*-isomers (effect of $I_2/h\nu$ on the *R. rubrum* compound as compared with ζ-carotene) but is entirely genuine, implying that the *R. rubrum* compound is not ζ-carotene but is a different, though closely related (no separation on t.l.c.) compound. This was confirmed by mass spectrometry. Whereas carrot ζ-carotene gave ions at M-69 (weak) and M-137 (strong) as expected (cleavage of the various doubly-allylic single bonds present; cf. p. 208), the *R. rubrum* compound (isomeric with ζ-carotene, $C_{40}H_{60}$, by precision m.s.) gave ions at M-69 and M-137 (both weak), and in addition at M-205, corresponding to cleavage of the 11,12 bond which must therefore be doubly allylic in this compound. This led to the structure below being proposed. The possibility that the more saturated end of the molecule is cyclic was eliminated by mass spectrometry (no ion at M-56, hence no α-ring present, p. 209) and t.l.c. (cyclisation to a β-end-group would have made the compound markedly less polar than ζ-carotene: Davies, 1970, loc. cit.). The *R. spheroides* compound probably has the same structure. For a possible further occurrence, see under θ-carotene below. In addition, Weeks and Andrewes (cf. Davies, 1970, loc. cit.; A. G. Andrewes, Ph.D. thesis, New Mexico State University, 1969, pp. 31—33) have found that diphenylamine-inhibited *F. dehydrogenans* (see Section 7h) produces traces of what is probably the same compound; in the absence of inhibitor, ζ-carotene is predominant:

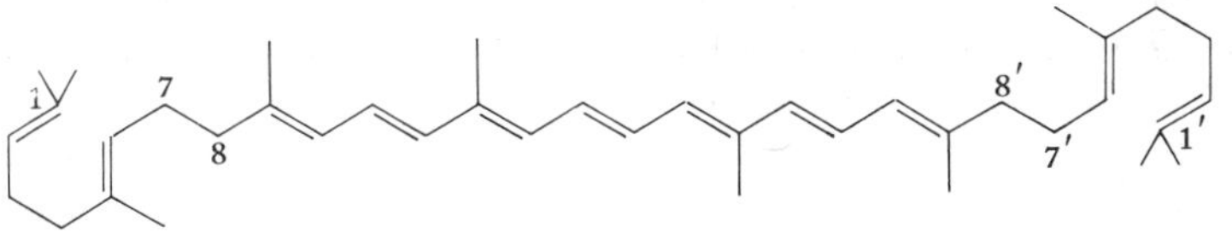

ζ-Carotene (ex carrots*); 7,8,7′,8′-tetrahydrolycopene: $\lambda_{max} \sim 425, 400, 378$ nm†

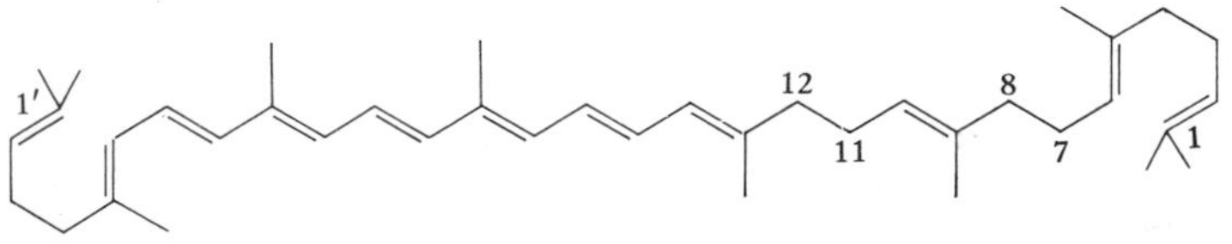

"Unsymmetrical-ζ-carotene" (ex *R. rubrum*, etc.): $\lambda_{max} \sim 419, 395, 374$ nm†

Unsymmetrical-ζ-carotene has been synthesised by Siddons and Weedon, using basically the same kind of route as used for ζ-carotene (2nd Edn., p. 275); the synthetic material had λ_{max} 420, 395, 374 nm (all-*trans*; light petroleum) in agreement with the data above.

For the biogenetic implications of the above discovery, see p. 342.

θ-**Carotene.** The carotenoid mentioned under this heading in the 2nd edition as having been isolated by Nakayama in 1958 (from *R. spheroides*, in fact) and tentatively identified by him as θ-carotene, was probably unsym-

---

* The compound from tomatoes, and most other sources cited in the 2nd Edn., p. 334, is probably also ζ-carotene.

† For the all-*trans* form in light petroleum (Davies, loc. cit.).

metrical-ζ-carotene. θ-Carotene itself, the substance detected in *Neurospora crassa* by Haxo in 1955, may also have been unsymmetrical-ζ-carotene.

New carotenoids. The following three compounds are acyclic at both ends and bear an obvious resemblance to neurosporene and lycopene, but show the important difference of having one end-group of the novel, isopropyl-(1,2-dihydro) type. They were discovered by H. C. Malhotra, G. Britton and Goodwin (Chem. Comm., 1970, 127) in *Rhodopseudomonas viridis*, one of the purple bacteria (for other carotenoids from allied bacteria, see Section 7c).

1′,2′-Dihydroneurosporene. Repeated chromatography of the fraction from the above source previously thought to be just neurosporene showed that although the latter was present in small amount, a second constituent (which was the major component) could be isolated. This had the same absorption spectrum (468, 439, 414 nm; petroleum) as neurosporene and very similar polarity but contained two extra hydrogens (m.s. molecular weight). The m.s. fragmentation pattern showed major ions at M-69 and M-139, presumed to be due to fission of doubly allylic single bonds (usually the major fragmentation pathway in carotenoid hydrocarbons; p. 208) which must therefore be at $C_{(3)}-C_{(4)}$ and $C_{(7')}-C_{(8')}$. The $-CH_2 \cdot CHMe_2$ end-group was confirmed by p.m.r. [6H doublet, $J = 6$, at τ 9.12 (Me_2CH-)]:

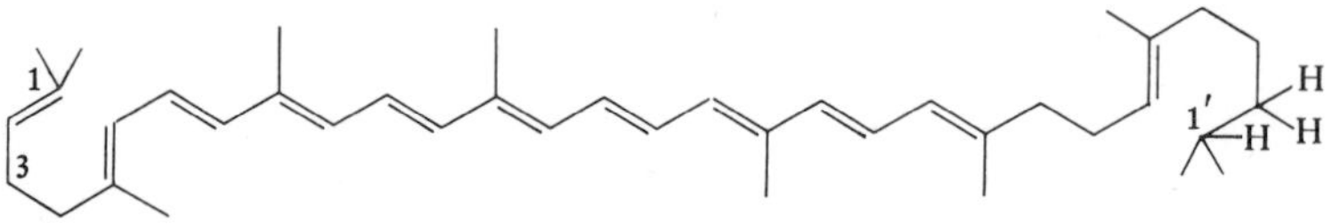

1,2-Dihydrolycopene. Similar treatment, to that above, of the "lycopene fraction" from the *R. viridis* showed the major constituent to have the same absorption spectrum as lycopene and similar polarity, but to have two more hydrogens. The main m.s. ions were at M-69 (strong) and M-71 (medium), corresponding to cleavage at the doubly allylic 3′,4′ and singly allylic 3,4 bonds, respectively. P.m.r. showed a doublet at τ 9.12, as above:

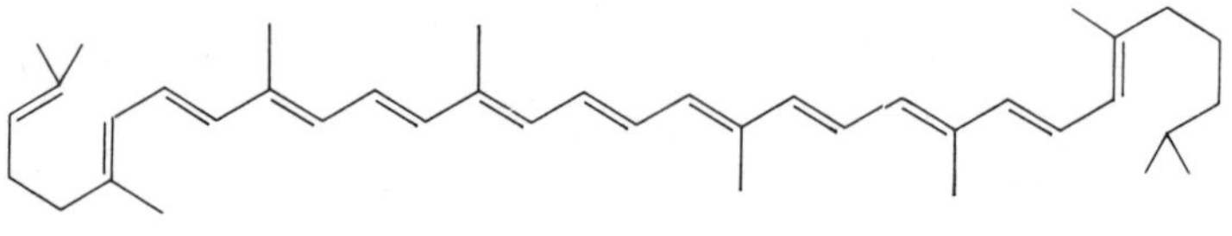

1,2-**Dihydro**-3,4-**dehydrolycopene** (probably). A minor fraction from the above source had λ_{max} values, 518, 483, 457 nm (petroleum) corresponding to a $(C=C)_{12}$ chromophore and a mass spectrum consistent with the following structure:

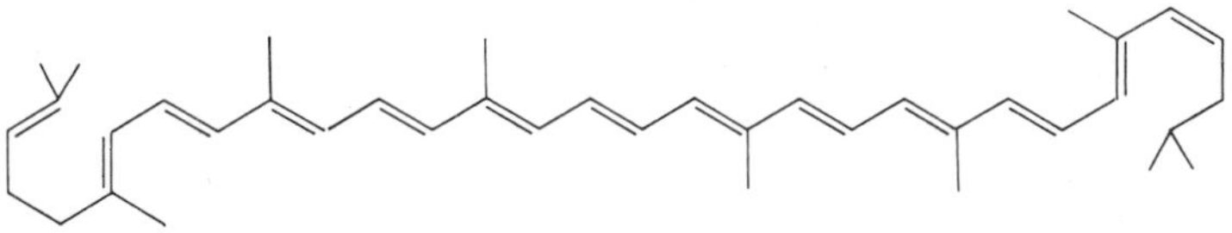

The mode of biogenesis of the isopropyl end-group was not studied by Malhotra et al., but the observation that the first two examples above co-occur with the analogous isopropylidene-end-group compounds suggest that they are derived from the latter by reduction in vivo.

3,4-**Dehydro**-β-**carotene** and 3,4:3′,4′-**bisdehydro**-β-**carotene** (compounds XXXVIII and XXXIX of 2nd Edn., p. 261) are produced by the fungus *Epicoccum nigrum* when grown under conditions of diphenylamine inhibition; UV-induced mutants of *E. nigrum* also produce these compounds (F. H. Foppen and O. G. Sassu, Biochim. biophys. Acta, 1969, 176: 357; Arch. Mikrobiol., 1970, 73: 216. Structures by λ_{max} values, polarity, i.r., and mixed chromatograms with specimens synthesised as in the 2nd Edn., loc. cit.).

β,γ-**Carotene**. Several $C_{40}H_{56}$ hydrocarbons with varying combinations of the ψ-ionone, β-ionone, and α-ionone type of end-groups (p. 193) have long been known in nature. Using the nomenclature system mentioned on p. 193, these are known as ψ,ψ-, β,β-, β,ε-, β,ψ-, ε,ψ-, and ε,ε-carotenes (i.e. lycopene, β-carotene, α-carotene, γ-carotene, δ-carotene, and ε-carotene). A new $C_{40}H_{56}$ (by m.s.) carotenoid first reported in 1968 by N. Arpin as one of the carotenoids of the fungus *Caloscypha fulgens* and designated P 444, has recently been studied in detail by Arpin, S. L. Jensen, and co-workers (Phytochem., 1971, 10: 1595). The compound had m.p. 176° (benzene– methanol), and λ_{max} values (478, 448, 420 nm in petroleum) similar to α-carotene's – and hence, probably, the same chromophore. However it was more polar than α-carotene on chromatography and its p.m.r. spectrum had two broad singlets near τ 5.4 (=CH_2) and nine, rather than ten, methyl groups (four of the "in-chain" type, three corresponding to the positions expected for β-end-groups*, and two near the positions found for the *gem*-methyls in carotenoids with α-end-groups*). This led to the following

* For definition, see footnote on p. 193.

structure being proposed for this carotenoid (since confirmed by synthesis; see below):

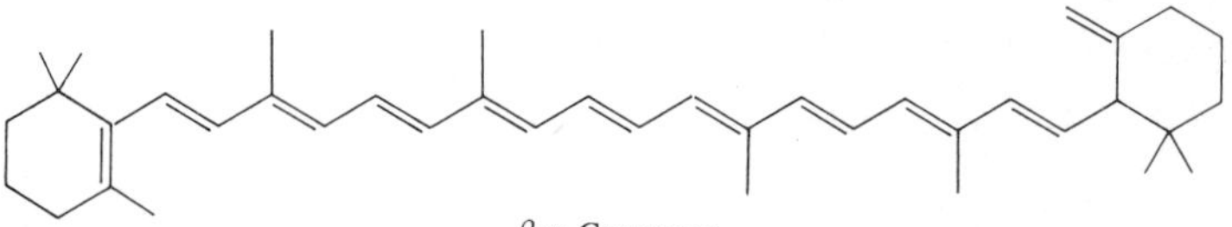

β,γ-Carotene

This represents the first reported occurrence of a carotenoid with an end-group containing an exocyclic methylene group (to be generally signified by "γ" in the new nomenclature). However, the occurrence of such an end-group in nature is not unexpected, assuming the suggested mode of carotenoid biogenesis outlined on p. 353 is correct, and β,γ-carotene's superficial similarity with α-carotene means that the former might be more widespread than it appears to be. It has also been found in an aphid (K. H. Weisgraber et al., Experientia, 1971, 27: 1017; Weisgraber, Jensen et al., Acta Chem. Scand., 1971, 25: 3878).

Synthesis (A. G. Andrewes and Jensen, ibid., 1971, 25: 1922): β- and γ-ionones were converted into XVa and XVb, respectively, by treatment with (i) $(EtO)_2 P(O) \cdot CH_2 \cdot CO_2 Et/NaOMe$ (Horner reaction); (ii) $LiAlH_4$; (iii) $Ph_3 PH^{\oplus} Br^{\ominus}$.

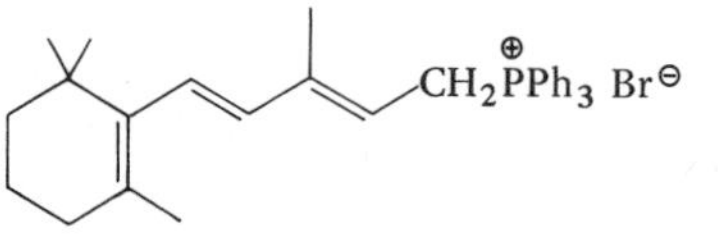

(XVa) 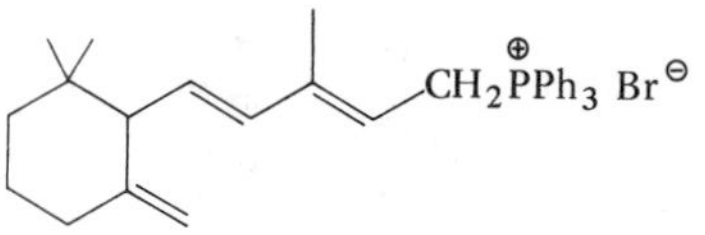   (XVb)

A two-stage Wittig synthesis (cf. p. 247 of 2nd Edn.) followed: XVa was treated with the C_{10}-trienedial shown on p. 275 of the 2nd Edn. and NaOMe to give the 1:1 condensation product (C_{25}) which reacted with XVb and NaOMe to give β,γ-carotene, identical with the natural material.

γ,γ-Carotene. One of the minor carotenoids accompanying the β,γ-carotene in the above-mentioned aphid has been shown (Weisgraber, Jensen, et al., 1971, loc. cit.) to be γ,γ-carotene (structure as for β,γ-carotene above but both end-groups of the exocyclic-methylene type) — by chromatographic comparison with β,γ-carotene and allied compounds, by its λ_{max} (469, 439, 414 nm in petroleum; nonaene chromophore) and m.s. data $(C_{40}H_{56})$, and by direct comparison with a synthetic sample.

Synthesis (Andrewes and Jensen, loc. cit.): By a Wittig condensation (NaOMe) of 2 moles XVb with one mole of the above-mentioned C_{10}-

trienedial. For full spectral and m.p. data on γ,γ-carotene, see Andrewes and Jensen. In both the above carotenoids, the $>C=CH_2$ groups exhibited a strong absorption at 889 cm^{-1} (KBr).

6. The vitamins A and related compounds

(a) Vitamin A₁: retinol*

Retinol (XVI)

(A different numbering system is occasionally used in which the $CH_2OH = C_{(1)}$).

Reviews covering retinol's physiological effects, and its essential rôle in health, have appeared (e.g. G. A. J. Pitt, Intern. J. Vitamin Res., 1966, 36: 249; G. Mackinney, in "Metabolic Pathways", 3rd Edn., Vol. II, ed. D. M. Greenberg, Academic Press, New York, 1968, p. 247; K. H. Lee et al., J. Pharm. Sci., 1970, 59: 1195, and refs. cited). This and related topics have been discussed in Ann. Rev. Biochem. (1971, 40: 501−512) and reviewed by T. Moore and others on pp. 245−303 and p. 153 et seq. of "The Vitamins" (2nd Edn., Vol. I, eds. W. H. Sebrell and R. S. Harris, Academic Press, New York, 1967); this work also includes (p. 113 et seq.) a compilation of the vitamin A, and "pro vitamin A" (see 2nd Edn., p. 283 for definition), content of various foods.

Concerning the utilisation of this air/light-sensitive substance as an essential additive in certain foodstuffs, see p. 110 of the last-mentioned work; A. J. Forlano et al. (J. Pharm. Sci., 1970, 59: 121) have shown that esterification of retinol with α,α-dimethylpalmitic acid, rather than the traditional palmitic or acetic acids, converts it into a form which undergoes significantly less deterioration on prolonged storage in foods.

For p.m.r. and ^{13}C-n.m.r. data on retinol and various derivatives and cis-isomers, cf. p. 207.

(i) The conversion of carotenoids into retinol

This reaction is known to take place, in mammals, as the β-carotene (or other pro-vitamin A) in the diet is absorbed through the wall of the intestine,

---

* The official (I.U.P.A.C.) present-day name, but the traditional term "vitamin A" is still often used, especially in pharmacological literature. The corresponding aldehyde is retinal (old names retinene₁ or simply retinene) and the corresponding acid, retinoic acid (vitamin A- or A₁-acid).

but whether this occurs through (1) central cleavage of the β-carotene molecule so yielding two molecules of retinol or (2) stepwise degradation from one end thereby yielding only one, was very uncertain. Opinion has now hardened in favour of (1). Thus, R. Blomstrand and B. Werner (C.A., 1967, 67: 114705x) studied the β-carotene → retinol conversion in vivo by following the fate of ingested β-carotene labelled at the 15,15' positions with ^{14}C or ^3H and of similarly labelled retinol, and concluded in favour of scheme (1). Similarly, D. S. Goodman et al. have continued their in vitro experiments using enzyme preparations from rat, and now hog, intestine and have shown that in such systems, each molecule of β-carotene furnishes approximately *two* of retinol (isolated as the aldehyde (retinal) owing to the lack of essential co-factors in such systems which are present in vivo and which reduce the aldehyde to the vitamin) [J. biol. Chem., 1966, 241: 1929 (includes a brief discussion of the possible mechanism of the cleavage reaction, which is O_2-dependent); 1967, 242: 3543; Biochem. J., 1969, 114: 689]. A crude concentrate of the enzyme responsible for the cleavage reaction, *"carotene 15,15'-dioxygenase"*, has been prepared from rabbit, and other, tissues, and some of its properties studied; its specificity for the *central* double bond of the carotenoid molecule may be due to this particular bond's steric accessibility (J. A. Olson and co-workers, J. Lipid Res., 1972, 13: 477: and refs. therein). The observation that comparative biological assays of β carotene and retinol indicate a conversion ratio nearer to 1:1 is not necessarily inconsistent with scheme (1). The β-carotene may, on ingestion, undergo prior absorption and/or degradation, or be less efficiently absorbed than retinol itself (cf. B. Stoecker et al., J. Nutrit., 1973, 103: 1112). The retinal → retinol step, mentioned above, has also been further studied (N. H. Fidge and Goodman, J. biol. Chem., 1968, 243: 4372).

It is now known that once in the bloodstream, the retinol circulates therein in the form of a 1:1 complex with a specific protein "carrier" (*"retinol-binding protein"*; $M \sim 21,000$) whose nature is currently under investigation by Goodman, P. A. Peterson and others (cf. J. biol. Chem., 1970, 245: 1903; 1971, 246: 6638; 1973, 248: 4009, 4698; Biochemistry, 1972, 11: 264, 4526).

"α-Vitamin A", λ_{max} 311 nm (ethanol) is the α-ionone analogue of vitamin A (i.e. C=C at $C_{(4)}$—$C_{(5)}$), and is formed along with it by mammals on ingestion of α-carotene (J. S. McAnally and C. D. Szymanski, Nature, 1966, 210: 1366); it was previously obtained synthetically (C. D. Robeson et al., J. Amer. chem. Soc., 1955, 77: 4111). The (synthetic) acyclic analogue of vitamin A is sometimes known, for convenience, as *"γ-vitamin A"* (Rüegg et al., J. chem. Soc., 1965, 2019).

(ii) *Synthesis of retinol (vitamin A) and related compounds*

Isler, Schwieter and co-workers (in Sebrell and Harris, op. cit., pp. 26, 104), have surveyed pre-1966 progress in this field.

G. Köbrich et al. (Ann., 1967, 704: 51), have described a method of building up the retinol skeleton, carrying a chloro- (or bromo-) substituent at $C_{(10)}$, from β-ionone:

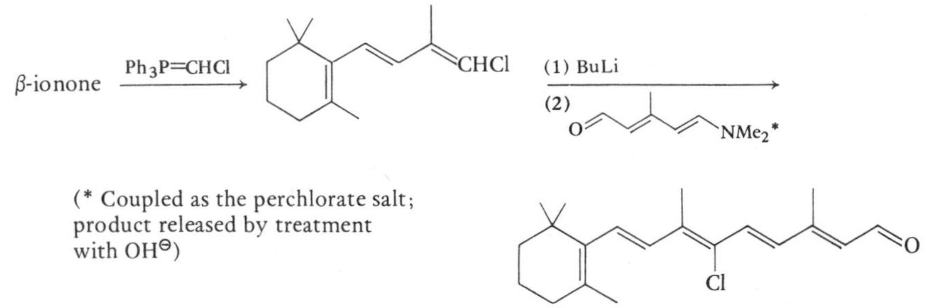

(* Coupled as the perchlorate salt; product released by treatment with $OH^{\ominus}$)

Retinoic acid; vitamin A₁ acid. (1) Full details of Pattenden et al.'s synthesis (2nd Edn., p. 287) of the 11,13-di-*cis* isomer of retinoic acid (using the Wittig-reaction based process outlined on p. 200) have appeared (Pattenden, Garbers et al., J. chem. Soc., C, 1968, 1984, 1982). A low-yield synthesis of the important (involved in the visual process?) 11-*cis* isomer, as its methyl ester, has also been reported (M. Mousseron-Canet et al., Bull. Soc. chim. Fr., 1969, 3242).

(2) An improved method of obtaining retinoic acid from the corresponding aldehyde (retinal) has been described. A. B. Barua et al. (Biochem. J., 1969, 113: 447) have applied Corey's method of oxidising allylic alcohols or aldehydes to the corresponding acid esters [treatment with cyanide and manganese dioxide in, e.g., methanol: $R \cdot CHO \rightarrow R \cdot CHOH \cdot CN \rightarrow R \cdot CO_2Me$ (J. Amer. chem. Soc., 1968, 90: 5616)] to retinal, and have thereby obtained retinoic acid, as its methyl ester, from retinal in > 90% yield; treatment with ethanolic alkali yielded the acid.

(3) Method "2" of 2nd Edn., p. 287, has been streamlined by A. B. Barua and M. C. Ghosh (Tetrahedron Letters, 1972, 1823) by applying the above reaction to retinol itself, whereupon retinoic acid, as its methyl ester, is obtained direct in 70% yield: retinol → [retinal] → methyl retinoate.

Treatment of methyl retinoate with manganese dioxide in hexane leads to oxidation of the allylic $C_{(4)}$ with the formation, 60%, of methyl 4-oxo-retinoate (Barua and Ghosh, 1972, loc. cit.; cf. 2nd Edn., p. 289 for an allied observation). With chromium trioxide in acetone, cleavage of the 5,6-bond occurs instead — as with β-carotene (2nd Edn., p. 260) (Schwieter et al., Helv., 1971, 54: 2447).

(b) *Vitamin A$_2$: 3,4-dehydroretinol**

Barua et al. (1969, loc. cit.) have also applied Corey's method, above, to 3,4-dehydroretinal* [obtainable from 3,4-dehydroretinol (2nd Edn., p. 288) by manganese dioxide oxidation] and have so obtained vitamin A$_2$ (3,4-dehydroretinoic) acid (LXXXIV of 2nd Edn., p. 287) by a new, and high-yield, route.

(c) *Retinal† and 3,4-dehydroretinal, and the visual process*

The p.m.r. spectra of all-*trans*-, 9-*cis*-, 11-*cis*-, and 13-*cis*-retinals have been recorded at 220 MHz (to expand the spectra to allow direct analysis), assignments being based on previous carotenoid p.m.r. data (D. J. Patel, Nature, 1969, 221: 825). Considerable changes occurred in the position of the various olefinic hydrogen, and to a smaller extent the methyl group, signals between one isomer and another as the changing shape of the molecule forced the atoms or groups in and out of the shielding/deshielding zones (cf. P. V. Demarco, J. W. ApSimon et al., Tetrahedron, 1967, 23: 2357; 1970, 26: 119) of the various double bonds. Coupling constants (*J* values) could also be extracted from the spectra. For example in the all-*trans* isomer ~15–16 Hz for *trans*-CH=CH's and ~11–12 Hz for the transoid protons (i.e. H's *trans*-coplanar to one another) about the $C_{(10)}-C_{(11)}$ single bond. In a CH–CH residue J_{H-H} is normally highly dependent on the dihedral angle between the two C–H bonds (Karplus equation), being maximal for the *trans*-coplanar conformation. Since $J_{10,11}$ in the sterically hindered and important (see below) 11-*cis* isomer showed no reduction from the value seen in the all-*trans* isomer, the 10,11-bond in the 11-*cis* isomer is still essentially transoid. It is inferred that the conformation taken up by the 11-*cis* molecule [at least in deuterochloroform solution], to relieve the considerable steric strain inherent in such a structure, is that in which there has been twisting around the 12,13 single bond rather than around the 10,11 bond. This is supported by the observed upfield shift undergone by the $C_{(11)}$- and $C_{(12)}$-H's on switching over from the all-*trans* to 11-*cis* isomer. The orientation taken up, in solution, by the cyclohexene ring relative to the polyene chain in the 11-*cis*- and the all-*trans*-retinals has also been inferred from p.m.r. data (Overhauser effects between the $C_{(5)}$-methyl and $C_{(8)}$-H;

---

* Strictly 3,4-didehydroretinol and 3,4-didehydroretinal.
† See also pp. 213, 226; and cf. p. 288 of Rodd, 2nd Edn., Vol. II B for earlier work.

measurement of long-range coupling constants). It appears the 6,7-bond is closer to being s-cis than s-trans — as with other carotenoids (p. 194) — with the ring at an angle of ~50° to the chain (B. Honig et al., Proc. Natl. Acad. Sci., U.S., 1971, 68: 1289). X-ray diffraction measurements by R. Gilardi et al. (Nature, 1971, 232: 187) have led to parallel conclusions regarding the conformation of these isomers in the solid state.

Patel (Nature, 1969, 224: 799) also used p.m.r. to study the stereo-isomerisation of all-*trans*-retinal by trifluoroacetic acid in CDCl_3 solution, using his observation, above, that the various *cis—trans* isomers of retinal give distinctive spectra. The 13- and 9-*cis* isomers were formed but not the 11-*cis* isomer.

L. Rimai et al. (J. Amer. chem. Soc., 1971, 93: 1353, 6288) have recorded the solution r.e.r. spectra (p. 212) of the 9-, 11-, and 13-*cis* and the all-*trans* forms of retinal. The C=C band occurs in all cases near 1580 cm^{-1} but there are marked differences in the 950—1400 cm^{-1} region of the spectra — sufficient for this technique to be of use as a method of identification (for an alternative method — viz. linear dichroism following irradiation at ~400 nm — of identification, and estimation, of these isomers, see J. Horwitz and J. Heller, J. biol. Chem., 1973, 248: 1051). For r.e.r. data on Schiff-base derivatives of retinal, cf. p. 213.

The visual process

The retina, a very thin (0.05 mm in humans) layer of tissue at the back of the eye*, contains two types of light-receptor: rods (for vision in dim light) and cones (for bright light and for colour vision).

Most of the work outlined below has been carried out on the pigments isolated from the outer segments* of the rods of cattle retinas, but the systems responsible for sight in other animals, including Man, are thought to be broadly similar (G. Wald, Nature, 1968, 219: 800; cf. also p. 241). In the dark, a very photosensitive carotenoid—protein complex (rhodopsin, or "visual purple") accumulates in the rods. On extracting this pigment (which is purple) and illuminating it, it is bleached being broken down into all-*trans*-retinal (yellow) and the protein.(opsin). Attempts to reproduce the first of these processes in vitro using opsin isolated from rods and various synthetic *cis—trans* isomers of retinal showed that only those isomers with a certain molecular shape would couple with the opsin, these being the 9-*cis*

---

* For a description of the constitution and physiology of the retina, cf. S. L. Bonting, Current Topics in Bioenergetics, 1969, 3 : 351—359, and for theories on how the chemical changes therein (discussed in the following pages) might give rise to the sense of sight, see articles by R. A. Weale and others in Brit. Med. Bull., 1970, 26 : No. 2 and Bonting, 1969, loc. cit., pp. 391—407.

and the (hindered*) 11-*cis* forms. Of these, it is the latter which is actually present in the eye. The entirely artificial pigment from 9-*cis*-retinal is known

All-*trans*-retinal (old name = all-*trans*-retinene$_1$) (XVII)

as *"isorhodopsin"* (R. Hubbard, Wald and co-workers, J. gen. Physiol., 1952/53, 36: 269; J. biol. Chem., 1956, 222: 865; Vitamins and Hormones, 1960, 18: 417). Thus it appeared that the bleaching process simply involved the isomerisation of the protein-bound 11-*cis*-retinal (the *"prosthetic group"*) to the all-*trans* form which, now being of the wrong shape to remain bound to the protein, would dissociate from it. However, although this is the overall effect, it gradually became apparent (by carrying out the above process at low temperatures and carefully monitoring changes in the absorption spectrum) that a series of intermediary pigments is involved (Hubbard, Wald, P. K. Brown and co-workers, Nature, 1959, 183: 442; 1964, 201: 340; J. gen. Physiol., 1963, 47: 215). Thus on taking rhodopsin in aqueous glycerol, freezing it to −190° (λ_{max} 502 nm) and illuminating it, a new pigment ("prelumirhodopsin") absorbing at considerably longer wavelengths (543 nm) is formed. On raising the temperature, this disappears and is replaced, in turn and at specific temperature-levels, by pigments with absorption maxima at 497, 478, 380, and finally 387 nm, as follows (refs. as above) (Scheme 1).

The changes up to and including metarhodopsin II formation, by which stage the absorption maximum has shifted as far as it is going to, occur very rapidly at normal physiological temperatures†. This correlates with the observation that, in practice, visual excitation in the eye is virtually

---

* i.e. of the type

$$
\begin{array}{c}
cis \\
\mathrm{HC{=}CH} \\
\diagup \qquad \diagdown \\
\mathrm{{=}C} \qquad \mathrm{C{=}} \\
\diagdown \qquad \diagup \\
\mathrm{H} \qquad \mathrm{Me}
\end{array}
$$

Considerable steric hindrance between the −**H** and the **Me**−, which leads to such bonds isomerising to *trans* particularly readily under the influence of light (cf. 2nd Edn., Vol. II B, p. 250).

† These very fast reactions can now in fact be followed to some degree, using recently developed rapid-response (laser pulse) spectroscopic methods operating at fixed wavelengths: cf. R. A. Cone, Nature New Biol., 1972, 236: 39; G. E. Busch et al., Proc. Natl. Acad. Sci., U.S., 1972, 69: 2802; T. Rosenfeld et al., Nature, 1972, 240: 482.

Scheme 1

THE "BLEACHING PROCESS" IN OUTLINE
(Additional intermediates may be formed transiently at other stages:
cf. Morton and Pitt, Adv. Enzymol., 1969, 32: 131)

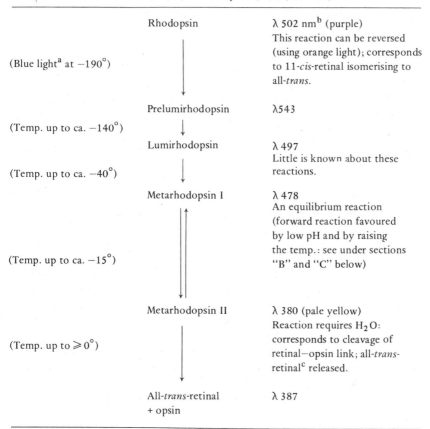

Rhodopsin   λ 502 nm[b] (purple)
This reaction can be reversed
(using orange light); corresponds
(Blue light[a] at −190°)   to 11-*cis*-retinal isomerising to
all-*trans*.

Prelumirhodopsin   λ543
(Temp. up to ca. −140°)

Lumirhodopsin   λ 497
Little is known about these
(Temp. up to ca. −40°)   reactions.

Metarhodopsin I   λ 478
An equilibrium reaction
(forward reaction favoured
by low pH and by raising
the temp.: see under sections
(Temp. up to ca. −15°)   "B" and "C" below)

Metarhodopsin II   λ 380 (pale yellow)
Reaction requires H_2O:
corresponds to cleavage of
(Temp. up to ⩾ 0°)   retinal−opsin link; all-*trans*-
retinal[c] released.

All-*trans*-retinal   λ 387
+ opsin

[a] Only this step requires light; all subsequent reactions are dark (i.e. thermal) reactions, apparently triggered off by the change in shape (from bent to straight) undergone by the retinal during the rhodopsin → prelumirhodopsin step.

[b] λ_{max} at −190°; 497 nm at normal temperatures.

[c] In vivo, the final product is, instead, the corresponding alcohol, retinol (vitamin A): see p. 240.

instantaneous (Wald, 1968, loc. cit.). The main object of recent research has been to discover:

(A) the means by which the retinal is linked to the opsin in each pigment;

(B) the reason for the surprisingly long wavelength position of the absorption maximum of the rhodopsin itself (cf. the 11-*cis*-retinal basically responsible has λ_{max} ~ 380 nm: Brown and Wald, J. biol. Chem., 1956, 222: 865) and of the following three pigments in the series, and for the considerable differences in λ_{max} between them;

(C) the nature of the retinal binding site and of the conformational and other changes undergone by the opsin during the bleaching process.

A. Properties of rhodopsin: The nature of the retinal—opsin link

Rhodopsin as isolated from cattle retinas has M ~ 30,000 and contains one molecule of 11-*cis*-retinal per molecule rhodopsin (J. Heller, Biochemistry, 1968, 7: 2906; J. E. Shields et al., Biochim. biophys. Acta, 1967, 147: 238; H. Shichi et al., J. biol. Chem., 1969, 244: 529). It gives a single broad absorption peak in the visible region, which occurs at ca. 500 nm in aqueous solvents (containing 2% w/v digitonin or other detergent as solubiliser), ϵ ~ 42,000* (Shichi, Biochemistry, 1970, 9: 1973; C. D. B. Bridges, Nature, 1970, 227: 1258; S. L. Bonting et al., ibid., p. 1259).

For methods of isolating rhodopsin, and in particular freeing it of lipid and other impurities by repeated aqueous chromatography on Sephadex and other adsorbents, see Shichi, 1970, loc. cit.; K. Hong et al., Proc. Natl. Acad. Sci., U.S., 1972, 69: 2617; Bonting et al., Arch. Biochem. Biophys., 1972, 151: 1. Studies on rhodopsin have been hampered by the difficulty of obtaining it pure and of finding reliable tests for purity: electrophoresis, chromatography on various adsorbents, E_{max}/E_{trough} ratio of the absorption curve, and estimates of the percentage retinal present (cf. the footnote) are commonly used.

It has long been thought (R. A. Morton et al., Biochem. J., 1955, 59: 122, 128) that, in rhodopsin, the retinal might be linked to the opsin through the agency of a Schiff base ($-CH=N-$) bond to one of the presumably multitudinous $-NH_2$ groups present in the opsin molecule. It was realised that if this were in fact so, it should be possible to stabilise the acid-labile retinal—opsin linkage by borohydride reduction ($-CH=N- \rightarrow -CH_2-NH-$), and degradation of the product might then yield a fragment containing retinal still attached to part of the opsin molecule. The first definite evidence on this came from the discovery (M. Akhtar, P. T. Blosse and P. B. Dewhurst, Life Sci., 1965, 4: 1221; D. Bownds and Wald, Nature, 1965, 205: 254) that although the supposed $-CH=N-$ linkage in rhodopsin itself

---

* Determined, e.g., by treating purified rhodopsin with ethanol, and estimating the retinal thereby released by adding thiobarbituric acid (which undergoes aldol-coupling yielding a compound, λ_{max} 530 nm, of known ϵ). Much lower values have been claimed by J. Heller (Nature, 1970, 225: 636) but have been discounted by Shichi, Bridges, Bonting et al., above. In addition, the value of M remains in some dispute with values approaching 40,000 claimed by Bownds et al. and H. Heitzmann (Nature New Biol., 1972, 235: 112, 114).

is apparently unaffected by borohydride treatment*, the mixture of pigments obtained (Scheme 1) on irradiating rhodopsin is, in contrast, quite susceptible to attack giving a compound with an absorption maximum at a markedly shorter wavelength (viz. 333 nm). In addition, this compound was found to be insoluble in organic solvents, suggesting that the opsin molecule was still attached, and its λ_{max} value was consistent with its containing a chromophore of the type:

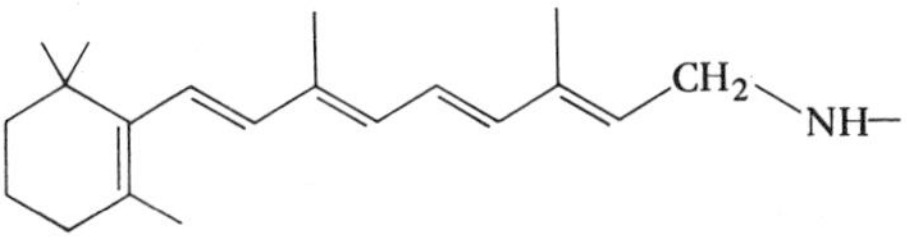

The same authors then showed that on borohydride treatment of, in turn, metarhodopsin I and metarhodopsin II (prepared by irradiation of rhodopsin at $0°$; see above), the latter was rapidly converted into a compound with λ_{max} 333 nm, whilst the former was relatively unaffected*, suggesting that the λ_{max} 333 compound mentioned above was the dihydro ($-CH=N- \rightarrow$ $-CH_2-NH-$) derivative of metarhodopsin II:

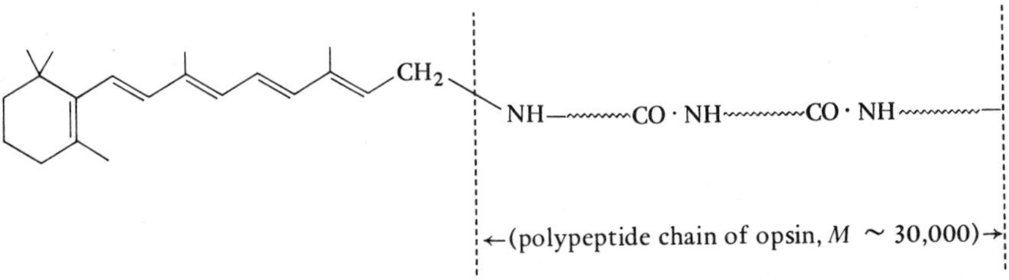

$\leftarrow$(polypeptide chain of opsin, $M \sim 30,000$)$\rightarrow$

λ_{max} 333 nm compound

Bownds (Nature, 1967, 216: 1178) and Akhtar et al. (Chem. Comm., 1967, 631; Biochem. J., 1968, 110: 693) later showed that on basic or enzymatic cleavage of the $-CO \cdot NH-$ links in the above compound, there could be isolated† from the complex mixture of amino-acids so obtained, a fragment consisting of the retinyl chromophore linked to the first amino-acid residue

---

* These compounds may of course still contain a $-CH=N-$ link, but buried sufficiently deeply in the protein part of the molecules to be immune from attack; see Section C below.

† By chromatography. The isolation was monitored either by using the strong u.v. fluorescence of the retinyl chromophore (Bownds) or radiochemically by using tritium-labelled rhodopsin, made by coupling tritiated 11-cis-retinal (R $\cdot$ CMe=CH $\cdot$ C*HO, from NaB^3H_4 reduction of 11-cis-retinal to [15-^{3}H]retinol followed by MnO_2 oxidation) with opsin, to make the λ 333 compound (Akhtar et al.).

of the polypeptide chain. This fragment was compared (t.l.c.) with a series of N-retinyl-aminoacids prepared by coupling retinal with the $-NH_2$ group of various amino-acids:

Scheme 2

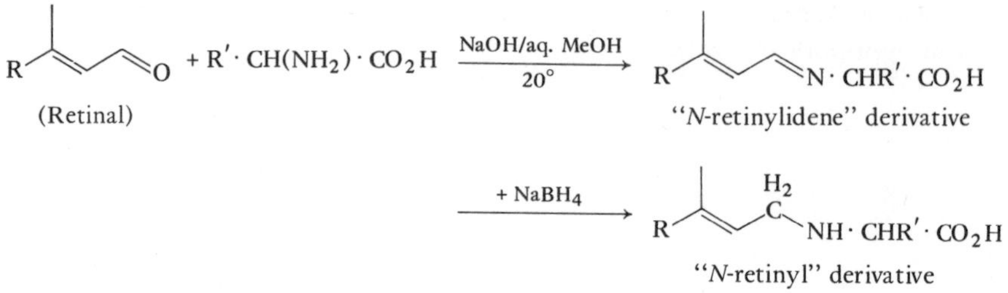

(Retinal) $+ R' \cdot CH(NH_2) \cdot CO_2H \xrightarrow[20°]{NaOH/aq.\ MeOH}$ "N-retinylidene" derivative $R \cdots N \cdot CHR' \cdot CO_2H$

$\xrightarrow{+\ NaBH_4}$ "N-retinyl" derivative $R \cdots \overset{H_2}{C} NH \cdot CHR' \cdot CO_2H$

In lysine, $H_2N(CH_2)_4 \cdot CH(NH_2) \cdot CO_2H$, the ϵ-NH_2 is markedly more reactive than the α-, so both ϵ- (by allowing brief reaction in the NaOH step) and α- (by first protecting the ϵ-NH_2 group) N-retinyl-lysines could be made.

This showed that the above-mentioned fragment was the N-retinyl derivative of the ϵ-NH_2 of a lysine residue, so that in metarhodopsin II the retinal is, apparently, joined to the opsin part of the molecule by a $-CH=N-$ linkage to the ϵ-NH_2 of a lysine residue therein:

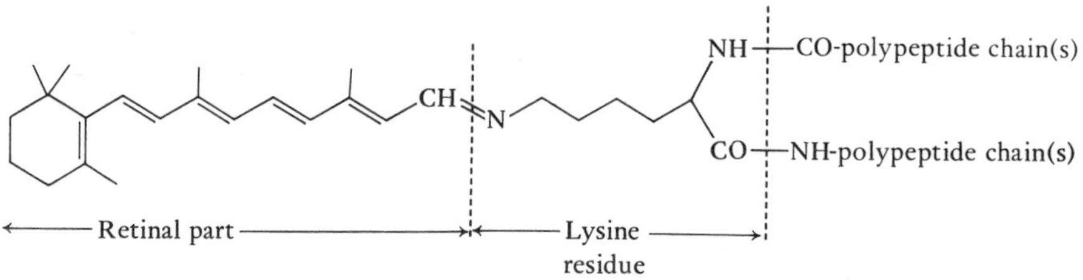

NH—CO-polypeptide chain(s)

CO—NH-polypeptide chain(s)

Retinal part ← → Lysine residue

Metarhodopsin II, probable part structure

However, the above arrangement is not necessarily the one occurring in rhodopsin itself, and experiments by Akhtar and M. D. Hirtenstein (Biochem. J., 1969, 115: 607) and R. P. Poincelot et al. (Biochemistry, 1970, 9: 1809) indicated that the (11-cis) retinal in rhodopsin might be attached, still through a $-CH=N-$ link, to the $-NH_2$ group of phosphatidylethanolamine ("PE"*), it being envisaged that at some stage during the subsequent series of conversions (Scheme 1) the retinyl moiety would migrate from the PE

* Approved IUPAC name, ethanolamine phosphoglyceride, EPG (J. Lipid Res., 1967, 8: 523).

$$CH_2O \cdot CO \cdot R$$
$$|$$
$$CHO \cdot CO \cdot R$$
$$|$$
$$\overset{\displaystyle O}{\overset{\displaystyle \|}{CH_2-O-P \cdot O \cdot CH_2 \cdot CH_2NH_2}}$$
$$|$$
$$OH$$

Phosphatidylethanolamine

R = fatty acid residue, e.g. oleic

$$CH_2O \cdot CO \cdot R$$
$$|$$
$$CHO \cdot CO \cdot R$$
$$|$$
$$CH_2O \cdot PO \cdot O \cdot CH_2 \cdot CH_2 \cdot N\underset{CH}{\diagdown}$$
$$|$$
$$OH$$

(retinyl residue)

N-RPE

amino-group to the lysine ϵ-amino-group. This retinal–phosphatidyl-ethanolamine compound (N-retinyl–phosphatidylethanolamine, N-RPE) would only account for a part of the rhodopsin molecule's molecular weight, and it would have to be attached to the rest of the opsin in some way (cf. Poincelot and Abrahamson, ibid., 1970, 9: 1823 for possibilities). In support of the above theory were the points:

(a) retinal couples readily with the $-NH_2$ group of PE requiring no catalyst other than the reagent (F. J. M. Daemen, Bonting et al., Nature, 1969, 222: 879; P. A. Plack et al., Biochem. J., 1969, 115: 927);

(b) PE (= "EPG") had been found to be a constituent of, or at least associated with, rhodopsin (R. G. Adams, J. Lipid Res., 1969, 10: 473);

(c) internal protonation of the $-CH=N-$ link by the phosphoric acid group in the PE might explain the surprising position of the rhodopsin absorption maximum (cf. below).

However the above theory was soon disputed by M. O. Hall and A. D. E. Bacharach (Nature, 1970, 225: 637) and by R. E. Anderson et al. (Biochemistry, 1970, 9: 3628; Nature New Biol., 1971, 229: 249) who, whilst accepting that *some* of the retinal in rod outer segments (cf. p. 229) is present in the form of a Schiff base with PE (N-RPE), concluded that this is only there as a *reserve* supply of retinal and is not a part of the rhodopsin molecule itself. Bonting et al., having recently succeeded in purifying rhodopsin until it contained negligible quantities of PE (or indeed any form of phosphorus), now support this view (Arch. Biochem. Biophys., 1972, 151: 1). Further work is now needed on this problem.

B. The position of the absorption maximum in rhodopsin and prelumi-rhodopsin, etc.

Various theories have been put forward to explain why rhodopsin, whose basic chromophore is 11-*cis*-retinal ($\lambda_{max} \sim 380$ nm) or, probably, a derived Schiff's base (λ_{max} of model Schiff bases, ~ 360 nm), absorbs at such long wavelength (497 nm; all data for aqueous or alcoholic solutions). Most of these theories assume that at least part of this difference is due to protonation of the Schiff base link:

$$R' \cdot CMe{=}CH \cdot CH{=}N{-}R \xrightarrow{\ H-X\ } R' \cdot CMe{=}CH \cdot CH{=}\overset{\oplus}{N}H{-}R\ X^{\ominus}$$

Thus, with model compounds (R = alkyl instead of R = opsin as would be the case in rhodopsin), this shifts the λ_{max} to ~440/450 nm in hydroxylic solvents and to ~460/485 nm in non-hydroxylic solvents such as chloroform, the position being markedly dependent on the nature of the counter-ion (ca. 20 nm longer wavelength for large ions like $I^{\ominus}$ and $ClO_4^{\ominus}$ as compared with small ones like $Cl^{\ominus}$) (Akhtar et al., Biochem. J., 1968, 110: 693; P. E. Blatz and J. H. Mohler, Chem. Comm., 1970, 614). This however does not appear to explain the whole of the λ_{max} shift seen in rhodopsin, still less that seen in some of the *other* visual pigments (λ_{max} values up to 562 nm: cf. C. D. B. Bridges in "Comprehensive Biochemistry", Vol. 27, eds. M. Florkin and E. H. Stotz, Elsevier, Amsterdam, 1967, p. 46) based on retinal. Akhtar et al. (Life Sci., 1965, 4: 1221; 1968, loc. cit.) have ascribed this additional λ_{max} shift to the formation of an internal charge-transfer complex between the $N^{\oplus}$ and the acidic site (A−H) in the opsin molecule which had protonated the nitrogen, and which would have acquired a negative charge in so doing ($-CH{=}N- + A{-}H \rightarrow -CH{=}\overset{\oplus}{N}H{-} + A^{\ominus}$). The size of the λ_{max} shift due to this effect would depend on (increase with) the $-\overset{\oplus}{N}H \cdots A^{\ominus}$ distance, and this would change markedly as the molecule underwent various contortions during the 11-*cis*-retinal → 11-*trans*, and subsequent, transformations outlined in Scheme 1 (a possibility which has since been supported − and elaborated on − by P. E. Blatz et al., Biochemistry, 1972, 11: 848*). However, others have ascribed the whole of the shift (i.e. to 497 nm) to either an extension of the effect noted above whereby the λ_{max} of model protonated Schiff's bases is roughly proportional to the size of the counter-ion ($X^{\ominus}$), that is to the size of the electrostatic interaction between the $N^{\oplus}$ and the $X^{\ominus}$ in the $R' \cdot CMe{=} CH \cdot CH{=}\overset{\oplus}{N}HR\ X^{\ominus}$ ion pair (effects which might be even larger with opsin$^{\ominus}$ as the counter-ion than with $I^{\ominus}$), or to the particular properties of the medium in which this protonated Schiff's base is lying within the rhodopsin molecule and which, therefore, is acting as the compound's "solvent" (Blatz and Mohler, 1970, loc. cit.; C. S. Irving et al., Biochemistry, 1970, 9: 858).

Meanwhile, Daemen and Bonting (Nature, 1969, 222: 879) suggested that the only grouping likely to be present in a protein molecule such as opsin capable of protonating a −CH=N− link would be a phosphoric acid residue, and showed that on coupling retinal with phosphatidylethanolamine (see

---

* Also elaborated on (M.O. treatment) by A. Waleh et al., Arch. Biochem. Biophys., 1973, 156: 261.

$$CH_2O \cdot CO \cdot R$$
$$CHO \cdot CO \cdot R$$
$$CH_2-O-\overset{O}{\underset{O^{\ominus}}{\overset{\|}{P}}}-O \cdot CH_2 \cdot \underset{\overset{|}{\oplus}NH=CH}{CH_2}$$

(XVIII)

above), what appeared to be an internal salt (XVIII), with λ_{max} 450 nm, was indeed formed. Adams et al. (Nature, 1970, 226: 270) went on to show that on repeating this reaction in a polar solvent, a second pigment was — eventually — produced, with λ_{max} 500 nm as for rhodopsin itself. This pigment was produced only if the fatty acid residues in the phosphatidyl-ethanolamine used contained C=C unsaturation, suggesting that the +50 nm shift seen in this compound is due to a $\pi-\pi$ interaction between the retinal chromophore and this unsaturation. This then seemed to offer a possible explanation of rhodopsin's λ_{max} value.

Recent work, however, has shown the above ideas must be modified. Thus it is now known (p. 235) that there is not sufficient phosphorus in rhodopsin to accommodate a structural unit of the XVIII type, and so it must now be assumed that the protonation of the —CH=N— link is caused by an acidic site within the opsin, *other than* phosphate (cf. M. Sundaralingam et al., Proc. Natl. Acad. Sci., U.S., 1972, 69: 1569 for analogies) with the resulting anion site on the opsin being held close to the $N^{\oplus}$ by the confining action of the opsin, thereby giving a λ_{max} of ~450 nm; and with the additional shift, to ~500 nm, being due to a $\pi-\pi$ interaction between the retinal chromophore and C=C, or other*, unsaturation in the surrounding opsin moiety (cf. the compound above; Sundaralingam et al., loc. cit., give additional analogies).

It is thought that the —CH=N— bond linking the retinal to the opsin is protonated not only in rhodopsin but also up to and including the metarhodopsin I stage, but that in the *metarhodopsin I → II step* the molecule undergoes a considerable change in form (see also later) which includes the deprotonation of this bond (Wald and co-workers, J. gen. Physiol., 1963, 47: 215).

R.e.r. spectroscopy on the pigment (lumirhodopsin?) present in intact bovine retina at $-70°$, indicates that the retinal is linked to the opsin at this stage by a protonated Schiff base bond (see p. 212).

Finally, B. Honig and M. Karplus (Nature, 1971, 229: 558) have pointed out that 11-*cis*-retinal, particularly in the restricted environment it occupies

---

* E.g. one of the Ar-amino acid residues: cf. Mendelsohn, Nature, 1973, 243: 22. A charge-transfer complex?

in rhodopsin, might exist in distinguishable *s—cis* and *s—trans* (about the $C_{(12)}-C_{(13)}$ single bond) forms. The presence of a *s—cis* bond within a polyene chromophore causes a marked shift of the λ_{max} towards longer wavelengths (refs. cited by Honig and Karplus). Thus if the retinal in rhodopsin were to undergo the sequence 11-*cis*, 12-*s—cis* → 11-*trans*, 12-*s—cis* → 11-*trans*, 12-*s—trans*, this could explain at least some of the dramatic changes in λ_{max} seen in the early stages of Scheme 1. Heller's visible-region c.d. measurements on a rhodopsin preparation in aqueous glycerol (Biochemistry, 1971, 10: 1402) at room temperature and at $-190°$, and then after irradiating the latter for a short while (giving prelumi-rhodopsin?) and for longer (giving?) — all of which operations yielded a different c.d. curve — may throw some light on the nature of these transformations. See also Honig et al., ibid., 1973, 12: 1637 in this respect.

C. The nature of the opsin part of the rhodopsin molecule

The opsin accounts for the bulk of the rhodopsin molecule and has a molecular weight of ∼30,000. Hydrolysis of opsin and amino-acid analysis of the product has shown opsin to contain 17 of the common amino acids in varying amounts (the major ones being alanine, glutamic acid, threonine, glycine, leucine, and phenylalanine), and a total of ca. 250 amino acid residues per molecule of opsin (Shields et al., 1967, loc. cit.; Shichi et al., J. biol. Chem., 1969, 244: 529; Bownds and A. C. Gaide-Huguenin, Nature, 1970, 225: 870).

The amino acids in the vicinity of the retinal binding-site at the metarhodopsin II stage were identified by Bownds (1967, loc. cit.) who, following enzymatic cleavage of the peptide links in the "λ_{max} 333 nm compound"*, isolated a series of fragments containing retinal linked to various small peptides the biggest being a decapeptide (ϵ-N-retinyl—lysine, thr, pro, ile, ala$_3$, phe$_3$). Information on the nature of this site has also been inferred from studying the structural features necessary in a prosthetic group for it to form, on mixing with opsin, a complex of the visual pigment type. It has already been mentioned that the 9-*cis* and the 11-*cis* isomers of retinal will react but that the all-*trans* will not. Blatz et al. (J. Amer. chem. Soc., 1969, 91: 5930), R. Nelson et al. (Proc. Natl. Acad. Sci., U.S., 1970, 66: 531), and D. R. Lewin and J. N. Thompson (Biochem. J., 1967, 103: 36P) tried coupling a wide variety of molecules of the retinal type with opsin and found that the 9-*cis* and the 11-*cis* isomers of retinal lacking the 9-methyl group, the 13-methyl group, or even both, would all couple, as would both 9- and 11-*cis*-retinal with the 5,6-bond saturated or epoxidised, or with an additional (3,4) double bond in the ring. However, in certain cases the

---

* Cf. p. 233.

reaction was much slower (and/or incomplete) than with 11-*cis*-retinal itself, and in no case did the all-*trans* isomer react. Also, extending or contracting the polyene side-chain or removing the ring, prevented reaction in both the *cis* and the *trans* series.

The retinal binding site in the opsin is, then, very discriminating in that it will only accept polyene aldehydes of a certain quite definite shape. In summary, the polyene side-chain must be the same length as in retinal and it must be bent, by a *cis* bond, at the $C_{(9)}$ or $C_{(11)}$ position; there must be a trimethylcyclohexyl ring (or possibly any similarly bulky group) attached to it; and although the 9- and 13-methyl groups are relatively unimportant, the 13-methyl group, at least, is necessary for the reaction to be as efficient as in the natural system.

On going through the sequence from rhodopsin to retinal (Scheme 1), the opsin part of the molecule undergoes several important changes which may be real changes in structure or just major changes in its conformation. Thus two additional −SH groups (probably towards the end of the sequence: cf. S. E. Ostroy et al., Biochim. biophys. Acta, 1966, 126: 409) and a new proton-binding group (Wald and co-workers, J. gen. Physiol., 1963, 47: 234) are exposed. In rhodopsin the retinal is so buried within the protein molecule for it to be inert to oxidative attack (Wald and Hubbard, in "The Enzymes", 2nd Edn., Vol. 3, eds. P. D. Boyer et al., Academic Press, New York, 1960, p. 382) and for the −CH=N− link, which binds it to the protein, to be unaffected by borohydride (see above) and to show no gain or loss of $H^{\oplus}$ over a wide range of pH (Bownds and Wald, 1965, loc. cit.). This kind of situation prevails up to and including the metarhodopsin I intermediate but in the subsequent step (conversion into metarhodopsin II) considerable changes occur. The −CH=N− link becomes sufficiently exposed to make it susceptible to borohydride reduction and the opsin–retinal chromophore interaction, which has, it is thought, helped to give rise to the long wavelength λ_{max} values seen in the first four pigments of the series, is broken. Measurements on the effect of temperature on the metarhodopsin I ⇌ II equilibrium indicate that the reaction is accompanied by a large increase in entropy (Wald and co-workers, J. gen. Physiol., 1963, 47: 215). All this suggests that a considerable "loosening up" or unfolding of the opsin occurs at this stage.

Shichi et al. (J. biol. Chem., 1969, 244: 529) have recorded o.r.d. and c.d. curves for purified rhodopsin in the 270−200 nm region (for a review of earlier results, mostly on rather less pure samples of rhodopsin, see Shichi et al., loc. cit., p. 535; and for c.d. measurements in the visible region, cf. p. 238, and Shichi, Biochemistry, 1970, 9: 1973). A marked drop in the amplitude of the c.d. curve was seen on subjecting this rhodopsin to the "bleaching" process (Scheme 1), corresponding to a drop in the proportion of the opsin part of the molecule existing in the highly ordered α-helix form. However, this effect was not seen in the rhodopsin present in a suspension of rod outer segments, suggesting that in its natural environment rhodopsin does not undergo the marked changes in conformation which it apparently undergoes when isolated from that environment's restraining influence.

The bleaching process in vivo; the final step in the bleaching process; the retinal cycle

Most of the work described so far was carried out on rhodopsin removed from its natural environment (the retina). However, the use of techniques such as fundus reflectometry (developed by Weale and Rushton: cf. Bonting, 1969, loc. cit., p. 375; W. A. Hagins, Nature, 1956, 177: 989), microspectrophotometry (developed Brown: cf. H. J. A. Dartnall, Brit. med. Bull., 1970, 26: 175), and r.e.r. spectroscopy (see p. 212), which allow the direct study of the visual pigments *as they are in the intact retina*, has demonstrated the presence of rhodopsin-like pigments in the retina of many animals and has shown that certain, and possibly all, of the intermediates formed on illuminating rhodopsin in vitro (Scheme I) are probably also formed in the intact retina. Thus although Schichi's o.r.d./c.d. measurements (above) suggest some differences in detail between in vitro and in vivo processes, it appears that, in general, the latter does parallel the former to a high degree. The important exception is that the final product of the "bleaching process" (Scheme I) in vivo is all-*trans*-retinol rather than all-*trans*-retinal (for refs. cf. Bonting et al., below). This is due to the presence in the intact retina, in close association with the rhodopsin, of an enzyme "retinoldehydrogenase" (which is normally removed during the isolation of the rhodopsin for in vitro experiments) which with the help of a co-factor also present in the retina (NADPH, a biological H-donor) reduces the $-CH=N-$ bond linking the all-*trans*-retinyl chromophore to the opsin in metarhodopsin II direct to all-*trans*-retinol (with simultaneous cleavage of the bond): i.e. $-CH=N- \rightarrow -CH_2OH$ (+ $H_2N-?$) direct (rather than $-CH=N- \rightarrow -CHO \rightarrow -CH_2OH$) (Bonting et al., Arch. Biochem. Biophys., 1970, 140: 275).

The mechanism by which the all-*trans*-retinol is converted into 11-*cis*-retinal, for coupling to the opsin thereby regenerating rhodopsin and completing the cycle, remains uncertain; the reactions involved appear to occur only "on demand" from the opsin: that is when there are opsin molecules present devoid of a prosthetic group (Wald, 1968, loc. cit.). The sequence is probably: (a) oxidation (retinoldehydrogenase; + NADP as biological H-acceptor?) to all-*trans*-retinal; (b) isomerisation of the latter to 11-*cis*-retinal. Thus, S. Amer and Akhtar (Biochem. J., 1972, 128: 987) have recently isolated a soluble enzyme system from the bovine retina which will catalyse reaction "b" in vitro (they also demonstrated the presence of a further enzyme, in the insoluble fraction from the same preparation, which catalyses the (faster) reaction of 11-*cis*-retinal so formed with the opsin to give rhodopsin).

Any wastage of retinal during the cycle is made up by incorporating small amounts of retinol, from the animal's stores, at the all-*trans*-retinol stage.

The visual process in o'her species

The rod vision in other animals has been less investigated but apparently also involves 11-*cis*-retinal as the prosthetic group, and a similar set of transformations to those described in the preceding pages. However, the absorption maximum of the pigment varies quite markedly from animal to animal (λ_{max} from ~478 to 528 nm; cf. Dartnall, 1970, loc. cit.) presumably due to subtle differences in the protein part of the molecule and/or to the mode of binding of the retinal to same.

Cone (colour) vision also depends on pigments formed from the coupling of 11-*cis*-retinal (usually, but see below) to protein molecules like opsin but is more complicated since for the perception of colour a complex of pigments is required. Thus in Man there are three pigments (all based on 11-*cis*-retinal but coupled to different protein molecules) sensitive to red, green, and blue light, respectively (absorption maxima at ~565, 540 and 435 nm, respectively) (Wald, 1968, loc. cit.).

The visual pigments (for both rod and cone vision) of land animals and most marine fish are based on retinal (and retinol) (old names, retinene$_1$ and vitamin A$_1$) but are replaced in freshwater fish by 3,4-dehydroretinal (and 3,4-dehydroretinol) (retinene$_2$ and vitamin A$_2$) (Wald, 1968, loc. cit.; Dartnall, 1970, loc. cit.).

For a wide-ranging discussion of some of the problems in the visual-process field, see Exptl. Eye Res., 1969, 8: 241; Adv. Enzymol., 1969, 32: 97.

(d) Kitol

C. Giannotti (Canad. J. Chem., 1968, 46: 3025) has reinvestigated the nature of the product obtained on irradiating retinol (vitamin A$_1$) or its acetate in vitro with light of visible/near-u.v. wavelengths, and has disputed the earlier claims by Kaneko (2nd Edn., p. 290) and Mousseron-Canet et al. (Bull. Soc. chim. Fr., 1966, 3043) that a single compound identical with natural kitol (ex fish liver oils) is produced. Giannotti found instead that the apparently single product could, on certain t.l.c. systems, be resolved into several closely similar compounds. The four major constituents were dimers isomeric with, but distinguishable from, natural kitol by their m.s. fragmentation patterns. Some may have been *cis/trans*, or optical, isomers. Natural kitol may also be a mixture of compounds. The structure given before, obtained on a highly purified sample of the natural material, stands, and the full version of Garbers et al.'s 1965 paper has appeared: J. chem. Soc., Perkin I, 1973, 590.

(e) *Properties of the vitamins A and related compounds*

(i) *Physical properties*

J. Kahan has summarised references to papers describing the t.l.c. separation of the two vitamins, their *cis/trans* isomers, and their derivatives (J. Chromatog., 1967, 30: 506; see also S. R. Ames, J. Assoc. offic. agric. Chemists, 1966, 49: 1071). M. Vecchi, W. Vetter et al. (Helv., 1967, 50: 1243) have developed a quantitative method of determining the composition of a mixture of retinol (A_1) *cis/trans*-isomers by silylation (by treatment with excess N-trimethylsilylacetamide in pyridine) followed by g.l.c. on a silicone oil column. Clean, well-separated peaks, corresponding to each isomer, are obtained.

Further 60 MHz p.m.r. spectral data, on 11,12-dehydroretinal stereo-isomers (Mousseron-Canet et al., Bull. Soc. chim. Fr., 1969, 3247), the all-*trans* and 11,13-di-*cis* isomers of retinoic acid and allied synthetic compounds lacking the ring (Pattenden and Weedon, J. chem. Soc., C, 1969, 1984), have appeared. In addition, the spectra of the all-*trans*, and several of the possible *cis*/di-*cis* isomers of retinol, retinal, retinoic acid, and derivatives, have been recorded at 220 MHz (which simplified interpretation by expanding the spectrum: G. Englert and co-workers, Pure Appl. Chem., 1969, 20: 365, 386).

(ii) *Chemical properties*

The structure given for **anhydrovitamin A₁** (LXXXVIII, 2nd Edn., p. 291) has been confirmed from its 220 MHz p.m.r. spectrum. In contrast to the 60 or 100 MHz spectra, where there is much overlapping of signals, the spectrum is sufficiently expanded at 220 MHz to allow each multiplet in the complex olefinic-H region to be clearly seen, and so assigned to an individual proton (Englert and co-workers, 1969, loc. cit.).

Blatz and D. L. Pippert (J. Amer. chem. Soc., 1968, 90: 1296) have found that on treating retinol or its acetate, λ_{max} 325 nm, with trichloroacetic or sulphuric acids in an organic solvent at *low* temperature and running the spectrum *immediately* at 0°, a sharp band at ca. 600 nm is observed, corresponding to a λ_{max} shift of nearly 300 nm. This they tentatively ascribed to the formation of the retinyl carbonium ion, XIX. This band decays with time, even at $0°$, and is replaced by others (due to polymers?):

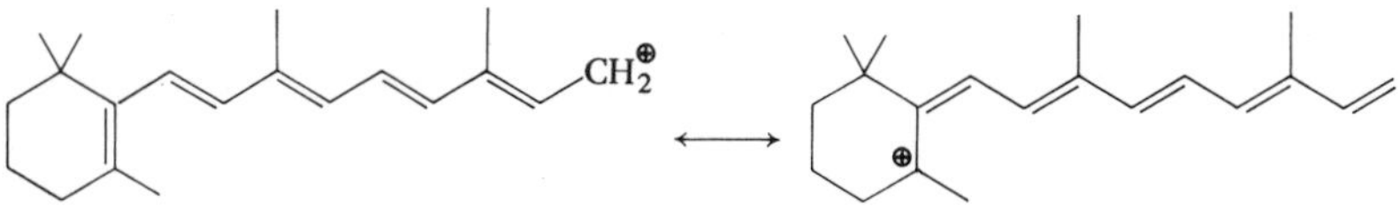

(XIX)

Similar λ_{max} shifts are observed on treating a colloidal suspension (essential: due to the presence of organised aggregates?) of retinol in water with *electron acceptors* (iodine, quinones). As in the above reaction, it is difficult to be certain as to the nature of the products formed (J. A. Lucy and F. U. Lichti, Biochem. J., 1969, 112: 221, 231).

The nature of **rehydrovitamin A_1** has been discussed by T. K. Murray and P. Erdody (Biochim. biophys. Acta, 1967, 136: 375) who conclude, in contrast to the assertion of some authors, that it differs from **retrovitamin A_1** (XX), the probable major-product from briefly treating retinol, as the acetate, with HBr in methylene dichloride (R. H. Beutel et al., J. Amer. chem. Soc., 1955, 77: 5166):

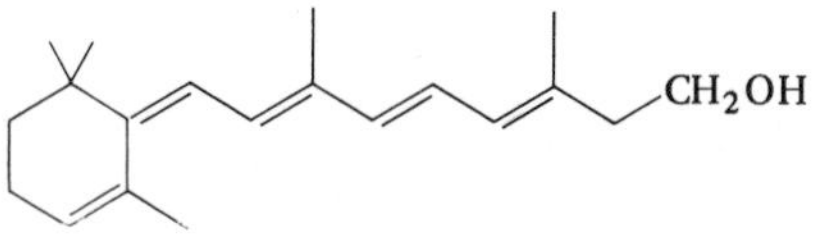

(XX)

The full paper describing the structure elucidation of the retinal ("retinene$_1$")–iron carbonyl adduct has appeared (R. Mason and G. B. Robertson, J. chem. Soc., A, 1970, 1229). It appears (X-ray crystallography) that the conformation about the 6,7-bond approximates closely to the *s–cis* form (XXI), rather than *s–trans* as drawn in their 1966 communication (Birch, Mason et al., Chem. Comm., 1966, 613):

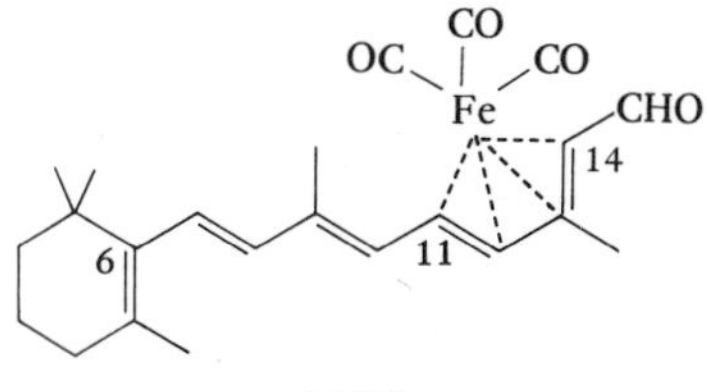

(XXI)

The $C_{(11)}-C_{(14)}$ portion of the molecule is not only curled round, as shown, but also far from planar. Similar adducts have since been prepared from lycopene and β-carotene (which see).

3,4-Dehydroretinol (vitamin A_2) has long been known to be very sensitive to atmospheric oxidation. In a study related to the problem of understanding why this should be so, Mousseron-Canet et al. (Bull. Soc. chim. Fr., 1969, 3252) have subjected the methyl ester derivative of 3,4-dehydroretinol and allied compounds containing a cyclohexadiene ring to sensitised photo-oxidation (O_2/sensitising dye/$h\nu$). The following exo-cyclic (as for ordinary β-ionone derivatives: p. 299) and endocyclic peroxides formed by 1,4 addition of O_2 to diene units, were the main products. The structures were deduced from the elemental analyses, u.v. (hence chromo-phore) and p.m.r. spectra of the compounds:

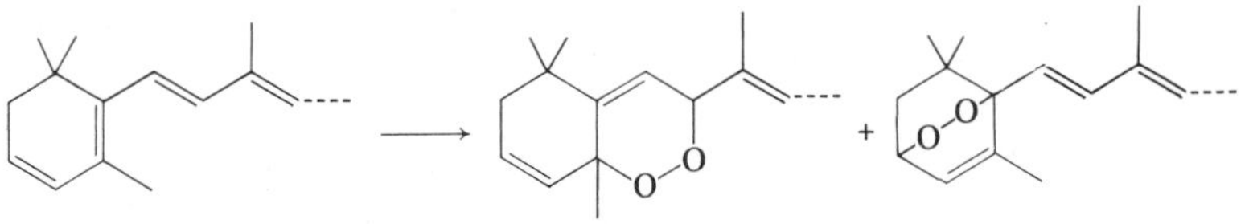

H. R. Cama and co-workers (Biochem. J., 1967, 103: 539; 1965, 95: 17) have prepared (by treatment of the parent compounds with per-acid, cf. p. 245, or as indicated) the 5,6-*epoxides* of retinyl acetate (XXII; λ_{max} 310 nm), of retinol itself (by saponification of the latter), of retinal (XXIII) (by manganese dioxide oxidation of the latter; λ_{max} 352 nm), of methyl retinoate (XXIV; λ_{max} 337 nm), and of retinoic acid (by saponification of the latter; λ_{max} 338 nm):

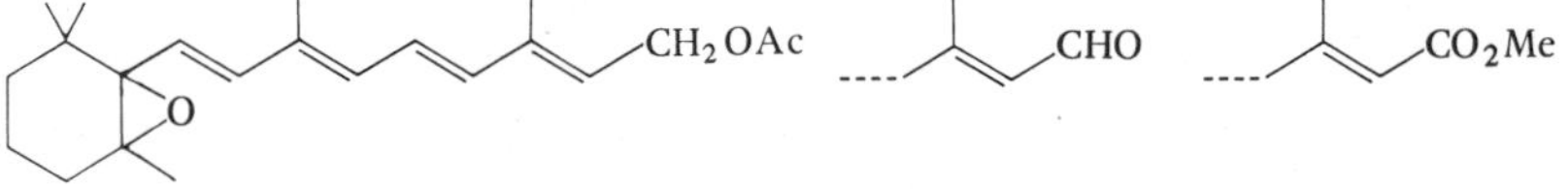

(XXII)   (XXIII)   (XXIV)

The corresponding furanoid oxides were also prepared, by treating the 5,6-epoxides with $H^{\oplus}$/EtOH (cf. p. 246); λ_{max} shift, ca. -30 nm.

All the above epoxides and furanoid oxides except those of retinol were stable, crystalline solids. The i.r. and/or p.m.r. spectra, and the biological activity and metabolic fate, of several were recorded. Lithium aluminium hydride, under mild conditions (cf. Section 7a), selectively reduces the

−CHO and −CO₂Me groups of 5,6-epoxyretinal and of methyl 5,6-epoxy-retinoate, to give 5,6-epoxyretinol:

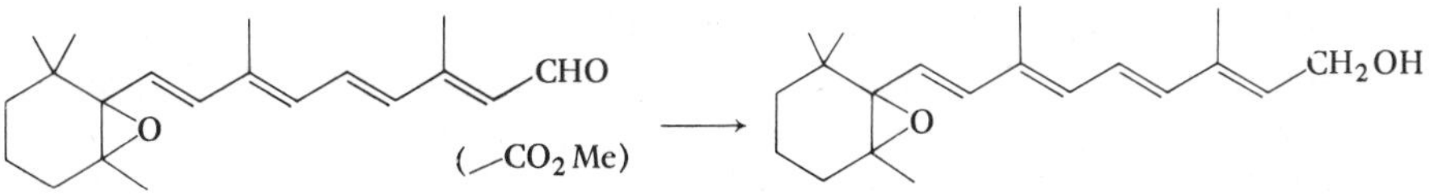

(⌐CO₂Me)

Parallel experiments in the 3,4-dehydroretinol (vitamin A₂) series by Cama et al. (Biochem. J., 1969, 111: 23; Ind. J. Biochem., 1969, 6: 26) led to the 5,6-*epoxides* of 3,4-dehydroretinyl acetate, 3,4-dehydroretinol, and 3,4-dehydroretinal; the 3,4-double bond remains intact. The structures were deduced from the u.v. spectra (which showed tetraene and tetraeneal chromophores), elemental analysis and p.m.r. spectra:

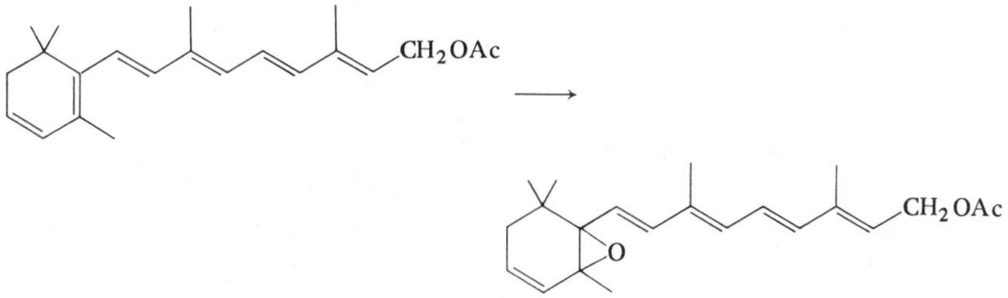

In contrast to the retinol series, none of these compounds rearranged to a furanoid form with acid. Biological activity was again recorded.

7. Carotenoid compounds containing oxygen ("xanthophylls"*)

(a) Carotenoid epoxides: general properties

The only compounds in this category previously reported, either "synthetic" or from natural sources, were the 5,6-epoxides (mainly of cyclic carotenoids with β-end-groups). Apparently, in the majority of the common cyclic carotenoids it is the 5,6 (and 5',6') double bond which is by far the most susceptible to epoxidation, both in vitro with per-acid and in vivo. Since then, however, C. Bodea and co-workers (Rev. Roum. Chim., 1970, 15: 965) have studied the effect of per-acid on a cyclic carotenoid containing

---

* The generic name for oxygen-containing carotenoids; hence, e.g., zeaxanthin is a xanthophyll. However, this term is occasionally used, especially in the older literature, for *a particular* carotenoid (viz. lutein).

conjugated carbonyl groups (canthaxanthin: p. 274) and have shown that in this case epoxidation occurs not in the ring at the 5,6 position(s) but in the chain, a mixture, separable by chromatography, of (probably) 9,10- and 13,14-monoepoxycanthaxanthins being produced. The structures of these compounds followed from their polarity and λ_{max} values (hence chromophore present), both before and after borohydride reduction to remove the complicating effect of a conjugated carbonyl group, and by the reactions outlined below:

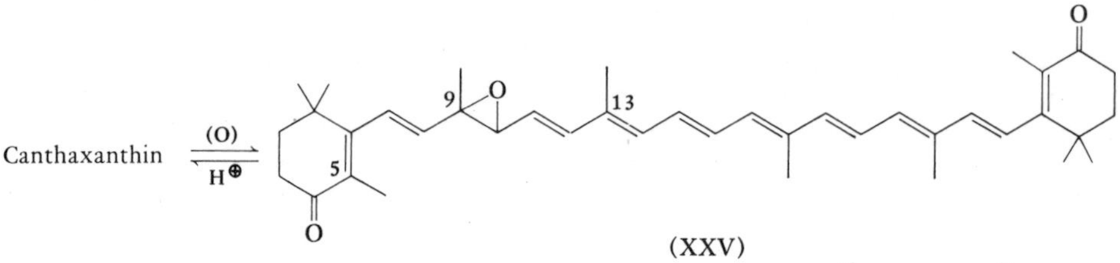

Canthaxanthin $\xrightarrow[H^{\oplus}]{(O)}$

(XXV)

+ the corresponding
13,14-mono-epoxide
(XXVI)

Compounds (XXV) and (XXVI) resembled the 5,6-epoxides in giving a blue colour with concentrated acid but differed in that on milder acid treatment, they did not rearrange to the furanoid form but reverted, with surprising facility, to the parent carotenoid. They also differed in being readily reduced by lithium aluminium hydride, yielding the corresponding 9- and 13-hydroxy compounds.

For a new type of natural epoxide see p. 355.

New work on the 5,6-epoxides

(1) *Reaction with acid.* This is normally thought of as being the characteristic reaction of the 5,6-epoxides (2nd Edn., p. 293), leading to the furanoid oxide (or "**5,8-epoxy-carotenoid**", e.g. XXVII) form. This remains generally true but Schwieter et al. (Helv., 1971, 54: 2447) have found that 5,6-epoxides with a short (diene) side-chain terminated by $-CO_2R$ do not react in this manner but, with aqueous acid, yield the glycol instead. Cama's A_2-epoxides, above, also failed to give furanoid derivatives.

(2) *Reaction with Lewis acids.* For the effect on a typical 5,6-epoxide of treatment with aluminium chloride, see under violaxanthin. The furanoid oxides (e.g. XXVII) react with titanium tetrachloride (cf. p. 213) in benzene at 20° to give, after a few minutes, a solid which, according to its λ_{max} value and $Cl^{\ominus}$ analysis, is a 1:1 complex between the furanoid oxide and the reagent. If the reaction is quenched with acetone soon after mixing the

reactants, the *retro* derivative of the parent carotenoid can be isolated in high yield (Tamas and Bodea, Rev. Roum. Chim., 1970, 15: 655):

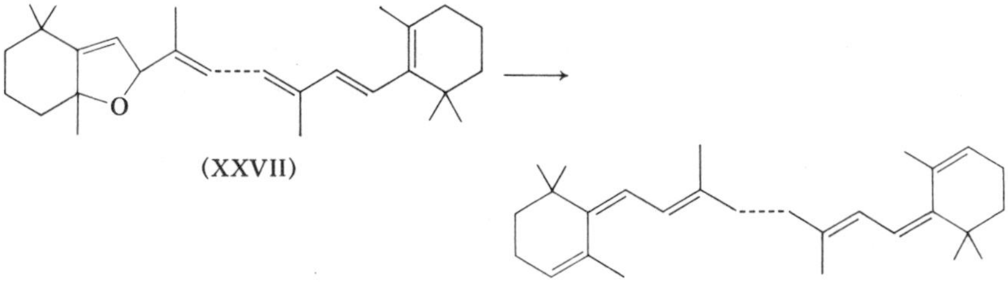

(XXVII)

With 3-hydroxy compounds, some of the initial product is transformed further, by allylic dehydration. Thus mutatoxanthin gives not only the expected product, eschscholtzxanthin, but also anhydro-eschscholtzxanthin:

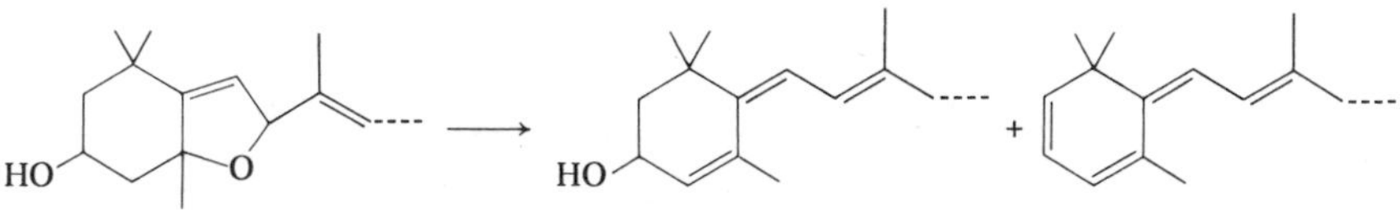

On heating with mercuric chloride, both 5,6-epoxides and their furanoid derivatives yield characteristic blue complexes (cf. Cama et al., Anal. Biochem., 1965, 12: 275).

(3) *Conversion to the parent carotenoid.* Full details of Cholnoky et al.'s method for carrying out this reaction (by prolonged treatment with a *large excess* of lithium aluminium hydride in ether under vigorous reflux: conditions important) have been published (Cholnoky et al., Ann., 1967, 708: 218). The earlier claim that simultaneous irradiation with i.r. light is necessary has been withdrawn (Cholnoky et al., 1967, loc. cit.; Hertzberg and Jensen, Phytochem., 1967, 6: 1119). Yields of 60–70% were reported for typical examples (e.g. violaxanthin → zeaxanthin; α-carotene-5,6-mono-epoxide → α-carotene *without* isomerisation into β-carotene). A. Hager and H. Stransky (Arch. Mikrobiol., 1970, 71: 132) have since advised using boiling tetrahydrofuran, rather than ether, as solvent; they also reported effecting this reaction enzymatically (cf. also Hager and Perz, Planta, 1970, 93: 314).

B. P. Schimmer and N. I. Krinsky (Biochemistry, 1966, 5: 3649) observed the same reaction occurring on subjecting antheraxanthin to similar conditions. The mechanism is unknown but apparently does not proceed simply via hydride reduction/dehydration (Schimmer and Krinsky, loc. cit.). This reaction has been of considerable value in determining the absolute configuration at $C_{(3)}$, $C_{(3')}$ of compounds such as violaxanthin (q.v.). Cholnoky et al. (1966, 1967, loc. cit.) and Hertzberg and Jensen (1967, loc. cit.) found that even the corresponding furanoid oxides could be reduced to the parent carotenoid by this method, and although the yield is usually low, it has provided a very useful method of relating the structures and absolute configuration of complex molecules such as neoxanthin and fucoxanthin (which see) to those of the relatively simple molecule, zeaxanthin:

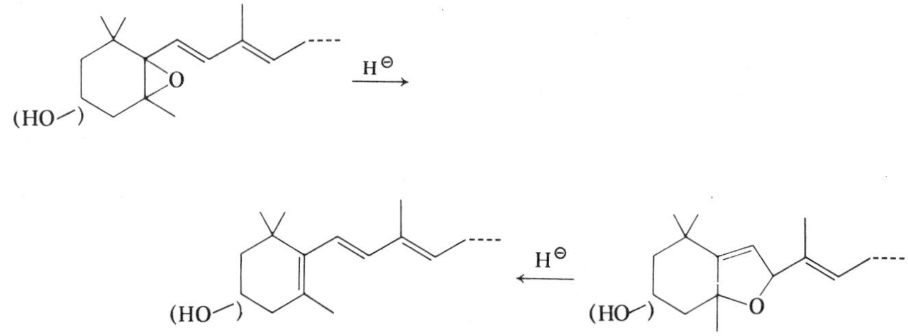

(4) *In vivo involvement.* The reversible sequence violaxanthin $\rightleftharpoons$ antheraxanthin $\rightleftharpoons$ zeaxanthin operating in green leaves, has received further attention from H. Y. Yamamoto et al. (Biochim. biophys. Acta, 1967, 141: 342; 1968, 153: 459; cf. also Hager and Stransky's use of an enzyme to reduce antheraxanthin to zeaxanthin, above). Both forward and reverse reactions are now thought to be basically dark reactions, triggered by an unidentified light-produced catalyst. They apparently involve (cf. p. 355) evolution and take-up, respectively, of molecular oxygen. The possibility that in this or some other way (e.g. protection of sensitive biochemical systems against oxidative-degradation), these compounds have a vital function in photosynthetic tissues, has been briefly discussed by Yamamoto and others* (1968, loc. cit., and refs. there cited). In addition, H. Kleinig (Planta, 1967, 75: 73) has discovered that the diadinoxanthin and diatoxanthin in *Vaucheria* sp. (misidentified at the time as antheraxanthin and zeaxanthin, respectively, pp. 305, 303) undergo an analogous interconversion:

$$\text{Diadinoxanthin} \underset{+O}{\overset{-O}{\rightleftarrows}} \text{diatoxanthin}$$

* And at length by Sapozhnikov, Pure Appl. Chem., 1973, 35: 47.

(b) Cyclic hydroxycarotenoids and their oxides, and the hydrocarbon oxides

The first group of compounds in this section (up to eschscholtzxanthin) are all well established and described in this Section of the 2nd Edition in varying degrees of detail. The last few compounds (from plectaniaxanthin onwards) are new additions to this section*.

Lutein, one of the most abundant xanthophylls in nature, sometimes known as "*xanthophyll*", especially in the older literature:

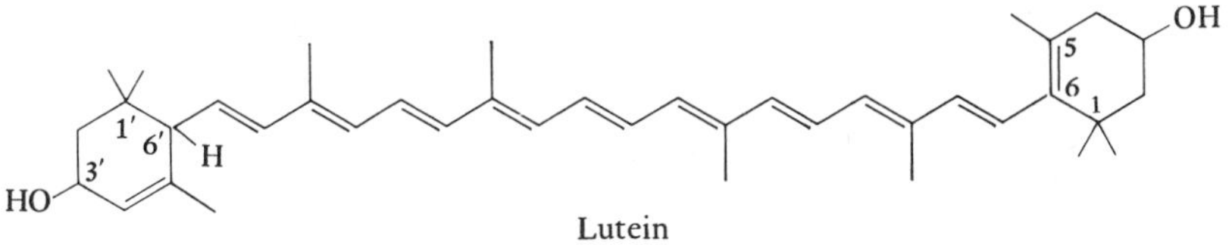

Lutein

O.r.d. data for lutein (various sources) have been recorded over the 200–400 nm range by Scopes, Weedon et al. (J. chem. Soc., C, 1969, 2527). The configuration at the three asymmetric centres could not be deduced from the curve due to a lack of reference data on similar compounds, but was inferred indirectly. Thus, D. Goodfellow, G. P. Moss and Weedon (Chem. Comm., 1970, 1578) showed that on removing the allylic-OH from lutein (using the method on p. 215) the product obtained was identical with α-cryptoxanthin (below) in all respects, including absolute configuration (comparison of c.d. curves). Hence the $C_{(3)}$ and $C_{(6')}$ positions in lutein must be R. As regards the $C_{(3')}$ position, the labelling experiments of Walton, Britton and Goodwin (p. 356) indicated that lutein's two hydroxyl groups are inserted into the molecule in nature in the same manner as the $C_{(3)}$-hydroxyl in β-cryptoxanthin (p. 254), which in turn suggests (Weedon et al., Chem. Comm., 1969, 1311) their configurations are the same. Hence lutein probably has the absolute configuration shown below:

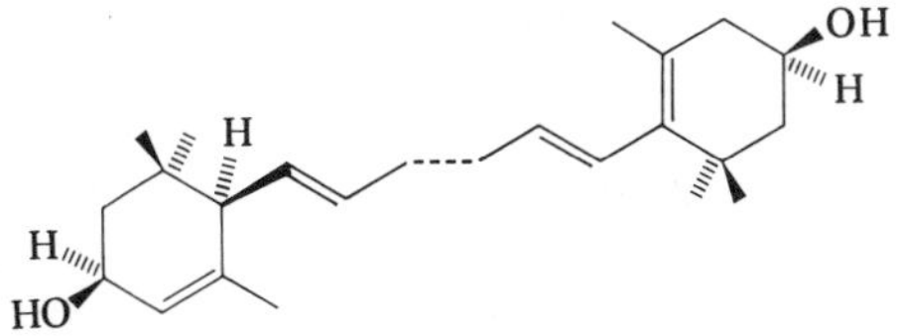

(= $3R$, $3'S$, $6'R$) [although both OH groups are of the "β" type (steroid nomenclature), the configuration at $C_{(3')}$ is S due to the manner in which the Sequence Rules (p. 212) operate in α-rings]

---

* Recent work has shown aleurixanthin (p. 333) should also be included in this section.

For a further method of differentiating the allylic and non-allylic OH groups in lutein, see p. 216.

Lutein-5,6-epoxide. It is now thought that many of the references to the presence of "taraxanthin" in nature were misidentifications, the compound in question being lutein epoxide; hence, the natural sources of the latter can be extended accordingly. For details see p. 333.

The o.r.d. curve of natural lutein epoxide has been recorded by Scopes, Weedon et al. (1969, loc. cit.). Comparison of the curve with those given by violaxanthin and lutein (using the approach outlined on p. 211), whose absolute configurations are known with a fair degree of certainty, indicated that lutein epoxide has the same configuration as these molecules at the common centres; that is probably as illustrated below (in which the epoxide ring is *trans* to the $C_{(3)}-OH$):

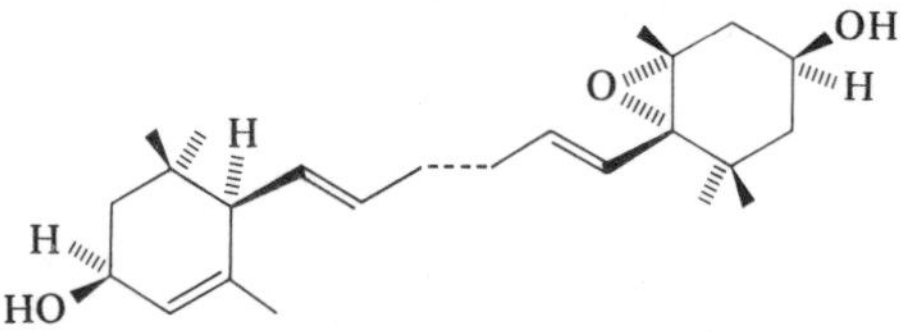

(= $3S, 5R, 6S, 3'S, 6'R$) [concerning the S notation given for the 3,3' positions, similar remarks to those given under lutein apply]

This contention is supported by the observation (Scopes, Weedon et al., loc. cit.) that *partially synthetic* lutein epoxide (by treatment of lutein with per-acid, a reaction which is thought (cf. Scopes et al.) to lead to the epoxide ring going in *cis* to the $C_{(3)}-OH$) has a completely different o.r.d. curve from natural lutein epoxide, and can be separated from it by chromatography.

Zeaxanthin. *Natural sources.* The variation in zeaxanthin content between one strain of maize (a traditional source of the pigment) and another has been studied by C. O. Grogan and C. W. Blessin (Crop Sci., 1968, 8: 730). K. Harashima and Y. Yajima (Agric. biol. Chem., Japan, 1969, 33: 1092) have found that zeaxanthin can be isolated in high yield from the berries of the shrub *Lycium chinensis*. *Delonix regia* flowers are also a good source (R. K. Barua and A. B. Barua, Biochem. J., 1966, 101: 250), as is Aasen et al.'s *Flexithrix* bacterium (Acta Chem. Scand., 1972, 26: 404). Certain reports referring to the occurrence of zeaxanthin in nature may be incorrect (cf. p. 300).

Regarding p.m.r. data on its olefinic-protons (only methyl groups mentioned before), cf. p. 207.

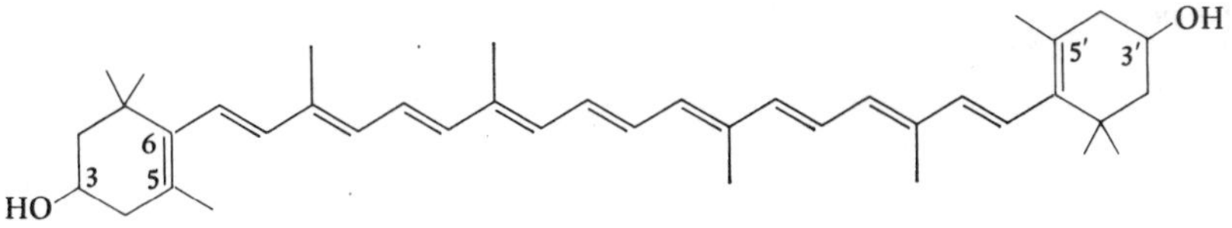

Zeaxanthin

Absolute configuration (at $C_{(3)}$, $C_{(3')}$). The o.r.d. (and c.d.) curves of zeaxanthin from various natural sources have been recorded by Scopes, Weedon et al. (1969, loc. cit.; extended by Aasen et al., 1972, loc. cit.) over the range 200—400 nm; all samples gave the same curve showing that they not only had the same gross structure but also the same configuration at $C_{(3)}$, $C_{(3')}$. It was on the basis of the presumed (biogenetically-inferred) absolute configuration at the two asymmetric centres of the zeaxanthin molecule and the o.r.d. curve that resulted therefrom, that the configuration at the $C_{(3)}$ and/or $C_{(3')}$ carbon atoms in various other xanthophylls was deduced by Scopes, Weedon et al. (see p. 211 for details). It was, therefore, important to check the key zeaxanthin configurations by an independent, and absolute, method. This was achieved as follows.

It is known that the $C_{(3)}$ and $C_{(3')}$ carbon atoms of zeaxanthin have the same configuration as the corresponding positions in fucoxanthin (details, p. 289), and in addition that they themselves must be of the same configuration (otherwise zeaxanthin, a symmetrical molecule, would be optically inactive). Since the configuration of the $C_{(3')}$-carbon of fucoxanthin has been shown to be as on p. 289 (formula XLVII), the full stereochemistry of zeaxanthin must be as shown below:

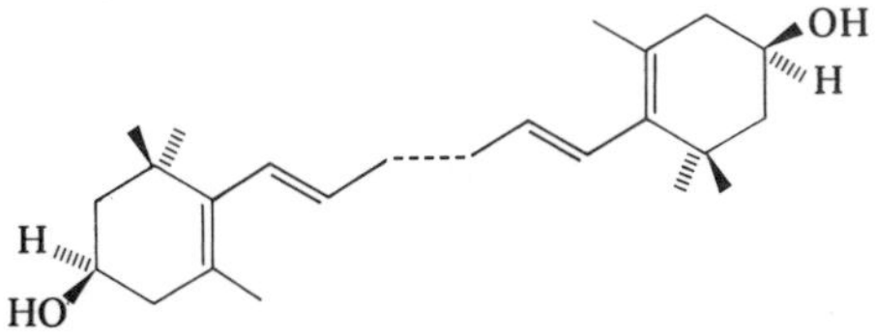

(= 3R, 3'R)

Synthesis. Two more total syntheses have been outlined by J. D. Surmatis et al. (Helv., 1970, 53: 974), one involving reduction of their synthetic (p. 271) rhodoxanthin and the other using the same starting materials as for the rhodoxanthin synthesis to build up 3,3'-dioxo-β-carotene which, again, can be reduced to zeaxanthin. In a further synthesis (Weedon and co-workers, J.

chem. Soc., C, 1971, 404), 3-hydroxy-β-ionone was built up from iso-phorone (p. 270) and then converted (CH_2=CHMgBr; $Ph_3PH^{\oplus}$ $Br^{\ominus}$) into (XXVIII). A (2 + 1) Wittig coupling* with the C_{10} dial (XXIX) gave zeaxanthin.

(XXVIII)   (XXIX)

All synthetic samples of zeaxanthin consist, presumably, of a mixture of SS and RS (*meso*) forms (at $C_{(3)}$, $C_{(3')}$) in addition to the naturally occurring (RR) isomer.

For the reaction of zeaxanthin with titanium tetrachloride, see Section 4(c).

Antheraxanthin (the 5,6-monoepoxide of zeaxanthin). Certain references to the occurrence of antheraxanthin in nature are now known to be incorrect: it appears that the pigment concerned was in fact the analogous 7,8-acetylenic compound, diadinoxanthin (q.v. for ref.).

Violaxanthin, one of the most abundant carotenoids in nature, is the 5,6:5',6'-diepoxide of zeaxanthin.

Absolute configuration. The o.r.d. curve over the 200–400 nm range has been recorded by Scopes, Weedon et al. (1969, loc. cit.). Removal of the epoxide rings (cf. Section 7a) yielded a sample of zeaxanthin giving the same o.r.d. curve as natural zeaxanthin, thereby showing the configuration at $C_{(3)}$, $C_{(3')}$ of violaxanthin to be as for zeaxanthin, above; that is R. The epoxide rings are thought to be *trans* to the ring-OH's, as in lutein epoxide above, since per-acid epoxidation of zeaxanthin (which is thought to go preferentially *cis* to the —OH: cf. Scopes, Weedon et al., loc. cit., p. 2531) gave as the major optically-active product a violaxanthin isomer whose o.r.d. curve was "enantiomeric" to that of natural violaxanthin. For a time, this assignment appeared to be inconsistent with the configuration which had been proposed, in 1967, for abscisic acid (cf. R. S. Burden and H. F. Taylor, Tetrahedron Letters, 1970, 4071); however it has been shown (S. Isoe et al., ibid., 1972, 2517; T. Oritani et al., ibid., p. 2521; G. Ryback, Chem. Comm., 1972, 1190) that this 1967 proposal was incorrect and that the

---

* Buddrus-type conditions (heating the salt and carbonyl compound together in an epoxy-alkane solvent) were found to be necessary for this reaction to proceed satisfactorily; NaOMe/MeOH failed.

$C_{(6)}$-oxygen of abscisic acid is (as written here: XXXIII) "α", rather than "β" as thought earlier. Since violaxanthin and abscisic acid have been chemically related (Burden and Taylor, loc. cit.), this must also be true for violaxanthin so supporting the above contention that, in each ring, the epoxide is *trans* to the —OH.

C. Costes et al. (Compt. rend., 1968, 266; C, 481) studied violaxanthin's reaction with aluminium trichloride. Molar quantities converted it into, probably, auroxanthin (as with H$^{\bullet}$) but a large excess of reagent produced a λ_{max} shift to 655 nm, presumably due to the formation of a compound containing a carbonium or oxonium ion.

Related natural products. Taylor and Burden (Nature, 1970, 227: 302) have shown that a growth inhibitor present in wheat and bean seedlings is identical (comparison of parent compounds and the corresponding acetates on t.l.c. and g.l.c.; comparison of biological activity) with the (XXX) + (XXXI) mixture known as xanthoxin previously obtained, along with several other compounds, by photo-oxidative degradation of violaxanthin [hv/H$_2$O containing dissolved O$_2$ (idem, Phytochem., 1970, 9: 2217); in practice (cf. Burden and Taylor, Tetrahedron Letters, 1970, 4071) zinc permanganate gave a better yield of this particular, and potentially useful, oxidation product]. **Loliolide** (XXXII), isolated from rye grass and elsewhere (R. Hodges and A. L. Porte, Tetrahedron, 1964, 20: 1463), also bears a resemblance to violaxanthin (and has been shown to be one of the products from the above photo-oxidation reaction: Taylor and Burden, Phytochem., 1970, 9: 2217). It seems likely that these compounds, and possibly also (Burden and Taylor, 1970, loc. cit.; S. Isoe et al., 1972, loc. cit.) the biologically-important compound **abscisic acid** (XXXIII), are all derived in nature by oxidative degradation of the violaxanthin commonly present in plant tissue.

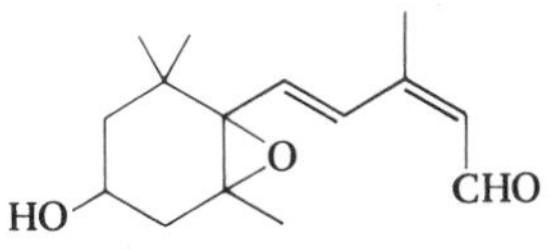

(XXX) 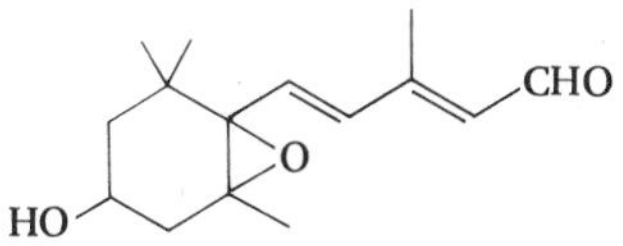   (XXXI) 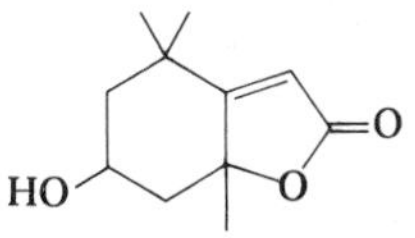   (XXXII)

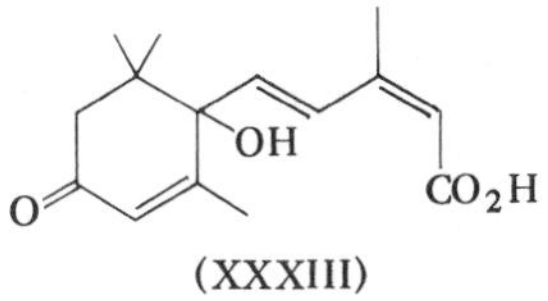

(XXXIII)

H. H. Strain's "violeoxanthin" (from pansy flowers: Arch. Biochem. Biophys., 1954, 48: 458) has been reinvestigated by Szabolcs and Toth (Acta Chim. Acad. Sci. Hung., 1970, 63: 229) who concluded (chemical studies, spectra, o.r.d.) that it is just a *cis*-isomer (*9-cis*?) of violaxanthin.

β-**Kryptoxanthin** (*β-cryptoxanthin, kryptoxanthin, cryptoxanthin*). *Absolute configuration.* The o.r.d. curve from 200 to 400 nm has been recorded by Scopes, Weedon et al. (1969, loc. cit.). It approximated to the zeaxanthin

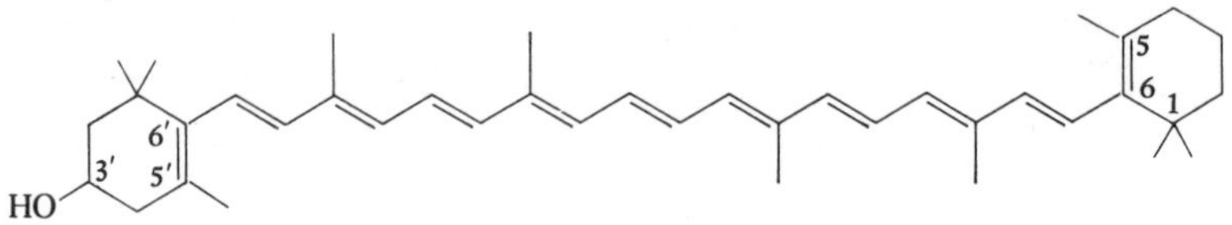

β-Kryptoxanthin

curve in shape but had about one half the intensity, showing that the (single) asymmetric centre, $C_{(3')}$, in β-kryptoxanthin has the same configuration, R, as the corresponding position in zeaxanthin.

Synthesis, of a mixture of isomers, R and S at $C_{(3')}$, has been achieved by a 1:1 Wittig coupling of (XXVIII) above and β-apo-12'-carotenal (C_{25}: prepared by Lindlar reduction of XXIII, 2nd Edn., p. 247) (Weedon and co-workers, 1971, loc. cit.).

Concerning β-kryptoxanthin's reaction with titanium tetrachloride, see Section 4(c).

β-Kryptoxanthin-5,6:5',6'-diepoxide has been tentatively identified in lemons (λ_{max} 470, 439, 416 nm) (Yokoyama et al., J. Food Sci., 1967, 32: 42; Curl, ibid., 1962, 27: 171).

α-**Kryptoxanthin** (*α-cryptoxanthin; zeinoxanthin*). The tentative structure given before (2nd Edn., p. 302) has been revised by J. Szabolcs and A. Rónai (Acta chim. Acad. Sci. Hung., 1969, 61: 309) who have shown that: (a) the compound does not give a positive test for allylic-OH using the trichloro-acetate (cf. p. 215) and $H^{\bullet}/CHCl_3$ tests; (b) on alkaline permanganate oxidation, the major product (according to t.l.c. comparisons with authentic samples and the pattern of *cis/trans*-isomers formed on $I_2/h\nu$ treatment) is *not* α-citraurin, as would have been required by the old formula, but the corresponding non-hydroxylated compound (α-apo-8'-carotenal; or "α-apo-2-carotenal", old name, cf. 2nd Edn., p. 236). This led to the following structure being proposed:

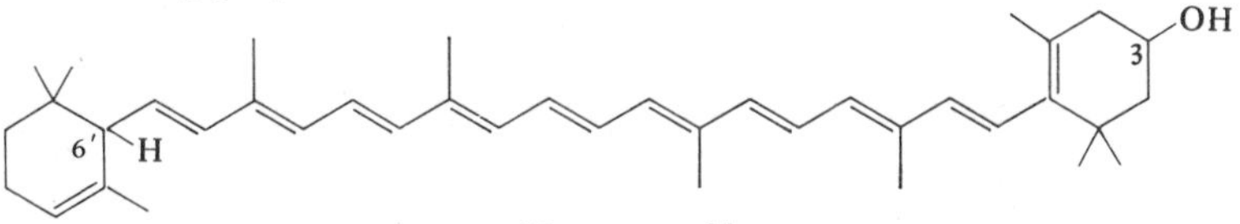

α-Kryptoxanthin

Both the p.m.r. spectrum (Szabolcs and Rónai; on the trichloroacetate) and mass spectrum (Scopes, Weedon et al., 1969, loc. cit., p. 2543: M^+ 552 and a strong peak at M-56 characteristic (p. 209) of an unsubstituted α-end-group) were consistent with the above formulation, as was the o.r.d. curve – below; further confirmation was provided by synthesis (below).

The above structure is identical with that tentatively suggested earlier for the corn (maize) pigment **zeinoxanthin**, and a direct comparison of the two compounds by M. Farkas (cf. Szabolcs and Rónai, loc. cit.) confirmed that they are identical.

Sources of α-k(c)ryptoxanthin (zeinoxanthin). Crogan and Blessin (Crop Sci., 1968, 8: 730) found it to be the *major* carotenoid in certain strains of corn (maize) which, are therefore, a convenient source; A. L. Livingston and R. E. Knowles (Phytochem., 1969, 8: 1311) have detected it in small amount in alfalfa.

Absolute configuration. Scopes, Weedon et al. (1969, loc. cit.) have recorded the o.r.d. curve of α-kryptoxanthin (from peppers) over the range 200–400 nm. Since the curve approximated to that obtained by summing, after halving their intensities, the curves given by zeaxanthin and α-carotene, it was inferred that the configurations at $C_{(3)}$ and $C_{(6')}$ in α-kryptoxanthin are as for the corresponding positions in the latter compounds; that is 3R, 6'R (as in lutein, above).

Synthesis. (1) Partially, from lutein by selectively removing the allylic-OH (see under lutein, above). (2) A total synthesis has been described by Weedon and co-workers (1971, loc. cit.). The phosphonium salt (XXVIII)

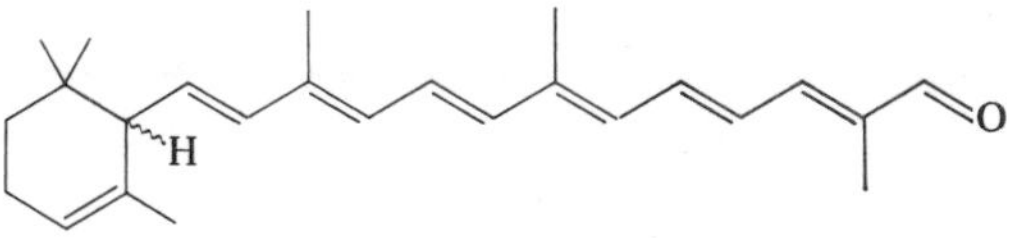

(XXXIV)

was coupled (Wittig reaction) with (±)-α-apo-12′-carotenal (C_{25}, XXXIV; prepared from LXIa, 2nd Edn., p. 274, by the 1:1 coupling with the trienedial (XXIX), p. 252). The product presumably contained in addition to the natural ($3R, 6'R$) isomer, the $3S, 6'R$ etc. forms.

β-Carotene monoepoxide is, probably, one of the major carotenoids in certain types of lemon (Yokoyama et al., J. Food Sci., 1967, 32: 42); see also under "flavacene" below.

"Flavacene" (*"flavacin"*). Direct comparison of the respective compounds by Hertzberg and Jensen (Phytochem., 1967, 6: 1119) has shown that "flavacene" probably has the same (gross*) structure as mutatochrome (the furanoid form of β-carotene monoepoxide).

Rubixanthin. The structure has been confirmed. The position of the hydroxyl group was the main uncertainty, only $C_{(4)}$ having been previously excluded as the point of attachment. However, the recently recorded p.m.r. spectrum (B. O. Brown and Weedon, Chem. Comm., 1968, 382; Arpin and Jensen, Phytochem., 1969, 8: 185) has essentially (cf. p. 205) now excluded the possibility of its being attached to $C_{(2)}$, and has also eliminated the possibility, not considered before, that it might be present as a $-CH_2OH$ group (as in the compounds described in Section 7i). In addition, the p.m.r. pattern was consistent with the presence of one zeaxanthin and one lycopene-type of end-group. The m.s. molecular formula and fragmentation pattern were also consistent with structure (XXXV):

(XXXV)

Absolute configuration. The o.r.d. curve (200–400 nm) has been recorded by Scopes, Weedon et al. (1969, loc. cit.), and the same remarks apply as those under β-kryptoxanthin, above. Arpin and Jensen (loc. cit.) have recorded the c.d. curve (250–500 nm).

Gazaniaxanthin. The similarity between this pigment and rubixanthin above has been noted before. A direct spectral comparison (i.r., p.m.r., m.s. fragmentation pattern) of rubixanthin and gazaniaxanthin (both $C_{40}H_{56}O$ by m.s.) by Brown and Weedon (1968, loc. cit.) has shown the two pigments

---

* The samples of "flavacene", probably formed by in vivo $H^{\oplus}$-catalysed rearrangement of β-carotene monoepoxide, and mutatochrome (prepared from β-carotene as in 2nd Edn., p. 303) may well have differed in the proportion of configurational isomers about $C_{(8)}$ (and possibly $C_{(5)}$ too).

to be almost indistinguishable spectrally, the only notable difference being in the allylic-CH_2 region of the p.m.r. In addition, the pattern of *cis/trans* isomers obtained on hv/I_2 treatment of each was identical, the various respective isomers being inseparable chromatographically, and the major product from the gazaniaxanthin experiment appeared identical (m.p., mixed m.p.) with rubixanthin. The possibility that the observed m.p. difference between rubixanthin and gazaniaxanthin is due to polymorphism or to a difference in configuration at $C_{(3)}$, has been excluded by crystallisation/mixed m.p. tests and by c.d./o.r.d. measurements, respectively (Brown and Weedon; Arpin and Jensen, loc. cit.). Apparently whereas rubixanthin is all-*trans*-(XXXV), gazaniaxanthin is merely a *cis*-isomer of same. The observation (Brown and Weedon) that the visible/u.v. absorption spectrum of the latter is indistinguishable from that of rubixanthin, as regards both peak positions *and* relative intensities, and that it lacks a *cis*-peak is most unusual for a *cis*-carotenoid and can only be explained satisfactorily by assuming that the *cis*-bond is at the end of the chromophore, namely at 5′,6′ (Brown and Weedon). This could also explain the p.m.r. differences noted above.

Absolute configuration. Both o.r.d. and c.d. curves have been obtained and found to be similar to those obtained, by the same authors, for rubixanthin (which see for refs.), thereby showing the configuration at $C_{(3)}$ to be as for rubixanthin (i.e. *R*).

Eschscholtzxanthin. *Biosynthesis.* The $T/^{14}C$ ratio of eschscholtzxanthin biosynthesised in the presence of a doubly-labelled $(T,^{14}C)MVA$ (cf. Section 8) suggests that it is formed in nature from a hydroxy-carotenoid containing normal end-groups (such as zeaxanthin, with which it co-occurs), rather than independently through a series of *retro*-type precursors (R. J. H. Williams, G. Britton and Goodwin, Biochim. biophys. Acta, 1966, 124: 200):

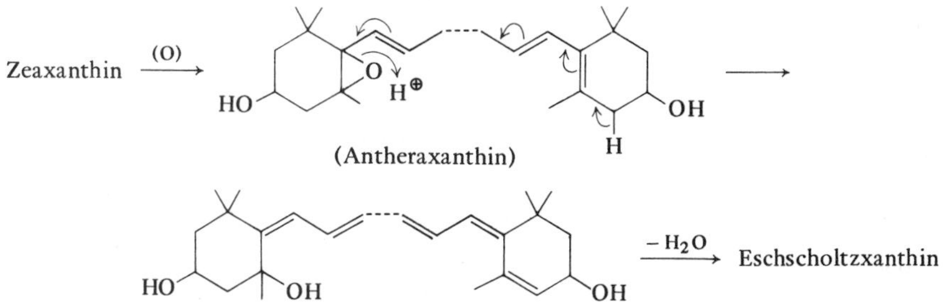

Zeaxanthin $\xrightarrow{(O)}$

(Antheraxanthin)

$\xrightarrow{-H_2O}$ Eschscholtzxanthin

Plectaniaxanthin, a carotenoid found by Arpin and Jensen (Phytochem., 1967, 6: 995), in the mushroom *Plectania coccinea.* It is present partly free, partly as a monoester (at $C_{(2')}$), and partly as a diester (at $C_{(1')}$, $C_{(2')}$) (the relationship between the three pigments being inferred from the similarity in their absorption spectra and the finding that base treatment of either the latter two pigments yielded the first). The free pigment gave a monoacetate on acetylation, which in turn gave a mono-trimethylsilyl-ether (*tert*-OH; cf. Section 3). Treatment with $H^{\oplus}/CHCl_3$ yielded a less polar compound, identified, by direct comparison, as 3,4-dehydrotorulene (for outline of structure, see footnote p. 260). This suggested the following structure, which was confirmed by comparison of the natural pigment with a sample of (XXXVI) synthesised by Schwieter and Isler (cf. Arpin and Jensen, loc. cit.):

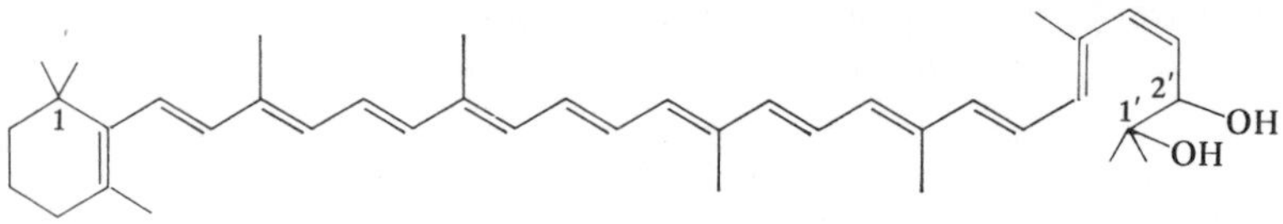

Plectaniaxanthin (XXXVI)

Arpin and Jensen found plectaniaxanthin to be rather less polar than expected for such a structure and ascribed this effect to intramolecular hydrogen bonding reducing the "availability" of the OH groups.

For another source (a yeast) see C. O. Chichester et al. (ibid., 1971, 10: 625). It has m.p. 172–173°, λ_{max} 502, 471, 445 nm. These authors also quote mass spectral and p.m.r. data. The $Me_2C(OH)-$ group gives two methyl peaks near τ 8.8 (rather than a single 6H peak as given by the *gem*-Me's of a spirilloxanthin-type end-group), due to its being attached to an asymmetric carbon atom (cf. p. 205).

2'-Dehydroplectaniaxanthin. Arpin and Jensen (loc. cit.) found that *P. coccinea* also contains a carotenoid, again present as an ester, with λ_{max} values at ca. 20 nm longer wavelength than plectaniaxanthin's; it was found to be identical with the *p*-chloranil oxidation product of plectaniaxanthin itself, thereby leading to the following structure (this also provides a total synthesis of 2'-dehydroplectaniaxanthin from synthetic (cf. above) plectaniaxanthin) :

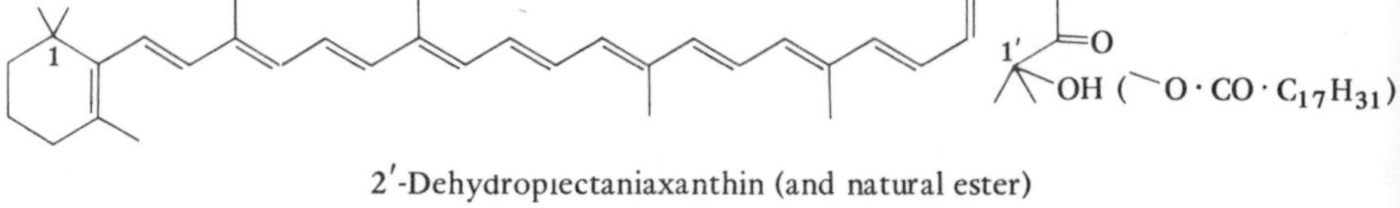

2'-Dehydroplectaniaxanthin (and natural ester)

For two further sources, see Arpin and Jensen (Bull. Soc. chim. Biol., 1967, 49: 527), Arpin et al. (Phytochem., 1969, 8: 897). In all three sources the pigment was found to be esterified at $C_{(1')}$. In the latter two, this esterification was almost entirely by a single acid, which was the same in each case [by t.l.c. comparison of the natural (esterified) pigments; g.l.c. of the methyl esters of the acids liberated by saponification] (Arpin et al.). Mass spectrometry of the liberated acid (as the methyl ester, and after epoxidation to aid clean fragmentation of the molecule) showed it to be linoleic ($C_{(18)}$, with C=C at $C_{(9)}$ and $C_{(12)}$). Similarly, the observation, in the spectrum of the natural (unsaponified) pigment, of fragments at m/e 350 and 322 (cleavage α to the C=O at $C_{(3')}-C_{(2')}$ and $C_{(2')}-C_{(1')}$ respectively) showed the $C_{(1')}$ position carries a $-O \cdot CO \cdot C_{17}H_{31}$ (linoleic) residue, as did the presence of ions at 280 and 263, corresponding to $[C_{17}H_{31} \cdot CO_2 H]^{\oplus}$ and $[C_{17}H_{31} \cdot CO]^{\oplus}$ (Arpin et al., 1969, loc. cit.). Four new 2-OH-β(α)-carotene derivatives have been reported (Quackenbush, ibid., 1973, 12: 2481; Jensen, Pure Appl. Chem., 1973, 35: 85).

Carotenoid glycosides. Certain of the very polar* carotenoids mentioned under the "Carotenoids of uncertain constitution" section in the 2nd Edn., p. 331, have now been shown to owe their high polarity to their occurring in nature linked to a sugar residue. These are described below. Further examples appear under the section on C_{50} carotenoids (p. 307), and under Section 7e. A related compound is *crocin,* in which the sugar unit, gentiobiose, is attached to the carotenoid (an acid) through an ester, rather than an ether, link.

Phlei-xanthophyll (and **4-keto-phlei-xanthophyll**). Phlei-xanthophyll was the name given earlier to the main carotenoidal constituent of the bacterium *Mycobacterium phlei* Vera (2nd Edn., p. 331). It has since been shown to be a mixture of two compounds, the name phlei-xanthophyll being retained for the main component (75%) of the mixture, the remainder being 4-keto-phlei-xanthophyll (Hertzberg and Jensen, Acta chem. Scand., 1967, 21: 15). The two compounds could only be separated by alumina chromatography of the mixture of the derived per-acetates, the native xanthophylls then being recovered by lithium aluminium hydride reduction and alkaline hydrolysis, respectively.

The phlei-xanthophyll so obtained crystallised from pyridine—petroleum as deep mauve needles, m.p. 209°; λ_{max} (CHCl$_3$) 522, 489, (465) nm.

---

* As inferred from measuring the pigment's partition ratio on distributing it between two immiscible solvents (usually petroleum and 90 or 95% methanol) or observing its behaviour on chromatography.

Hertzberg and Jensen inferred the nature of the chromophore, and the presence or otherwise of $Me_2 C(O)-$ (p.m.r.), allylic-OH (p-chloranil gave a conjugated ketone: i.r.), tert-OH, etc. groups in the usual manner. I.r. spectroscopy showed no $-CO_2 H$ group to be present. To account for the high polarity of the compound with only hydroxyl groups would require several to be present, and placing these in a C_{40} skeleton would have presented problems. However, mass spectrometry indicated a C_{46} rather than C_{40} formulation. Acid treatment (0.02 N HCl in $CHCl_3$/3 min/25°) of phlei-xanthophyll resulted in the formation of (a) a much less polar compound with absorption maxima at 14 nm longer wavelength, identified as 3,4-dehydrotorulene* (by λ_{max} values and direct comparison with the authentic compound), and (b) in the aqueous phase from the reaction work-up, a carbohydrate identified (paper chromatography and other standard tests) as D-glucose. This led to the realisation that phlei-xantho-phyll is a glycoside of a C_{40} carotenoid, and suggested the following structure, where R = D-glucose residue :

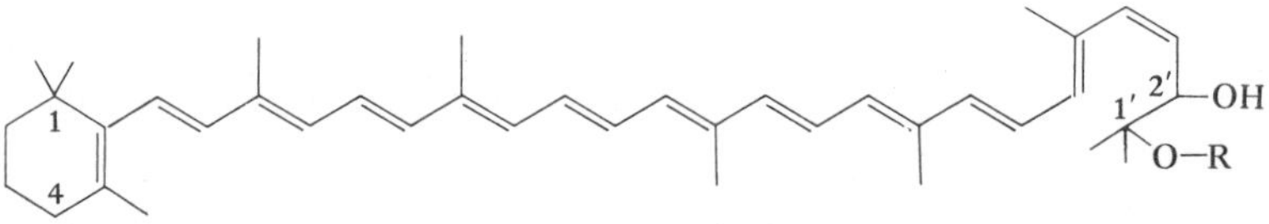

This was confirmed by total synthesis of the tetra-acetate of the allylic oxidation product of phlei-xanthopyll (structure as above except $>C=O$ at $C_{(2')}$ and R = glucose tetra-acetate residue) by a Königs–Knorr condensation ($AgCO_3$) of the corresponding synthetic C_{40} compound with tetra-acetoxy-bromoglucose, and direct comparison of the two samples.

The second compound obtained from the above separation of per-acetates had absorption maxima at slightly longer wavelengths than phlei-xantho-phyll, but after hydride reduction the spectra coincided, suggesting the

---

* Formed by

3,4-dehydrotorulene

presence of a conjugated-carbonyl group (confirmed by i.r.) but otherwise a similar chromophore to that of phlei-xanthophyll. Acid hydrolysis again converted the pigment into a much less polar derivative, this time identified as 4-oxo-torulene (by hydride reduction then acid treatment, which yields 3,4-dehydrotorulene: cf. above), and D-glucose. Finally, comparison of its p.m.r. spectrum with that of phlei-xanthophyll was consistent with the presence of a carbonyl group in the β-ring at $C_{(4)}$ (ring-methyls deshielded), thereby leading to the structure of this compound as being the 4-*oxo-derivative of phlei-xanthophyll*; i.e. as for phlei-xanthophyll but with $> C=O$ at $C_{(4)}$ (Hertzberg and Jensen, loc. cit.).

Myxobacton ester, the major pigment from three *Myxobacterales* species studied by H. Kleinig and H. Reichenbach (Naturwiss., 1970, 57 : 92; Arch. Mikrobiol., 1970, 74: 223), was shown by them (chemical studies; visible, i.r., and mass spectra) to be as for 4-oxo-phlei-xanthophyll above but lacking the —OH at $C_{(2')}$ and to have for its sugar unit at $C_{(1')}$ a glucose residue esterified with a mixture of fatty acids. The parent C_{40} structure is identical with the carotenoid deoxy-flexixanthin (2nd Edn., p. 318).

Myxobactin ester, a minor pigment from the same source, was considered to be identical with myxobacton ester at the acyclic end of the molecule but to have a trimethylcyclohexadiene (vitamin A_2-type) ring at the cyclic end.

The C_{46} structures resulting from removal of the esterifying acids are known as *myxobacton* and *myxobactin* respectively. The glucose residues could be removed by vigorous alkali or MeOH/HCl treatment, but some dehydration at $C_{(1')}/C_{(2')}$ also occurred.

An allied myxobacterium has been found to contain (same methods as above) a group of four glycosides consisting of (a) two compounds with structure as for phlei-xanthophyll but lacking the —OH at $C_{(2')}$ and with in one case rhamnose at $C_{(1')}$ and in the other an esterified glucose residue*, and (b) the corresponding pair of compounds with an additional —OH in the ring of the carotenoid skeleton, tentatively placed at $C_{(3)}$ (Kleinig et al., ibid., 1971, 78: 224).

Myxoxanthophyll and oscillaxanthin, first described, and briefly studied, about 30 years ago, are very polar compounds which occur in certain blue-green algae, including *Oscillatoria rubescens* which contains both. Hertzberg and Jensen have now confirmed their presence in *O. rubescens* (but report that an *Athrospira* sp. is a better source: Phytochem., 1966, 5: 557), and have carried out a detailed study of their structures using modern techniques (ibid., 1969, 8: 1259, 1281). Thus myxoxanthophyll was found

---

* The acyclic analogue since found, in an allied source (Kleinig et al., Phytochem., 1973, 12: 2483).

to be a C_{46} compound ($C_{46}H_{66}O_7$ by precision m.s.) rather than C_{40} as previously thought. The absorption spectrum was as for phlei-xanthophyll. Acetylation gave a tetra-acetate (m.s.) which still contained a *tert*-OH (silylation). Vigorous LiAlH$_4$ reduction gave a much less polar compound (removal of the sugar residue*) identified as saproxanthin (2nd Edn., p. 304), establishing the carbon skeleton. Similarly, oscillaxanthin ($C_{52}H_{76}O_{12}$; from precision m.s. on the acetate derivative, $M = 1144$) was found to have an acyclic tridecaene chromophore, to give a hexa-acetate, still containing two *tert*-OH's, and on reduction to give the C_{40} dihydroxy compound corresponding to spirilloxanthin (by comparison with the authentic compound)*. Both pigments, on treatment with $H^{\oplus}$/MeOH followed by aqueous $H^{\oplus}$, yielded, in the aqueous fraction, a sugar identified as L-rhamnose. The structures were confirmed by a study of the fragmentation patterns of the acetate derivatives (Hertzberg and Jensen, loc. cit.; discussed by Jensen, Pure Applied Chem., 1969, 20: 434–437). Thus myxoxanthophyll has the following structure, where "R" is mainly L-rhamnose (a small percentage of the pigment studied by Hertzberg and Jensen was attached to glucose (?) rather than rhamnose); the parent C_{40} structure is known as "*myxol*":

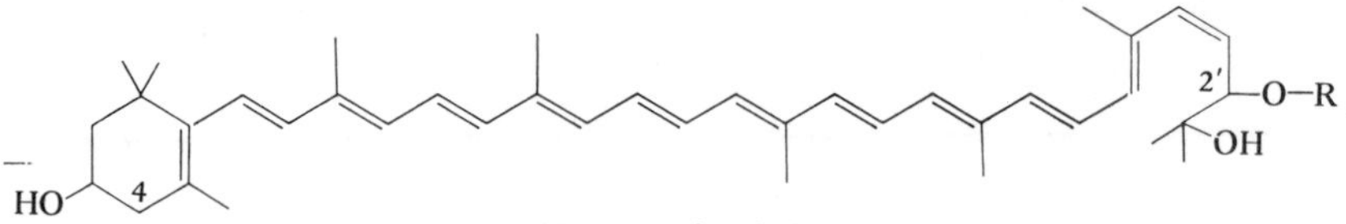

Myxoxanthophyll

The oscillaxanthin structure is given below under Section 7c, where it, formally, belongs.

O. *limosa* contains compounds with the same carotenoid skeletons as myxoxanthophyll and oscillaxanthin but with different sugar units attached (Francis et al., Phytochem., 1970, 9: 629). What appears to be a 2′-glycoside of 4-*oxo-myxol* has been detected in various blue-green algae, including O. *limosa*, by Jensen et al. (ibid., 1971, 10: 3121).

Aphanizophyll, mentioned briefly in the 2nd Edn. (p. 331) as being of uncertain constitution, has been reinvestigated using the techniques used for myxoxanthophyll, including precision-m.s. (on the acetate) which gave the correct formula ($C_{46}H_{66}O_8$) for the first time; this led to the formulation of the structure as (probably) the 4-hydroxy-derivative of myxoxanthophyll (Hertzberg and Jensen, ibid., 1971, 10: 3251).

The p.m.r. spectra of the above glycosides were generally run on the

---

* Displacement of $OR^{\ominus}$ by $H^{\ominus}$.

acetate derivatives. The *OAc* peaks complicate the in-chain methyl group region ($\sim \tau$ 8.0) but in general the only other peaks from the sugar units are the relatively weak *H*–C–OAc multiplets. However, in oscillaxanthin a doublet due to the *Me*CH< group of the rhamnose residues can be seen in the saturated-methyl group region. The methyls of the $Me_2 C(OH)$– groups, being non-equivalent due to the neighbouring asymmetry, usually give two separate 3H peaks (~8.8) instead of the more familiar 6H singlet. For the chromatographic resolution of mixtures of these very polar compounds, Kleinig and Reichenbach recommend first removing any fatty-acid residues by saponification and then converting them into the peracetates followed by chromatography on *magnesia* (J. Chromatog., 1972, 68: 270).

(c) Acyclic hydroxy-, methoxy-, and methoxy oxo-carotenoids

Lycoxanthin and **lycophyll.** The tentative structures given before (2nd Edn., p. 306) have been revised, see Section 7(i) (p. 320).

Oscillaxanthin, a very polar carotenoid of previously unknown constitution, has now been shown to have the following structure (for details, see above) :

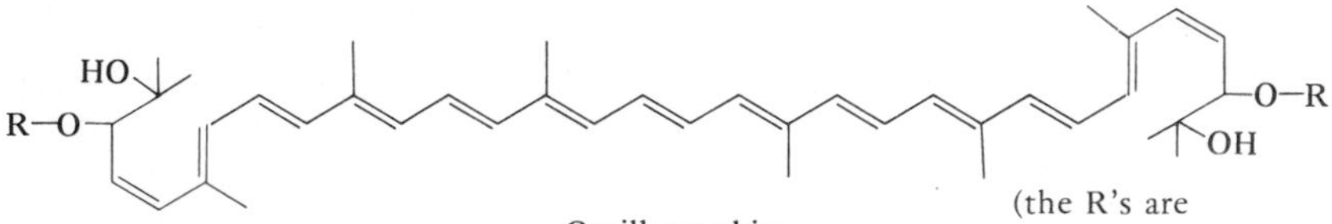

Oscillaxanthin

(the R's are
L-rhamnose residues)

Carotenoids from purple bacteria. New work on these compounds is outlined below.

Rhodopin. For new carotenoids based on the rhodopin skeleton but with one of the in-chain methyl groups oxidised to the $-CH_2OH$ or $-CHO$ levels, see Sections 7i (p. 320) and 7d(i) (p. 266), respectively.

3,4-Dehydrorhodopin. A total synthesis has been reported (D. F. Schneider and Weedon, J. chem. Soc., C, 1967, 1686) (Scheme 3):

Scheme 3

$$Me_2C=O + HC\equiv C \cdot CH_2 Br \xrightarrow{\text{Reformatsky}} Me_2 C(OH) \cdot CH_2 \cdot C\equiv CH \xrightarrow{\text{PhLi; mvk}}$$

$$Me_2 C(OH) \cdot CH_2 \cdot C\equiv C \cdot CMe(OH) \cdot CH=CH_2 \xrightarrow[\text{(ii) } Ph_3PH^{\oplus}Br^{\ominus}]{\text{(i) } LiAlH_4}$$

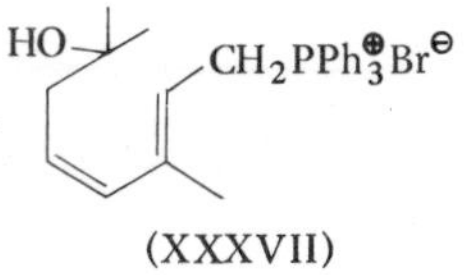

(XXXVII)

(mvk = methyl vinyl ketone, Me $\cdot$ CO $\cdot$ CH=CH$_2$)

A Wittig condensation (NaOMe/MeOH) of (XXXVII) with apo-8'-lycopenal (C_{30}; structure, 2nd Edn., p. 313; synthesis by a 1:1 Wittig condensation of compounds (XLVI) and (XXIV) of 2nd Edn., pp. 267, 248) yielded a pigment with properties identical to those reported for the natural compound (thereby also confirming the structure given in 2nd Edn., p. 307 for the latter).

3,4,3',4'-Tetrahydrospirilloxanthin, m.p. 175°, the major (96%) carotenoid of bacterium "RG3" (Hertzberg and Jensen, Acta chem. Scand., 1967, 21: 371). It was found to have the same chromophore as lycopene (absorption spectrum) but to be rather more polar and to contain two methoxy-groups (i.r., p.m.r.), apparently present as two $Me_2C(OMe)$– groupings (4Me singlet at $\tau 8.86$). The structure thereby inferred (XXXVIII; Hertzberg and Jensen) was proved by total synthesis and direct comparison of the natural and the synthetic material:

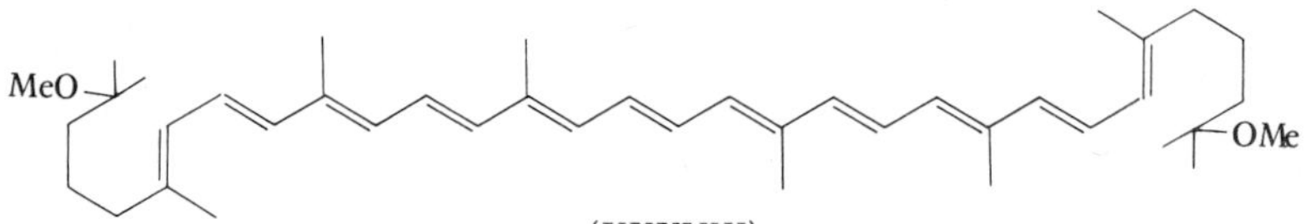

(XXXVIII)

Synthesis (Hertzberg and Jensen, loc. cit.). This was based on the first route used by Surmatis and Ofner in their syntheses of spirilloxanthin (2nd Edn., p. 309) except that (a) the NBS/CCl$_4$ dehydrogenation was omitted and (b) the second step of the sequence was effected by a Wittig (with $Ph_3P=CH \cdot CO_2Et$), rather than a Reformatsky, reaction.

See below for a further occurrence of this compound.

Spirilloxanthin. A new route to compound (CVIa; 2nd Edn., p. 309) used in one of the syntheses of this compound given before has been reported by Schneider and Weedon (1967, loc. cit.; same sequence as used for the synthesis of XXXVII above, except that the –OH of the Reformatsky product was methylated before carrying out the next step).

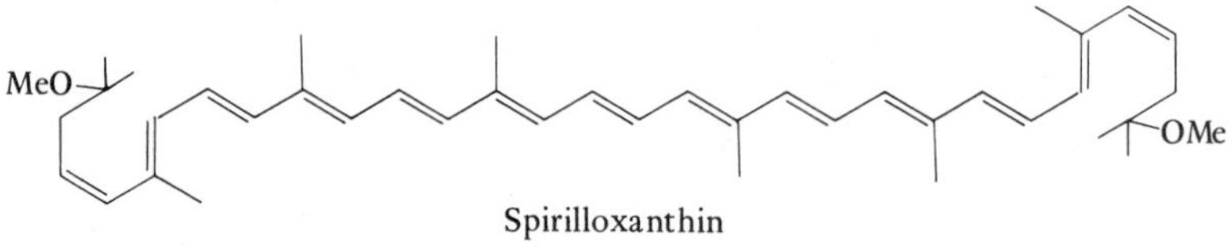

Spirilloxanthin

"OH-spirilloxanthin". A total synthesis of compound CVII (2nd Edn., p. 310; the structure tentatively assigned to natural "OH-spirilloxanthin") has been reported by Schneider and Weedon (1967, loc. cit.):

Compound XXXVII above, was condensed (1:1 Wittig reaction; NaOMe/ MeOH) with crocetindial (p. 267 for structure). The C_{30} product was then condensed (Wittig) with the methoxy-analogue of XXXVII (prepared using the same sequence as for XXXVII except that the —OH of the Reformatsky product was methylated before continuing the reaction sequence) to give the desired compound. This was found to have similar properties to those reported earlier for natural "OH-spirilloxanthin" by Jensen (2nd Edn., p. 309) thereby supporting the tentative formulation for the latter. A direct comparison of the synthetic and the natural material has not been carried out.

Bacterioruberin α. The structure which had tentatively been put forward for this pigment has been revised, see Section 7h (p. 313).

11',12'-Dihydrospheroidene and **3,4,11',12'-tetrahydrospheroidene.** Two pigments isolated from the bacterium *Rhodospirillum rubrum* grown in the presence of diphenylamine (which inhibits the formation of the carotenoids normally produced by the organism and leads to the appearance of compounds not normally detectable) have been identified by B. H. Davies et al. (J. chem. Soc., C, 1969, 1266; Biochem. J., 1970, 116: 101), as being the 11',12'-dihydro- and 3,4,11',12'-tetrahydro- analogues of the well-established carotenoid spheroidene (below); $\lambda\lambda_{max}$ (petroleum) 440, 414, 392 and 419, 395, 374 nm (as for "unsymmetrical ζ-carotene, p. 221), corresponding to octaene and heptaene chromophores, respectively. The molecular formulae were determined by precision m.s. This data in conjunction with the m.s. fragmentation patterns [ions formed by cleavage of doubly allylic single bonds, particularly that at 11',12' adjacent to the chromophore (cf. p. 209) and by loss of methanol] led to the proposed structures.

The same source has also yielded a pigment identified (tentatively by Davies et al., 1970, loc. cit.; confirmed by H. C. Malhotra, G. Britton and T. W. Goodwin, Phytochem., 1970, 9: 2369) as **3,4-dihydrospheroidene.**

Spheroidene has additionally been found in the above, diphenylamine-inhibited, source (Davies et al., 1969, 1970, loc. cit.; Goodwin and co-workers, Phytochem., 1969, 8: 1047):

Spheroidene

OH-Spheroidene, sometimes now known as 1'-*hydroxy*-1',2'-*dihydrospheroidene*, has also been obtained from the above-mentioned source (Davies et al., 1970, loc. cit.; Goodwin and co-workers, 1969, loc. cit.).

3,4-*Dihydro-anhydrorhodovibrin*, having the structure as for spheroidene but with C=C at $C_{(7')}$–$C_{(8')}$ instead of $C_{(3)}$–$C_{(4)}$, one of the minor pigments in the above source, was tentatively assigned this structure by both Davies et al. (1970, loc. cit.) and Malhotra, Britton and Goodwin (loc. cit.).

In addition to the above mono-oxygenated compounds, the above source was found to contain a series of dimethoxy-compounds based on the spirilloxanthin skeleton (see above) which were identified with varying degrees of certainty depending on the amount available (often very small) by Malhotra, Britton and Goodwin (loc. cit.) as: **3,4-Dihydrospirilloxanthin**; **3,4,7,8-tetrahydrospirilloxanthin** (also reported to occur in *Rhodopseudomonas spheroides*); 3,4,3',4'-tetrahydrospirilloxanthin (p. 264); and tentatively 3,4,3',4',7,8-*hexahydrospirilloxanthin* and 3,4,3',4',7,8,11,12-*octahydrospirilloxanthin*.

Phillipsiaxanthin, the major carotenoid of the dark red fungus *Phillipsia carminea*, probably has the following structure (Arpin and Jensen, Bull. Soc. chim. Biol., 1967, 49: 527), and therefore bears a close relationship to pigment "P518" (2nd Edn., p. 312).

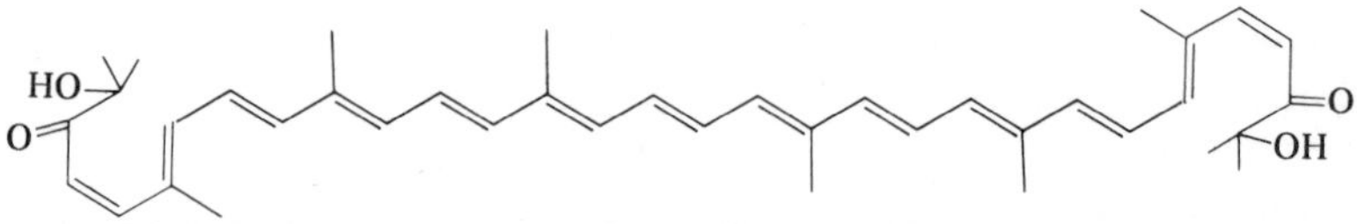

Phillipsiaxanthin

Thiothece-OH-484, earlier known as "demethylated okenone" (p. 331), a pigment accompanying the okenone (p. 330) in a *Thiothece* sp. studied by Jensen and co-workers first in 1968 but in greater detail more recently (Acta chem. Scand., 1972, 26: 2194; m.s., i.r., and chemical studies); this showed it not to be aromatic, as had been thought earlier, but an acyclic carotenoid with, probably, one end of the conjugated-carbonyl type in okenone and the other the hydroxy-end-group of chloroxanthin (2nd Edn., p. 310).

The latter of Kleinig's pair of canthaxanthin-like compounds noted on p. 275 would formally belong to this section.

(d) Aldehydo-, keto-, and hydroxyketo-carotenoids

(i) Aldehydes and hydroxyaldehydes

β-**Citraurin** (C_{30}). *Absolute configuration* (at $C_{(3)}$). Scopes et al. (J. chem. Soc., C, 1969, 2527) measured the o.r.d. (of the oxime derivative) and showed it to be similar to that of zeaxanthin in both general shape and positions of extrema but of about half the amplitude (expected as only one asymmetric centre present), and hence concluded that $C_{(3)}$ in natural β-citraurin is of the same configuration (R) as the $C_{(3)}$, $C_{(3')}$ positions in zeaxanthin.

K. Egger et al. have reported the African plant *Palisota barteri* as a new source of β-citraurin (Z. Naturforsch., 1968, 23b: 1105).

Violaxanthinals. Carotenoids tentatively identified as **apo-10'-violaxan-thinal** and **apo-12'-violaxanthinal** (structure as for 10' but with one less CH=CH) have been detected in the peel of certain varieties of orange and tangerine, respectively, by A. L. Curl [J. Food Sci., 1967, 32: 141; 1965, 30: 13; structures from polarity, absorption spectrum in hexane and in ethanol (loss of fine structure), effect of $H^{\oplus}$ and of $NaBH_4$ (2nd Edn., pp. 293, 245)] :

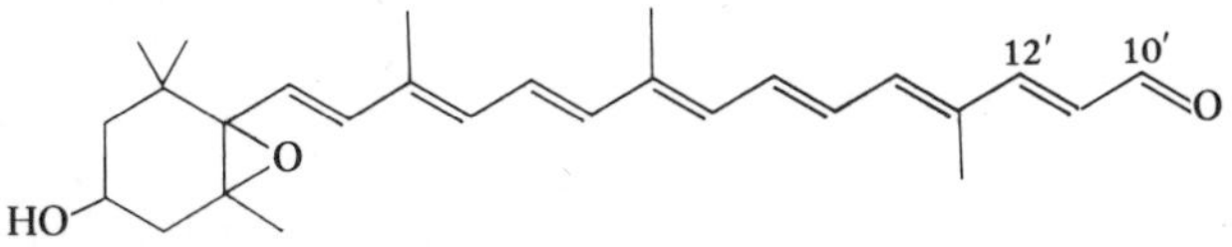

Apo-10'-violaxanthinal

Crocetindial (C_{20}). This well known key intermediate in one of the major routes used for the total synthesis of carotenoids (2nd Edn., p. 248), has now been found to occur in the flowers of *Jacquinia angustifolia* (C. H. Eugster et al., Helv., 1969, 52: 806). The natural sample had m.p. 192–193° (benzene–petroleum), λ_{max} 472, 445, 421 nm (benzene) and its structure was confirmed by direct comparison with a sample of synthetic crocetindial.

"Crocetin half-aldehyde". A very polar compound accompanying the above pigment in *J. angustifolia,* but present in much smaller amount, was shown to be the corresponding mono-acid (from comparison of its spectral properties with those of the corresponding synthetic ethyl ester). It was isolated by chromatography on a polyamide column using aqueous acetic acid as solvent and had λ_{max} 466, 439, 415 nm (benzene) :

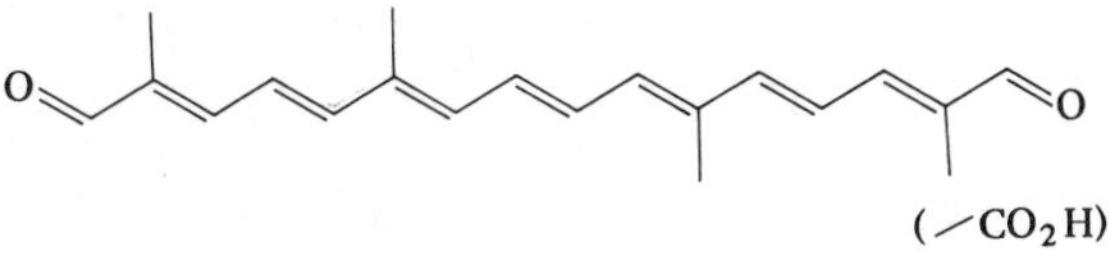

$(\diagup CO_2H)$

Crocetindial (and crocetin half-aldehyde)

The following four compounds are all thought to contain the novel feature of an in-chain —CHO group, presumably formed by biogenetic oxidation of the original —CH$_3$ group. In one, and possibly two, cases the corresponding —CH$_2$OH compound has also been found (see Section 7i, p. 322).

Rhodopin-20-al. This compound was referred to in the 2nd Edn. (p. 332) as "warmingone" (Jensen and Schmidt, Arch. Mikrobiol., 1963, 46: 138). It has also been named "Pigment 2" (Jensen and Schmidt, loc. cit.). It was first isolated by Jensen and Schmidt in 1963 (loc. cit.) from *Chromatium warmingii* strain Migula, and subsequently found (A. J. Aasen and Jensen, Acta chem. Scand., 1967, 21: 2185, wherein called "*rhodopinal*") in two other strains of *C. warmingii* and in a *Thiocystis* sp. It is the major carotenoid in each of the above sources, and is accompanied by the corresponding rhodopinol (Section 7i, p. 321), the corresponding lycopenal (below), rhodopin itself and lycopene. The latter has the basic skeleton from which presumably the others are derived biogenetically. Rhodopin-20-al is unusual in that its most stable form is a *cis*-isomer (as which it exists in nature; cf. below) which even on iodine-catalysed stereoisomerisation yields none of the all-*trans* form. The corresponding lycopenal mentioned below behaves similarly suggesting this is a characteristic property of carotenoids possessing an in-chain —CHO group. The naturally occurring *cis*-isomer has λ_{max} 495 nm (plus several strong *cis*-peaks at 360—400 nm) (petroleum). Although it could be obtained pure according to chromatographic and spectral criteria, it could not be crystallised (a precipitated sample, acetone—petroleum, had m.p. ca. 195°) and it showed rather lower stability than the average carotenoid (Aasen and Jensen, 1967, loc. cit.). The presence of a conjugated aldehyde was inferred from i.r. (2760, 1685 cm^{-1}) and p.m.r. (τ 0.5) data, and chemically (formation of oxime and other derivatives; reaction with Wittig reagents). No intermediates were detected during these reactions implying only one carbonyl group present. P.m.r. data also suggested the presence of the same end-groups as in rhodopin [Me_2C(OH)— at τ 8.78; Me_2C= at 8.39, 8.32], eliminated the possibility of methoxy groups, and showed that the —CHO group was present in place of one of the in-chain methyl groups (only three of which could be seen in the spectrum). The relationship to rhodopin was confirmed by converting the —CHO group into —CH$_3$ [(i) LiAlH$_4$, (ii) tosylation, (iii) LiAlH$_4$] and showing the product to be rhodopin. The position of the —CHO group in the chain was inferred from a consideration of the above data and of various chemical inter-relationships established with two allied compounds (rhodopinol, lyco-penal) and their derivatives (Aasen and Jensen, 1967, loc. cit.), coupled with the eventually successful determination of the mass spectrum of this unusually unstable carotenoid (G. W. Francis and Jensen, Acta chem. Scand., 1970, 24: 2705):.

Rhodopin-20-al

(For carotenoid numbering see p. 192)

Aasen and Jensen (1967, loc. cit.) tentatively placed the *cis*-bond present in the naturally-occurring compound at the point of attachment of the —CHO group to the polyene chain (i.e. 13-*cis* or 12-*s-cis*)* for the following reasons. The stability of the *cis*-form as compared with the all-*trans* is exceptional, suggesting that this is a direct result of, and dependent on, having a —CHO group attached to the chromophore. Also the *cis*-peak in the spectrum (above) is very strong, characteristic of a *cis*-bond near the centre of the chromophore (cf. 2nd Edn., p. 255). The cross-conjugated —CHO group has a marked bathochromic effect on the absorption spectrum (ca. +30 nm; Aasen and Jensen, 1967, loc. cit.).

Lycopen-20-al. This compound was previously called "**anhydro-warmingone**" (Schmidt et al., Arch. Mikrobiol., 1965, 52: 132) or "Pigment 1" (Jensen and Schmidt, 1963, loc. cit.); discovery (1963) and subsequent further sources (1967) as for the pigment above, but present in only trace amounts. It shows the same absorption spectrum, including the strong *cis*-peaks, as the preceding pigment. The proposed structure (as for the rhodopinal above but with both end-groups of the lycopene type) is mainly based on the finding that it appears to be identical with the $POCl_3$—pyridine dehydration product of rhodopinal (Aasen and Jensen, 1967, loc. cit.) and on its mass spectral molecular weight (550) and fragmentation pattern (Francis and Jensen, 1970, loc. cit.). It shows the same unusual *cis/trans* isomerisation behaviour (Aasen and Jensen) as the rhodopinal above, and is assumed to contain a *cis*-bond at the same position (i.e. 13-*cis* or 12-*s-cis*).

P500 (3,4,3′,4′-tetrahydrospirilloxanthin-20-al?) A minor carotenoid from another purple bacterium (*Thiococcus* sp.) (details, Aasen and Jensen, 1967, loc. cit., p. 2191), with an absorption spectrum as for the rhodopinal above but being rather less polar, and with a strong methoxyl band in the i.r. A series of small-scale reactions (reduction, acetylation, formation of aldehyde derivatives, etc.) coupled with m.s. results (Aasen and Jensen, 1967; Francis

---

* Definition of *s-cis*, p. 294.

and Jensen, 1970, loc. cit.) suggested the following structure, again with a stable *cis*-bond at $C_{(13)}$ (or *s-cis* at $C_{(12)}/C_{(13)}$):

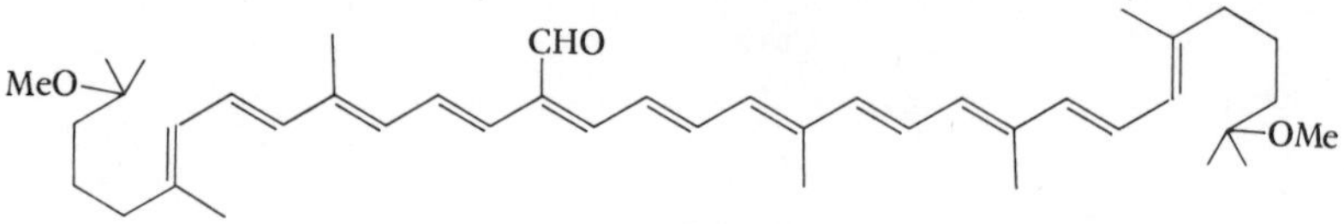

P500(?)

"*Monomethoxylycopenal*", a pigment from a *Lamprocystis* sp., has been tentatively assigned a structure similar to that of P500, but with one end (which one is unknown) of the lycopene type (Francis and Jensen, 1970, loc. cit.).

Thiothece-460 and Thiothece-425, two apo-carotenoids which accompany (and are degradation products of ?) the okenone in a *Thiothece* sp. studied by Andrewes and Jensen (Acta chem. Scand., 1972, 26: 2194) who found each to have one end-group of the okenone (non-aromatic end) type and to be, probably, apo-8' and apo-12' respectively.

(*ii*) *Keto- and hydroxyketo-carotenoids*

2'-*Dehydroplectaniaxanthin* and 4-*keto-phlei-xanthophyll*, both hydroxy-keto-carotenoids have, for simplicity, been described along with the corresponding non-ketonic compounds towards the end of Section 7b (p. 258).

Rhodoxanthin. Two total syntheses have now been described.

Scheme 4

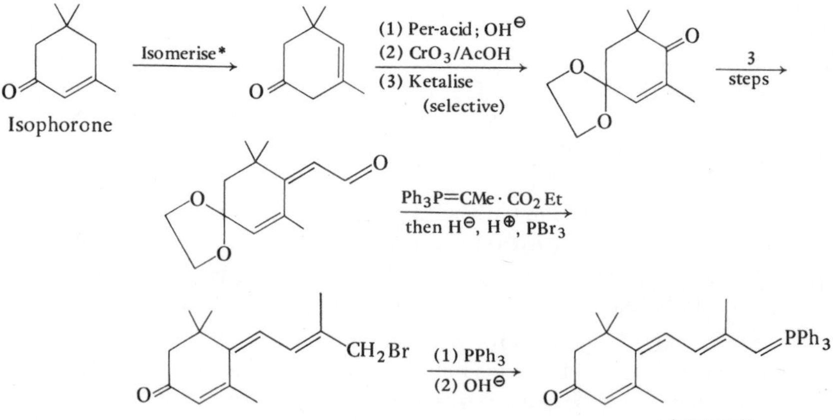

(XXXIX)

---

* Cf. J. Meinwald et al., J. org. Chem., 1971, 36 : 1446 for brief review.

(1) By O. Isler and co-workers (Helv., 1967, 50: 1606), using a C_{14} + C_{12} + C_{14} → C_{40} principle. The $C_{(3),(3')}$ keto groups required in the final molecule were derived from isophorone, which was built up into the C_{14} unit as in Scheme 4.

The C_{12} unit was prepared from compound LXXVII of Vol. II B, 2nd Edn. (p. 285) (an intermediate in a vitamin A synthesis) and then converted into rhodoxanthin (Scheme 5).

Scheme 5

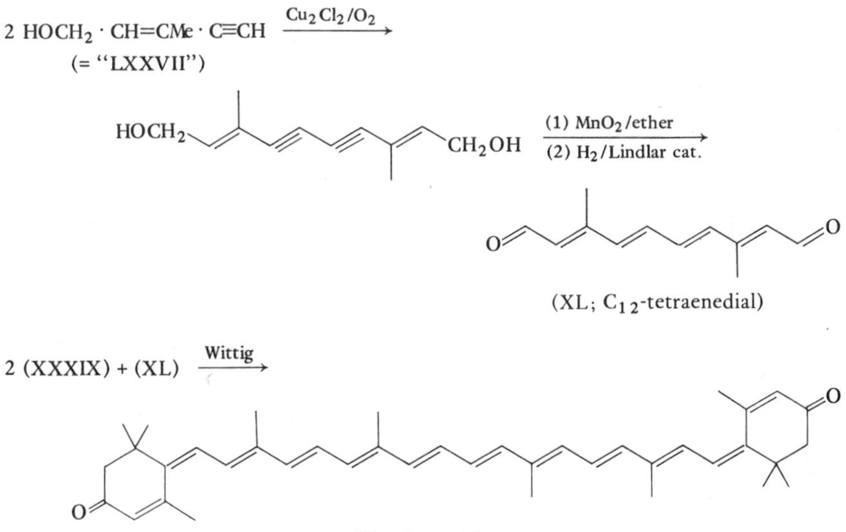

$$2 \; HOCH_2 \cdot CH{=}CMe \cdot C{\equiv}CH \xrightarrow{\;Cu_2Cl_2/O_2\;}$$

(= "LXXVII")

$$\xrightarrow[\text{(2) } H_2/\text{Lindlar cat.}]{\text{(1) } MnO_2/\text{ether}}$$

(XL; C_{12}-tetraenedial)

$$2 \; (XXXIX) + (XL) \xrightarrow{\;Wittig\;}$$

Rhodoxanthin

(2) By J. D. Surmatis et al. (Helv., 1970, 53: 974), starting from 3,3-ethylenedioxy-β-ionone (XLI) (Surmatis et al., J. org. Chem., 1970, 35: 1053) (Scheme 6).

Physical properties. For the p.m.r. spectrum of rhodoxanthin, see Isler and co-workers (loc. cit.). The *gem*-methyl groups give a singlet at the rather low-field position of τ 8.62, due to the deshielding influence of the double bonds of the conjugated polyene ketone chromophore in the immediate vicinity.

Biosynthesis. Pathway-blocking (p. 341) experiments on carotenoid formation by the mould *Epicoccum nigrum* suggest that the rhodoxanthin therein originates from β-carotene by oxidation, via zeaxanthin (Foppen and co-workers, Arch. Mikrobiol., 1970, 73: 216).

Scheme 6

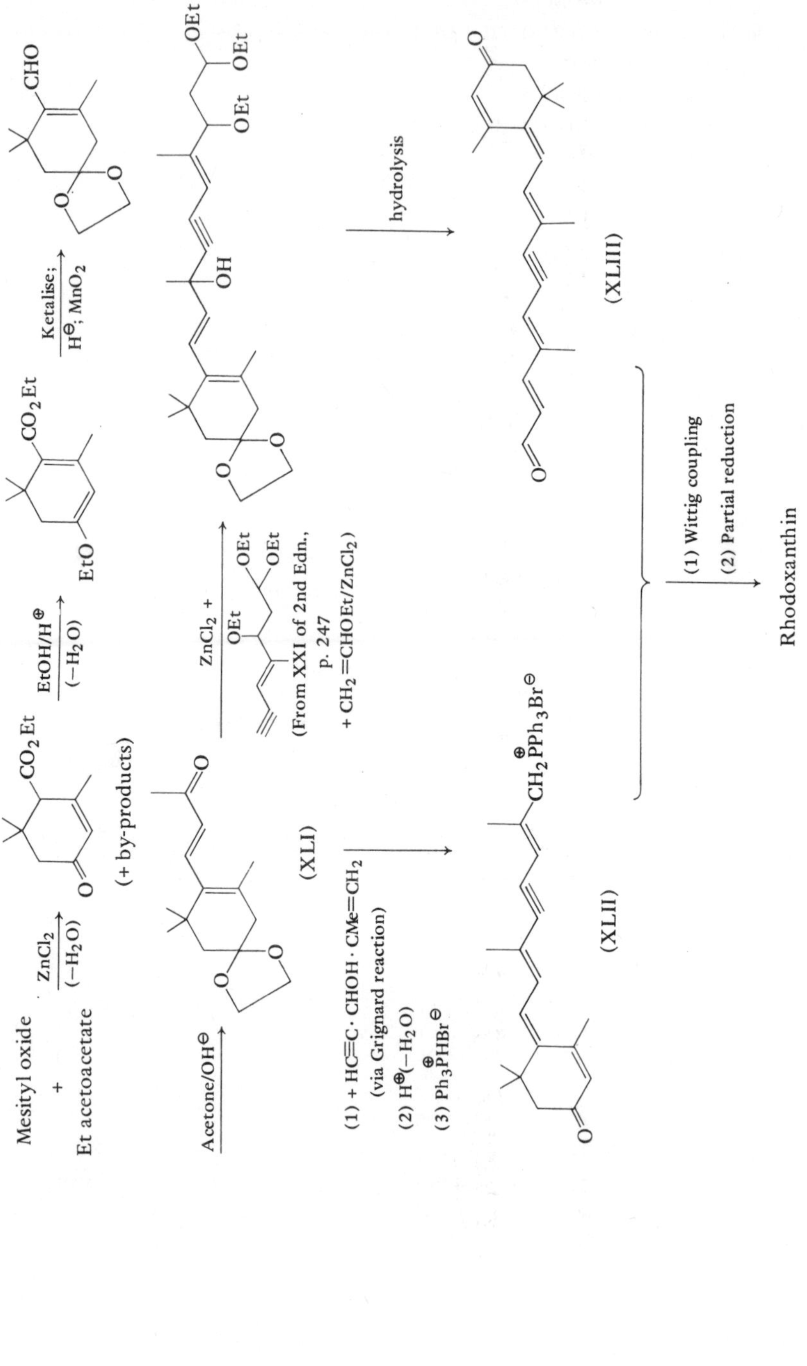
Mesityl oxide
+
Et acetoacetate
$ZnCl_2$
$(-H_2O)$
CO_2Et
(+ by-products)
$EtOH/H^{\oplus}$
$(-H_2O)$
EtO
CO_2Et
Ketalise;
$H^{\ominus}$; MnO_2
CHO
$Acetone/OH^{\ominus}$
(XLI)
$ZnCl_2$ +
OEt
OEt
OEt
(From XXI of 2nd Edn.,
p. 247
+ CH_2=CHOEt/$ZnCl_2$)
OH
OEt
OEt
OEt
hydrolysis
(XLIII)
(1) + HC≡C · CHOH · CMe=CH_2
(via Grignard reaction)
(2) $H^{\oplus}$($-H_2O$)
(3) $Ph_3\overset{\oplus}{P}HBr^{\ominus}$
$CH_2\overset{\oplus}{P}Ph_3Br^{\ominus}$
(XLII)
(1) Wittig coupling
(2) Partial reduction
Rhodoxanthin

Echinenone (4-oxo-β-carotene). A new (C_{20} + C_{20}) synthesis with an overall yield of 55%, based on the now readily available retinol (vitamin A; made on an industrial scale), has been described by Surmatis et al. (Helv., 1970, 53: 974).

(i) Retinol $\xrightarrow{\text{MnO}_2/\text{CH}_2\text{Cl}_2}$ retinal $\xrightarrow[\substack{\text{cf. 2nd Edn.,}\\ \text{p. 262}}]{\text{NBS/AcOH}}$ 4-acetoxyretinal

(ii) 4-Acetoxyretinal is then condensed (NaOMe/MeOH) with the Wittig salt from retinol (retinol + $Ph_3PH^{\oplus}HSO_4^{\ominus}$) to give 4-acetoxy-β-carotene; saponification followed by Oppenauer oxidation yields echinenone.

Phoenicopterone (4-oxo-α-carotene?) (see 2nd Edn., p. 317 for the γ-carotene analogue), a carotenoid isolated from the flamingo *Phoenicopterus ruber* by D. L. Fox's group (Compar. Biochem. Physiol., 1966, 17: 841). It was assigned the following structure from its absorption spectrum and comparison (partition ratio; t.l.c.) of both the pigment itself and of its borohydride reduction product with authentic (synthetic: see below) 4-oxo-α-carotene and its reduction product. However, the proliferation of carotenoid structures and the unreliability of simple mixed-t.l.c. tests mean that this assignment must be considered tentative:

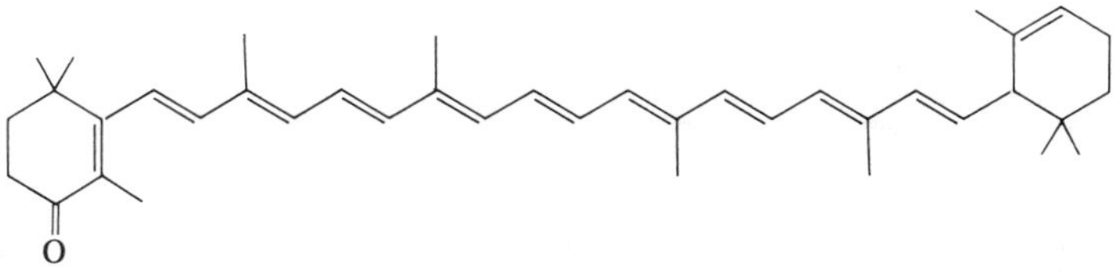

4-Oxo-α-carotene (= phoenicopterone?)

Partial synthesis (of 4-oxo-α-carotene). From (natural) α-carotene by brief treatment with *N*-bromosuccinimide in reagent (ethanol-containing: cf. 2nd Edn., p. 261) chloroform (cf. Zechmeister et al., J. Amer. chem. Soc., 1958, 80: 2991).

3′-Hydroxy-4-oxo-β-carotene. The two algae studied by Hertzberg and Jensen (Phytochem., 1966, 5: 557) whilst working on the structures of myxoxanthophyll and oscillaxanthin (p. 261) also yielded a series of other carotenoids, a minor one of which was assigned this structure on the basis of absorption spectrum, polarity, chemical interconversions, and biogenetic reasoning (co-occurrence with carotenoids having one or both end-groups of the same type):

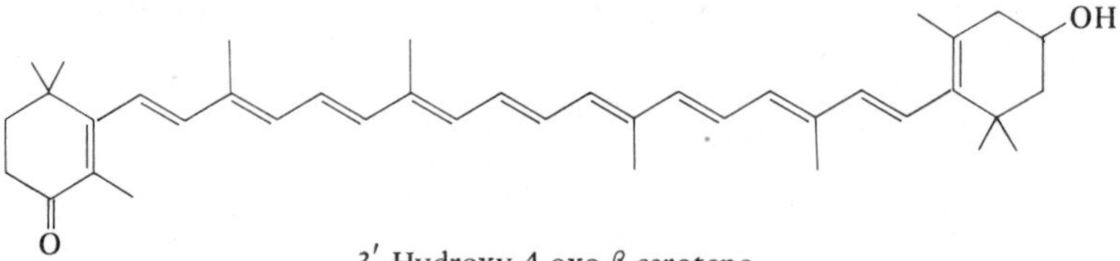

3′-Hydroxy-4-oxo-β-carotene

4-Hydroxy-4′-oxo-β-carotene. Kleinig and Czygan (Z. Naturforsch., 1969, 24b: 927) have assigned this structure, using similar reasoning to that above, to one of the minor pigments produced by a green alga grown under stressful conditions (nitrogen-deficiency; cf. "secondary carotenoids", p. 195). The same structure was assigned, tentatively, by Krinsky et al. (Compar. Biochem. Physiol., 1965, 16: 189) to a minor pigment isolated from the marine organism *Hydra littoralis* fed on a diet rich in canthaxanthin (the corresponding diketone).

There have also been other occasional reports (for refs. see S. A. Campbell, ibid., 1969, 30: 803) of compounds isolated in small amount which have behaved as expected for hydroxy-oxo-β-carotenes but until sufficient material is acquired for m.s. and/or other physical data to be obtained, these assignments, and even those above, should be treated with caution.

Thiothece-474, a minor pigment accompanying the okenone (p. 330) in a *Thiothece* sp. studied by A. G. Andrewes and Jensen (Acta chem. Scand., 1972, 26: 2194) probably has one end-group as for the acyclic end of okenone and the other of the β-carotene type (chemical transformations, spectra).

Canthaxanthin. A synthesis based on retinol (vitamin A) has been described by Surmatis et al. (Helv., 1970, 53: 974). The 4-acetoxy-β-carotene prepared as under echinenone, above, is converted into *retro*-dehydro-β-carotene by treatment with acid (HBr/—45°) and this on treatment with N-bromosuccinimide/acetic acid yields isozeaxanthin diacetate (4,4′-diacetoxy-β-carotene); saponification followed by Oppenauer oxidation gives canthaxanthin (20% overall from retinol):

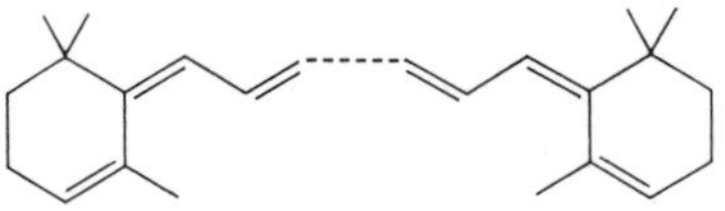

retro-Dehydro-β-carotene 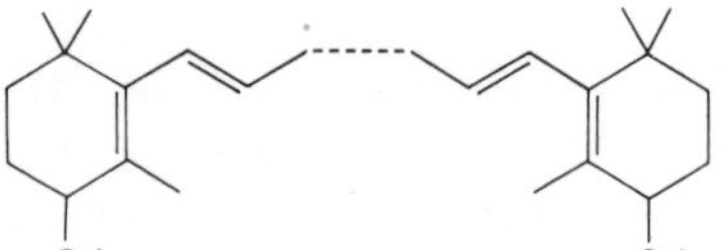   4,4′-Diacetoxy-β-carotene

Reaction with N-*bromosuccinimide.* P. Karrer et al. (Helv., 1961, 44: 1261) found that brief treatment of canthaxanthin with NBS in chloroform–acetic acid gave, after quenching with organic base, a complex mixture of products of uncertain structure. G. W. Francis et al. (Acta chem. Scand., 1970, 24: 3053) have now identified the major product (ca. 20%) of the reaction as being the central-acetylenic analogue of canthaxanthin, 15,15'-dehydro-canthaxanthin [by spectral studies and direct comparison with authentic (cf. 2nd Edn., pp. 319–320) material].

Canthaxanthin

Although partial reduction of 15,15'-dehydro-compounds to the corresponding polyene is a well known and frequently used reaction, the reverse reaction, as above, has not been described before (and is likely to be restricted to those carotenoids having carbonyl or similar groups terminating the polyene chain; for the more usual types of NBS reaction, see Vol. IIB, 2nd Edn., p. 261 ff.).

For canthaxanthin's reaction with per-acid, see Section 7a (p. 246); and for new p.m.r. data cf. p. 207.

Canthaxanthin, along with smaller amounts of astaxanthin and phoenicoxanthin (which see for ref.) below, is responsible for the characteristic pinkish colour of most flamingoes.

A pair of (unnamed) carotenoids with structures apparently corresponding to canthaxanthin but with (either one or both) ends acyclic, have been reported by Kleinig et al. (Arch. Mikrobiol., 1971, 76: 364).

Astaxanthin. *Absorption spectrum.* The report (by Kuhn and Sörensen 1938) that in pyridine astaxanthin shows two subsidiary maxima in addition to the main band, has been shown to be erroneous; only a single broad band is observed (Sörensen et al., Acta chem. Scand., 1968, 22: 344). The position of this band in a variety of solvents is given by M. Buchwald and W. P. Jencks [Biochemistry, 1968, 7: 834; normal dependence (2nd Edn., p. 254) on solvent seen]. These authors also note the effect of adding strong organic acids to such solutions [reversible formation of a species with λ_{max} shifted to longer wavelength (~840 nm), similar to the behaviour of other carotenoids] and of adding inorganic salts to aqueous dispersions of astaxanthin [considerable λ_{max} shift to shorter (down to ~400 nm),

Astaxanthin

unexpected, or longer (up to ~550 nm) wavelengths depending on the salt], and discuss these findings in relation to the problem of explaining the marked λ_{max} shift undergone by astaxanthin when it couples with protein to give a complex such as α-crustacyanin (2nd Edn., p. 321).

The o.r.d. curve of astaxanthin has been recorded over the range 200–400 nm by Scopes et al. (J. chem. Soc., C, 1969, 2527) and in the visible range by Buchwald and Jencks (1968, loc. cit.), but no deduction was made concerning the configuration of the —OH groups at $C_{(3)}$, $C_{(3')}$.

Astaxanthin has been found to be the sole carotenoid in a series of corals from Norwegian waters (Jensen and co-workers, Acta chem. Scand., 1970, 24: 3055). Actinioerythrin bears a close structural relationship to astaxanthin, and may be formed from it in nature. Violerythrin, a derivative, can be obtained from astaxanthin by manganese dioxide oxidation (Section 7j, p. 329, for details).

J. D. Surmatis and R. Thommen (J. org. Chem., 1967, 32: 180) have developed a further synthetic route to the astaxanthin skeleton, though this time to the dimethyl-ether rather than to astaxanthin itself. The —OMe function was introduced into the end-group at the start of the sequence:

$$\beta\text{-ionone} \xrightarrow{\text{NBS; base}} 3,4\text{-dehydro-}\beta\text{-ionone} \xrightarrow{\text{MeOH/H}^\oplus} 3\text{-methoxy-}\beta\text{-ionone}$$

Chain extension of the latter to the C_{19}-polyene aldehyde, followed by a $C_{19} + C_2 + C_{19} \rightarrow C_{40}$ coupling with BrMgC≡CMgBr gave the dimethyl ether of 15,15′-dehydrozeaxanthin (cf. Scheme 16 of 2nd Edn., p. 298). Oxygen functions, as —OAc, were introduced at $C_{(4)}$, $C_{(4')}$ using the method (NBS in $CHCl_3$—AcOH etc.; 30%) given in 2nd Edn., p. 262. This was followed by saponification, and Oppenauer oxidation of the $C_{(4)}$- >CHOH groups to >C=O.

The following five carotenoids all show the characteristic astaxanthin-type instability to oxygen in the presence of base, due to the facile oxidation undergone by the α-ketol function of such end-groups to the α-diketone oxidation level (flexixanthin, and its 2'-hydroxy derivative (see below) and some of the carotenoids on pp. 307–309 also come into this category):

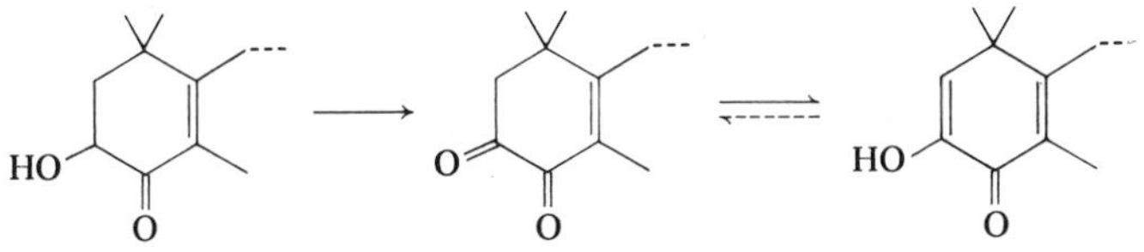

(first four compounds)

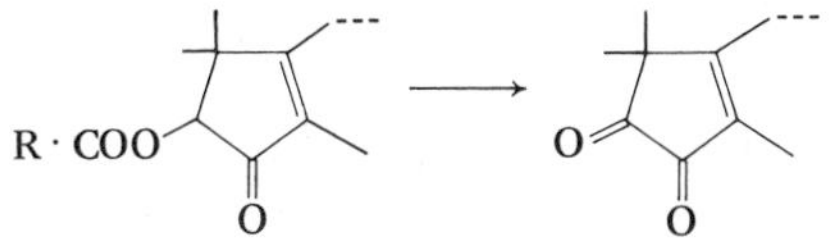

(actinioerythrin)

Adonirubin and adonixanthin, two of the minor* pigments found in the red flowers of *Adonis annua* by K. Egger (Phytochem., 1965, 4: 609) wherein they occur esterified by a mixture of fatty acids, mainly myristic (C_{14}, saturated) but also palmitic, etc. (Egger and Kleinig, ibid., 1967, 6: 437). Because saponification led to simultaneous autoxidation of the astaxanthin end-group present in each case (see above), much of the structure-elucidation work was carried out on these, autoxidised, derivatives (known as "*dehydro-adonirubin*" and "*dehydro-adonixanthin*"). A detailed study of the effect of carrying out various chemical transformations (borohydride reduction; subsequent etherification of any allylic-OH's so formed) on the absorption spectrum and polarity (various t.l.c. systems) of these pigments, in comparison with similar data on known compounds such as astaxanthin and echinenone, led to the suggestion that adonirubin is 3-*hydroxy*-4,4'-*dioxo-β-carotene* (i.e. one end-group of the astaxanthin type and one of the canthaxanthin type) and adonixanthin is 3,3'-*dihydroxy*-4-*oxo-β-carotene* (one end astaxanthin, and one zeaxanthin). The latter structure remains tentative but the former has essentially been confirmed by carrying out a direct comparison (Egger and Kleinig, ibid., 1967, 6: 903) of its dehydro-derivative with synthetic 3,4,4'-trioxo-β-carotene, prepared by treating canthaxanthin with O_2/KOBu^t as for the synthesis of astacene (2nd

---

* The major pigment is astaxanthin (also esterified) (Egger, 1965). C. Bodea et al. have surveyed the carotenoids in certain other *Adonis* species. The yellow varieties contain mainly zeaxanthin and lutein epoxides but the red, as above, mainly astaxanthin along with small amounts of compounds more of the above type (and also, apparently, 3,4-*dihydroxy-β-carotene*) (Rev. Roum. Biochim., 1966, 3: 305).

Edn., p. 321) but stopping the reaction half-way (cf. Weedon, Chem. Brit., 1967, 430). For a further source of adonirubin and adonixanthin, see under hydroxyechinenone, below.

One of the pigments in goldfish, *β-doradexanthin*, has been assigned the same (probable) structure as adonixanthin (Chichester et al., Internat. J. Biochem., 1970, 1: 438). An accompanying pigment, α-**doradexanthin**, was shown [m.s., p.m.r., i.r. spectra; chromophore from the λ_{max} values of the borohydride reduction product (to avoid the complication of $>C=O$ conjugation)] to have the isomeric structure in which the C=C is at $C_{(4')}-C_{(5')}$. Both pigments were present in the esterified form and most of the structural work was done on the O_2/base autoxidation derivatives, the β- and α-*doradecins*.

The structure assigned to adonirubin, which can be considered as 3-*hydroxycanthaxanthin*, was also deduced independently by D. L. Fox and T. S. Hopkins (Compar. Biochem. Physiol., 1966, 17: 841) for a pigment, which they named **phoenicoxanthin**, isolated by them from three species of flamingo, in which it occurs unesterified and so can be studied direct [structure: by applying similar methods to those used by Egger above; by biogenetic reasoning (co-occurs in all three sources with canthaxanthin and astaxanthin); by direct comparison of the O_2/base oxidation product with Weedon's trioxo-β-carotene as above]. A compound with properties very like those reported for phoenicoxanthin (= adonirubin) has been isolated from certain crustaceans (B. M. Gilchrist, ibid., 1968, 24: 123) and algae (Czygan and E. Kessler, Z. Naturforsch., 1967, 22b: 1085).

Hydroxyechinenone, a further minor pigment isolated from *A. annua* by Egger (1965, loc. cit.; cf. adonirubin and adonixanthin, above) and thought to be identical (Egger and Kleinig, 1967, loc. cit.) with the "hydroxy-echinenone" isolated earlier by Krinsky et al. from *Euglena gracilis* (Arch. Biochem. Biophys., 1960, 91: 271). The proposed structure, 3-*hydroxy-4-oxo-β-carotene* (= 3-*hydroxy-echinenone*), one end-group like astaxanthin, one like β-carotene, was inferred (Egger, 1965) from carrying out tests, of the kind Egger used above, on the oxygen/base autoxidation product, "*dehydro-hydroxyechinenone*"; and was essentially confirmed by showing (Egger and Kleinig, 1967, loc. cit.) that the latter is, probably, identical with synthetic (see below) 3,4-dioxo-β-carotene.

Synthesis (A. P. Leftwick and Weedon. Chem. Comm., 1967, 49). Echinenone (synthetic) was converted into 3,4-dioxo-β-carotene using similar conditions (O_2/KOBu^t) to those used for the synthesis of astacene (2nd Edn., p. 321); borohydride reduction, to the corresponding dihydroxy-compound, followed by selective oxidation (Oppenauer) of the allylic-OH, yielded 3-hydroxy-4-oxo-β-carotene.

Adonirubin, adonixanthin, and hydroxyechinenone also occur, again with —OH groups esterified, in the marine alga *Acetabularia mediterranea* (Kleinig and Egger, Phytochem., 1967, 6: 611).

Actinioerythrin, the last of this group of five astaxanthin-like carotenoids, is discussed under Section 7j (p. 327).

Flexixanthin. For a new source of this carotenoid and its deoxy-derivative (2nd Edn., pp. 322, 318) and the first report of its 2'-hydroxy-derivative, 2'-hydroxyflexixanthin (astaxanthin-like at one end, acyclic at the other), all in the same bacterium, see Jensen et al. (Acta chem. Scand., 1972, 26: 2528).

Philosamiaxanthin, an alkali-labile carotenoid isolated from *Philosamia cynthia* (silkworm) where it co-occurs with lutein (p. 249), is thought to be as for lutein but with the $-CMe=CH \cdot CHOH-$ grouping replaced by $-CMe=CH \cdot CO-$ (by small-scale chemical etc. tests as in 2nd Edn., p. 245, and direct t.l.c. comparison of the hydride reduction product with authentic lutein; not confirmed spectrally) (K. Harashima, Internat. J. Biochem., 1970, 1: 523). Partial synthesis: by nickel peroxide oxidation of lutein as on p. 217.

Capsanthin and Capsorubin. The stereochemistry of the two (identical) substituted cyclopentane rings in capsorubin was given earlier (2nd Edn., p. 324, formula CXXIV)*, and corresponds to an S configuration at $C_{(3)}$[†] and R at $C_{(5)}$. The (identical) cyclopentane-ring end-group of capsanthin was also shown to have that stereochemistry, but the configuration at $C_{(3)}$ at the other end of the molecule, a zeaxanthin-type end-group, was unknown. Scopes et al. (J. chem. Soc., C, 1969, 2527) have since recorded the o.r.d. curves of both pigments, and comparison of the capsanthin curve with those of capsorubin and zeaxanthin has shown (cf. p. 211) the zeaxanthin end of capsanthin to have the same configuration at $C_{(3)}$ as zeaxanthin itself; that is R[†].

K(c)ryptocapsin. The o.r.d. curve has been measured by Scopes et al. (loc. cit.), and comparison with that of capsorubin has confirmed that the $C_{(5)}$-ring end-group of kryptocapsin has the same configuration, that is $3S$, $5R$ (peaks and troughs in similar positions but with ca. one half the capsorubin amplitude).

"Capsanthin-5,6-epoxide", corresponding to the capsanthin structure but with a 5,6-epoxy group at the "zeaxanthin end". A carotenoid with properties (partition ratio; λ_{max} values; effect of $H^{\oplus}$) consistent with its having this structure was first reported by A. L. Curl (in a variety of pepper;

---

* This, the natural optical isomer, has recently been synthesised (Weedon, Pure Appl. Chem., 1973, 35: 117).

† In both cases this corresponds to a "β"—OH (steroid nomenclature), and "α"—H, at those positions.

J. agric. Food Chem., 1962, 10: 504). B. H. Davies et al. (Phytochem., 1970, 9: 797) have since detected what is probably the same compound in another pepper, again in association with capsanthin itself and allied compounds. In neither case was the substance isolated crystalline or full tests carried out, but such a structure is to be expected as an intermediate in the rearrangement of violaxanthin to capsorubin in the (suggested) biogenetic sequence leading to the latter compound in vivo (cf. 2nd Edn., p. 345).

The following compounds have all been found in various citrus and related fruits, and are responsible, at least in part, for their colour.

Semi-β-carotenone. In 1968, H. Yokoyama and M. J. White (Phytochem., 1968, 7: 1031) isolated the major pigment from the ripe (bright red) fruit of *Murraya exotica* and found it to be a carotenoid with, after chromatography of the crude acetone extract of the fruit and crystallisation of the major component from hexane, m.p. 115—117°, λ_{max} 495, 467, 440 nm (hexane). Elemental and m.s. analysis gave the formula ($C_{40}H_{56}O_2$), the absorption spectrum suggested a capsanthin-type chromophore, and i.r. indicated the presence of both saturated (unusual in carotenoids) and conjugated C=O (the latter was confirmed by showing that borohydride reduction causes a λ_{max} shift of ca. −25 nm). The p.m.r. spectrum was consistent with the presence of a β-type end-group and four in-chain methyls; a 3H singlet at τ 7.89 suggested —CO·Me. The remaining methyl group signal, a 2Me singlet at τ 8.82, can be assigned to the $C_{(1')}$ gem-methyl group in the structure below (at significantly different position to the ring gem-methyls, at τ 8.95). These results led Yokoyama and White to conclude that the pigment was identical with a substance obtained from (natural) β-carotene, by chromic acid oxidation, by Kuhn et al. in the 1930's and named by them "semi-β-carotenone" (2nd Edn., p. 260). This was confirmed by carrying out a direct comparison (t.l.c., spectra) of the natural pigment with a sample of synthetic (see below) semi-β-carotenone.

Semi-β-carotenone was also found to be the major carotenoid in the ripe fruit of *Triphasia trifolia* (Yokoyama and White, loc. cit.).

Synthesis (Yokoyama and White, loc. cit.; based on Kuhn's method, above). From synthetic (cf. p. 218) β-carotene by treating the latter, in benzene, with chromic—acetic acid, and chromatographic removal of the various other products formed simultaneously (yield, 15%).

β-Carotenone. Yokoyama and White (1968, loc. cit.) isolated, by chromatography, a second pigment, m.p. 176°, from *T. trifolia* and, using the methods outlined above for semi-β-carotenone, showed it to be β-carotenone [analysis gave $C_{40}H_{56}O_4$; λ_{max} values indicated a capsorubin-type chromophore; and the p.m.r. spectrum was markedly simpler than that above, characteristic of a symmetrical carotenoid, and still showed a band (this time of 2Me intensity) attributable to —CO · *Me*]. Confirmation was by comparison with a synthetic sample.

Synthesis (Yokoyama and White, loc. cit.). As for semi-β-carotenone above. β-Carotenone is one of the later oxidation products and can be isolated chromatographically from the crude reaction product.

Semi-β-carotenone   β-carotenone

Triphasiaxanthin. A further pigment (cf. above) isolated from *T. trifolia* by Yokoyama and co-workers (J. org. Chem., 1970, 35: 2080); m.p. 95–97°, λ_{max} as for semi-β-carotenone; more polar than the above two compounds and showed —OH absorption in the i.r. in addition to saturated and conjugated C=O; other tests as for the compounds above. The —OH was shown to be non-allylic (H$^{\oplus}$ test negative; attempted allylic-oxidation failed), and was placed at $C_{(3)}$ rather than $C_{(2)}$ on p.m.r. grounds (*gem*-methyl groups gave a single 2Me peak; might be expected to be non-equivalent if —OH were at $C_{(2)}$). It was concluded that triphasiaxanthin is the 3-*hydroxy derivative of semi-β-carotenone*.

The m.s. fragmentation pattern was also reported, each ion being identified by precise mass measurement. An $M-C_9H_{15}O$ ion was one of those observed, corresponding to loss of a hydroxylated β-end-group, as required by the above formulation.

Semi-α-carotenone. A minor carotenoid isolated by Yokoyama and H. C. Guerrero (Phytochem., 1970, 9: 231) from *M. exotica* (cf. above) was found, using the same methods as used for semi-β-carotenone above, to be identical with the known α-carotene derivative, semi-α-carotenone (structure as for semi-β-carotenone but with the ring double bond at $C_{(4)}-C_{(5)}$ rather than $C_{(5)}-C_{(6)}$; first prepared in the 1930's by Karrer by chromic acid oxidation of natural α-carotene, 2nd Edn., p. 260). The main differences, spectrally, from semi-β-carotenone were in the λ_{max} values (5 nm shorter wavelength) and certain of the p.m.r. signals. The ring-*gem*-methyl groups gave two separate 3H peaks (τ 9.00, 9.15) and the $C_{(5)}$-methyl was no longer of the end-of-chain type in agreement with the presence of an α-, rather than a β-, type of end-group (cf. 2nd Edn., p. 257 and L. M. Jackman et al., J. chem. Soc., 1960, 2870). Confirmation of the structure was by comparison with authentic semi-α-carotenone prepared from (natural) α-carotene as above: this constitutes a partial synthesis of the pigment. Semi-α-carotenone is also one of the minor carotenoids in *T. trifolia* (see below).

Absolute configuration of natural semi-α-carotenone (at $C_{(6)}$) (R. Buchecker et al., Helv., 1970, 53: 1210). Comparison of the c.d. curve of the borohydride-reduced pigment (to remove the complicating effect of the carbonyl groups) with that of synthetic 6*S*,6′*S*-ε-carotene showed the natural carotenone to have the opposite configuration at the 6-position: that is to be 6*R* the same as natural α-carotene.

Biogenesis of the above four carotenones. Yokoyama and White (Phytochem., 1970, 9: 1795) studied the build-up of these four compounds, and the fluctuations in the level of other carotenoids present, in the fruit of *T. trifolia* during ripening from the dark green to the fully ripe (crimson) stage. β-Carotene, and smaller amounts of α-carotene and 3-hydroxy-β-carotene (cryptoxanthin), appeared early on but these later waned rapidly as the semi-β-carotenone, and soon afterwards the other three carotenones, appeared (and, subsequently, steadily increased). This strongly suggests the carotenones arise by in vivo oxidation of the corresponding ring-compounds:

β-carotene → semi-β-carotenone → β-carotenone
α-carotene → semi-α-carotenone
cryptoxanthin → triphasiaxanthin

The following four compounds are also from various citrus fruit, but are *apo* (< C_{40}) carotenoids.

Reticulataxanthin (C_{33}), first isolated (from tangerine: *Citrus reticulata*) by Curl (J. Food Sci., 1962, 27: 537) who inferred the approximate structure. Subsequently, Yokoyama and White (J. org. Chem., 1965, 30: 2482) found the same compound in another *Citrus* sp., isolated it crystalline

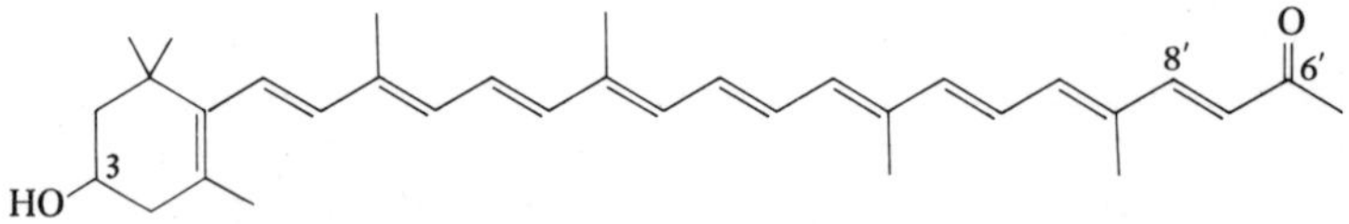

Reticulataxanthin

(m.p. 172°), carried out a full structural study using similar methods to those outlined above, and arrived at the above structure. This was confirmed by showing that reticulataxanthin with hot ethanolic alkali yields acetone and β-citraurin by retro-aldol cleavage at $C_{(8')}-C_{(7')}$.

Absolute configuration. The o.r.d. curve has been recorded by Scopes et al. (J. chem. Soc., C, 1969, 2527); the general shape and the position of the extrema were as for zeaxanthin, but of less amplitude, showing that the single asymmetric centre present, at $C_{(3)}$, is as in zeaxanthin (now known to be R).

Citranaxanthin (C_{33}) and **sintaxanthin** (C_{31}), both isolated (m.p.'s 156, 145°) from a citrus hybrid by Yokoyama and White (J. org. Chem., 1965, 30: 2481, 3994). Similar structural studies to those used for reticulata-xanthin led to the conclusion that citranaxanthin is as for reticulataxanthin except it lacks the —OH at $C_{(3)}$, whilst sintaxanthin is the corresponding compound with one less conjugated C=C (i.e. C=O at $C_{(8')}$, instead of $C_{(6')}$).

Synthesis of citranaxanthin has been achieved by aldol condensation (EtOH—KOH) of acetone with β-apo-8'-carotenal (2nd Edn., p. 313). Sintaxanthin has been synthesised by treating the same carotenal with lithium methyl in ether (R · CH=CMe · CHO → R · CH=CMe · CHOH · Me) followed by manganese dioxide oxidation of the allylic-OH.

8'-Hydroxy-7',8'-dihydrocitranaxanthin, a further carotenoid isolated from the above citrus hybrid by Yokoyama and White (J. org. Chem., 1966, 31: 3452); m.p. 146—147°; [structure from spectra, as above, and effect of $H^{\oplus}$ (gave citranaxanthin)]:

8'-Hydroxy-7',8'-dihydrocitranaxanthin

Synthesis. As for citranaxanthin but the reaction was carried out at lower temperature to suppress the spontaneous dehydration of the β-hydroxy-ketone formed initially.

The mode of biogenesis of these rather unusual (not truly isoprenoid) structures remains unknown. The possibility that they are artefacts formed by aldol condensation of the corresponding (natural) carotenals and the acetone used during isolation, has been dismissed by Yokoyama (1966, unpubl.).

Synthetic analogues. J. D. Surmatis, R. Thommen et al. (J. org. Chem., 1969, 34: 3039; Helv., 1970, 53: 974) have synthesised three compounds containing a cross-conjugated carbonyl group within the polyene chromo-phore (partly so as to study the effect on the visible absorption spectrum of inserting $>C=O$ within a polyene system). The example below is based on the β-carotene skeleton:

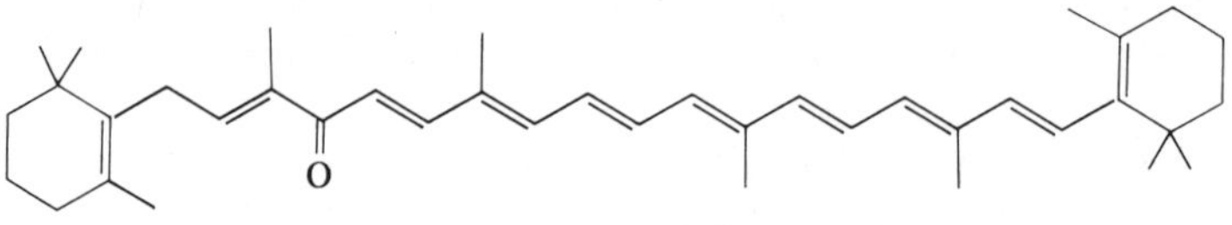

The fucoxanthin derivative "isofucoxanthin" belongs to this class (2nd Edn., p. 330); cf. also the synthetic analogues noted under Section 7e (p. 287).

(e) Carotenoid carboxylic acids

Torularhodin (C_{40}). The *methyl ester* of torularhodin has now been found in nature, in the fungus *Cookeina sulpices* (N. Arpin and Jensen, Compt. rend., 1967, 265: D, 1083) and a further source of the acid itself reported (O. G. Sassu and F. H. Foppen, Phytochem., 1967, 6: 907; 1968, 7: 1605). Further biosynthetic studies on the acid have shown that the $-CO_2H$ group arises by biological oxidation of, *specifically*, one of the two methyl groups in the $RCH=CMe_2$ end-group of the precursor, torulene [by showing that the $-CO_2H$ of the $>C=C(Me)CO_2H$ end-group of torularhodin, biosynthesised by a yeast dosed with $C_{(2)}$-labelled mevalonic acid contained *all* the ^{14}C label, rather than sharing it half and half with the methyl group (K. L. Simpson et al., Biochem. J., 1970, 117: 921)]. Further evidence that torularhodin arises in nature from γ-carotene as shown below, has been provided by the observation that all five compounds probably co-occur in two *Rhodotorula* (yeast) species, and by measuring the change in level of three of these compounds during growth of the yeasts (R. Bonaly and J. P. Malenge, Biochim. biophys. Acta, 1968, 164: 306):

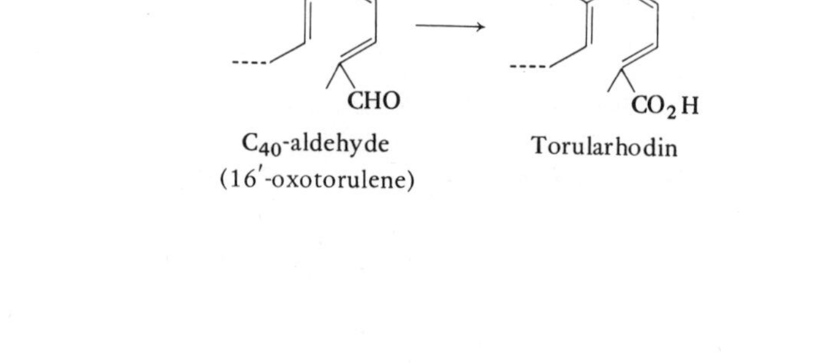

γ-Carotene ⟶

Torulene

C_{40}-alcohol

C_{40}-aldehyde
(16'-oxotorulene)

Torularhodin

Neurosporaxanthin (C_{35}). The *methyl ester* has been found in *Nectria cinnabarina*, a red, woodland fungus [structure by absorption spectrum, effect of alkali, hydride, etc., and direct comparison with the synthetic compound (2nd Edn., p. 326)] (J. L. Fiasson and M. P. Bouchez, Phytochem., 1970, 9: 1133).

Apo-6′-lycopenoic acid (C_{32}). The *methyl ester* (*methyl apo-6′-lycopenoate*), has been found to occur in the berries of the plant *Sheperdia canadensis*. The other main carotenoid present was found to be lycopene, from which the ester is presumably derived in vivo by oxidative degradation. The structure of the pigment (m.p. ca. 145°, λ_{max} 503, 471 (448) nm, in petroleum) was inferred from i.r. and p.m.r. spectra and the usual chemical interconversions (effect of hydride reduction, alkaline hydrolysis, etc.), and was confirmed by total synthesis (H. Kjösen and Jensen, ibid., 1969, 8: 483):

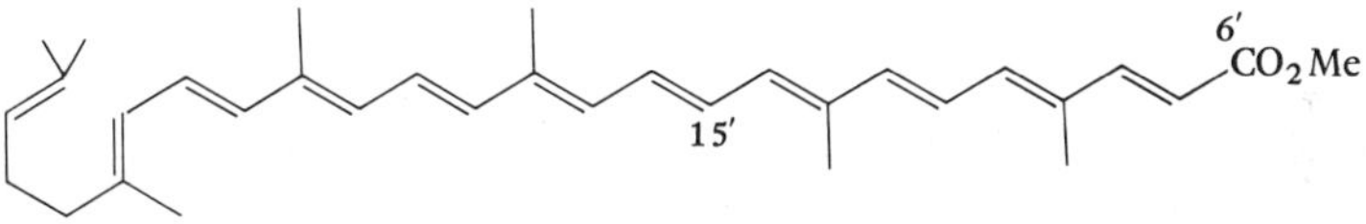

Synthesis (Kjösen and Jensen, loc. cit.). By a 1:1 Wittig condensation of geranylidene-triphenylphosphorane (compound XLVI; 2nd Edn., p. 267) and crocetindial (2nd Edn., p. 248) to give a mixture from which could be isolated the desired 1:1 condensation product, the C_{30} aldehyde apo-8′-lycopenal. To this was added the phosphorane derived from methyl bromoacetate, $Ph_3P=CH \cdot CO_2Me$, in a second Wittig reaction.

A (C_{30}) *3-dehydro-apo-8′-lycopenoic acid derivative* with the following structure has been found to be the major carotenoid (m.p. 182°) in a bacterium studied by Aasen, Francis and Jensen (Acta chem. Scand., 1969, 23: 2605). The absorption spectrum, and probably therefore the chromophore, was as for the preceding compound and a conjugated $-CO_2Me$ was present (i.r., p.m.r.). However, it was found to be very polar (chromatography) and to contain four acetylatable hydroxyl groups (m.s. on the per-acetate), suggesting a glycoside. Application of the methods used for myxoxanthophyll (acid hydrolysis to release the sugar residue; m.s. fragmentation patterns; etc.: p. 262) led to the structure shown below, where R is a sugar residue, probably mannose:

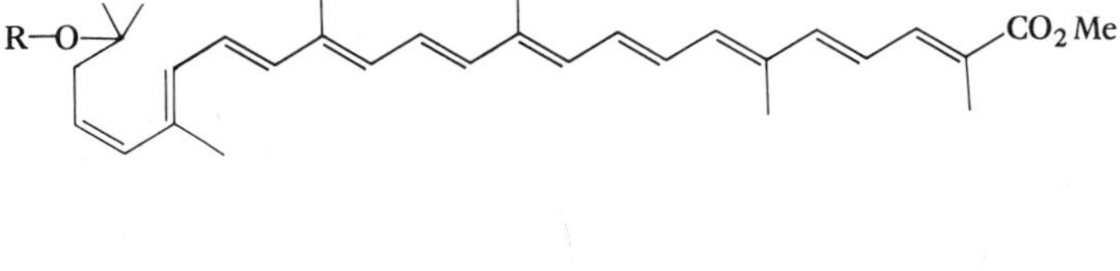

β-Apo-8'-carotenoic acid (C_{30}: structure as for neurosporaxanthin but with a shorter polyene chain, the $-CO_2H$ group being at $C_{(8')}$), is reported to be the major carotenoid present in a strain of maize (J. Baraud et al., Compt. rend., 1965, 260: 7045). It was identified by comparison with authentic material synthesised by the method used for neurosporaxanthin (2nd Edn., p. 326).

Bixin. Natural bixin is a *cis*-compound (at $C_{(16)}-C_{(17)}$), and several syntheses of the all-*trans* isomer (as its dimethyl ester, "methylbixin") are

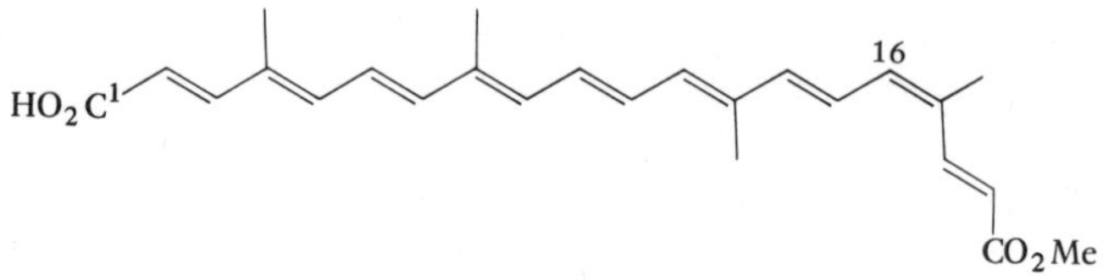

Natural bixin (XLIV)

known (2nd Edn., p. 327). The first synthesis of the 16-*cis* isomer has been accomplished by Pattenden and Weedon (J. chem. Soc., C, 1970, 235), using a method for introducing *cis*-bonds developed by Weedon's group described in Section 3b (p. 200):

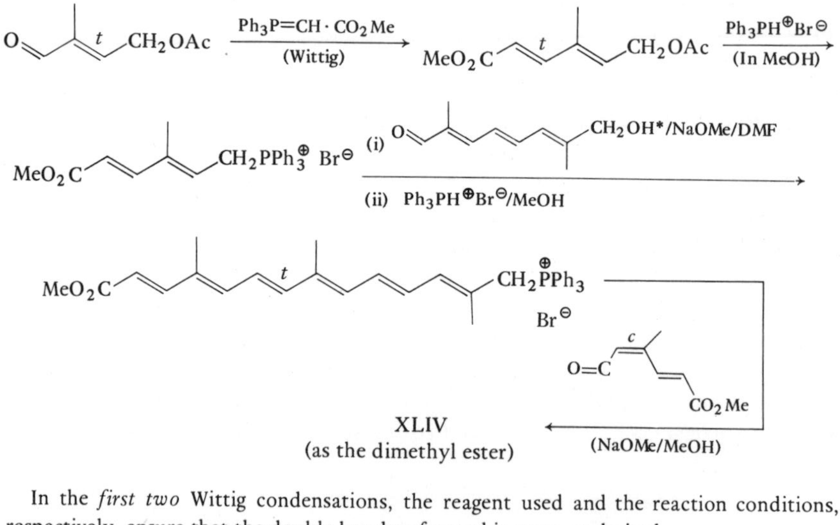

XLIV
(as the dimethyl ester)

In the *first two* Wittig condensations, the reagent used and the reaction conditions, respectively, ensure that the double bond so formed is *trans*, as desired.

---

* Useful bi-functional synthetic intermediate, prepared by partial reduction with $NaBH_4/MeOH$ (reaction followed by change in absorption spectrum) of the corresponding dial (XXIX, p. 252).

The presence of a *cis* bond at $C_{(16),(17)}$ is readily detected by p.m.r. spectroscopy, the protons α and β to the $-CO_2$Me group at the *cis* end of the 16-*cis* molecule giving an AB quartet ($J \sim 16$; *trans*-CH=CH) with the doublets centred at $\sim \tau$ 4.1 (normal for the α-H in such a system) and, distinctively, at τ 1.98 (β-H; much lower than in the corresponding *trans*-arrangement (cf. below) due to deshielding by the nearby $C_{(14)}-C_{(15)}$ double bond). For comparison, the α,β-protons at the *trans*-end of the 16-*cis* molecule, and at both ends of the all-*trans* molecule, give an AB quartet, $J \sim 16$, at ca. 4.1 and 2.6 (Pattenden and Weedon, loc. cit.).

   Crocetin half-aldehyde, described under Section 7d, i (p. 267), is also a carotenoid acid.

   Synthetic analogues. A series of entirely synthetic compounds of the type XLV has been prepared by H. H. Haeck and T. Kralt (Rec. Trav. chim., 1968, 87: 709) from polyene aldehydes and ketones (synthetic, cf. 2nd Edn., p. 247) using the Stobbe condensation:

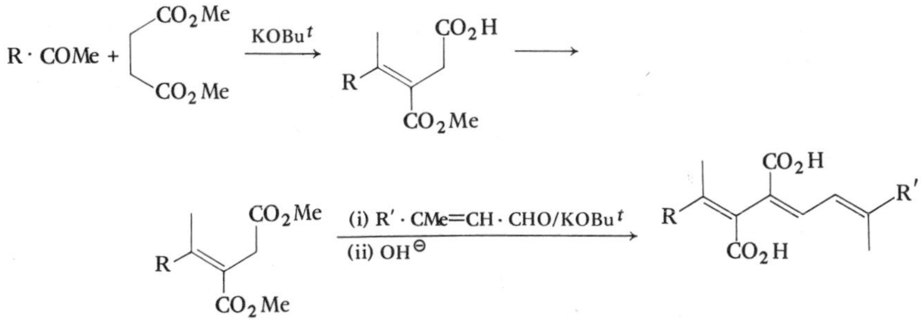

The interesting feature of these compounds is that the steric conflict between the $-CO_2$H and the methyl groups effectively destroys through-conjugation in the polyene chain, presumably by twisting the two halves of the molecule around the $=C(CO_2H)-C(CO_2H)=$ single bond. Thus XLV does not have a modified β-carotene spectrum but exhibits two peaks at much shorter wavelengths corresponding to a conjugated pentaene acid and a conjugated hexaene acid:

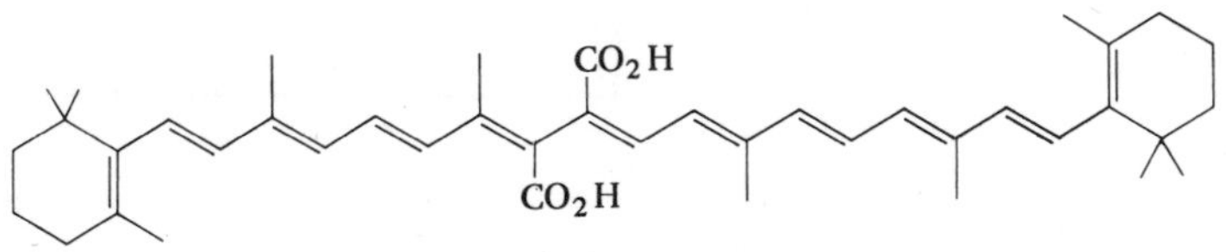

(XLV)

(f) Allenic carotenoids

Several further examples of this type of carotenoid have been discovered. In addition, further details of the work leading to the structure of neoxanthin (foliaxanthin), which appeared in the form of preliminary communications in time to be mentioned briefly in the 2nd Edn. (p. 330), have appeared, and novel chemical transformations of the molecule have been reported. The names "*foliaxanthin*" and "*foliachrome*" (2nd Edn., pp. 329, 330) have been abandoned in favour of "*neoxanthin*" and "*neochrome*": see under neoxanthin below.

The terminal allenic bond in these compounds essentially behaves in the absorption spectrum as a single (terminal) C=C. Thus neoxanthin has λ_{max} values similar to those of neurosporene, with a linear nonaene chromophore. In the p.m.r. spectrum, the in-chain methyl group adjacent to the C=C=C grouping occurs at markedly different position from a normal in-chain methyl group (ca. τ 8.2 in both neoxanthin and fucoxanthin, as compared with a normal value of $\sim \tau$ 8.05). Further, two of the three end-group methyl groups occur at markedly lower field (e.g. τ 8.67 in neoxanthin, the other being at ca. 8.9) than in most* of the more usual carotenoid end-groups (cf. refs. given below).

The interesting conversion shown below occurs with either $H^{\oplus}/CHCl_3$ (along with competing reactions; yield only ca. 20%) or excess $LiAlH_4$ (almost quantitatively)†, in the case of two of the allenic compounds

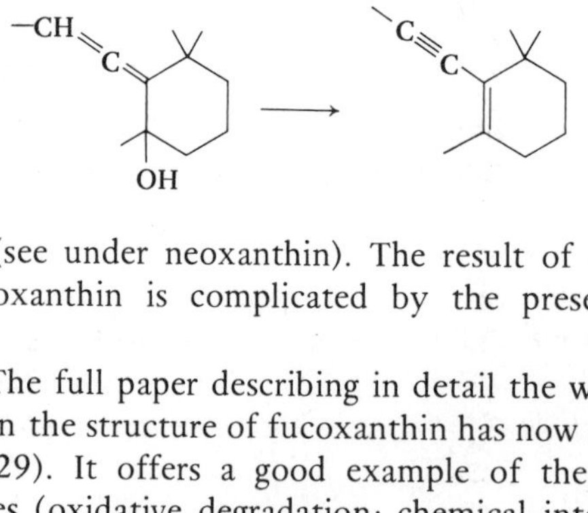

described below (see under neoxanthin). The result of carrying out these reactions on fucoxanthin is complicated by the presence of additional functional groups.

Fucoxanthin. The full paper describing in detail the work carried out by Weedon's group on the structure of fucoxanthin has now appeared (J. chem. Soc., C, 1969, 429). It offers a good example of the combined use of classical techniques (oxidative degradation; chemical interconversions) and

---

* The *gem*-methyl groups of rhodoxanthin (and of the artificial pigments astacene, dehydroadonirubin, etc.) also occur $\sim$ 8.6/8.7, but most others are in the range 8.8 to 9.1.

† Mild conditions; vigorous treatment causes gradual reduction to the —CH=CH— level (cf. p. 293).

of modern physical methods (in particular to determine the structures of the many fragments obtained from oxidative degradation) in the elucidation of the structure of a complex and reactive carotenoid.

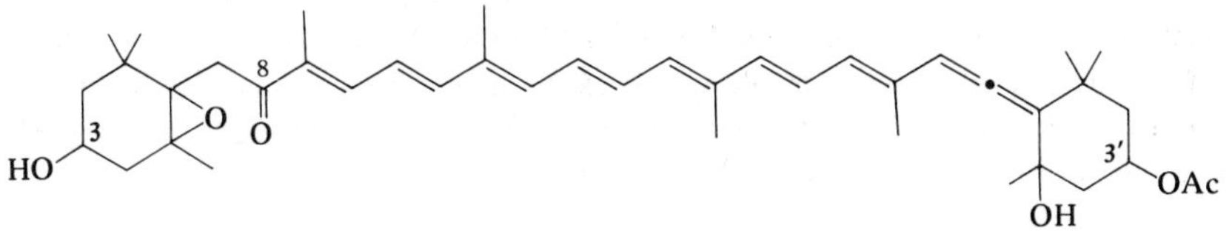

Fucoxanthin

Absolute configuration. Fucoxanthin has six asymmetric centres and once the gross structure of the molecule had been elucidated, attempts were made to determine the stereochemistry of these centres.

Lithium aluminium hydride reduction of fucoxanthin (to the "*fuco-xanthols*", a pair of epimers about $C_{(8)}$) followed by brief treatment with a trace of HCl in chloroform (allylic dehydration followed by epoxide rearrangement) leads to a mixture of epimeric (at $C_{(8)}$) furanoid oxides, the *fucochromes* (gross structure as for neochrome, below)*. Vigorous treatment of the mixed fucochromes with LiAlH$_4$ (as for neochrome) gave, in small yield, zeaxanthin (Weedon et al., 1969, loc. cit.). The latter gave (Scopes et al., J. chem. Soc., C, 1969, 2527) the same o.r.d. curve as natural zeaxanthin, showing that the $C_{(3)}$,$C_{(3')}$—OH's in the fucochromes, and hence in fucoxanthin, have the same configuration as those in natural zeaxanthin (p. 251). The full stereochemistry at the allenic end of the molecule was deduced by taking one of the fragments (XLVI) isolated from the oxidative degradation of fucoxanthin (zinc permanganate; cf. Weedon et al., 1969, loc. cit.), converting it into the *p*-bromobenzoate (via LiAlH$_4$; MnO$_2$; *p*-Br·C$_6$H$_4$·COCl), and determining its stereochemistry (XLVII)

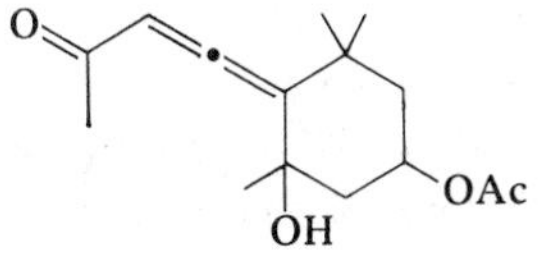

(XLVI) 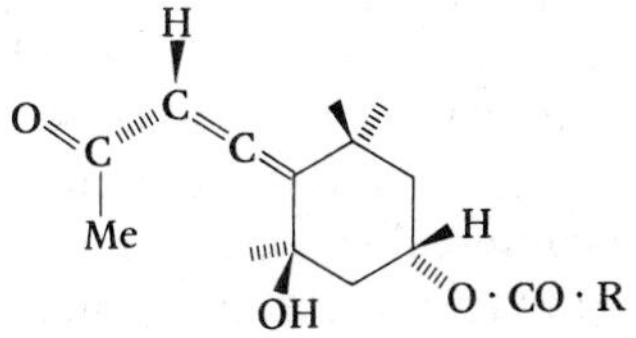   (XLVII)

---

* The fucochromes are themselves unstable to acid (cf. neochrome, below) and the use of slightly more vigorous conditions leads to further reaction giving compounds (XLVIII), (XLIX), and *diadinochrome* (p. 305) by loss of the tertiary —OH (Nitsche, Tetrahedron Letters, 1970, 4913).

by X-ray crystallography (Weedon, S. W. Russell and co-workers, Chem. Comm., 1969, 1311).

At the other end of the molecule, the configuration of the epoxide group is uncertain although it is thought to be *trans* to the $C_{(3)}$-OH (Weedon et al., 1969, loc. cit., p. 439; Weedon, Russell and co-workers, 1969, loc. cit.). This is the only doubt remaining regarding fucoxanthin's stereochemistry; if the epoxide is in fact *trans* to the OH then the full stereochemistry would be as shown below:

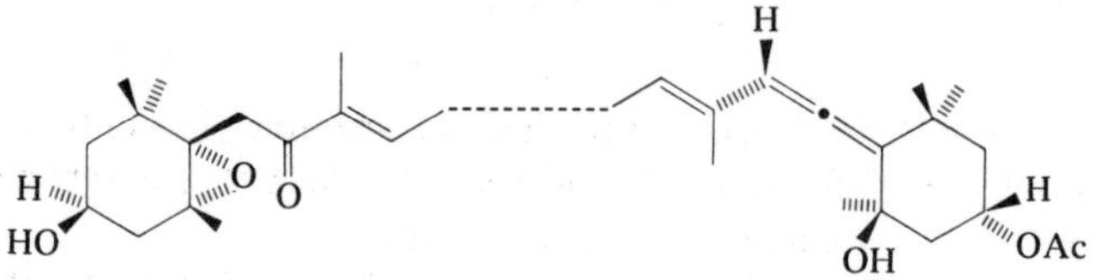

Reactions undergone by fucoxanthin

(1) Conversion (LiAlH$_4$; H$^{\oplus}$; LiAlH$_4$) into the *fucoxanthols, fucochromes*, and *zeaxanthin*, and under certain conditions, into *diadinochrome*: see under Absolute Configuration above.

(2) Regarding methods for (a) converting the $C_{(3')}$-OAc into —OH and (b) removing the epoxide end-group by cleaving the $C_{(6)}$—$C_{(7)}$ bond, see under fucoxanthinol and paracentrone, below.

Neoxanthin (and "foliaxanthin"). Neoxanthin is one of the commonest carotenoids in nature, being one of the major pigments in all green leaves so far investigated and in many algae. In addition, a pigment known as *"foliaxanthin"* isolated by L. Cholnoky in 1955 from paprika and, subsequently, other sources (cf. Cholnoky, Weedon et al., Chem. Comm., 1966, 404 for refs.), was shown in 1966/1967 to be identical in all respects, including absolute configuration at the asymmetric centres, with neoxanthin by direct comparison of samples of Cholnoky's "foliaxanthin" with samples of authentic neoxanthin (C. O. Chichester, Cholnoky, Weedon and co-workers, ibid., 1966, 807; 1967, 484; full paper: J. chem. Soc., C, 1969, 1256). The names *foliaxanthin* (for the natural pigment) and *foliachrome* (for the corresponding furanoid oxide, formed on H$^{\oplus}$-catalysed rearrangement) have therefore been abandoned in favour of *neoxanthin* and *neochrome*.

Neoxanthin was first isolated pure in 1938 by H. H. Strain ("Leaf Xanthophylls", Carnegie Inst., Washington, 1938). Chemical studies by Krinsky, Curl and others over the years led to various suggested structures but it was not until 1966/1967 that the following, and correct, structure was finally proposed on the basis of previous chemical data and i.r./p.m.r. (allene group) spectra (Chichester et al., Chem. Comm., 1966, 807):

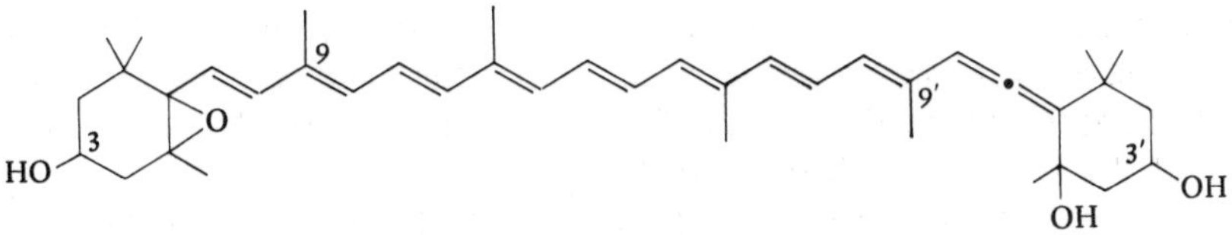

Neoxanthin

This structure was identical with that simultaneously, and independently, deduced (Cholnoky, Weedon et al., 1966, loc. cit.) from spectral and chemical data, including LiAlH$_4$ reduction to zeaxanthin (cf. below) thereby establishing the gross skeleton of the compound, for Cholnoky's "folia-xanthin". As mentioned above, the latter was soon afterwards shown to be identical with neoxanthin, thereby providing strong support for the above structure. The structure was then confirmed by a full spectral study [p.m.r., i.r., mass spectrum (molecular formula and fragmentation pattern on the furanoid-oxide form)], using comparisons with known compounds, on samples of both "neoxanthin" and "foliaxanthin" (Cholnoky, Weedon et al., Chem. Comm., 1967, 484). Full details of this study have since appeared (idem, J. chem. Soc., C, 1969, 1256).

Physical properties. Most samples of neoxanthin isolated from nature have had m.p. ca. 128° (for chromatographically pure but uncrystallised samples) or ca. 134° (after crystallisation, from acetone—petroleum) and λ_{max} 467, 439, 416 nm (ethanol) (cf. refs. given above). This is probably a *cis*-isomer since iodine-catalysed stereoisomerisation yields a higher melting (ca. 143°; from acetone—petroleum) form with λ_{max} at slightly longer wavelengths, 470, 441, 422 nm (ethanol), presumed to be all-*trans*-neoxanthin (Cholnoky, Weedon et al., 1969, loc. cit.). It was the latter isomer which was isolated by Strain in 1938 (loc. cit.; m.p. 143—145°), and which Cholnoky, Weedon et al. (1969, loc. cit.) have since shown can be isolated from maple leaves providing the saponification step, which apparently causes stereo-isomerisation of the natural isomer, is omitted from the isolation procedure.

Stereochemistry (cf. Cholnoky, Weedon et al., 1969, loc. cit.). The above two isomers have identical i.r. and mass spectra but quite different o.r.d. curves, as expected for *cis/trans* isomers of the same compound. In the p.m.r. spectrum, they show slight, but distinct, differences in the positions of the methyl group bands corresponding the the 5,6-epoxide end of the molecule, suggesting that it is at this end the difference in C=C stereo-chemistry lies. The smallness of the *cis*-peak in the *cis*-isomer suggests that the *cis*-bond therein lies near the end of the chromophore (cf. 2nd Edn., p. 255), probably at $C_{(9)}$.

Absolute configuration. Cholnoky, Weedon et al. (1967, 1969, loc. cit.) found that the zeaxanthin obtained from neochrome as below had the same o.r.d. curve as natural zeaxanthin, and so has the same absolute configuration as natural zeaxanthin at $C_{(3)}, C_{(3')}$. The other two centres at the allene end of the molecule probably have the same stereochemistry as "synthetic" (*ex* fucoxanthin: see below) fucoxanthinol (on the grounds that the $C_{(1')}$, $C_{(5')}$, and $C_{(9')}$ methyl group signals occur at very similar positions in the two compounds: Cholnoky, Weedon et al., 1969, loc. cit.; supported by work on the "grasshopper ketone", below). At the other end, the epoxide ring is believed, but not proved, to be *trans* to the $C_{(3)}$—OH (cf. idem, ibid., p. 1261). Neoxanthin would then have the absolute stereochemistry given below:

Reactions

(i) *With acid.* Brief treatment (at 25°) with a trace of HCl in chloroform or ether yields mainly the corresponding furanoid oxide, *neochrome*, m.p. 148°, λ_{max} shift ca. −16 nm, in the usual way. However, prolonged treatment with $H^{\oplus}/CHCl_3$, or use of larger quantities of acid, leads to the formation of secondary products formed by further reaction of the neochrome. Both Curl and Krinsky (refs. in 2nd Edn., p. 330) carried out intensive studies of the reactions occurring, and inferred something of their nature, but it was not until after neoxanthin's structure had been elucidated that real progress was made. Thus K. Egger, A. G. Dabbagh and H. Nitsche (Tetrahedron Letters, 1969, 2995) found that $H^{\oplus}/CHCl_3$ on neoxanthin for 4 hours gave three products, all of which has the same absorption spectrum as neochrome*. Two were, apparently, the expected dehydration products produced by removal of the tertiary-OH (structures by comparison of polarity with xanthophylls of known structure, test for allylic-OH, etc. and comparison with the reaction on the diacetate below). The third product

---

* The double bonds of the allene group, being at 90° to one another, do not transmit conjugation effects.

was, interestingly, *diadinochrome* (p. 305) (confirmed by comparison with authentic diadinochrome). For the mechanism of formation of the latter, see p. 309.

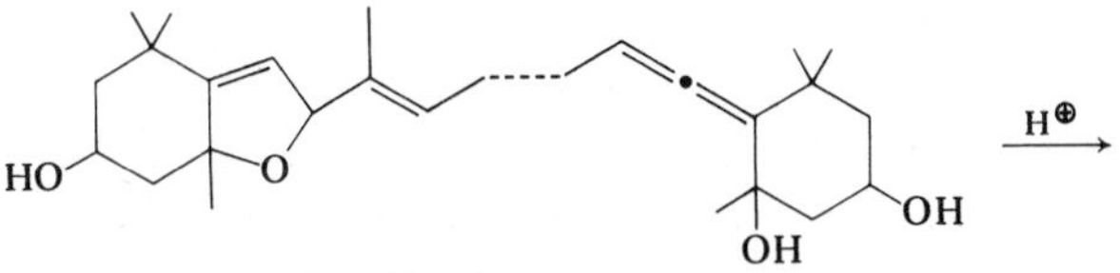

Neochrome

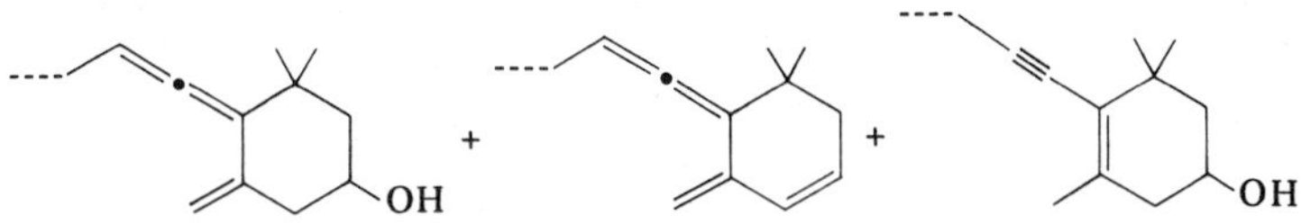

(XLVIII)   (XLIX)   Diadinochrome
Major product

Using the ($C_{(3)}$, $C_{(3')}$) diacetate of neoxanthin, formation of the middle of the above three compounds, which would involve elimination of allylic-OAc, is suppressed, the other two products then being obtained, as their acetates, in the proportions of ca. 75% (dehydration product) and 25% (diadino-chrome).

De-epoxyneoxanthin, below, undergoes a similar reaction.

(ii) *With lithium aluminium hydride.* (a) Neoxanthin on treatment with excess $LiAlH_4$ in ether (20°) gives diatoxanthin (p. 303; identified by mixed t.l.c.'s with natural diatoxanthin and i.r.) in high yield (Nitsche, ibid., 1970, 3343). This involves the epoxide end of the neoxanthin molecule being reduced to the corresponding cyclohexenyl moiety in the usual way (p. 247) and the allene end undergoing a rearrangement with the formation of a triple bond, rather as seen with acid above. The mechanism presumably involves a similar electron flow but triggered by removal of the $C_{(8')}$-proton by $H^{\ominus}$ ion:

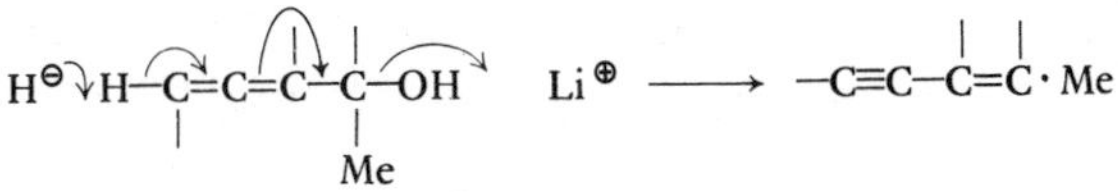

De-epoxyneoxanthin also yields diatoxanthin on treatment with $LiAlH_4$; see below.

(b) Neochrome on vigorous treatment (vigorous reflux; negligible reaction at 0°) with excess $LiAlH_4$ gives, in small yield, zeaxanthin (Cholnoky,

Weedon et al., 1969, loc. cit.). This reaction was of considerable value in the studies, above, on the structure and absolute configuration of neoxanthin. The low-yield reduction of the furanoid to the parent compound is usual (p. 248), and presumably the more severe conditions used here explain the reduction in this reaction of the allene end to the polyene, as opposed to the acetylene (cf. above), level.

A C_{13} allenic ketone isolated from the grasshopper *Romalea microptera* and shown to have the structure below (J. Meinwald et al., Tetrahedron Letters, 1968, 2959), bears an obvious resemblance to neoxanthin, and is thought to be formed by oxidative degradation of neoxanthin in ingested food.

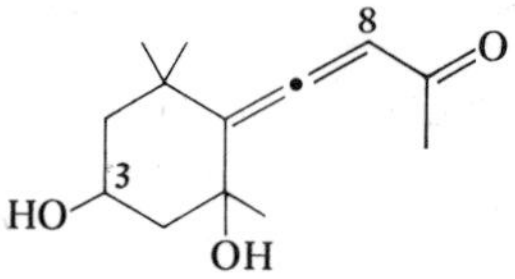

"Grasshopper ketone"

Hopefully-stereospecific syntheses of the ketone have been reported by both Meinwald (ibid., 1969, 1657; starting from isophorone) and Weedon's group (Chem. Comm., 1969, 85; from a trimethylcyclohexanedione derivative used in carotenoid synthesis)*. Both groups inserted the allenic grouping using a sequence of the type

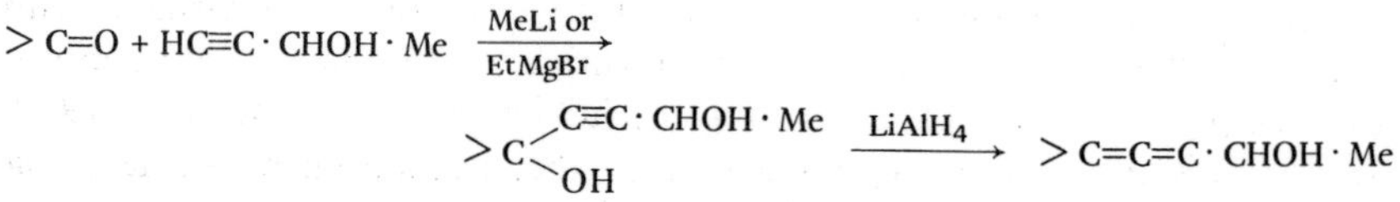

$$> C{=}O + HC{\equiv}C \cdot CHOH \cdot Me \xrightarrow[\text{EtMgBr}]{\text{MeLi or}}$$

$$> C \underset{OH}{\overset{C{\equiv}C \cdot CHOH \cdot Me}{<}} \xrightarrow{\text{LiAlH}_4} > C{=}C{=}C \cdot CHOH \cdot Me$$

Weedon's product had the same configuration (X-ray study on the $C_{(3)}$-p-bromobenzoate; Weedon et al., ibid., 1969, 754) at $C_{(3)}$, $C_{(5)}$, $C_{(8)}$ as that proposed for the allenic end of the neoxanthin molecule. Natural "grasshopper ketone" has since been shown to have the same stereo-chemistry at these positions, thereby providing support for the proposed neoxanthin (allenic end) stereochemistry (idem, ibid., 1970, 1231). The same ketone (as the $C_{(3)}$-acetate) is one of the products obtained on oxidative degradation of fucoxanthin (see above).

De(s)epoxyneoxanthin (de-epoxyneoxanthin) is the major carotenoid in the flowers of *Mimulus guttatus* and apparently occurs in two other *Mimulus* species (Nitsche, Egger and Dabbagh, Tetrahedron Letters, 1969, 2999). This compound's structure, as neoxanthin but lacking the epoxide group, was

---

* Also, since then, by K. Mori, Tetrahedron Letters, 1973, 723.

inferred from a study of its polarity ($\sim$violaxanthin), absorption spectrum (chromophore), the effect of acetylation, and of MeOH/H$^{\oplus}$ on the diacetate so formed (gave a monomethyl ether; hence tertiary allylic OH present), and the effect of H$^{\oplus}$ (no λ_{max} shift characteristic of conjugated epoxide, but gave three other products, by dehydration etc. of the t-OH, with unchanged λ_{max} values; cf. neoxanthin and neochrome). One of the latter three products was found to be *diatoxanthin* (p. 303) (identified by direct comparison on t.l.c. etc. with authentic material) formed in the same way as diadinochrome is from neoxanthin, above. Nitsche (ibid., 1970, 3343) has, since, also shown that the sole product obtained on treating de-epoxyneo-xanthin with excess lithium aluminium hydride at 20° is diatoxanthin. This parallels the reaction seen above in the neoxanthin series.

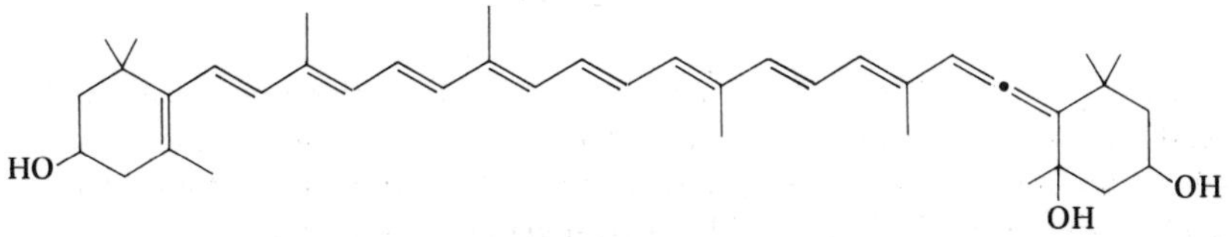

De-epoxyneoxanthin

De-epoxyneoxanthin has since been isolated from other plants (Nitsche and Pleugel, Phytochem., 1972, 11: 3383), and in addition may be identical with Krinsky's *"trollein-like carotenoid"* (p. 334).

   Mimulaxanthin. The source which yielded de-epoxyneoxanthin above, and which also contains neoxanthin itself, has been found (Nitsche, Phytochem., 1972, 11: 401) to contain a third allenic carotenoid, "mimulaxanthin". Its structure was studied using similar techniques as were used for de-epoxyneo-xanthin and it was concluded that mimulaxanthin is, probably, as for neoxanthin but with the epoxide ring at $C_{(6)}$–$C_{(5)}$ replaced* by –CH(–C=C$\sim$)–CMe(OH)–.

   Fucoxanthinol. One of the three pigments isolated (the others were fucoxanthin, above, and paracentrone, below) from the sea urchin (*Para-centrotus lividus*) by Lederer, Weedon and co-workers (J. chem. Soc., C, 1969, 1264) showed obvious spectral similarities to fucoxanthin [including allene and conjugated C=O bands in the i.r.; and an AB quartet (J, 18 Hz) near τ 7, characteristic of –CH$_2$–CO in an asymmetric environment], but was more polar than fucoxanthin, was only a C$_{40}$ compound (m.s.), and showed no acetate absorption (i.r., p.m.r.). It appeared to be very similar to a derivative of fucoxanthin, known as *"fucoxanthinol"*, which had been

---

* The same, or similar, pigment since reported in avocado: J. Gross et al., Phytochem., 1973, 12: 2259.

prepared from fucoxanthin during structural work on that compound and which was known to have the structure below. This supposition was then confirmed (Lederer, Weedon and co-workers, loc. cit.) by carrying out a direct comparison (t.l.c., various systems; spectra) of the pigment from the sea urchin with a sample of fucoxanthinol prepared from fucoxanthin as below.

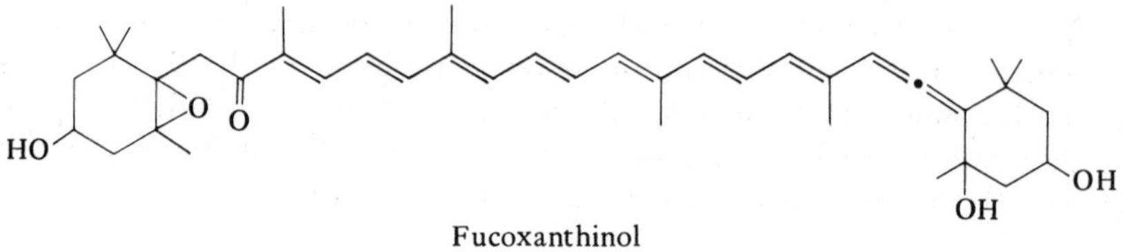

Fucoxanthinol

Partial synthesis. From fucoxanthin by LiAlH$_4$ reduction followed by oxidation of the resulting allylic −OH with dichlorodicyanobenzoquinone (Weedon and co-workers, J. chem. Soc., C, 1969, 429; direct base- or acid-catalysed hydrolysis yields rearrangement products such as isofuco-xanthin instead).

Pure fucoxanthinol has m.p. 146−148°, λ_{max} 476, 448, 425 nm (petroleum) (Weedon and co-workers, loc. cit.). Weak base converts it into the corresponding "iso"-compound, *isofucoxanthinol* (as for fucoxanthin → isofucoxanthinol; 2nd Edn., p. 330).

The fucoxanthinol in *P. lividus* probably arises by enzymatic hydrolysis of the fucoxanthin present in the algae of its diet.

Paracentrone, one of the pigments, m.p. 147−149°, isolated from *P. lividus* by Lederer, Weedon and co-workers (see under fucoxanthinol, above), was found by them to have λ_{max} values and i.r. bands (C=C=C, OH, conj. CO) similar to those of fucoxanthinol, but to be only a C$_{31}$ compound (C$_{31}$H$_{42}$O$_3$, by precision m.s.). The p.m.r. spectrum showed certain similarities to fucoxanthinol's but lacked the −CH$_2$·CO− quartet and had an additional band at τ 7.65 (probably Me·CO−). The following structure was proposed by the above authors in 1969 (loc. cit.) and subsequently confirmed by a partial synthesis (see below):

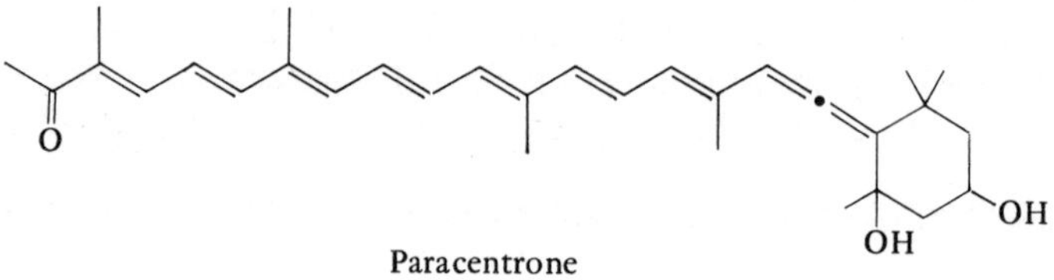

Paracentrone

Partial synthesis (J. Hora, T. P. Toube and Weedon, J. chem. Soc., C, 1970, 241). Oppenauer oxidation of fucoxanthin results in: (a) oxidation at $C_{(3)}$; and (b) base-catalysed cleavage of the $C_{(6)}-C_{(7)}$ bond. Subsequent removal of the fucoxanthin acetate group by mild alkaline hydrolysis (K_2CO_3 in wet methanol; 20°) yields paracentrone (ca. 7% overall; confirmed by comparison with the natural pigment — mixed m.p., t.l.c., spectra).

Fucoxanthin $\longrightarrow$

The paracentrone in *P. lividus* probably arises from the fucoxanthin present in its diet, possibly by biological oxidation of the $C_{(3)}-OH$ to carbonyl followed by a rearrangement like that outlined above.

Vaucheriaxanthin, one of the major carotenoids (heteroxanthin and diadinoxanthin are two others) of three *Vaucheria* spp. and a *Botrydium* sp. (algae) studied by Kleinig and Egger (Z. Naturforsch., 1967, 22b: 868) where it exists esterified. The free carotenoid has $\lambda_{max} \sim 470, 442, 419$ nm (ethanol; as for a nonaene) and formula $C_{40}H_{56}O_5$ (precision m.s.). It contains an allene group (i.r.), three acetylatable and one tertiary hydroxyls, and a conjugated epoxide ($H^\oplus$ test; λ_{max} shift, ca. -20 nm) (Kleinig and Egger, loc. cit.; Strain et al., Phytochem., 1968, 7: 1417). H. Nitsche and Egger (Tetrahedron Letters, 1970, 1435) found that two of the OH's are allylic (methylated with $MeOH/H^\oplus$) including the *t*-OH (effect of $MeOH/H^\oplus$ on the acetylated pigment), and that on treatment of the triacetate, above, with $H^\oplus/CHCl_3$ there occurred a rearrangement analogous to that seen with neoxanthin diacetate (above), indicating a similar relationship of *t*-OH and allene group. The presence of an in-chain $-CH_2OH$ (cf. Section 7i, p. 320) was inferred from the effect of *p*-chloranil (formation of a conjugated aldehyde), and it was placed at $C_{(9')}$ rather than $C_{(9)}$ since on acid rearrangement of the epoxide there occurred a λ_{max} shift corresponding to a change of chromophore from octaeneal to heptaeneal (no change if at $C_{(9)}$). The secondary OH's were placed at $C_{(3),(3')}$ by analogy.

Vaucheriaxanthin (?)

Caloxanthin and **nostoxanthin**, are a pair of pigments occurring together, along with zeaxanthin (p. 250) and other known carotenoids, in certain blue-green algae (*Cyanophyceae*) examined by H. Stransky and A. Hager, and others (cf. Arch. Mikrobiol., 1970, 72: 84). They were assigned the following structures by Stransky and Hager (loc. cit.) from a consideration of their λ_{max} values, co-occurrence with zeaxanthin, polarity, i.r. (C=C=C band), acetylation/silylation tests (two secondary, no *tert*-hydroxyls), and rearrangement on treatment with hydride ion to a compound identified as zeaxanthin (probably; via an intermediate, identified as caloxanthin, in the case of nostoxanthin); but doubt has recently been cast on these structures by Jensen (Phytochem., 1971, 10: 3121).

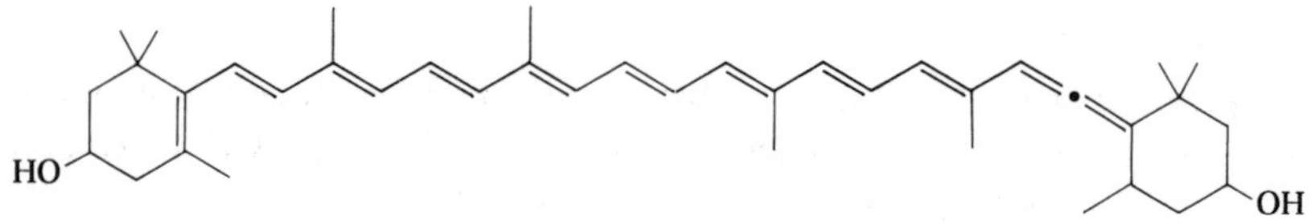

Caloxanthin (?)

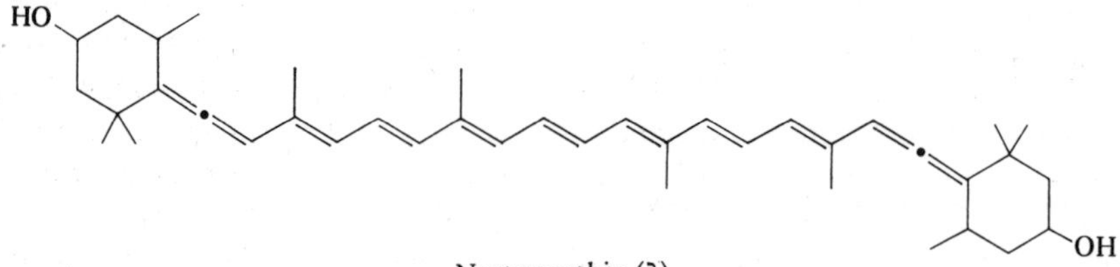

Nostoxanthin (?)

Peridinin has also been assigned an allenic structure: p. 335.

Biogenesis of allenic carotenoids (excluding fucoxanthinol and paracentrone, which are probably derived directly from pre-formed fucoxanthin). The allenic end of the allenic carotenoids so far discovered conforms to the same pattern (in the proposed caloxanthin and nostoxanthin structures, the $C_{(5,5')}$—OH's have been lost). Various theories have been put forward as to how this system might originate in nature.

(a) That the $C_{(5)}$—$C_{(8)}$ part of the violaxanthin molecule*, which usually accompanies at least fucoxanthin and neoxanthin in nature, undergoes rearrangement and/or hydrolysis to the α-hydroxyallene grouping (Weedon et al., J. chem. Soc., C, 1969, 429, p. 442; Chem. Comm., 1969, 754; B. H. Davies et al., Phytochem., 1970, 9: 797).

(b) From the corresponding acetylenic compound, but the reverse process (see p. 293) appears more likely (cf. e.g., Cholnoky, Weedon et al., J. chem. Soc., C, 1969, 1256).

---

* Or antheraxanthin for de(s)epoxyneoxanthin.

(c) A possibility that has some experimental backing follows from the discovery, by S. Isoe et al., M. Mousseron-Canet et al. and C. S. Foote et al. (Tetrahedron Letters, 1968, 5561, 6037, 6041; also Mousseron-Canet et al., Bull. Soc. chim. Fr., 1970, 1968), that dye-sensitised photo-oxidation of β-ionone derivatives (L) leads to the formation (via L, arrows) of allenic hydroperoxides (LI), amongst other products (e.g. LII):

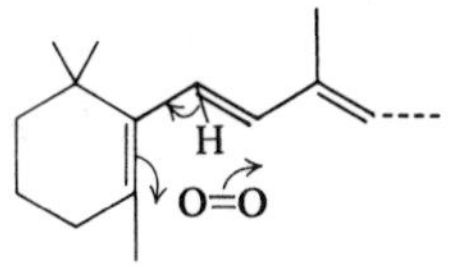

(L) 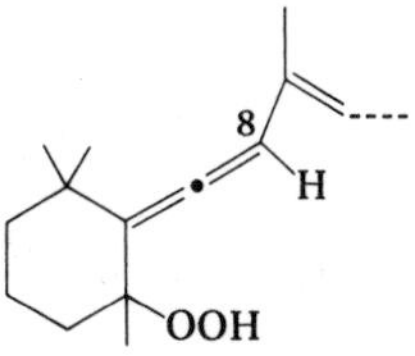   (LI)

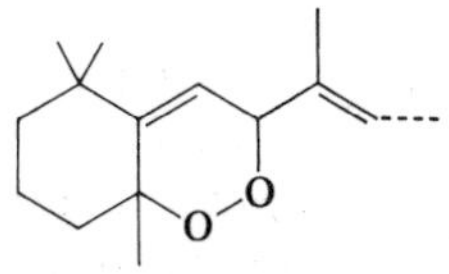

(LII) 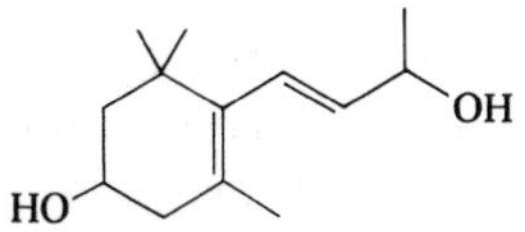   (LIII)

This suggests that the allenic carotenoids might originate by a parallel reaction occurring in vivo on, for example, zeaxanthin. Weedon's finding (Chem. Comm., 1970, 1231) that the stereochemistry at $C_{(8)}$ of the hydroperoxide formed from (LIII) is opposite to that found in nature in neoxanthin and fucoxanthin apparently casts doubt on this hypothesis, but the stereochemistry at this point might well depend on the length of the attached chain in the starting material (very short in LIII; much longer in zeaxanthin). Similarly, K. Tsukida et al. (Chem. pharm. Bull., 1969, 17: 1755) found that chlorophyll-sensitised photo-oxidation of β-carotene (oxygen bubbled through an irradiated solution of β-carotene containing a trace of chlorophyll) gave, amongst other products, a small amount of LIV (the stereochemistry at $C_{(8')}$ was not elucidated):

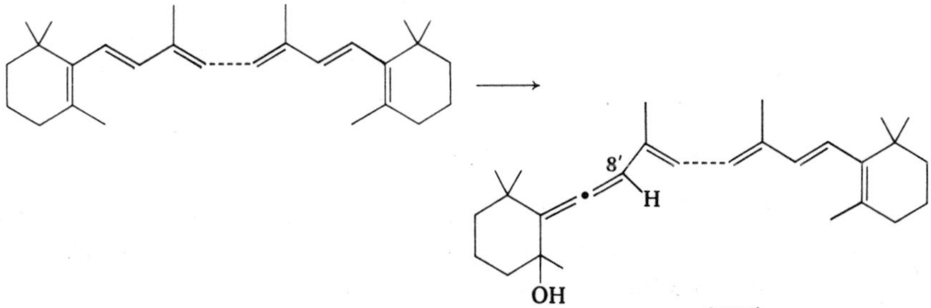

(LIV)

In addition it is conceivable that the "other" end-group of fucoxanthin originates from a compound like (LII); e.g. from zeaxanthin:

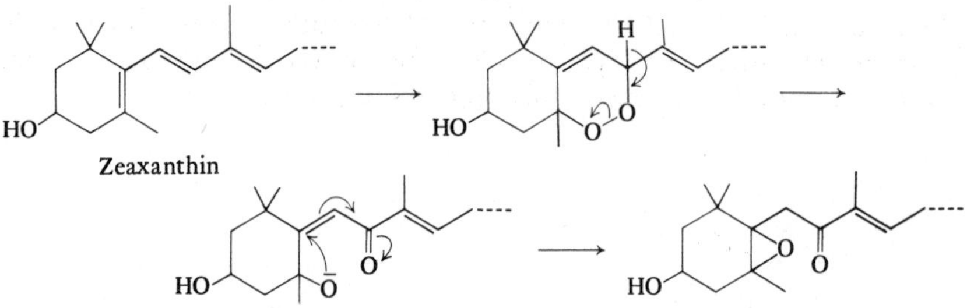

Zeaxanthin

(g) *Acetylenic carotenoids*

General properties. Although the presence in nature of carotenoids containing one or more acetylenic bonds was only recognised in 1967 (see under alloxanthin, below), they may in fact be relatively common since an acetylenic bond within the unsaturated chain of a carotenoid (particularly if located near one end) has very little effect on either the visible absorption spectrum or polarity (and in much of the earlier work it was frequently just on a combination of these two characteristics that the nature of a carotenoid isolated from a new source was inferred). Thus, the acetylenic carotenoids alloxanthin and diatoxanthin have visible absorption spectra and polarities (t.l.c. R_f values; position on column chromatogram) very similar to those of the common all–C=C carotenoid, zeaxanthin (for data, see Egger, Nitsche and Kleinig, Phytochem., 1969, 8: 1583; D. J. Chapman, ibid., 1966, 5: 1331, and the relevant sections of this publication), and so certain of the pigments tentatively identified as zeaxanthin, or described as "zeaxanthin-like", in pre-1967 reports may in reality have been alloxanthin or diatoxanthin (for specific examples of where this is now known to have occurred,

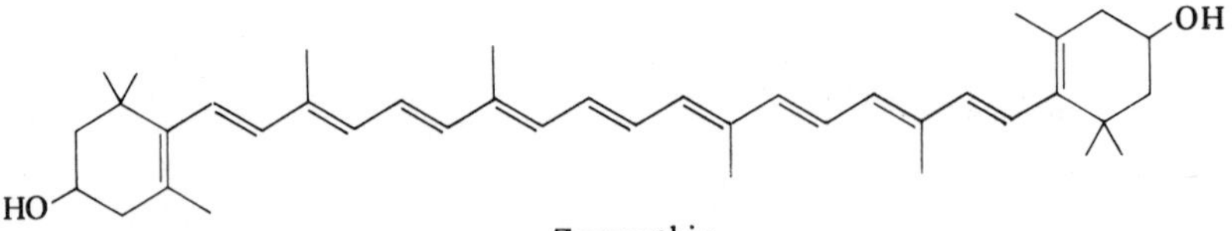

Zeaxanthin

see under alloxanthin and diatoxanthin below). Similarly, the pigment now known to be diadinoxanthin (q.v. for refs.) was, earlier on, occasionally reported as antheraxanthin (5,6-epoxyzeaxanthin).

The main characteristics of acetylenic carotenoids which allow them to be distinguished from the corresponding "all—C=C" compounds are given below.

(i) The molecular formula (by precision mass spectrometry on the molecular ion) has more double-bond equivalents than can be easily accounted for knowing the chromophore (visible absorption spectrum) and number of carbonyl-containing groups (i.r.).

(ii) *I.r.* This is of less help than might be expected since only a relatively weak C≡C band is observed, particularly for a mono-acetylenic compound (except in the carbonyl-conjugated example, "asterinsäure").

(iii) ^1H-*n.m.r.* (*p.m.r.*). The 7,8 (and 7',8') olefinic-H's in most all—C=C cyclic carotenoids fortuitously occur at the same position (τ 3.85) in the spectrum and so are seen as a characteristic 4H singlet protruding above the complex olefinic-H multiplets in the region (see, e.g. the spectra reproduced in Pure Appl. Chem., 1969, 20: pp. 403—404). This peak is lacking (or halved in intensity) in di- (or mono-) acetylenic carotenoids [all acetylenic carotenoids so far discovered in nature have had the triple bond at the 7,8-position(s)] (cf. also p. 302). ^{13}C-n.m.r. measurements should also be of value: cf. p. 207.

(iv) Certain of the acetylenic carotenoids show increased fine structure in their visible absorption spectra as compared with the corresponding all—C=C compounds suggesting the end-group double bond (forced by steric hindrance effects to lie at an angle to the polyene chain in the all—C=C compounds, p. 194) lies more nearly in the plane of the polyene chain in the acetylenic compounds (cf. Scopes et al., J. chem. Soc., C, 1969, 2527; N. A. Sørensen, S. L. Jensen and co-workers, Acta chem. Scand., 1968, 22: 344).

(v) *Susceptibility to 5,6-epoxidation.* The 5,6-double bond in ordinary carotenoids is readily epoxidised on treating with per-acid (p. 245); it appears that the 5,6-double bond in the unit shown below is inert (at least in diatoxanthin and diadinoxanthin*: cf. Egger, Nitsche and Kleinig, 1969, loc. cit.):

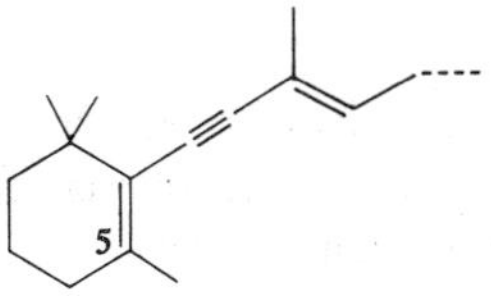

Alloxanthin. In 1966, D. J. Chapman (Phytochem., 1966, 5: 1331) isolated what appeared to be a new carotenoid from the alga *Cryptomonas ovata*, and

---

* = "*Vaucheria*-zeaxanthin" and "*Vaucheria*-antheraxanthin", respectively.

named it alloxanthin. It was the major carotenoid present and had m.p.
186–188° (corr.; from acetone–petroleum). Its absorption spectrum (λ_{max}
480, 451 nm, in hexane) and polarity on t.l.c. were almost identical with
those of zeaxanthin. It had been detected in *C. ovata* by F. T. Haxo and
D. C. Fork in 1959 and reported as a "zeaxanthin-like carotenoid" (Nature,
1959, 184: 1051). The structure of alloxanthin was inferred (A. K. Mallams,
E. S. Waight, Weedon et al., Chem. Comm., 1967, 301) from a consideration
of its obvious similarity to zeaxanthin; its molecular formula ($C_{40}H_{52}O_2$ by
precision m.s.; = zeaxanthin-4H); the formation of a diacetate (two
non-tertiary −OH groups); the presence of a weak band at 2167 cm^{-1} in the
i.r. (C≡C), and the lack of the τ 3.85 singlet in the p.m.r. spectrum alluded
to above.

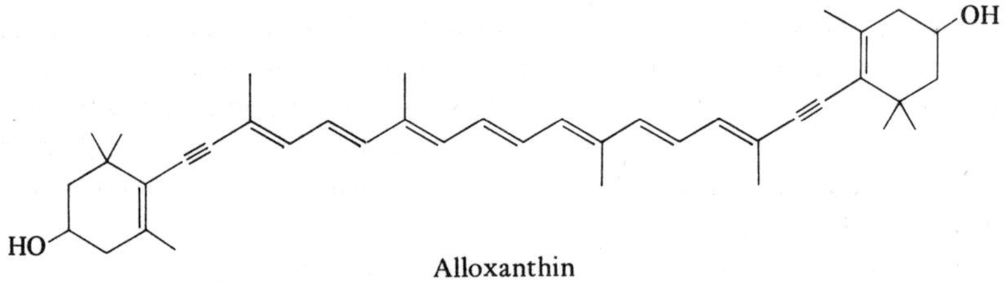

Alloxanthin

P.m.r. of such compounds; effect of the 7,8 triple bond. The acetylenic bond
has a marked effect on the position of the signal due to the methyl groups in
the cyclic end-group as compared with the corresponding CH=CH compound:

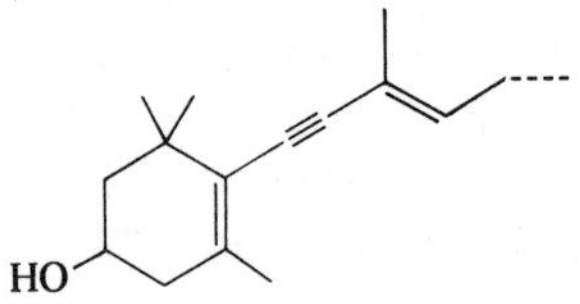

gem-Me's at τ 8.86, 8.81; C$_{(5)}$-Me at 8.10

(Mallams et al., loc. cit.)

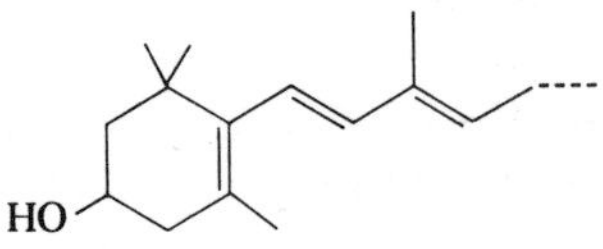

gem-Me's at 8.92 (both); C$_{(5)}$-Me at 8.26

(M.S. Barber et al., J. chem. Soc., 1960, 2870: cmpd. 26)

The triple bond has relatively little effect on the signal due to in-chain
methyl groups, although they are made distinguishable by its presence being
at τ 8.01 and 8.06 in alloxanthin (probably C$_{(9),(9')}$ and C$_{(13),(13')}$-
methyls, respectively: all four at 8.05 in zeaxanthin).

Absolute configuration (at C$_{(3)}$ and C$_{(3')}$). Both zeaxanthin (structure
above) and alloxanthin on catalytic hydrogenation to the corresponding
fully saturated compound give a product which is laevorotatory, and hence

alloxanthin has been assigned the same stereochemistry at these positions as zeaxanthin (viz. 3R, 3'R: see p. 251) (Scopes et al., 1969, loc. cit.):

Alloxanthin

Occurrence elsewhere. Chapman (1966, loc. cit.) found that alloxanthin was also the major carotenoid in the algae *Rhodomonas* D3 and *Hemiselmis virescens*, and it also occurs in *Mytilus edulis* and *M. californianus* (molluscs) and crab (probably *ex* ingested algae) (S. A. Campbell, Compar. Biochem. Physiol., 1970, 32: 97; 1969, 30: 803). In addition, the pigment known as *"cryptomonaxanthin"* (reported by others in two of Chapman's sources) was probably alloxanthin (Scopes et al., 1969, loc. cit., pp. 2530, 2540). Similarly, the carotenoids *"pectenoxanthin"*, isolated by E. Lederer from the scallop *Pecten maximus* (Bull. Soc. chim. biol., 1938, 20: 567) and *"cynthiaxanthin"*, from *Halocynthia papillosa* [Lederer, 1938, loc. cit.), have been shown to be identical with alloxanthin (by re-isolation of the compounds and comparison of their (or their diacetate's) m.p.'s, visible, i.r., p.m.r., and mass spectra, and t.l.c. behaviour with those of alloxanthin (Lederer, Waight, Weedon and co-workers, Chem. Comm., 1967, 941; Scopes et al., 1969, loc. cit)].

Diatoxanthin. This is a relatively well-known carotenoid having been found in diatoms in 1944 (H. H. Strain et al., Biol. Bull., 1944, 86: 169; Strain, 32nd Ann. Priestley Lecture, Penn. State Univ., 1958; cf. also K. Egger et al., Phytochem., 1969, 8: 1583) and reported in various other algae since then (mainly in members of the Chrysophyceae and Xanthophyceae) [Strain, 1958, loc. cit.; M. B. Allen, S. W. Jeffrey and co-workers, J. gen. Microbiol., 1960, 23: 93; 1964, 34: 259; 1964, 36: 277; Chapman, 1966, loc. cit.; Egger et al., 1969, loc. cit.; Strain et al., Phytochem., 1970, 9: 2561; *but,* the pigment referred to as diatoxanthin in some of the pre-1966 reports (e.g. Allen et al.'s second paper) was probably alloxanthin, from which it is almost indistinguishable using simple chromatographic methods]. Diatoxanthin was also detected in the two molluscs studied by Campbell (see under alloxanthin, above) and, as mentioned above, certain of the "zeaxanthin-like" pigments reported before 1967 may also have been diatoxanthin (e.g. Egger et al., 1969, loc. cit., found that the "zeaxanthin" in three *Vaucheria* spp. is really diatoxanthin). Diatoxanthin has m.p. 201°; λ_{max} 479, 450 (hexane); 479, 452, (428) nm (ethanol) (Chapman, 1966, loc. cit.; Egger et al., 1969, loc. cit.). Its similarity to zeaxanthin, with regard to polarity and absorption spectrum, was noted early on and Chapman (loc. cit.) in turn noted its similarity to alloxanthin. Its structure was deduced in 1967 from a consideration of these observations, its molecular formula ($C_{40}H_{54}O_2$ by precision mass spectrometry; = zeaxanthin-2H), and a comparison of its spectral properties with those of alloxanthin, above (Mallams, Waight, Weedon et al., 1967, loc. cit.):

Diatoxanthin

Absolute configuration. Deduced to be *3R,3'R* as for alloxanthin by comparison (cf. p. 211) of its o.r.d. curve with that predicted for a molecule made up of half zeaxanthin plus half alloxanthin, both known to have the *R* configuration at the $C_{(3)},C_{(3')}$ positions (see relevant sections) (Scopes et al., 1969, loc. cit.).

Careful chromatography alongside authentic samples will distinguish diatoxanthin from zeaxanthin fairly readily but from alloxanthin only partially or indistinctly. For the latter it is better to observe the effect of iodine-catalysed stereoisomerisation (viz. exposure of the carotenoid in a solvent containing a trace of iodine to the effect of light). The relative amounts, and R_f values, of the mixture of *cis/trans* isomers produced from the two compounds are markedly different (Egger et al., 1969, loc. cit.; Chapman, 1966, loc. cit.).

Partial synthesis, from the abundant natural carotenoid neoxanthin (q.v.).

Crocoxanthin, one of the minor carotenoids in each of the three algae from which Chapman isolated alloxanthin (q.v.), has m.p. 163–165° (from benzene–methanol), λ_{max} 475, 445, 422 nm (hexane) (Chapman, 1966, loc. cit.). Chapman noted that its polarity and absorption spectrum were barely distinguishable from those of α-cryptoxanthin. This coupled with its molecular formula ($C_{40}H_{54}O$, by precision m.s.) and a comparison of its spectral properties with those of alloxanthin, led to its structure (Mallams, Waight, Weedon et al., 1967, loc. cit.).

Crocoxanthin's o.r.d. curve was measured by Scopes et al. (1969, loc. cit.) but from this it was not possible to infer with certainty its absolute

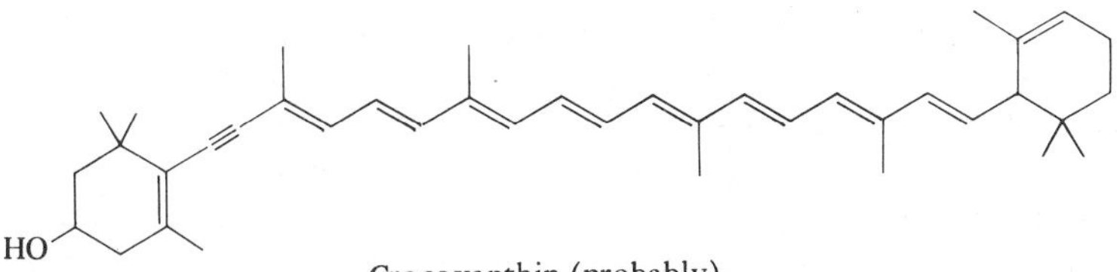

Crocoxanthin (probably)

configuration, althougn it did seem likely that the α-carotene end of the molecule has the same configuration at $C_{(6')}$ as α-carotene itself.

Monadoxanthin, one of the minor carotenoids in *Rhodomonas* D3 (one of the above-mentioned three algae), has m.p. 165° (acetone–petroleum), λ_{max} as for crocoxanthin. It was found to be slightly less polar than alloxanthin, and to have polarity and absorption spectrum similar to, but just distinct from, those of lutein (Chapman, 1966, loc. cit.). Its molecular formula $(C_{40}H_{54}O_2)$ and a probable structure were inferred in the same way as for crocoxanthin above (Mallams, Waight, Weedon et al., 1967, loc. cit.).

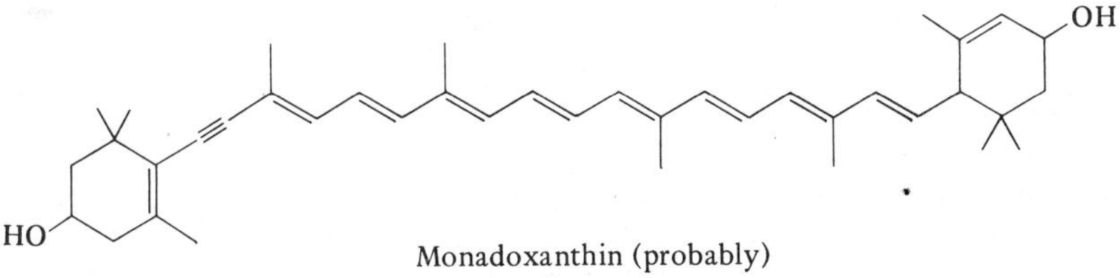

Monadoxanthin (probably)

Scopes et al. (1969, loc. cit.) determined monadoxanthin's o.r.d. curve but were unable to deduce its absolute configuration (3 centres) since the contribution to the o.r.d. curve to be expected from the right hand side of the above formula was not known (cf. p. 211).

Diadinoxanthin, a pigment isolated from diatoms and certain other algae (Strain et al., 1944, 1958, 1970, loc. cit.; Chem. Comm., 1968, 32; Egger et al., 1969, loc. cit.) and which, as mentioned on p. 300, has occasionally been reported as antheraxanthin with which it is easily confused. The "antheraxanthin" in *Euglena gracilis* and in three *Vaucheria* species (algae) has recently been shown to be diadinoxanthin (Strain et al., 1968, loc. cit.; Egger et al., 1969, loc. cit.). Diadinoxanthin has m.p. ca. 160° (benzene–petroleum), λ_{max} 478, 448, (424) nm (ethanol) and gives a diacetate (Strain et al., 1968, 1970, loc. cit.). With acid, a λ_{max} shift corresponding to the rearrangement of a conjugated 5,6-epoxide to the furanoid derivative occurs. The product, *diadinochrome* (cf. below), has λ_{max} 458, 431, (409) nm (ethanol). Diadinoxanthin's structure was inferred (Strain et al., 1968, loc. cit.) from its close resemblance (absorption spectrum, polarity; cf. above) to antheraxanthin (p. 252), its formula ($C_{40}H_{54}O_3$ by m.s.; = antheraxanthin-2H), the presence of a weak $C\equiv C$ band in the i.r., its p.m.r. spectrum (ten methyls present, none having been hydroxylated) and the mass spectrum which showed one OH group at each end of the molecule and also ions at m/e 181 and 221 corresponding to pyrylium and homopyrylium ions, as expected for an end-group of the type shown below (cf. p. 209):

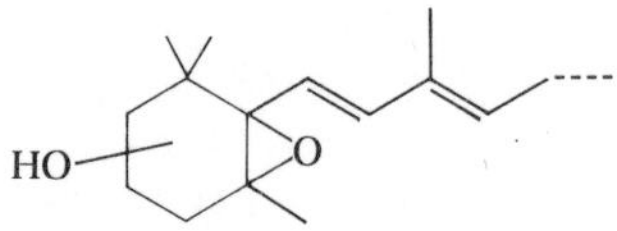

The hydroxyl groups were shown to be secondary (diacetate formed) and not allylic (no reaction with $MeOH/H^{\oplus}$) but their placing at $C_{(3)}, C_{(3')}$ was only by analogy with other carotenoids. However, the subsequent finding that diadinochrome can be obtained in one step from neoxanthin (q.v.) has provided support for this placing.

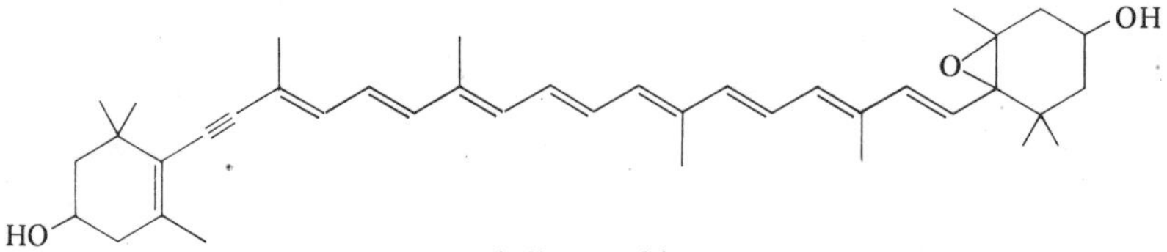

Diadinoxanthin

Diadinoxanthin is the 5,6-epoxide of diatoxanthin, with which it sometimes co-occurs in nature (Strain et al., 1970, loc. cit.; Egger et al., 1969, loc. cit.).

Heteroxanthin. A very polar carotenoid (more polar than neoxanthin, for example) isolated from a *Vaucheria* sp. and from *Tribonema aequale* by Strain et al. (Phytochem., 1968, 7: 1417; 1970, 9: 2561), and apparently present in all other members of the Xanthophyceae (algae) which have been investigated (cf. Strain et al., 1970, loc. cit.). It has m.p. ca. 180° (benzene–petroleum), λ_{max} 478, 448, (423) nm (ethanol) and readily gives a diacetate (Strain et al., 1970, loc. cit.; Strain, K. Aitzetmüller et al., Chem. Comm., 1970, 876). The following structure was proposed by the last-named authors from a study of the compound's formula ($C_{40}H_{56}O_4$: precision m.s.), its visible absorption spectrum (which indicated the same chromophore as in diadinoxanthin, above), lack of conjugated 5,6-epoxide, i.r. spectrum (weak $C\equiv C$ band; no $>C=O$ functions, hence all oxygens probably –OH's), p.m.r. spectrum [Me groups at the positions found for the acetylenic half of the diadinoxanthin molecule and for azafrin (2nd Edn., p. 328)] and mass spectral fragmentation pattern (comparison with data obtained on hydroxy-carotenoids of known structures; C. R. Enzell et al., Acta chem. Scand., 1969, 23: 727):

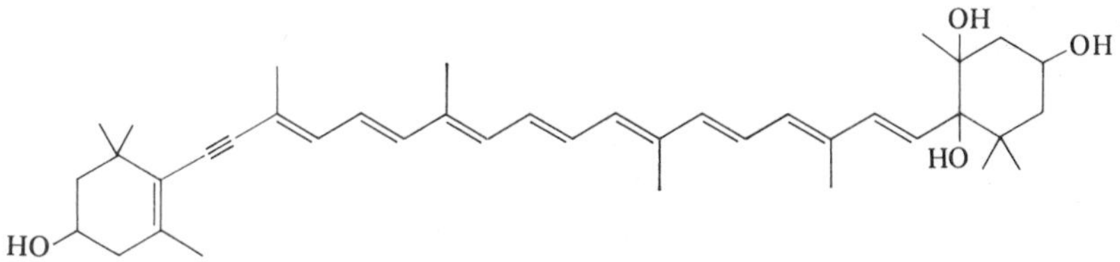

Heteroxanthin (probably)

Meanwhile, H. Nitsche (Tetrahedron Letters, 1970, 3345) put forward a different structure (which included a rather improbable enolic −OH function at $C_{(8)}$) for a pigment isolated from a *Vaucheria* sp. (*"Vaucheria-hetero-xanthin"*) with similar properties to Strain's heteroxanthin. Strain et al. (ibid., 1971, 733) have since shown that Nitsche's heteroxanthin is identical with theirs and have dismissed Nitsche's structure as being inconsistent with the p.m.r. and m.s. data obtained on Strain's material.

Pectenolone is a pigment found in the small marine animals *Pecten maximus* and *Halocynthia papillosa* by Lederer, Waight, Weedon and co-workers (1967, loc. cit.). It gave (and was isolated as) a diacetate, λ_{max} 463 nm (ethanol), molecular formula $C_{44}H_{56}O_5$ (precision m.s.), ν_{max} 2175 (C≡C), 1674 (conj. C=O). Reduction (KBH$_4$), to remove the complicating effect on the absorption spectrum of C=O conjugation, gave a product with λ_{max} values (478, 451 nm; ethanol) consistent with its containing a (7,8-) dehydro-β-carotene chromophore. This led to the following structure being proposed:

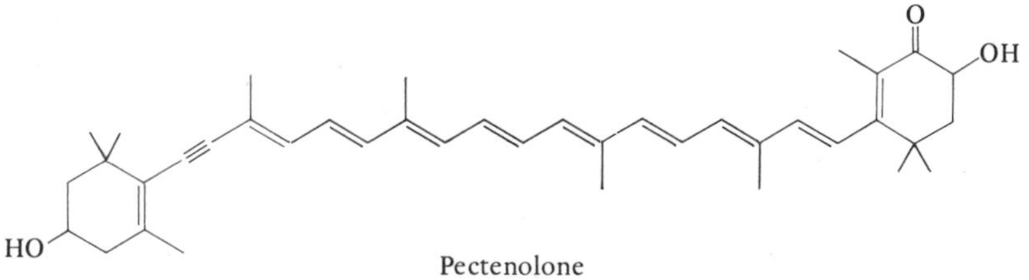

Pectenolone

In confirmation, pectenolone's p.m.r. spectrum was consistent with this half alloxanthin plus half astaxanthin formulation. Interestingly, pectenolone co-occurs with these two carotenoids in both of the above-mentioned sources.

"Asterinsäure" ("*Asterin(ic)-acid*"), was first isolated in 1934 from the starfish *Asterias rubens* (H. von Euler and H. Hellström, Z. physiol. Chem., 1934, 233: 89) where it occurs as a carotenoid–protein complex. It was tentatively identified as astaxanthin (or astacene) in the 1930's but more recently it was re-isolated by N. A. Sørensen, S. L. Jensen and co-workers (Acta chem. Scand., 1968, 22: 344) and studied by modern techniques, which showed it to be different. Their sample, m.p. 216° (pyridine–water), λ_{max} (522), 495, (478) nm (pyridine) (similar to astaxanthin, p. 276, but more fine structure), was slightly more polar than astaxanthin on chromatography ($CaCO_3$-impregnated paper), and showed a C≡C band in the i.r. at 2150 cm^{-1} which was quite intense (suggesting conjugation with a >C=O group). Treatment with alkali in air gave a product with acidic properties (cf. the astaxanthin → astacene reaction: 2nd Edn., p. 320) suggesting the presence of an asta-xanthin-type end-group. Reduction ($NaBH_4$) followed by acetylation gave a compound resembling (λ_{max}; t.l.c.) 3,4,3',4'-tetra-acetoxy-β-carotene. The mass spectrum of the pigment was complex suggesting the presence of two components, one with M$^{\oplus}$ at 594 and one with M$^{\oplus}$ at 592, corresponding respectively to 2H and 4H less than astaxanthin; the latter ion being too intense for it to be likely to be entirely due to an M-2 ion (cf. p. 210) from the 594 constituent. This suggested that the pigment was a mixture of LV (*M*, 594) and the corresponding (7,8:7',8'-) bis-acetylenic compound (*M*, 592):

(LV)
7,8-Dehydroastaxanthin

Recent work by the same group (Francis et al., ibid., 1970, 24: 3050) has confirmed this supposition, the two compounds being partially separated by repeated chromatography on cellulose. Both underwent the above asta-xanthin → astacene type reaction. They also found that the proportion of these two carotenoids in *A. rubens* varied from batch to batch. One sample contained only the bis-acetylenic compound and it proved possible thereby to obtain this pigment pure (m.p. 210°, from ether; λ_{max} similar to that quoted above; M$^{\oplus}$, 592).

Triophaxanthin and *hopkinsiaxanthin* are both apo(C$_{31}$)-carotenoids isolated from shell-less molluscs and thought (visible, i.r., and mass spectra; chemical tests) to be, respectively, as for sintaxanthin, p. 283, but with C≡C at C$_{(7)}$,C$_{(8)}$ and an −OH at C$_{(3)}$; and, apparently, a 4-oxo derivative of same (hence an astaxanthin-type end-group) (J. W. McBeth, Compar. Biochem. Physiol., 1972, 41B: 55, 69). As regards the biogenesis of these compounds: formation from another marine carotenoid, paracentrone (cf. p. 296), or by C$_{(6)}$,C$_{(7)}$-cleavage (cf. p. 297) (etc.) of the H$_2$O-addition product of bis-dehydroastaxanthin, above, would appear possible. *Mytiloxanthin* (q.v.) is also, probably, acetylenic.

Biogenesis of acetylenic carotenoids. Nothing definite is known concerning the mode of biogenesis of the acetylenic bond. However, the tendency for acetylenic carotenoids to occur in nature in association with allenic carotenoids (cf. e.g. Chapman, 1966, loc. cit.; Strain, Aitzetmüller et al., 1970, loc. cit.; Strain et al., 1944, 1968, loc. cit.) such as fucoxanthin and neoxanthin suggests they may be derived from these:

Neoxanthin (fucoxanthin)
end-group(s)

An analogy for such a transformation has been provided by the observation that the allenic end-group of neoxanthin undergoes this reaction in vitro on treatment with acid or hydride-ion (p. 293).

(b) C$_{50}$ (and C$_{45}$) carotenoids (cf. comments on p. 192)

"Dehydrogenans-P439" (decaprenoxanthin). *Dehydrogenans*-P439 was the name* given by S. L. Jensen, O. B. Weeks and co-workers in 1966/1967 (Acta chem. Scand., 1967, 21: 1972) to a pigment isolated by Weeks et al. (Arch. Biochem. Biophys., 1967, 121: 35) from the bacterium *Flavobacterium dehydrogenans*. It was found to be the major carotenoid therein and to have (Jensen, Weeks et al., Acta chem. Scand., 1968, 22: 1171) m.p. 153−155° (red needles from acetone−petroleum), λ_{max} 470, 439, 416 nm (petroleum) indicative of an acyclic nonaene chromophore. Preliminary

---

* The P439 designation is based on the position, in petroleum, of the compound's central peak in the visible absorption spectrum.

structural studies in 1966/1967 (loc. cit.) showed the presence of two hydroxyl groups (chromatographic and partition properties; ready formation of what was shown to be, by polarity, p.m.r., and mass spectrum, a diacetate); precision mass spectrometry on the molecular ion gave a formula of $C_{50}H_{72}O_2$. This C_{50} formulation could not be accounted for by the compound being a normal C_{40} carotenoid carrying, for example, two C_5 sugar units (cf. p. 259) or esterified by a C_{10} fatty acid (xanthophylls frequently occur esterified in nature) since there were no available oxygen atoms to attach such units. It was concluded that the compound must have a C_{50} carbon skeleton, thereby becoming the first example in nature of such a carotenoid. The usual chemical and physical methods were then applied to elucidating the full structure of the compound.

The p.m.r. spectrum showed the usual peaks due to in-chain methyl groups (4 Me at $\sim \tau$ 8.05) and also peaks consistent with the presence of two α-type rings (6H singlets at τ 9.25, 9.05, 8.47); however, there were *additional* peaks at τ 8.32 (6H; 2 $\times$ Me$-$C$=$?) and 5.98 (4H; 2 $\times$ $=$C $\cdot$ CH$_2$OH?), and also increased general absorption in the olefinic-H and the CH$_2$ regions. The latter peaks were similar to those, due to the $=$CMe $\cdot$ CH$_2$OH grouping, seen in lycoxanthin and lycophyll (p. 320). The point of attachment of these "extra" groupings was deduced from the m.s. fragmentation pattern. As mentioned on p. 209, an α-type end-group gives a strong fragment ion at M-56 due to a retro-Diels—Alder cleavage. In the present case, a strong peak appeared at M-140 instead (and at M-182 in the diacetate), thereby strongly suggesting that the two "extra" C_5 units, each terminating in $=$CMe $\cdot$ CH$_2$OH, must be attached to the pigment's end-groups. In addition, the point of attachment must, presumably, be $C_{(2)}$ (and $C_{(2')}$) unless a rather unusual rearrangement had occurred in the mass spectrometer:

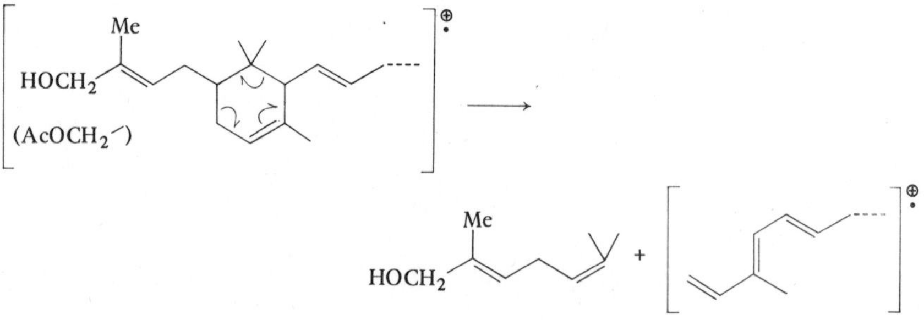

m/e = 140
(182 for the acetate)

This led to the following structure for *dehydrogenans-P439* being proposed by Jensen, Weeks et al. (1967; loc. cit.):

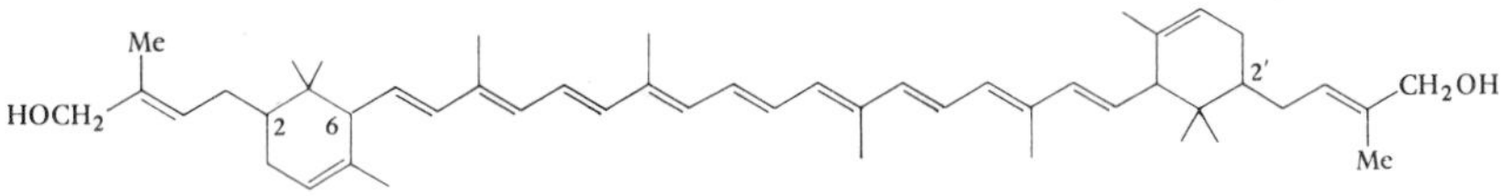

Dehydrogenans-P439; decaprenoxanthin

Full details of the structure determination appeared in 1968 (Jensen, Weeks et al., loc. cit.). The allylic—CH$_2$OH groups could be oxidised by nickel peroxide (cf. p. 216) to give the corresponding dial but, surprisingly, failed to give a methyl ether with MeOH/H$^{\oplus}$; other chemical studies were also reported. Once the full structure had been elucidated the P439 designation was replaced by the name *decaprenoxanthin*, since it contains ten isoprene units (Weeks et al., Nature, 1969, 224: 879).

The configuration about the $>$C=C(Me)CH$_2$OH double bond is unknown but a consideration of the position of the Me group in the p.m.r. spectrum in conjunction with the argument given under lycoxanthin and lycophyll (p. 321) suggests that it is as for those pigments; i.e. as written above. The position observed for the —CHO proton in the p.m.r. spectrum of the above dial, taken in comparison with data from related compounds, supports this (Schwieter and Jensen, Acta chem. Scand., 1969, 23: 1057). P. M. Scopes et al. (J. chem. Soc., C, 1969, 2527) have since measured decaprenoxanthin's o.r.d. curve (in dioxan; 200—400 nm) but the stereochemistry at the 2- and the 6-positions could not be inferred, due to a lack of data on sufficiently closely related compounds of known stereochemistry. Concerning the possible mode of biogenesis of decaprenoxanthin, see the end of this section.

Decaprenoxanthin has also been found, free or in glycoside form, in *Corynebacterium erythrogenes* and in an *Arthrobacter* sp. (see under corynexanthin, below).

Once the above novel structure had been established, several other carotenoids whose structures were unknown were also found to be C$_{50}$ compounds, and other examples were discovered by searching new sources for such compounds. These are listed below. In general, similar methods of structure elucidation were used to those outlined above.

Corynexanthin, a very polar carotenoid first detected in *C. erythrogenes* and another *Corynebacterium* sp. by W. Hodgkiss et al. in 1954; recently re-investigated by Weeks and A. G. Andrewes (Arch. Biochem. Biophys., 1970, 137: 284), who showed it to have the same chromophore as decaprenoxanthin (λ_{max} values) and formula $C_{56}H_{82}O_7$ (m.s.; corresponding to $C_{50}H_{72}O_2 + C_6H_{10}O_5$). Like decaprenoxanthin, it gave an M-140 peak in its mass spectrum, indicating that at least one end of the molecule was of the decaprenoxanthin type, but also showed strong peaks at M-162 and M-178 [corresponding to loss of $C_6H_{10}O_5$ and $C_6H_{10}O_6$; due to cleavage of the (carotenoid)O—glycoside and the carotenoid—O(glycoside) links, respectively]. Acid hydrolysis of the pigment (aqueous HCl/CHCl₃/100°; sealed tube), isolation of the water-soluble component of the reaction mixture, and comparison (t.l.c.) with authentic glucose showed this component, corresponding to the $C_6H_{10}O_5$ unit above, to be glucose. This led to the structure of corynexanthin as being as for decaprenoxanthin but with an —O-glucose unit in place of one of the two —OH groups. A mono- and a di-glycoside of decaprenoxanthin (as well as decaprenoxanthin itself) have since been found in an *Arthrobacter* sp. by Arpin et al. (Acta chem. Scand., 1972, 26: 2524). The sugar unit is probably D-glucose and the linkage of the β type, in each case; the monoglycoside thus closely resembles, and may be fully identical with, corynexanthin. For further examples of carotenoids occurring as glycosides, see Sections 7b (p. 259) and 7c (p. 263).

Weeks and Andrewes (loc. cit.) found that the above bacteria also contain decaprenoxanthin itself and a further C_{50} carotenoid, tentatively identified as deshydroxydecaprenoxanthin (see below).

Deshydroxydecaprenoxanthin. Weeks et al. (Arch. Biochem. Biophys., 1967, 121: 35) found that *Flavobacterium dehydrogenans* contains, in addition to *dehydrogenans*-P439 (above), a series of minor carotenoids; these were designated (cf. footnote p. 309) *dehydrogenans*-P422, *dehydrogenans*-P373, and *dehydrogenans*-P439 mono-OH(?). Preliminary studies by Weeks et al. were followed by a full p.m.r. and mass spectral study (formulae by precision m.s.; comparison of fragmentation patterns with those of decaprenoxanthin and other known compounds) by Weeks, Weedon and co-workers (Nature, 1969, 224: 879). All three pigments had visible absorption spectra showing good fine structure, characteristic of acyclic chromophores, and had λ_{max} positions corresponding to eight, six, and nine conjugated double bonds, respectively. The first two were found to be C_{45} compounds (described below) and the latter, $C_{50}H_{72}O$. A consideration of the latter's fragmentation pattern (presence of M-140, as for decaprenoxanthin, and also M-124, ascribed to retro-Diels—Alder cleavage of a decaprenoxanthin-like

"*Dehydrogenans*-P439 mono-OH" = deshydroxydecaprenoxanthin

end-group lacking the oxygen) and its other spectral properties, led to the structure given below and the pigment was therefore renamed **deshydroxy-decaprenoxanthin**. This compound has since been tentatively identified elsewhere (see under corynexanthin, above).

Bacterioruberin (formerly known as α-bacterioruberin or **bacterioruberin-α**). This pigment, the major carotenoid of the bacterium *Halobacterium salinarium**, was formulated by Jensen (in 1960; Vol. IIB, 2nd Edn., p. 310) as the di-hydroxy analogue of spirilloxanthin (C_{40}, acyclic). However, subsequent comparison of bacterioruberin with an authentic (synthetic) sample of the latter compound showed this suggestion to be incorrect (D. F. Schneider and Weedon, J. chem. Soc., C, 1967, 1686) and a new investigation of the pigment was carried out, this time with the benefit of p.m.r. and mass spectrometry (M. Kelly, S. Norgard and Jensen, Acta chem. Scand., 1967, 21: 2578; 1970, 24: 2169). Thus it immediately became apparent that bacterioruberin was another example of a C_{50} carotenoid ($C_{50}H_{76}O_4$, by precision m.s.). It did not give an acetate but did react with the silylating reagent (Me_3SiCl–pyridine–hexamethyldisilazane; cf. p. 199) to give a product having $M^{\oplus}$ 1028, corresponding to a tetra-trimethylsilyl ether (hence four *tert*-hydroxyls). The p.m.r. spectrum had the simplicity expected of a symmetrical structure. It showed six in-chain methyl groups (τ 8.03–8.09), four Me_2CO– methyls at τ 8.77 and another four at τ 8.82 (thereby *excluding* structures with end-groups of the type wherein the "extra" C_5 units are attached at $C_{(1)}$, $C_{(1')}$:

$$Me_2C(OH) \cdot CH_2 \cdot CH_2 - CH_2 \cdot C_{(1/1')}Me(OH) \cdot CH_2 \cdot CH = \cdots)$$

Treatment with $POCl_3$/pyridine gave, by the expected dehydration at the *tert*-hydroxyl positions, a mixture, separable by chromatography, of all the four possible dehydration products: loss of water occurring firstly from the $Me_2C(OH)$-groups of the "extra" C_5 units, and only after this at $C_{(1)}$ and $C_{(1')}$. This followed from the chromophore length of each product as inferred from its visible absorption spectrum and an examination of its m.s. fragmentation pattern (cf. the anhydrobacterioruberins mentioned below):

Bacterioruberin

---

* Cf. Jensen and co-workers (1960, 1970) for method of isolation; and Gochnauer et al. (Arch. Mikrobiol., 1972, 84: 339) for the effect of altering culture conditions on bacterioruberin production by other *Halobacterium* spp.

End-groups carrying *tert*-OH groups at $C_{(1)}$ and/or $C_{(1')}$ are common in carotenoids of the normal (C_{40}) type found in bacteria (cf. Vol. IIB, 2nd Edn., p. 307ff). The asymmetry at $C_{(2)}$,$C_{(2')}$ gives a c.d. effect, and the curve has been recorded by G. Borch et al. (Acta chem. Scand., 1972, 26: 402).

Bacterioruberin, a derived *monoglycoside*, and what is probably the corresponding (symmetrical) *diglycoside*, have been found in another halophilic (salt-loving) bacterium by Arpin et al. (ibid., 1972, 26: 2526). The sugar units of the glycosides are a mixture of glucose (mainly) and mannose, and appear (m.s. fragmentation pattern) to be attached to the "extra" C_5 unit(s) rather than at $C_{(1)}$,$C_{(1')}$.

β-Bacterioruberin, one of several pigments of unknown structure isolated by various groups from *H. salinarium* or related organisms (Jensen, 1960, loc. cit.; Kelly et al., loc. cit.). It may be merely a *cis*-isomer of bacterioruberin (R. M. Baxter, Canad. J. Microbiol., 1960, 6: 417).

Monoanhydrobacterioruberin. Kelly, Norgard and Jensen (1970, loc. cit.) found the bacterioruberin in *H. salinarium* to be accompanied by small amounts of other, less polar C_{50} carotenoids. One of these gave a tritrimethylsilyl ether (only) [from m.s.: $M^{\oplus}$ corresponding to $C_{50}H_{71}$-$(OSiMe_3)_3$] and exhibited p.m.r. peaks at τ 8.34, 8.39 (=CMe_2) and was shown, by direct comparison, to be identical with the first product formed in the bacterioruberin dehydration sequence alluded to above.

Monoanhydrobacterioruberin

Bisanhydrobacterioruberin, a further minor carotenoid in the above-mentioned bacterium (Kelly, Norgard and Jensen, 1970, loc. cit.) and a major carotenoid ("*C.p. 496*") in *Corynebacterium poinsettiae** (from which it was isolated crystalline, m.p. 170°, λ_{max} 527, 493, 465 nm, in petroleum) (Norgard, A. J. Aasen and Jensen, Acta chem. Scand., 1970, 24: 2183), was shown to be identical with the second-formed dehydration product mentioned under bacterioruberin, and its structure was deduced from p.m.r.,

---

* The carotenoids of this bacterium were studied some years ago by M.P. Starr et al. but most of their tentative identifications were subsequently proved to be incorrect (cf. Norgard et al., 1970).

m.s., dehydration, and silylation studies (Kelly, Norgard and Jensen, 1970; Norgard, Aasen and Jensen, 1970):

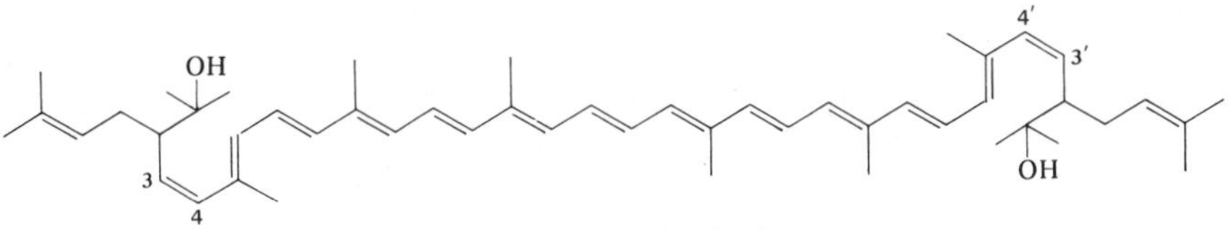

Bisanhydrobacterioruberin

The c.d. curve (of a sample *ex C. poinsettiae*) has been recorded by Borch et al. (1972, loc. cit.). It was similar to that obtained with bacterioruberin, showing that the configuration at $C_{(2)}, C_{(2')}$ is the same as in bacterioruberin (but still unknown, due to a lack of data on related compounds of known stereochemistry).

Pigment C.p. 470* (3,4,3',4'-tetrahydrobisanhydrobacterioruberin ?), a minor carotenoid in *C. poinsettiae* with a markedly shorter chromophore than the above compounds (λ_{max} values). It has been tentatively identified (m.s., λ_{max} values, silylation studies, and by analogy with the preceding structure) as the 3,4,3',4'-tetrahydro-derivative of the preceding carotenoid (Norgard, Aasen and Jensen, 1970, loc. cit.).

Pigment C.p. 473*, a further carotenoid from *C. poinsettiae*, having one acetylatable hydroxyl and one *tert*-hydroxyl (silylation) and λ_{max} 504, 473, 446 nm (petroleum). A full p.m.r., i.r., and m.s. study led to the following suggested structure (Norgard, Aasen and Jensen, loc. cit.). The main doubt concerns the point of attachment of the "extra" C$_5$ unit at the β-ring end. In decaprenoxanthin, the M-140 ion in the mass spectrum, from a retro-Diels–Alder cleavage of the α-ring, provided the clue, but such rearrangements rarely occur in compounds containing β-end groups as present here. However, the p.m.r. τ values observed for this end-group, and analogy with the above compounds, suggested the $C_{(2)}$ position, as shown:

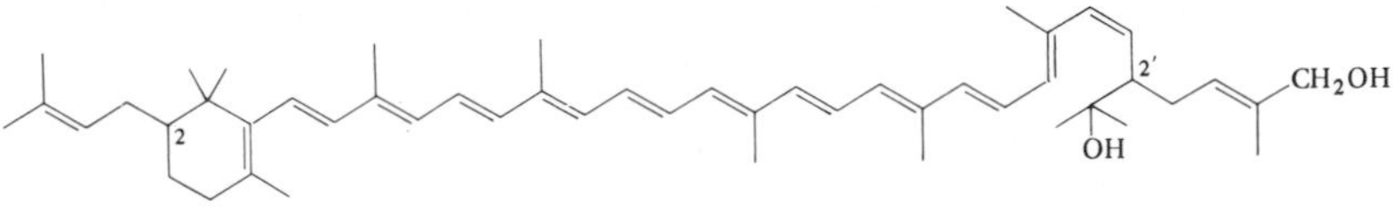

C.p. 473 (probably)

---

* These designations are based on the position of the central peak in the visible absorption spectrum (petroleum).

Pigment C.p. 450*, from *C. poinsettiae*, having a polarity and visible absorption spectrum very similar to those of zeaxanthin, but shown to be a C_{50} compound by mass spectrometry. The presence of two allylic primary hydroxyl groups was inferred from p.m.r. (4H at τ 5.97) and acetylation studies; no *tert*-hydroxyl was detected by silylation. The other p.m.r. signals were as found for β-carotene but with additional peaks attributable to a =CMe$_2$ grouping, leading to the proposed structure (Norgard, Aasen and Jensen, loc. cit.):

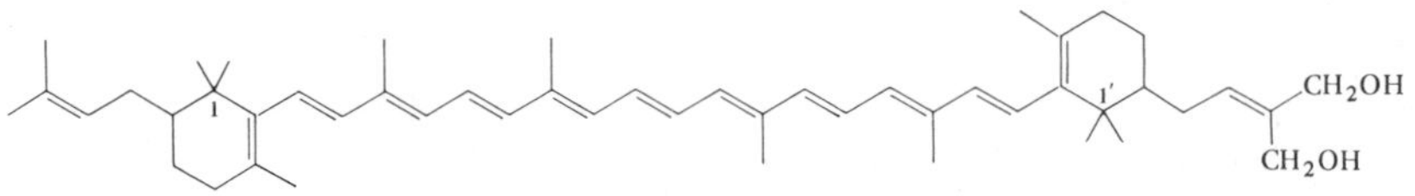

C.p. 450 (probably)

The point of attachment of the "extra" C_5 units is again uncertain, resting mainly on analogy and biosynthetic reasoning (cf. p. 318).

Sarcinaxanthin, from the bacterium *Sarcina lutea* (W. Hodgkiss et al., J. gen. Microbiol., 1954, 11: 438; 2nd Edn., p. 331), has recently been recognised to have the same absorption spectrum and formula ($C_{50}H_{72}O_2$: precision m.s.) as decaprenoxanthin (Jensen, Weeks et al., Nature, 1967, 214: 379), and was tentatively identified with that compound by the latter authors. However, a re-investigation by Jensen et al. (Pure and Applied Chem., 1969, 20: 428; Acta chem. Scand., 1970, 24: 1460) showed that the characteristic M-140 peak in the fragmentation pattern of decaprenoxanthin was not present in the sarcinaxanthin spectrum, and they tentatively re-formulated sarcinaxanthin as shown below:

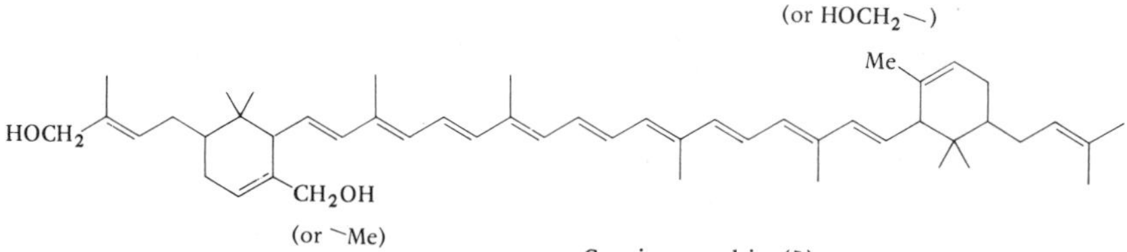

Sarcinaxanthin (?)

A *mono-glycoside* (the mono-D-glucoside) of a C_{50} diol, possibly sarcinaxanthin, has also been isolated from *S. lutea*. The m.s. fragmentation pattern of this glycoside showed that it has the two oxygen functions at *opposite* ends of the molecule (cf. the two suggested sarcinaxanthin structures above) (Jensen et al., 1970, loc. cit.).

---

* Cf. footnote p. 315.

S. flava contains sarcinaxanthin (Jensen, Weeks et al., 1967, loc. cit.) and at least one other C_{50} carotenoid, $C_{50}H_{70}O$ by m.s. (D. Thirkell, R. H. C. Strang and J. R. Chapman, J. gen. Microbiol., 1967, 49: 157; the C_{50} diol also reported was, presumably, mainly sarcinaxanthin).

Thus C_{50} carotenoids appear to be of relatively common occurrence in bacteria (see refs. above and Weeks and Andrewes, Arch. Biochem. Biophys., 1970, 137: 284).

C_{45} carotenoids. During the above searches for examples of C_{50} carotenoids in nature, several C_{45} compounds were discovered. In all cases so far examined, one end-group is of the ordinary C_{40} type and the other carries an "extra" C_5 unit of the type present in the C_{50} compounds.

Pigment C.p. 482*, m.p. 153°, is one of a series of carotenoids (the others were mainly of the C_{50} type, see above) isolated from the bacterium *Corynebacterium poinsettiae* by Jensen and co-workers (Acta chem. Scand., 1969, 23: 1463; 1970, 24: 2183). The following structure was inferred from its absorption spectrum (acyclic dodecaene chromophore), mass spectrum ($M^{\oplus}$ corresponding to $C_{45}H_{64}O$), p.m.r. data (twelve methyl signals at positions expected; 2nd Edn., p. 257), and acetylation/silylation studies (cf. techniques used for C_{50} carotenoids). These results do not prove unequivocally the position at which the "extra" C_5 and the $Me_2 C(OH)$- units join but this is thought, on biosynthetic grounds, to be as shown:

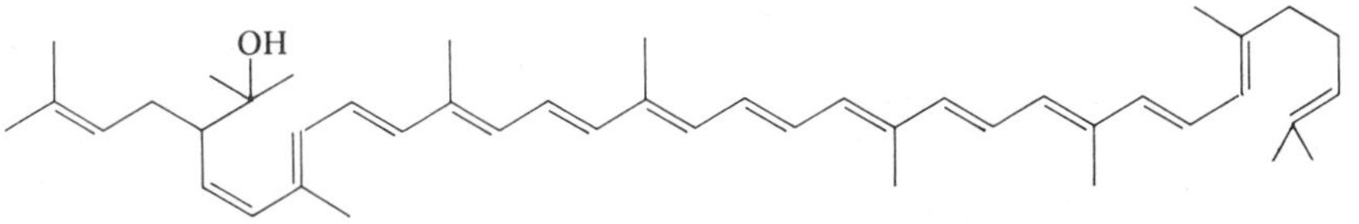

C.p. 482 (probably) (= "2-isopentenyl-3,4-dehydrorhodopin")

Dehydrogenans-P373 (nonaprenoxanthin) and **dehydrogenans-P422***, two of the minor carotenoids found along with deshydroxydecaprenoxanthin (q.v. for details; C_{50}) and decaprenoxanthin (also C_{50}) in *Flavobacterium dehydrogenans*. Their absorption spectra showed them to have acyclic hexaene and octaene chromophores respectively and precision mass spectrometry gave formulae $C_{45}H_{68}O$ and $C_{45}H_{66}O$ respectively. Their chromatographic properties implied the presence of one hydroxyl group. Comparison of the fragmentation pattern of the former with that of decaprenoxanthin and phytoene suggested it has a structure which is a hybrid of these two molecules [significant peaks at M-140, from the decaprenoxanthin end (cf. p. 310) and at M-205, by cleavage of the doubly allylic $C_{(11')}-C_{(12')}$

---

* Cf. footnote p. 315.

bond; cf. p. 209]. It was named *nonaprenoxanthin* (since it has nine isoprene units) (Weeks, Weedon and co-workers, Nature, 1969, 224: 879):

Dehydrogenans-P373; nonaprenoxanthin

The mass spectrum of the second compound exhibited strong peaks at M-140 (cf. above) and M-137 (loss of $C_{10}H_{17}$; corresponding to cleavage of the doubly-allylic $C_{(7')}-C_{(8')}$ bond) and, in conjunction with the data noted above, was formulated as 11',12'-dehydrononaprenoxanthin.

Biogenesis of C_{50} (and C_{45}) carotenoids

Weeks and R. J. Garner (Arch. Biochem. Biophys., 1967, 121: 35) found that the decaprenoxanthin in *F. dehydrogenans* (p. 309) is accompanied, particularly under conditions of limited aeration, by a series of more saturated compounds with absorption spectra very similar to those of the well-known C_{40} carotenoid precursors, phytoene, phytofluene, ζ-carotene, and neurosporene (cf. p. 335). Because it was recognised that there was a possibility that they were in fact the C_{50} versions of phytoene etc., they were designated "*dehydrogenans*-phytoene", "*dehydrogenans*-phytofluene", "*dehydrogenans*-ζ-carotene", etc. However, Weeks, Weedon et al. (Nature, 1969, 224: 879) have since shown (comparison of m.s. fragmentation patterns and of t.l.c. behaviour) that at least two, and probably all four, of these "*dehydrogenans*"-compounds are the normal C_{40} carotenoid precursors. This suggests that in *F. dehydrogenans* at least (and conceivably in other systems producing C_{50} carotenoids), the normal Porter–Lincoln pathway (phytoene → phytofluene → ζ-carotene → neurosporene → lycopene; cf. p. 338) operates at least up to the ζ-carotene stage, and that alkylation by dimethylallyl pyrophosphate ($C_{(5)}$; Vol. IIB, 2nd Edn., p. 341) then occurs*, with concomitant ring closure in certain cases (cf. the $H^{\oplus}$-triggered ring-closure reaction thought to be responsible for the formation of cyclic end-groups in the normal (C_{40}) compounds; p. 353, Scheme 14):

---

* The fact (p. 315) that two of the C_{50} carotenoids give c.d. curves, due to asymmetry at $C_{(2)},C_{(2')}$, indicates the C_5 units are added stereospecifically (at least in those compounds).

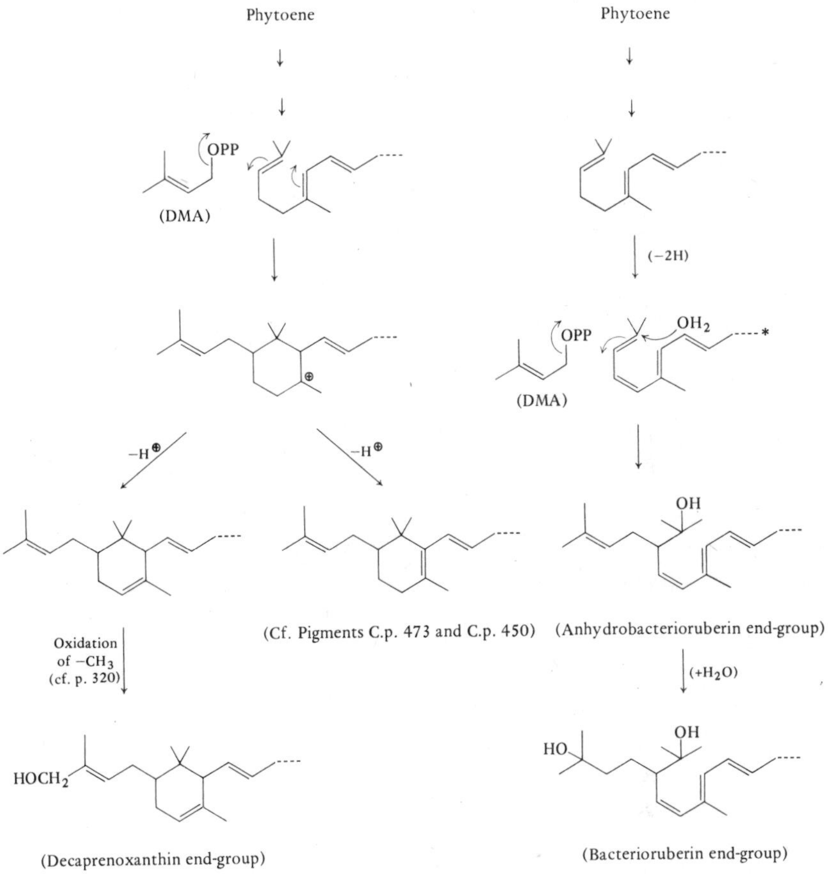
Phytoene
Phytoene
OPP
(DMA)
(−2H)
OPP
OH_2
(DMA)
$-H^{\oplus}$
$-H^{\oplus}$
OH
(Cf. Pigments C.p. 473 and C.p. 450)
(Anhydrobacterioruberin end-group)
Oxidation of $-CH_3$ (cf. p. 320)
$(+H_2O)$
$HOCH_2$
HO
OH
(Decaprenoxanthin end-group)
(Bacterioruberin end-group)

Notes: −OPP = pyrophosphate group; DMA = dimethylallyl pyrophosphate: see text.

* Cyclisation prevented by its being rigid and straight and so unable to coil like the other end-group? (real shape of a fully-unsaturated end-group:

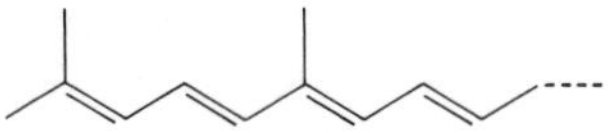

not as drawn above; cf. footnote p. 217).

(i) Carotenoids with partially oxidised methyl groups

Until recently almost all carotenoid structures elucidated had all ten skeletal methyl groups intact. Of late, however, several structures containing one or more of these groups oxidised to the $-CH_2OH$ level have come to light and most are discussed here. Other carotenoids possessing this feature but referred to elsewhere include vaucheriaxanthin (allenic, p. 297) and certain of the C_{50} carotenoids (Section 7h, p. 309). Those compounds in which oxidation through to the $-CHO$ or $-CO_2H$ level has occurred (e.g. 16'-oxo-torulene and torularhodin, p. 284) are referred to in Sections 7d(i) (p. 268) and 7e (p. 284) respectively.

Lycoxanthin and **lycophyll**. The tentative structures originally assigned to these carotenoids (Vol. IIB, 2nd Edn., p. 306) had been inferred from rather restricted structural studies in 1936. They had the hydroxyl group(s) placed at the $C_{(3),(3')}$ position(s) mainly by analogy with the position of the hydroxyl-groups in most of the then known carotenoids. These structures have been revised. Thus, in 1968, Cholnoky, Szabolcs and Waight (Tetrahedron Letters, 1968, 1931) and M. C. Markham and Jensen (Phytochem., 1968, 7: 839) independently re-isolated samples of the two pigments and subjected them to examination by modern physical techniques and to additional chemical tests. Mass spectrometry confirmed the molecular formulae (cf. p. 208) and the presence of one and two hydroxyl groups respectively (M-18 and M-18 plus M-36 peaks, respectively). The p.m.r. spectra were particularly revealing. Thus the lycoxanthin spectrum showed only nine methyl groups, four of the in-chain ($\sim\tau$ 8.0) and two of the acyclic end-of-chain type ($\sim\tau$ 8.2), a pair of isopropylidene methyl groups, as in lycopene (τ 8.39, 8.32), and a further single methyl signal at τ 8.32 (=ĊMe). In addition there was a 2H peak at $\sim\tau$ 6.0, consistent with there being a =Ċ $\cdot$ CH$_2$OH group in lycoxanthin. This assignment was confirmed by the marked downfield shift of the peak (to $\sim\tau$ 5.5) on acetylation, as is usual for $-CH_2OH \rightarrow -CH_2OAc$. Nickel peroxide oxidation (specific for allylic-OH) gave a product with decreased polarity and two less hydrogens (m.s.) but with an unchanged visible absorption spectrum. This confirmed the presence of a C=C $\cdot$ CH$_2$OH grouping and showed it to be isolated from the main chromophore of the molecule. Lycophyll gave a similar result on oxidation and its p.m.r. spectrum was also similar to that of lycoxanthin, except that the peaks assigned in the lycoxanthin spectrum to the $-CH=CMe \cdot CH_2OH$ end-group were, relatively, twice as intense suggesting the presence of *two* such groupings in lycophyll. Neither lycoxanthin nor lycophyll showed a

λ_{max} shift on treating with $H^{\oplus}/CHCl_3$ in the cold (test for $-OH$ allylic to the chromophore) but, on vigorous treatment, two compounds tentatively identified as mono- and bis-dehydrolycopenes (Vol. IIB, 2nd Edn., p. 280) were obtained, in agreement with the following structures:

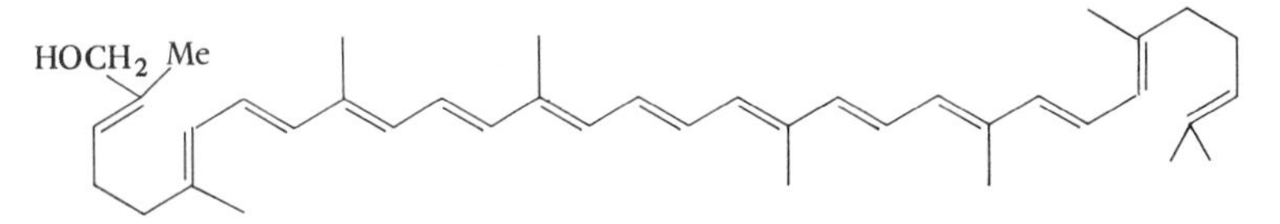

Lycoxanthin

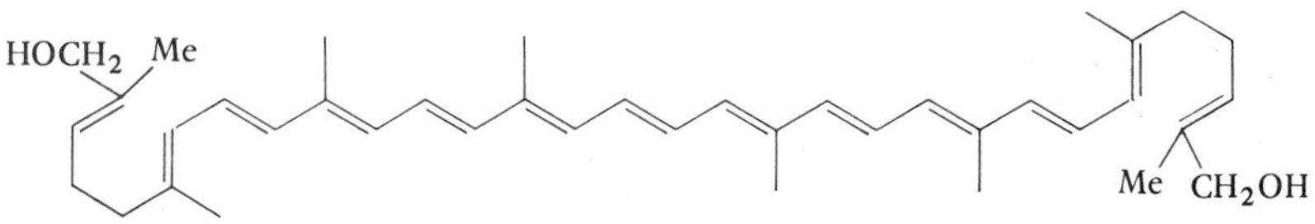

Lycophyll

Concerning the configuration of the substituents about the double bond in the hydroxylated isopropylidene group(s); a consideration of the position of the methyl group in the p.m.r. spectrum (τ 8.32) in comparison with the τ value of allied compounds (8.39 and 8.32 for the methyl groups *trans* and *cis* respectively to the olefinic-H in a $HRC=CMe_2$ grouping) and of the deshielding influence that a nearby oxygen usually has, suggests that the configuration is as shown*.

Synthesis: By attaching appropriate synthetic C_{10} units to crocetindial (cf. p. 267) by, respectively, a two-stage Wittig reaction (lycoxanthin) and a one-stage (2 : 1) Wittig reaction (lycophyll), the terminal oxygen function(s) being in the $-CO_2Me$ form until the final (LiAlH$_4$-reduction) step (Kjøsen and Jensen, Acta chem. Scand., 1972, 26: 4121).

Rhodopin-20-ol (*Rhodopinol*), previously called "*warmingol*" (K. Schmidt et al., Arch. Mikrobiol., 1965, 52: 132) or "*Pigment 3*" (Jensen and Schmidt, ibid., 1963, 46: 138). This was the first carotenoid recognised (Aasen and Jensen, Acta chem. Scand., 1967, 21: 2185) to have one of the in-chain methyl groups oxidised to the $-CH_2OH$ level. For its discovery in *Chromatium warmingii* (Jensen and Schmidt, 1963, loc. cit.) and for subsequent further sources, see under rhodopin-20-al (p. 268) with which it

---

* Since confirmed by p.m.r. studies with model compounds and by total synthesis, the stereochemistry of the terminal group(s) being checked at each step by having them in the $-CH=C(Me)CO_2Me$ form until the final (hydride-reduction) step [the $-CH=$ proton of $-CH=C(Me)CO_2Me$ occurs at $\sim\tau$ 3.3 if *cis* to the $-CO_2Me$ but at $\sim\tau$ 4.1 if *trans*] (cf. Kjøsen and Jensen, Acta chem. Scand., 1972, 26: 4121).

co-occurs. It occurs in nature mainly as a *cis*-isomer (neo-B), and has m.p. ca. 182° (after chromatography and repeated precipitation using acetone + petrol); λ_{max} 495, 466, (440) (+ *cis*-peaks ~355) nm in acetone (Aasen and Jensen, 1967, loc. cit.). The proposed structure is based mainly on the finding (cf. Aasen and Jensen) that it is, apparently, identical with the compound obtained on $LiAlH_4$ reduction of the above-mentioned rhodo-pinal. The mass-spectral molecular weight (570) and fragmentation pattern of the pigment lent support to this proposed structure (Francis and Jensen, ibid., 1970, 24: 2705), as did p.m.r., i.r., and polarity data and the results of a number of chemical tests (acetylation, NiO_2 oxidation, etc.) (Aasen and Jensen, loc. cit.):

Rhodopin-20-ol

The unusual stability of a *cis*-configuration in the corresponding rhodopinal (cf. above) is retained to some degree in this compound. The position the *cis*-bond occupies in the naturally-occurring compound is uncertain but presumed to be $C_{(13)}$ (or 12-*s*-*cis*) as for the rhodopinal.

The "*lycopenol*" detected in a *Lamprocystis* sp. by Jensen et al. in 1968 is a further example of the above type of compound (structure as above but MeO— in place of HO—?) (cf. Francis and Jensen, 1970, loc. cit., p. 2710 for details).

Loroxanthin. Numerous chromatographic studies over the years have shown that whereas most green algae (Chlorophyceae) consistently produce the same (three) major xanthophylls as the green parts of "higher" plants, certain members contain an additional compound (see Aitzetmüller, Strain, W. A. Svec and co-workers, Phytochem., 1969, 8: 1761 for refs.). In 1969, the latter group isolated samples of this compound, which they named loroxanthin, from the algae *Chlorella vulgaris* (readily accessible and a good source) and *Scenedesmus obliquus*. It had λ_{max} 474, 446 nm (in ethanol: curve identical with that of lutein, p. 249, suggesting the same chromophore present), was about as polar as neoxanthin (which could, therefore, have obscured it in certain of the earlier studies on the algae), and was found to have the formula $C_{40}H_{56}O_3$ (precision m.s.). It gave a triacetate with acetic anhydride and a di-methyl ether with $MeOH/H^\oplus$, indicating three non-tertiary hydroxyls present two of which being allylic. Of the two allylic hydroxyls, one was found to be primary [p.m.r. signal at ~τ 5.5

$(=\overset{|}{C} \cdot CH_2OH)$ which moved downfield on acetylation, as for lycoxanthin; oxidation with p-chloranil gave an aldehyde (i.r., p.m.r. data), λ_{max} 480 nm in ethanol]. Comparison of the p.m.r. spectrum with that of lutein and an examination of the fragments obtained on mass spectrometry of loroxanthin and the above derivatives indicated that each end-group carries an hydroxyl (one methylatable, one not), these being at $C_{(3')}$ at the α-type end (since known to be allylic) and probably at $C_{(3)}$ at the β-end. The p.m.r. spectrum also showed that loroxanthin contains only nine methyl groups, six of which roughly corresponded in τ value to the end-group methyl groups of lutein whilst the other three were close to the position expected for in-chain methyl groups (cf. Vol. IIB, 2nd Edn., p. 257). This confirmed that the

(A)   (B)

$=\overset{|}{C} \cdot CH_2OH$ group in loroxanthin is an oxidised in-chain methyl. This was also revealed by the m.s. fragmentation patterns of loroxanthin and its triacetate wherein the usual carotenoid M-106 ion (loss of m-xylene from the polyene chain) was weaker than usual and new ions were seen at M-122 and M-164, respectively (corresponding to loss of m-methyl-benzyl alcohol and acetate, respectively). These patterns also suggested that the oxygen function was at $C_{(9)}$ or $C_{(9')}$ rather than at the more central $C_{(13)}$ or $C_{(13')}$ locations. In the di-methyl ether mass spectrum, there was an ion $C_{27}H_{36}O$ (by precise mass determination), corresponding to loss of $C_{15}H_{24}O_2$; the latter's composition indicated it must contain both the unmethylated (i.e. non-allylic) hydroxyl and one of the methylated hydroxyls (presumably that originating from the in-chain $-CH_2OH$). Thus the C_{15} fragment would have structure A (above) and the in-chain $-CH_2OH$ would be at $C_{(9)}$. However, m.s. data can be misleading (unsuspected in-the-source rearrangements or demethylations) and proof of the structure of loroxanthin had to await the degradative work of Britton, Goodwin and co-workers (Phytochem., 1970, 9: 2545). Thus permanganate degradation of loroxanthin (ex *S. obliquus*) triacetate yielded a series of aldehyde fragments (corresponding to cleavage of various double bonds in the polyene chain) which were separated (t.l.c.), saponified, reduced (borohydride), and subjected to mass spectrometry (a) as they stood and (b) after treatment with MeOH/H$^{\oplus}$ (as a means of detecting, and estimating the number of, allylic hydroxyls in each molecule). Thus it was shown, for example, that one of the aldehydes isolated had structure (B).

These results confirm the structure for loroxanthin:

Loroxanthin

For other sources of loroxanthin, cf. Aitzetmüller, Strain, Svec et al., 1969, loc. cit. and Ricketts, ibid., 1971, 10: 161, and for a *partial synthesis* from the carotenoid siphonaxanthin, see below.

Siphonaxanthin (classified as of uncertain constitution in Vol. IIB, 2nd Edn., p. 331). Whereas most green algae contain a fairly constant spectrum of xanthophylls, Strain found that those of the order Siphonales contain, in relatively large amounts, an additional carotenoid, siphonaxanthin, or the derived ester *siphonein* (Strain in "Manual of Phycology", ed. G. M. Smith, Chronica Botanica, Waltham, Mass., 1951, p. 243). More recently, H. Kleinig (J. Phycol., 1969, 5: 281) has carried out a further survey of the occurrence of siphonaxanthin and siphonein, and has discussed the *chemotaxonomic* significance of their presence or otherwise in the various Siphonales species studied (the pigment composition of an alga is sometimes of value in classifying it in cases where the more traditional botanical criteria give an ambiguous indication: cf. p. 195).

Small-scale experiments on siphonaxanthin (λ_{max} 448 nm, in ethanol), using the change in t.l.c. R_f values and in visible absorption spectrum to follow changes in structure (Kleinig and Egger, Phytochem., 1967, 6: 1681), indicated the presence of a linear octaeneone chromophore [from lack of spectral fine structure in ethanol; cf. Vol. IIB, 2nd Edn., p. 254); hydride reduction yielded a compound showing good spectral fine structure (showing that its octaene chromophore did not extend into a ring; cf. Vol. IIB, 2nd Edn., p. 253) and a λ_{max} shift of ~27 nm, as expected for $(C=C)_n \cdot CO- \rightarrow (C=C)_n \cdot CHOH-]$, a primary and probably two secondary hydroxyl groups, and no conjugated epoxide ($H^\oplus$ test). The marked shift seen in the absorption maximum (448 → 455 nm; ethanol) on selectively acetylating the primary hydroxyl was taken to indicate close proximity (H-bonding destroyed on acetylation) of this hydroxyl to the carbonyl group. Kleinig and Egger proposed a structure on this evidence which was later revised to that below when they found (Kleinig, Nitsche and Egger,

Tetrahedron Letters, 1969, 5139) that siphonaxanthin could be converted into loroxanthin (identity confirmed by direct comparison with an authentic sample) by the sequence: acetylation to the triacetate (to protect the hydroxyl groups); $NaBH_4$ reduction of the C=O group; dehydration ($POCl_3$ —pyridine); and removal of the acetate groups by alkaline hydrolysis.

Siphonaxanthin

This structure was confirmed by Britton, Goodwin and co-workers (1970, loc. cit.) by confirming the tentative structure for loroxanthin (see above) on which the siphonaxanthin structure was based and by carrying out a thorough investigation on siphonaxanthin, its di-methyl ether (from $MeOH/H^{\oplus}$ treatment), and the derived acetates by p.m.r. (only three in-chain methyl groups; 2H peak at τ 5.54, ascribed to $=\overset{|}{C} \cdot CH_2OH$), and mass spectrometry (ions due to cleavage α to the keto-group; M-56 ions characteristic of an α-ionone ring). The same authors also converted siphonaxanthin into loroxanthin in two steps by treating the borohydride reduction product of siphonaxanthin with $H^{\oplus}/CHCl_3$. The unusual feature of a carbonyl group at $C_{(8)}$ is reminiscent of the fucoxanthin structure.

Siphonein has λ_{max} 455 nm (ethanol) (Kleinig and Egger, 1967, loc. cit.) as for siphonaxanthin monoacetate rather than as for siphonaxanthin itself, suggesting that it is the (primary) hydroxyl group adjoining the C=O of siphonaxanthin which is esterified in siphonein. This was supported by the observed lower reactivity towards acetylation of the remaining hydroxyl groups in ·siphonein as compared with the primary hydroxyl in siphonaxanthin (and confirmed by Ricketts' observation that the 2H peak mentioned above is 0.6 τ lower in siphonein, indicating the adjoining hydroxyl is acylated: Phytochem., 1971, 10: 155). The esterifying acid in the siphonein from Kleinig and Egger's source was apparently dodecanoic (lauric) acid, and that from Britton, Goodwin et al.'s source was a mono-unsaturated analogue of same.

Two of the xanthophylls ("KI" and "KIS") isolated from various members of the Prasinophyceae (algae) by Ricketts in 1967/70 have since been identified as siphonein and siphonaxanthin, respectively, and hence constitute a new source (ibid., 1970, 9: 1835; 1971, 10: 161); and an alga studied by Weber and Czygan (Arch. Mikrobiol., 1972, 84: 243) is, apparently, rich in (unesterified) siphonaxanthin.

Pyrenoxanthin, isolated from *Chlorella pyrenoidosa*, a green alga, by H. Y. Yamamoto, H. Yokoyama and H. Boettger (J. org. Chem., 1969, 34: 4207), and probably identical with a polar xanthophyll previously detected therein by Allen et al. (J. gen. Microbiol., 1960, 23: 93), was found to have m.p. 148–149° (from ether–petroleum), λ_{max} 472, 448, 420 nm (petroleum) (similar to lutein, p. 249), and formula $C_{40}H_{56}O_3$ (m.s.). The presence of an in-chain $-CH_2OH$ group was inferred from the p.m.r. spectrum (τ 5.46: cf. above) and by NiO_2 oxidation, which gave a polyene aldehyde. The smallest fragment containing two oxygen atoms, seen in the mass spectrum was $C_{20}H_{27}O_2$ (formula by precision m.s.), indicating that the in-chain $-CH_2OH$ must be at a position remote from either end-group. It was suggested that the compound probably has one of the following structures and that it might be identical with Krinsky's "trollein-like carotenoid" (p. 334):

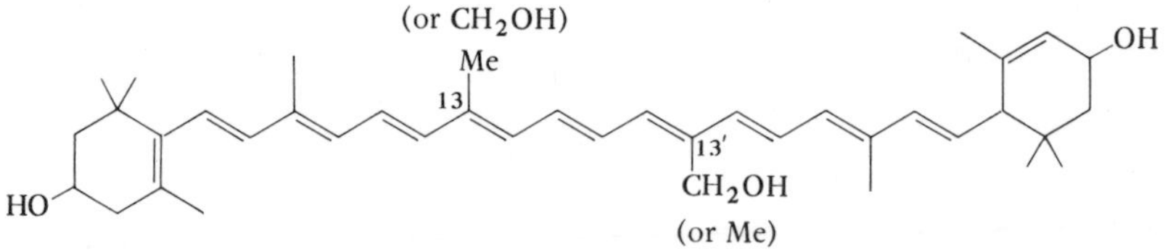

16′-Hydroxytorulene, the C_{40} polyene alcohol corresponding to torulene (Vol. IIB, 2nd Edn., p. 280) was, probably, detected by R. Bonaly and J. P. Malenge (Biochim. Biophys. Acta, 1968, 164: 306) in two *Rhodotorula* yeasts.

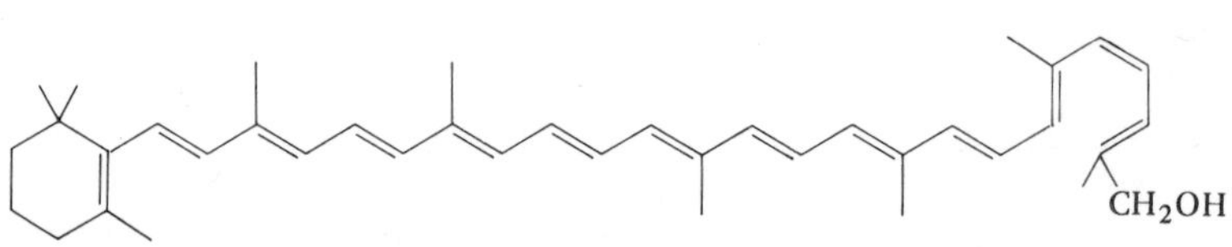

16′-Hydroxytorulene

The corresponding $-CH_3$ (torulene), $-CHO$ (16′-oxotorulene), and $-CO_2H$ (torularhodin) compounds (Vol. IIB, 2nd Edn., pp. 280, 313, 325 for structures) were also detected in the organism and are presumably linked biogenetically: $R-CH_3 \rightarrow R-CH_2OH \rightarrow R-CHO \rightarrow R-CO_2H$.

The oxidation of a terminal $-CH_3$ group to $-CH_2OH$, and subsequently to $-CHO$ and $-CO_2H$, may represent the first step in the oxidative degradation of C_{40} carotenoids. This apparently occurs in certain organisms, and leads to compounds such as neurosporaxanthin containing fewer than 40 carbon atoms (Vol. IIB, 2nd Edn., p. 326).

(j) Nor-carotenoids: C_{38} and allied compounds

Actinioerythrin is the main carotenoid of a variety of the sea anemone (*Actinia equina*) and responsible for its red or mauve colour. It was first isolated by Lederer and co-workers (Bull. Soc. chim. Biol., 1934, 16: 105) and shown by I. M. Heilbron et al. (Biochem. J., 1935, 29: 1384) to be converted, on saponification in air, into a blue crystalline compound which they named *violerythrin*. Recently S. Hertzberg, S. L. Jensen and co-workers (Acta chem. Scand., 1969, 23: 3290) have carried out a thorough examination of actinioerythrin, which included an extensive series of chemical interconversions with spectroscopic methods being used to elucidate structures. Their sample of the pigment had m.p. 91°, λ_{max} 529, 496, (470) (petroleum); (550), 518 nm (chloroform). Examination of actinioerythrin itself was complicated by its occurring as a mixture of esters and much of the structural work was carried out on violerythrin (m.p. 236°; λ_{max}, one broad band, at 580 nm in chloroform). Violerythrin gave a bis-quinoxaline derivative with *o*-phenylenediamine, indicating two α-diketone groups present, and had M 564 (m.s.; corresponding to $C_{38}H_{44}O_4$), and ν_{max} 1675 (conj. C=O) and 1750 cm^{-1} (suggestive of a 5-ring ketone). The C_2 deficiency could not be accounted for by there being two less methyl groups than usual, since ten were visible in the p.m.r. spectrum of actinioerythrin (four in-chain at $\sim\tau$ 8.0, two end-of-chain at $\sim\tau$ 8.1, and four others at $\sim$8.6 and 8.8, presumably 2 x *gem*-Me). This, in conjunction with the i.r. spectrum and the obvious resemblance of the above-mentioned reaction of actinioerythrin with alkali in the presence of air to the astaxanthin → astacene conversion under similar conditions (Vol. IIB, 2nd Edn., p. 320), led to the following suggestion for the structure of actinioerythrin (violerythrin being the corresponding bis-α-diketone):

Actinioerythrin (a 2,2′-bis-*nor*-astaxanthin derivative)

This was confirmed by a series of chemical interconversions (stepwise reduction with $NaBH_4$; acetylation; oxidation; etc.) a few of which are illustrated below (Scheme 7). Thus on prolonged treatment with borohydride (or briefly, with $LiAlH_4$), the corresponding tetra-ol, *violerythrol*, was obtained as a separable mixture of isomers, presumably due to the new hydroxyls (at $C_{(4)}$ and $C_{(4')}$) being introduced either *cis* or *trans* to the hydroxyls already present (at $C_{(3)}, C_{(3')}$) thereby producing compounds of differing polarities due to varying H-bonding effects. This on selective oxidation of the allylic hydroxyl groups (air/I_2 catalyst/hv) gave *actinioerythrol*, the parent carotenoid corresponding to actinioerythrin. By *brief* treatment of actinioerythrin with borohydride, the reduction could be stopped at the intermediate stage (reduction of conj. C=O only) and hydrolysis then yielded the esterifying fatty acids (shown by g.l.c.-mass spectrometry, of the methyl esters, to be mainly the saturated unbranched C_{10}, C_{11}, and C_{12} acids):

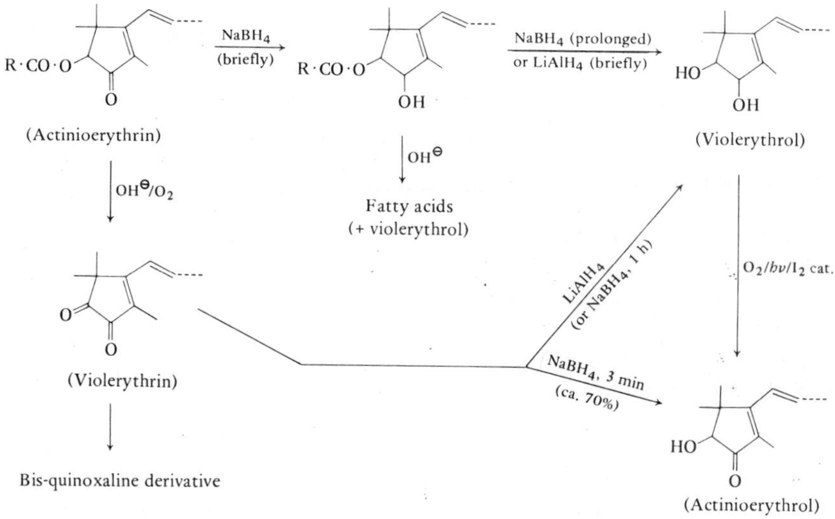

Scheme 7.

(R = various fatty-acid chains)

Actinioerythrin has since been found in another sea anemone (Jensen et al., Acta chem. Scand., 1970, 24: 3055).

Synthesis of actinioerythrol (and of violerythrin). R. Holzel, A. P. Leftwick and Weedon (Chem. Comm., 1969, 128), found that on treating astacene* (which has already been synthesised) with manganese dioxide in acetone, there was obtained a mixture from which violerythrin could be isolated. Since the latter can be converted (see above) into actinioerythrol, the parent of the naturally-occurring pigment, this constitutes a synthesis of that compound also. Astaxanthin (structure below) also yielded some violerythrin on treating with MnO_2 as above.

The mechanism of the above reaction is thought to be similar to that postulated for the formation of actinioerythrin in vivo (see below).

Biogenesis. The presence of small amounts of astaxanthin esters along with the actinioerythrin in both the above-mentioned sources of the latter, coupled with the widespread occurrence of astaxanthin in the whole marine environment and the obvious structural resemblance between the two compounds, led to the suggestion that actinioerythrin is formed in nature from astaxanthin, possibly as in Scheme 8 (Hertzberg and Jensen, Acta

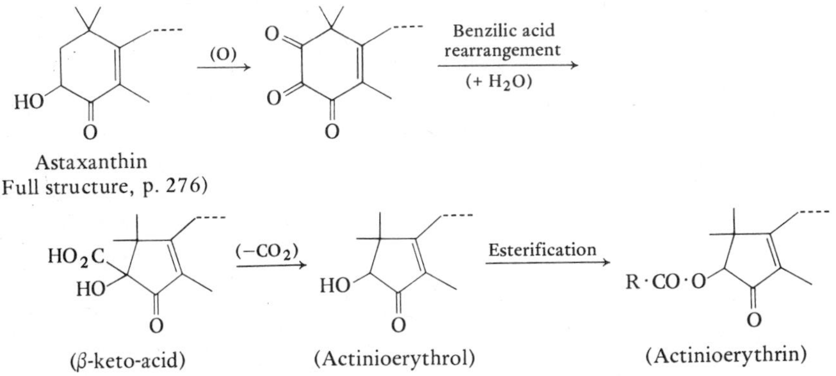

Scheme 8.

chem. Scand., 1968, 22: 1714). This suggestion has received support from the recent discovery that one of the minor carotenoids in *A. equina* has one end-group of the actinioerythrin type and the other of the (esterified) astaxanthin type. Treatment of this compound with alkali removes the ester groups and causes the expected autoxidation of the end-groups to, respectively, the α-diketone and diosphenol levels, this derivative being named

---

* Astacene is the diosphenol corresponding to astaxanthin, i.e. with the following end group:

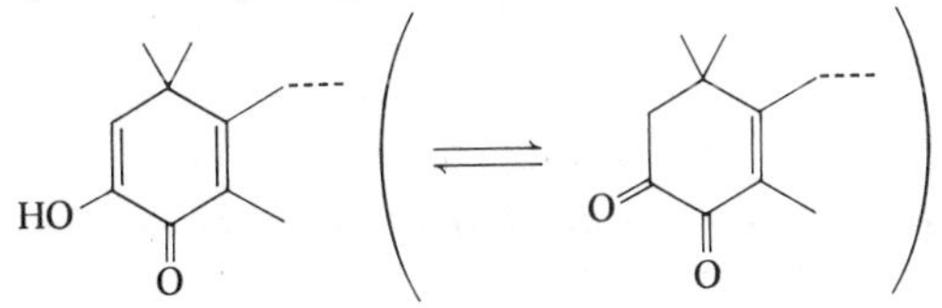

roserythrin. Roserythrin has been synthesised, being an intermediate, isolable by t.l.c., in the above-mentioned astacene + MnO_2 reaction (Francis et al., ibid., 1972, 26: 1097).

(k) Aryl carotenoids

Five carotenoid hydrocarbons and one xanthophyll with either one or both end-groups of the aryl (2,3,6- or 2,3,4-trimethylphenyl) type were mentioned in the main work (Vol. IIB, 2nd End., pp. 279 and 305). Although these compounds were then separately classified under the "hydrocarbons" and "xanthophylls" sections, the distinctive presence of an aryl end-group, which requires a special mode of biogenesis (Vol. IIB, 2nd Edn., pp. 346, 234), makes it more logical to group all such compounds together. Recent investigations have added four more examples of this type of structure.

Okenone. The full structure of this compound has now been elucidated (S. L. Jensen, Acta chem. Scand., 1967, 21: 961). The p.m.r. spectrum was particularly revealing, Me-Ar (3 Me at τ 7.7--7.8), $Me_2 C(O)$, and —OMe being plainly visible. An i.r. band at 800 cm^{-1} indicated an aryl ring with two adjacent hydrogens. A series of chemical interconversions (hydride reduction (to "okenol") followed by H$^{\oplus}$-catalysed dehydration, etc.) was also carried out. The aromatic substitution pattern was tentatively inferred from the λ_{max} position of okenol (to avoid the complicating effect of CO-conjugation; the steric hindrance caused by a 2,3,6-substitution pattern would have resulted in a shorter wavelength position for the absorption maxima than is actually observed; cf. Vol. IIB, 2nd Edn., p. 278).

Okenone

Synthesis (A. J. Aasen and Jensen, Acta chem. Scand., 1967, 21: 970). A 1 : 1 Wittig condensation of the 2,3,4-trimethylbenzyl Wittig reagent and crocetindial (C_{20}: cf. p. 267) followed by a second Wittig condensation of the product with the Wittig reagent from $Me_2 C(OMe) \cdot CH_2 \cdot CH_2 \cdot CH_2 \cdot$ CMe=CH $\cdot$ CH$_2$Br (cf. p. 264) gave (LVI). Introduction of —OAc at the allylic ($C_{(4')}$) position with NBS/AcOH followed by alkaline hydrolysis and *p*-chloranil/I_2 oxidation (Vol. IIB, 2nd Edn., pp. 262, 266), yielded a

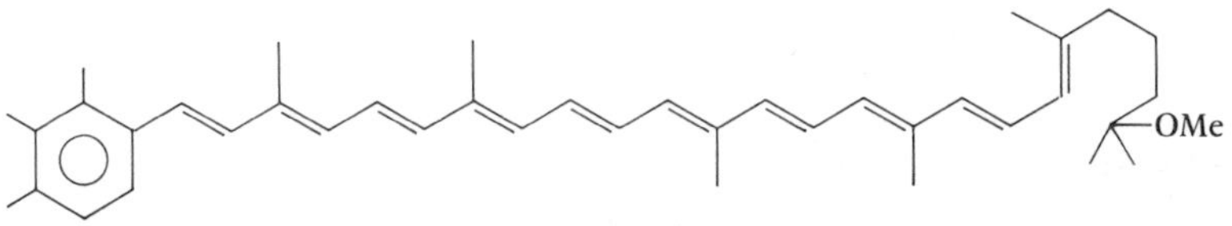

(LVI)

compound whose properties agreed in all respects with those of natural okenone. For a further natural source of okenone, see below.

"Demethylated okenone" was the name given by N. Pfennig, M. C. Markham and Jensen (Arch. Mikrobiol., 1968, 62: 178) to a minor pigment accompanying the okenone in a *Thiothece* sp. (one of the purple bacteria). It showed similarities to okenone but was more polar, and apparently contained a *tert*-hydroxyl rather than a *tert*-methoxy group, and so its structure was inferred to be as for okenone, but having $-OH$ in place of $-OMe$; hence the name. However a recent more thorough (chemical and spectral) study on larger quantities of the pigment, has shown it to be non-aromatic and to belong instead to Section 7c (p. 266).

Thiothece-484, λ_{max} 513, 484, 458 nm, a minor pigment accompanying the okenone in the above *Thiothece* sp., appears, from chemical and spectral work, to be as for okenone but with one (which is unknown) of the aryl-methyls oxidised to $-CO_2 Me$ (Andrewes and Jensen, Acta chem. Scand., 1972, 26: 2194).

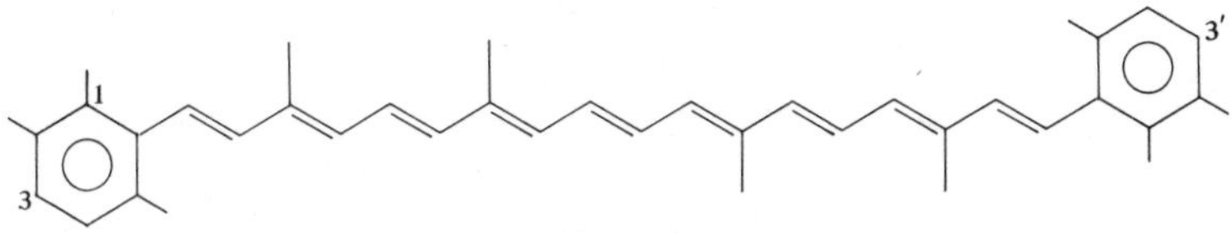

Isorenieratene

Phenolic carotenoids. Two examples of this new type of carotenoid, both related to the known isorenieratene (also present in the organism) have been discovered in *Streptomyces mediolani* by F. Arcamone et al. (Experientia, 1969, 25: 241; Gazz. chim. ital., 1970, 100: 581). Both compounds showed hydroxyl absorption (i.r.), behaved as phenolic compounds (soluble in ethanolic KOH), and had similar λ_{max} values to those of isorenieratene; they gave a di- and mono-acetate, respectively. Chromic acid oxidation of the latter derivatives gave, respectively, crocetindial (cf. p. 267) and 4-acetoxy-2,3,6-trimethylbenzaldehyde, and these plus 2,3,6-trimethylbenzaldehyde. This led to their formulation as 3,3'-dihydroxyisorenieratene and 3-hydroxy-isorenieratene, respectively.

Synthesis (Arcamone et al., loc. cit.). This was effected as for isorenie-ratene and renieratene respectively (Vol. IIB, 2nd Edn., p. 278, method ii) but 4-hydroxy-2,3,6-trimethylbenzaldehyde and a mixture of this and 2,3,6-trimethylbenzaldehyde, respectively, were used as starting materials. The hydroxyl groups were protected as their tetrahydropyranyl ethers throughout the reaction sequence, and then released by treatment with aqueous acid.

(*l*) *Carotenoids of uncertain constitution*

The structures of many of the compounds noted under this section before (Vol. IIB, 2nd Edn., p. 330; and called Section 7*g* therein rather than 7*l* as here) have now been fully, or almost fully, elucidated. These compounds now appear under the appropriate sections (see the Subject Index for individual page-number references). This applies to zeinoxanthin, myxoxan-thophyll, aphanizophyll, oscillaxanthin, phlei-xanthophyll, pectenoxanthin, siphonaxanthin, sarcinaxanthin, diadinoxanthin, diatoxanthin, hydroxy-echinenone, adonirubin, adonixanthin, okenone, warmingone, reticulataxan-thin, citranaxanthin and sintaxanthin. This list also includes certain compounds whose structures were announced too late to be described fully before and also those where there appeared to be some doubt, which has since been removed, as to the correctness of the structure.

In addition, Lederer, Weedon et al. (J. chem. Soc., C, 1969, 1264) have shown that Lederer's (1934, 1938) "**pentaxanthin**" mentioned in this section before (on p. 331) was probably the base-catalysed rearrangement product ("*isofucoxanthinol*") of the fucoxanthinol since shown by them (1969, loc. cit.) to be present in the same source (sea urchins) as investigated by Lederer earlier. It can therefore be considered an artefact. S. L. Jensen and A. Jensen (Prog. chem. Fats Lipids, 1965, 8: 162) pointed out that **sarcinene** (Vol. IIB, 2nd Edn., p. 331), and also Karrer and Solmssen's "*flavorhodin*" (Helv., 1935, 18: 1306), resemble, and may be identical with, the well-known compound neurosporene. The former might, alternatively, be the hydrocarbon analogue of sarcinaxanthin (C_{50}; p. 316), with which it co-occurs; a spectral study is required to clarify this.

The following progress has been made in elucidating the nature of the other compounds mentioned in this section in the second edition (Vol. IIB).

Trollixanthin, the major pigment in the bright yellow petals of the garden flower *Trollius europaeus*. Egger and Dabbagh (Tetrahedron Letters, 1970, 1433) have shown that Karrer's proposed structure is incorrect, only *one* allylic-OH being detected by the $H^{\oplus}$/MeOH test, and that trollixanthin is indistinguishable from neoxanthin (p. 290) by mixed chromatograms (of the pigments themselves, of their acetates and of the set of *cis/trans* isomers obtained from each on $I_2/h\nu$ treatment) and in its reaction with $H^{\oplus}/CHCl_3$. However, the corresponding furanoid derivatives (p. 246 for definition), *trollichrome* and *neochrome*, have very different reported m.p.'s ($206°$ and $148°$), suggesting that trollixanthin and neoxanthin do differ in some respect, possibly in the relative configuration of the $C_{(3)}$-hydroxyl and the 5,6-epoxy-group (Weedon et al., J. chem. Soc., C, 1969, 429).

    Taraxanthin, first reported by Kuhn and Lederer in 1931 as a constituent of dandelion (*Taraxacum officinale*) petals, and later, by others, of various other plants (Vol. IIB, 2nd Edn., p. 332; Karrer and Jucker, "Carotenoids", Elsevier, Amsterdam, 1950, p. 320). Egger (Planta, 1968, 80: 65) and Cholnoky, Weedon et al. (J. chem. Soc., C, 1969, 1256) have since surveyed the major alleged sources of this pigment, and report that the compound described as "taraxanthin" is, in most cases, lutein-5,6-epoxide (p. 250). However, the latter authors also ran the mass spectrum of a sample of Kuhn's original material and found it to be different from lutein epoxide ($C_{40}H_{56}O_4$, which is as for neoxanthin, instead of $C_{40}H_{56}O_3$). Also, Nitsche and Pleugel (Phytochem., 1972, 11: 3383) found that their sample of *T. officinale* (and two other plants) did contain, in addition to lutein epoxide, a pigment which analysed for $C_{40}H_{56}O_4$ (m.s.), and which gave a fragmentation pattern rather similar to that given by neoxanthin (9-*cis*-form, p. 291). Hence, "taraxanthin" may simply be a stereoisomer of neoxanthin, although it could still be a genuine carotenoid of restricted and erratic occurrence (the carotenoids in a plant sometimes vary widely in amount, and even in type, depending on the location in which the plant was grown).

    The pigment isolated by Strain (1954) from *T. officinale* petals and named by him *"tareoxanthin"* has been re-investigated by Szabolcs and Toth (Acta Chim. Acad. Sci., Hung., 1970, 63: 229) who conclude that it is a mixture, not resolved under Strain's conditions, of two *cis* isomers of lutein epoxide. **Aleurixanthin's** structure has now been fully defined by Jensen et al. as being, as γ-carotene but with a (novel) $-CHOH \cdot C(Me)=CH_2$ end-group (Phytochem., 1973, 12: 2751).

The following compounds were not mentioned in the 2nd edition (Vol. IIB), mainly due to a lack of definite characterisation at that time.

    Trollein, a rather ill-defined carotenoid first mentioned by A. L. Curl and G. F. Bailey (Food Res., 1955, 20: 371) as a minor constituent of Valencia oranges. It has the same chromophore as lutein (λ_{max} values) but a polarity corresponding to at least one more hydroxyl group. It was later resolved into trolleins *a* and *b*. Tests ($H^{\oplus}$; $H^{\oplus}/MeOH$) for epoxide and allylic- or *tert*-hydroxyls, were negative (Curl, J. agric. Food Chem., 1960, 8: 356; J. Food Sci., 1965, 30: 426). What initially appeared to be the same compound was detected by Krinsky (Anal. Biochem., 1963, 6: 293; Plant Physiol.,

1964, 39: 680) and others (cf. T. R. Ricketts, Phytochem., 1967, 6: 669, 1375) in *Euglena gracilis* and various other algae, but Krinsky's material was later found to differ from Curl's and was renamed *"trollein-like carotenoid"* (Schimmer and Krinsky, Biochemistry, 1966, 5: 1814). Yokoyama et al. (J. Food Sci., 1967, 32: 42) detected a similar compound in lemons. Yamamoto et al. have suggested that Krinsky's material is identical with pyrenoxanthin (q.v. for ref.) and Nitsche et al. that it is identical with de-epoxyneoxanthin (q.v.). A trollein-like carotenoid has also been found in a bacterium (the sole carotenoid present) by Czygan and Heumann (Arch. Mikrobiol., 1967, 57: 123), and a structure was tentatively suggested on the basis of chemical tests, polarity, i.r., λ_{max} values, etc. (as for zeaxanthin but with H—OH added across one of the $C_{(6)}$—$C_{(5)}$ bonds).

Tunaxanthin, a xanthophyll with a nonaene chromophore isolated from tuna and various other fish (S. Hirao et al., C.A., 1961, 55: 9703h; 1969, 71: 120889j; R. E. Torregrosa, C.A., 1966, 65: 20563d; N. Tsukuda et al., C.A., 1968, 68: 66759q; G. F. Crozier, Comp. Biochem. Physiol., 1967, 23: 179), has been tentatively identified as 3,3'-dihydroxy-ε-carotene (cf. Crozier, loc. cit.).

Crustaxanthin, also from the marine world (C. Bodea et al., C.A., 1968, 68: 13214b; Czygan, Z. Naturforsch., 1968, 23b: 1367; Herring, J. mar. biol. Assoc., U.K., 1969, 49: 766; for o.r.d. data, see Scopes et al., J. chem. Soc., C, 1969, 2527), has been formulated as the tetra-hydroxy compound corresponding to astaxanthin (common in the marine environment and suggested to be its precursor). However, substantiation of this, and clarification of the differences (ascribed to differing hydroxyl group configurations) noted by Herring, awaits a full spectral study. Concerning the allied compound *3,4-dihydroxy-β-carotene*, see footnote p. 277.

Idoxanthin, from a crustacean, has been assigned a tentative structure (one end as for astaxanthin, the other as for crustaxanthin; with both of which it co-occurs) on the basis of λ_{max} values, polarity, allylic-methylation studies, and comparison with the compound of that structure obtained, as a mixture of epimers, by partial hydride reduction of astaxanthin (Herring, 1969, loc. cit.); again, a full spectral study is needed.

Mytiloxanthin is one of the major pigments in the two *Mytilus* species (molluscs) mentioned under alloxanthin, the other pigment therein (S. A. Campbell, Comp. Biochem. Physiol., 1970, 32: 97). It resembles zeaxanthin in polarity, and λ_{max} values before and after borohydride reduction suggest the presence of a conjugated carbonyl group (apparently cross-conjugated: $C=C \cdot CO \cdot (C=C)_n$)*. **Micronone**, from *Micromonas pusilla* (Ricketts, Phyto-

---

* Structure recently assigned; based on ½ alloxanthin + ½ capsorubin (Weedon, Pure Appl. Chem., 1973, 35: 125).

chem., 1966, 5: 571) and related algae (idem, ibid., 1967, 6: 669; 1970, 9: 1835) also contains conjugated carbonyl, but only one hydroxyl group. The *pirardixanthins*, a group of four carotenoids isolated by Krinsky and H. M. Lenhoff (Compar. Biochem. Physiol., 1965, 16: 189) from *Hydra pirardi* when fed on a diet (of brine shrimp, as in nature) containing canthaxanthin as virtually the sole carotenoid, are apparently metabolic products of the latter. The major compound, **diketopirardixanthin**, apparently has an acyclic nonaene chromophore lacking conjugated carbonyl groups, and Krinsky and Lenhoff suggest that its structure is as for canthaxanthin (p. 275) but with the 5,6 and 5', 6' double bonds saturated. Two of the other compounds, **dihydroxypirardixanthin** and **hydroxyketopirardixanthin**, might be the corresponding dihydroxy- and hydroxyoxo-compounds, but all three assignments are tentative. The fourth compound, **hydroxypirardixanthin**, was present in trace amounts. **Guaraxanthin**, from the scarlet ibis (Fox and Hopkins, ibid., 1966, 19: 267), has similar properties to astaxanthin but a shorter chromophore, and is possibly as for astaxanthin, but with $-CH_2-CH_2-$ or $-CH_2CO-$ (cf. fucoxanthin, siphonaxanthin) at $C_{(7)}-C_{(8)}$.

Peridinin, probably identical with *"sulcatoxanthin"*, has been assigned (Strain, Jensen et al., J. Amer. chem. Soc., 1971, 93: 1823; Pure Appl. Chem., 1973, 35: 92), a structure containing two completely novel features for a carotenoid, a shortened polyene chain and a butenolide ring:

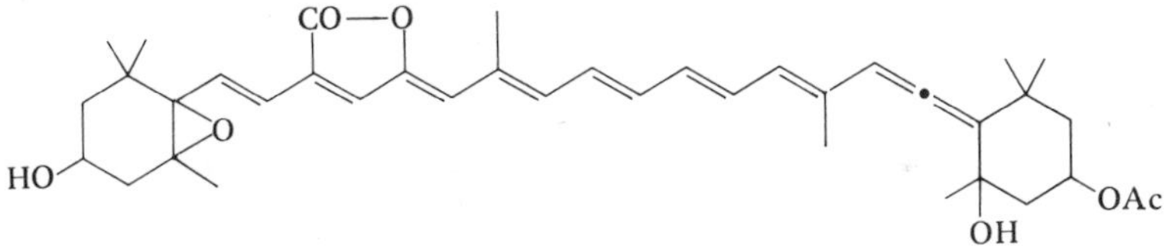

Peridinin

8. The biosynthesis of carotenoids

(a) Biosynthesis of hydrocarbons

(i) General

The state of knowledge in this field at the beginning of the period under review (Vol. IIB, 2nd Edn., p. 333) is summarised by the following postulates.

(1) The major tomato carotenoid, lycopene, is formed from more saturated precursors with the same skeleton (phytoene, phytofluene, ς-carotene, neurosporene; structures below) by a series of dehydrogenation

reactions, each of which brings into conjugation a previously isolated double bond:

This theory was originally put forward in 1950 by J. W. Porter and R. E. Lincoln following an investigation of the carotenoid content of various strains of tomato, and the reaction sequence became known as the *"Porter–Lincoln pathway"*. Later, others showed that the lycopene in certain other natural *carotenogenic* (carotenoid-producing) systems (notably various purple bacteria) is probably formed in the same way and showed that analogous dehydrogenation reactions occur during the conversion of lycopene to the still more unsaturated compound spirilloxanthin.

(2) The C_{40} skeleton originates from mevalonic acid by a series of reactions like that leading to the C_{30} compound squalene, with the difference that instead of the final step involving the tail-to-tail condensation of two C_{15} units with concomitant reduction to give a

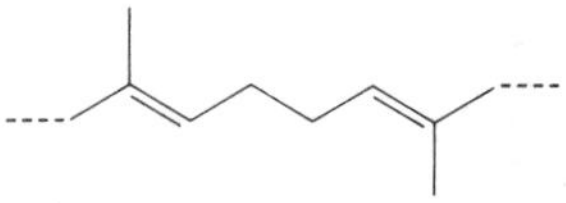

central unit, two C_{20} units unite *without reduction* occurring to give

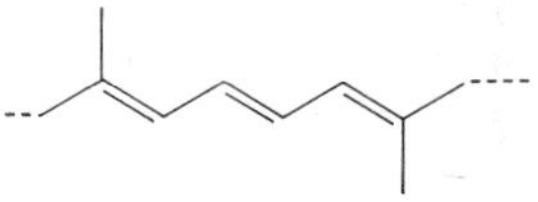

(phytoene)

This was postulated because various searches for the C_{40} analogue of squalene, "lycopersene" (structure as for phytoene but with $=C-CH_2-CH_2-C=$ at the centre), had failed to reveal it in systems in which phytoene and the later members of the Porter–Lincoln series of compounds were being formed in large quantities, and because of co-factor requirements (Vol. II B, 2nd Edn., p. 343).

These two postulates are summarised in Scheme 9. However, the evidence for certain parts of this scheme of lycopene biosynthesis was at that time of a circumstantial nature. Considerable doubt existed as to whether the lycopene produced by natural systems other than tomato fruit and purple bacteria (e.g. green (photosynthetic) tissues) originated in the same way. In addition, there was relatively little convincing experimental evidence for the involvement of geranylgeranyl pyrophosphate (C_{20}), or even for the individual steps of the Porter—Lincoln pathway. This was mainly due to the experimental difficulties associated with incorporating a (labelled) labile hydrocarbon oil such as phytoene into an aqueous carotenoid-synthesising system such as tomato fruit or green tissues, and afterwards, of isolating the various polyenes from such a system *free of radioactive contaminants* (cf. below). Some of the recent work relevant to these problems is outlined below. In most cases, ^{14}C-labelled compounds were used so as to be able to trace the progress of the very small amounts of the substrate and the various intermediates being studied. To aid manipulation of the products, the corresponding unlabelled compounds were usually added as carriers.

(ii) The conversion of mevalonic acid into phytoene and lycopene

In 1967, F. B. Jungalwala and Porter (Arch. Biochem. Biophys., 1967, 119: 209) succeeded in extracting and partially purifying, in a water-soluble form, the enzyme system responsible for lycopene formation in tomato fruit*. They showed that on incubating this enzyme concentrate with ^{14}C-labelled IPP (biosynthetic; cf. Scheme 9) in a buffered aqueous medium containing various co-factors, ^{14}C-labelled phytoene was produced. They also showed that a mixture of [^{14}C] farnesyl pyrophosphate (biosynthesised from MVA) and unlabelled IPP was converted into [^{14}C] phytoene under these conditions, and that each system also produced, simultaneously, a ^{14}C-labelled compound identified with fair certainty (hydrolysis to the corresponding alcohol, and g.l.c. comparison with authentic geranyl-geraniol) as geranylgeranyl pyrophosphate†. Later, when [^{14}C]geranylgeranyl pyrophosphate itself became available (biosynthetic: M. O. Oster and C. A. West, ibid., 1968, 127: 112), Porter and co-workers showed (ibid., p. 124) that the same tomato enzyme system would also convert this compound into phytoene, only Mg$^{2\oplus}$ and an −SH antioxidant being required as co-factors (as opposed to the several required for the MVA → C$_{20}$ pyrophosphate sequence). A repeat of the above experiment with [^{14}C] IPP in 1969 (Suzue and Porter, Biochim, Biophys. Acta, 1969, 176: 653) showed that, in addition to the ^{14}C-labelled phytoene, small amounts of labelled phytofluene, neurosporene, and lycopene were also produced. At the same time, Lee and Chichester (Phytochem., 1969, 8: 603) prepared [^{14}C] geranyl-geranyl pyrophosphate, and demonstrated that an enzyme system prepared from the carotene-synthesising fungus *Phycomyces blakesleeanus* was capable of converting it (same co-factor requirements as found by Porter above) into labelled phytoene, phytofluene, ζ-carotene, neurosporene, and

---

* By extracting the plastids (cell constituents) present in the tissue with acetone, evaporating to dryness, extracting the residue with (aqueous) phosphate buffer, and selective precipitation of various fractions by addition of ammonium sulphate, the most active fraction being selected for use.

† Supported by Suzue and Porter, 1969 (cf. Porter et al., J. biol. Chem., 1969, 244: 3641).

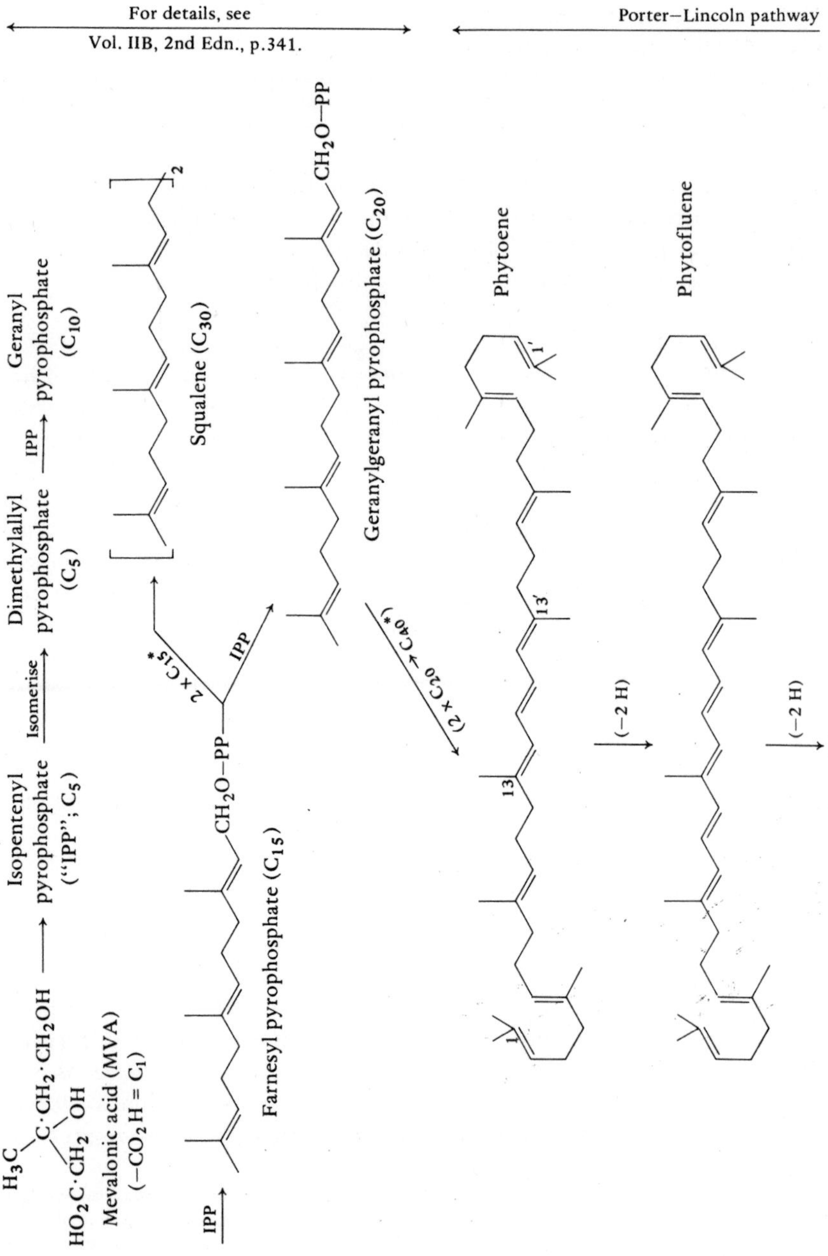

For details, see
Vol. IIB, 2nd Edn., p.341.
Porter–Lincoln pathway
H_3C
$C \cdot CH_2 \cdot CH_2OH$
$HO_2C \cdot CH_2$ OH
Mevalonic acid (MVA)
$(-CO_2H = C_1)$
Isopentenyl pyrophosphate ("IPP"; C_5)
Isomerise
Dimethylallyl pyrophosphate (C_5)
IPP
Geranyl pyrophosphate (C_{10})
IPP
CH_2O–PP
Farnesyl pyrophosphate (C_{15})
$2 \times C_{15}^*$
IPP
Squalene (C_{30})
2
CH_2O–PP
Geranylgeranyl pyrophosphate (C_{20})
$(2 \times C_{20} \rightarrow C_{40}^*)$
1
13
13'
1'
Phytoene
(−2 H)
Phytofluene
(−2 H)

ζ-Carotene

Neurosporene

Lycopene

(−2 H)

(−2 H)

Scheme 9. The central double bond in both phytoene and phytofluene has long been thought to be *cis* (from i.r. and thiourea-adduct tests: cf. Vol. II B, 2nd Edn., p. 335), and all isolated double bonds *trans* (from the p.m.r. position of attached methyl groups: Davies et al., J. chem. Soc., C, 1966, 2154), with the subsequent three polyenes each all-*trans*. It has now been shown (B. H. Davies et al., Phytochem., 1972, 11: 3187; R. Herber et al., Biochim. Biophys. Acta, 1972, 280: 194) that by careful chromatography the phytoene from various natural sources (tomatoes, carrots, various fungi) can be separated into two isomers, the major one (> 95%) apparently (from i.r., u.v., p.m.r. spectra; and effect of $I_2/h\nu$) being the central-*cis* isomer, and the minor one (with the same u.v. max. values but sharper peaks; more polar) the all-*trans*. That this was so, and that the *other* double bonds in the central triene unit are *trans* in both isomers, was demonstrated (Weedon et al., Chem. Comm., 1972, 996; Herber et al., loc. cit.) by p.m.r. spectrometry: (a) by comparing the position of the methyl groups attached to the triene unit (at ~τ 8.26 in both natural isomers) with the position predicted for various phytoene C=C configurations, as inferred from data on a series of synthetic trienes with known C=C configurations (viz. *Me* expected at ~τ 8.26 for *Me-trans*-C=C−*cis*-C=C or for *Me-trans−trans*; at ~τ 8.23 for *Me-cis−trans*, and at ~τ 8.18 for *Me-cis−cis*); and (b) by high-field p.m.r. of the olefinic region, which expands the pattern sufficiently to extract $J15,15'$ values (found to be as expected for, respectively, *cis*-C=C and *trans*-C=C at this position).

Certain organisms grown under stress (inhibitors present or excessive illumination) produce relatively large amounts of the all-*trans* isomer (Davies et al.; Herber et al.; Valadon et al., Phytochem., 1973, 12: 161).

* Concerning the mechanism of these two (tail-to-tail) condensations, see pp. 346−350.

$$ \text{−PP} = \text{pyrophosphate residue,} \quad -P(O)\cdot O\cdot P(O)\cdot OH $$

lycopene. As the experiment proceeded, the ^{14}C-activity moved steadily along the series from phytoene to lycopene.

The above reports show that geranylgeranyl pyrophosphate is definitely an intermediate on the biosynthetic pathway leading to lycopene, and provide strong support for the "Porter–Lincoln pathway" concept, in two widely different organisms. Further support for various parts of the sequence of reactions outlined in Scheme 9 has come from the work with ^{3}H, ^{14}C-labelled mevalonic acids described below and from the following observations.

(a) Incubation of ^{14}C-labelled phytoene (biosynthetic: using Suzue and Porter's method above) with the above tomato enzyme system in aqueous buffer containing $Mg^{2\oplus}$ and other co-factors, and an emulsifying agent to solubilise the phytoene, yielded (practical details below) all four subsequent members of the Porter–Lincoln series. Enzyme systems prepared in a similar manner from three other strains of tomato behaved in the same way (Suzue, Porter et al., J. biol. Chem., 1970, 245: 4708).

(b) An enzyme system prepared from spinach leaves converted [^{14}C] IPP into [^{14}C] phytoene, and [^{14}C] phytoene (biosynthetic: see above) into phytofluene and lycopene (Porter and co-workers, Arch. Biochem. Biophys., 1970, 137: 547).

(c) The conversion MVA → phytoene has been demonstrated in a soluble enzyme system from peas by J. E. Graebe (Phytochem., 1968, 7: 2003) and in bean leaves by J. M. Charlton et al. (Biochem. J., 1967, 105: 205), again using ^{14}C-labelling.

(d) Lycopene in various fungi (*Neurospora* and *Rhodotorula* spp.) apparently also originates from phytoene via the Porter–Lincoln series of compounds (J. Villoutreix, J. P. Malenge et al., Biochim. Biophys. Acta, 1967, 136: 459; 1968, 164: 306).

Taken in conjunction with the evidence outlined before (Vol. II B, 2nd Edn., p. 334) and elsewhere (e.g. in the review by Porter and Anderson, Ann. Rev. Plant Physiol., 1967, 18: 197, 209 et seq.; cf. also below), it seems virtually certain that the *lycopene in a wide range of organisms originates from MVA (and so ultimately, acetate) by the sequence of reactions given in Scheme 9* (or in certain organisms via a slight modification of same: see below). However, there have been reports that the phytoene in some organisms appears to be formed in disproportionately large amounts and is therefore perhaps a by-product rather than an obligatory intermediate on the pathway (e.g. F. W. Quackenbush et al., Arch. Biochem. Biophys., 1966, 114: 326, and refs. given by Porter and Anderson, loc. cit., p. 210). Such anomalies may however be due to the difficulty of isolating, from a complex system, the small amounts of labelled intermediates completely free of radioactive contaminants. In this respect, the isolation method used by Porter and his co-workers would appear the most reliable so far devised, being based on both adsorption and volatility characteristics. Thus, following incubation, the mixture was saponified, the hydrocarbons extracted with petroleum, the appropriate carriers added, and the mixture chromatographed. The absorption spectrum (hence the identity, and amount, of polyene therein) and radioactivity of each fraction was measured; usually the activity of each fraction closely paralleled its polyene content. Fractions containing the same polyene were then combined, evaporated, and each polyene separately hydrogenated, the product ("*lycopersane*", fully reduced lycopersene, in each case) being subjected to gas–liquid radiochromatography and the activity of the lycopersane peak measured. Most of the other authors cited above used either an allied method to check purity or (for ζ-carotene to lycopene) crystallised the product to constant specific activity. However, Villoutreix, Malenge et al. (loc. cit.) avoided the problem associated with radioactive assaying by using the technique whereby phytoene is made to accumulate by adding an inhibitor* to the organism and then, following removal of the inhibitor, its conversion into phytofluene etc. is followed by taking a sample at intervals, resolving it into its constituents chromatographically, and assaying each by u.v./visible absorption spectroscopy. A series of curves (see e.g. p. 464 of their 1967 paper) showing the build-up of phytofluene etc., and decline of phytoene, is thereby obtained.

　　Further arguments have been put forward for the thesis that "*lycopersene*" (p. 336) is not produced as an intermediate between geranylgeranyl pyrophosphate and phytoene during carotenoid biosynthesis. Thus Charlton et al. (1967, loc. cit.) were unable to detect such a compound in their bean leaf system. Similarly Scharf and Simpson (Biochem. J., 1968, 106: 311) showed that a lycopersene-like compound extractable from a carotenoid-producing red yeast is a contaminant derived from the medium on which the yeast is commonly grown; and it seemed that the earlier claims by others of lycopersene occurring in similar sources (Scharf and Simpson for refs.) could possibly be explained in the same way. Finally, Chichester et al. (Phytochem., 1972, 11: 681) could not detect [^{14}C]lycopersene on incubating [^{14}C]MVA with an enzyme preparation from a *Phycomyces* mutant known to produce large amounts of phytoene.

---

* In this case, a trace of diphenylamine (cf. Vol. II B, 2nd Edn., p. 334). Other reagents of value in biosynthetic studies include CPTA (2-(*p*-chlorophenylthio)triethylamine hydrochloride: Yokoyama et al., Phytochem., 1972, 11: 1721, 2985), and nicotine (Howes and Batra, Biochim. Biophys. Acta, 1970, 222: 174; Goodwin et al., Pure Appl. Chem., 1973, 35: 29), both of which inhibit end-group cyclisation, and dimethylsulphoxide (p. 350). CPTA also *stimulates* overall carotenoid production in certain cases (Yokoyama et al., loc. cit.), as do certain allied compounds (idem, Phytochem., 1973, 12: 2665).

Recently, however, this whole question has been re-opened by the discovery (Porter et al., J. biol. Chem., 1972, 247: 6730; 1973, 248: 2755) that an enzyme preparation from yeast ("squalene synthetase") which had previously been shown (Beytia, Qureshi and Porter, ibid., 1973, 248: 1856) to produce squalene on being incubated with farnesyl (C_{15}) pyrophosphate, would also catalyse the self-condensation of the C_{20} analogue of the latter, geranylgeranyl pyrophosphate (^{14}C-labelled, biosynthetic; cf. p. 337), giving a C_{40}-hydrocarbon identified as lycopersene [by t.l.c. (radioactive spot at the same R_f as authentic (synthetic) lycopersene); gas-liquid radiochromatography (radioactive peak at same position as authentic lycopersene); m.s. of the g.l.c. effluent ($M^\oplus$ at 546 as required; ions corresponding to loss of 69, 137, and 205 mass units (successive cleavage of bis-allylic C—C's); pattern identical to that given by authentic lycopersene)]. A modified version of Jungalwala and Porter's tomato enzyme-preparation (p. 337) effected the same conversion (Porter and co-workers, ibid., 1973, 248: 2768). NADPH and $Mg^{2\oplus}$ were essential co-factors in each system. In addition, the ^{14}C-labelled lycopersene so formed was, apparently, converted into phytoene and lycopene on incubation with the tomato enzyme-preparation, though only slowly. Whether *all* the phytoene in *all* (including *natural*) carotenoid-producing systems is produced via lycopersene remains to be seen (cf. B. H. Davies, Chimia, Switz., 1973, 27: 174); but it might be that it is, the reason for its non-detection by others being technical. Thus, Porter's group found lycopersene to be even more prone to (oxidative?) degradation than phytoene, column chromatography, rather than the more usually used t.l.c., being essential for separation work if reasonable recoveries are to be obtained. In addition, phytoene, often present in relatively large amounts, is difficult to separate from lycopersene and so can easily overwhelm it; and earlier attempts to detect it in incubation experiments (by Porter and others, 1962, 1963; see Vol. II B, 2nd Edn., p. 343 for refs.) may have been thwarted by a lack of sufficient NADPH co-factor or by the experiments being run for too long (thereby allowing any lycopersene formed to be converted into phytoene). However, it remains a fact that the above systems are to some extent artificial, and the various constituents therein* may have activated a pathway not normally operating in nature.

Modified Porter—Lincoln pathway. The discovery that the heptaene hydrocarbon in *Rhodospirillum rubrum* and possibly certain other organisms (p. 220 for refs.) is not, as had previously been assumed, ζ-carotene (in which the chromophore is situated centrally in the molecule) but an isomer ("unsymmetrical-ζ-carotene") with the chromophore situated to one side,

---

* and/or the *absence* of others, present in the natural system.

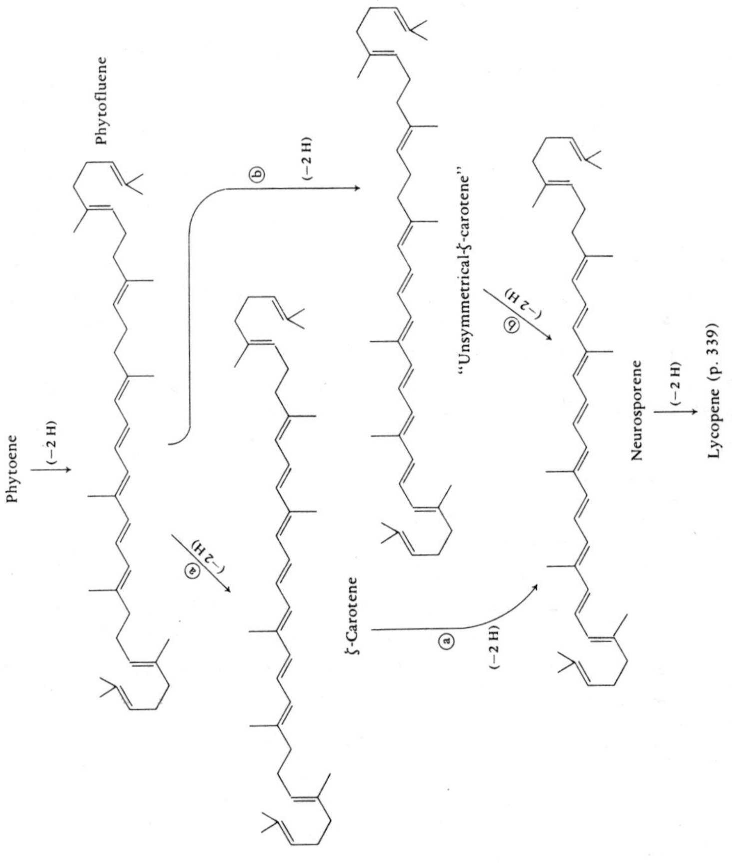

Scheme 10. Transformations marked ⓐ used by most organisms; those marked ⓑ by *R. rubrum*, etc.

has led to the suggestion that the lycopene in these organisms is formed via a slightly modified Porter—Lincoln sequence, as in Scheme 10 (Davies et al., J. chem. Soc., C, 1969, 1266; Biochem. J., 1970, 116: 93).

Other sequences involving dehydrogenation reactions. It was mentioned before (Vol. II B, 2nd Edn., pp. 336—338) that there is evidence that several different biosynthetic sequences (Schemes 22 and 23 of 2nd edn.) operate in bacteria. These sequences involve dehydrogenations of the Porter—Lincoln type, alternating with two other types of transformation: (a) hydration of the terminal isopropylidene groups, $-CH=CMe_2 \rightarrow -CH_2-C(OH)Me_2$; (b) methylation of the resulting $-OH$ group.

The discovery (pp. 265, 266) of still more pigments of the spheroidene, spirilloxanthin, etc. type in *R. rubrum* when cultured in the presence of diphenylamine (to inhibit the formation of the normal carotenoids) means that at *each level of dehydrogenation* from "unsymmetrical-ζ-carotene" to spirilloxanthin, there are now known molecules with almost all possible permutations of the three types of end-group (viz. $-CH=CMe_2$, $-CH_2-C(OH)Me_2$, $-CH_2-C(OMe)Me_2$). Consequently, there must be several alternative pathways leading from the unsymmetrical-ζ-carotene stage to the final, single product spirilloxanthin, obtained when the organism is allowed to grow normally. Some of these are outlined by B. H. Davies (Biochem. J., 1970, 116: 101) but following Malhotra et al.'s more recent work (p. 266), additional pathways need to be formulated so as to include their dimethoxy-compounds. An indication of the types of transformation probably occurring is given in Scheme 11. It may be that the enzymes catalysing the above-mentioned three basic reactions have very low specificity and will transform any molecule of approximately the required shape into the corresponding dehydro-, hydrated, or methylated derivative, the reactions occurring until the skeleton is fully unsaturated, and both end-groups are at the methylated stage, as in spirilloxanthin.

Stereochemistry. Over recent years several stereospecifically labelled forms of mevalonic acid (MVA) have been synthesised (initially by Cornforth, Popják et al., Proc. roy. Soc., 1966, 163B: 465, 492; Biochem. J., 1966, 101: 553). These have been added to systems producing squalene, and more recently phytoene and allied compounds, in an attempt to determine the fate of each of the atoms of MVA when it is converted (as in Scheme 9) into these compounds. Two different radioactive atoms, ^{14}C and tritium (^{3}H or "T"), are normally built into the MVA molecule so that the amount of tritium incorporated can be measured* *relative to* the ^{14}C. The ^{14}C count

---

* The disintegrations due to ^{3}H and to ^{14}C are sufficiently different in energy for them to be distinguished and measured separately by standard scintillation techniques.

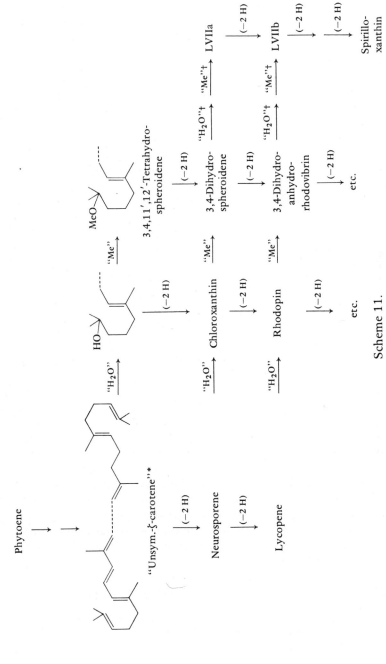

Scheme 11.

* For full structure, see p. 343; for other structures see p. 338, and Section 7c (p. 263) of this volume and of Vol. II B (2nd Edn.).

† Reactions occurring at the "other end" of the molecule [to give 3,4,3′,4′,7,8-hexahydro- and 3,4,3′,4′-tetrahydro-spirilloxanthin (LVIIa, b) (cf. Malhotra et al., loc. cit.) which on stepwise dehydrogenation will yield spirilloxanthin].

thereby acts as an *internal standard* and so problems associated with there being only partial incorporation of the labelled substrate are eliminated. The following compounds, (4R)- and (4S)-[2-^{14}C,4-T]MVA, were two of the first to be used (Goodwin and R. J. H. Williams, Proc. roy. Soc., 1966, 163B: 515):

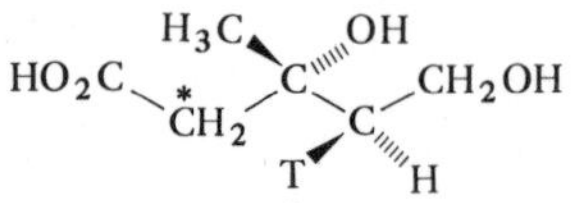

$$\overset{*}{C} = {}^{14}\text{C-labelled carbon}$$

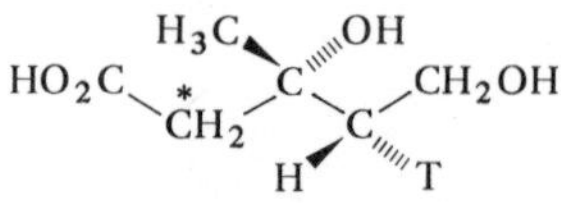

$$(-\text{CO}_2\text{H} = \text{C}_{(1)})$$

(LVIII)   (LIX)

Goodwin and Williams (loc. cit.) found that the phytoene produced by carrot slices incubated with LVIII had a T/^{14}C ratio of 8:8 whereas that produced in the presence of LIX contained no tritium at all. Hence in the building up of phytoene from eight molecules of MVA (Scheme 9), it is the "α"-H (or T)† on C$_{(4)}$ in the MVA molecule, as drawn here, which is removed at each step involving proton loss (cf. Vol. II B, 2nd Edn., p. 341). Similarly,

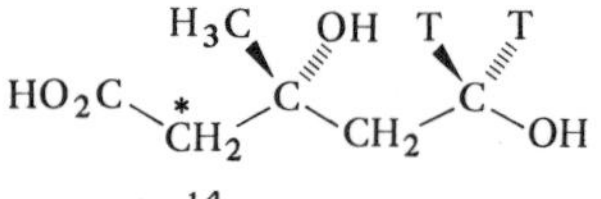

[2-^{14}C,5-T$_2$]MVA   (5R)-[2-^{14}C,5-T]MVA

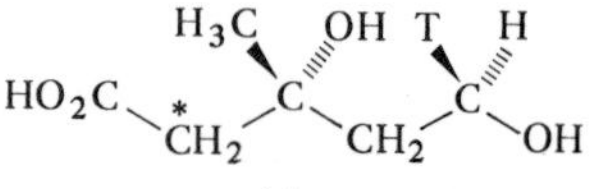

(LX)   (LXI)

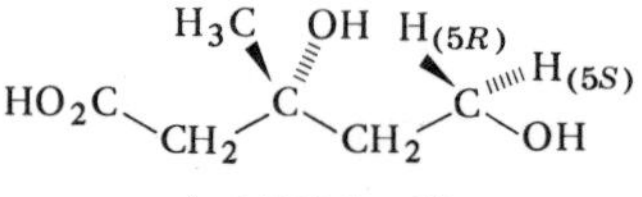

(= MVA itself)

(LXII)

comparison of the T/^{14}C ratio in phytoene (C$_{40}$) biosynthesised in a bean-leaf system dosed with (a) LX and (b) LXI, with the T/^{14}C ratio of the geranyl-geranyl pyrophosphate (C$_{20}$) formed simultaneously, showed that in the 2 x C$_{20}$ → C$_{40}$ link-up involving these two compounds (Scheme 9), two hydrogens ultimately originating from the C$_{(5)}$ of (two molecules of) MVA are lost stereospecifically and that these are the C$_{(5)}$ *pro-S*-H's (i.e. of the H$_{(5S)}$ type in LXII); as in Scheme 12 (M. J. Buggy, G. Britton and Goodwin, Biochem. J., 1969, 114: 641).

---

† Using steroid nomenclature; that is the "*pro-S*" hydrogen using Hanson's extension of the standard R,S system of nomenclature (J. Amer. chem. Soc., 1966, 88: 2731). The two hydrogens on C$_4$ of MVA (also those on C$_5$ and C$_2$) are cleanly differentiated by the enzymes responsible for the biogenesis of the carotenoids and other terpenoids.

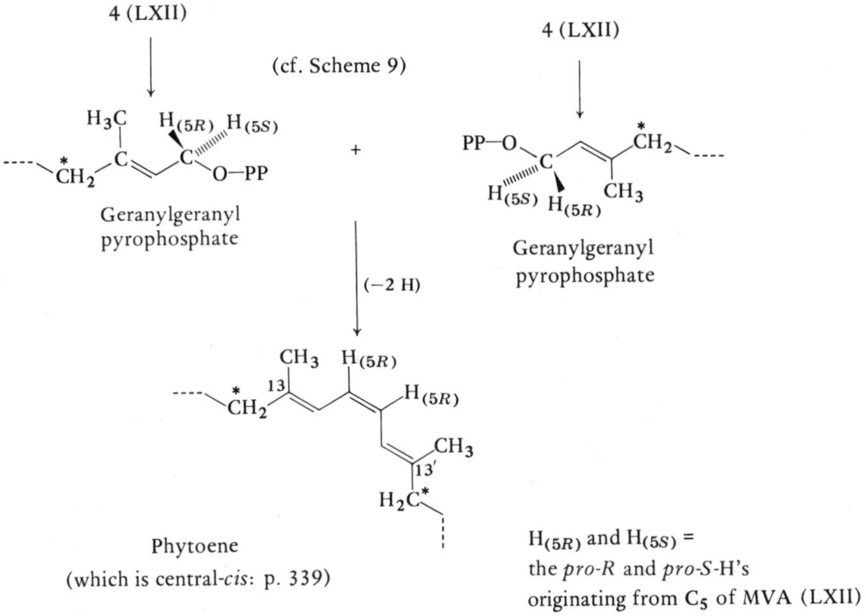

4 (LXII)

4 (LXII)

(cf. Scheme 9)

H_3C $H_{(5R)}$ $H_{(5S)}$

----$\overset{*}{C}H_2$—C=C—C—O—PP

+

PP—O—C=C—$\overset{*}{C}H_2$----

$H_{(5S)}$ $H_{(5R)}$ CH_3

Geranylgeranyl
pyrophosphate

Geranylgeranyl
pyrophosphate

(−2 H)

CH_3 $H_{(5R)}$

----$\overset{*}{C}H_2$—$\overset{13}{C}$=C—C—$H_{(5R)}$

CH_3

$13'$

$H_2\overset{*}{C}$

Phytoene
(which is central-*cis*: p. 339)

$H_{(5R)}$ and $H_{(5S)}$ =
the *pro-R* and *pro-S*-H's
originating from C_5 of MVA (LXII)

Scheme 12. The mechanism of the above reaction is currently under discussion (see text): the formation of a *cis* bond is obviously significant; R. J. H. Williams et al. (Biochem. J., 1967, 104: 767) have speculated on this point. This (immediate) formation of a rigid triene system was previously considered to be a possible explanation of why in the C_{40}-series one has a succession of dehydrogenation reactions to give the (essentially linear) carotenoids, rather than, as in the C_{30}-series, coiling up (of squalene) to give a series of polycyclic compounds (triterpenes, etc.). However, if it transpires (cf. p. 342) that lycopersene is in fact formed as an intermediate in the above transformation *in nature*, this explanation is invalidated.

Further experiments with LX and LXI, and with samples of geranylgeranyl pyrophosphate biosynthesised from (a) (2R)- and (b) (2S)-[2-^{14}C,2-T]MVA (LXIII) and (LXIV) respectively, suggest (cf. Goodwin, Biochem. J., 1971, 123: 297) that the subsequent desaturation steps undergone by phytoene

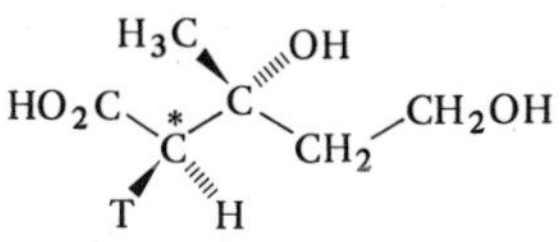

H_3C OH

HO_2C—$\overset{*}{C}$—C—CH_2—CH_2OH

T   H

(2R)-[2-^{14}C,2-T]MVA
(LXIII)

H_3C OH

HO_2C—$\overset{*}{C}$—C—CH_2OH

H   T

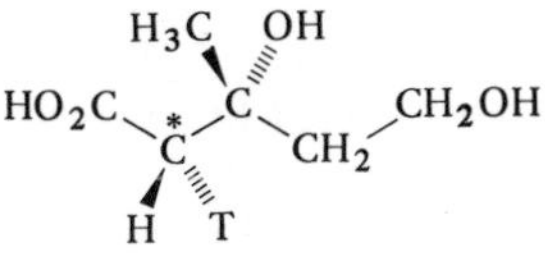

(2S)-[2-^{14}C,2-T]MVA
(LXIV)

(Scheme 9) involve a stereospecific *trans*-elimination of two hydrogens at each stage:

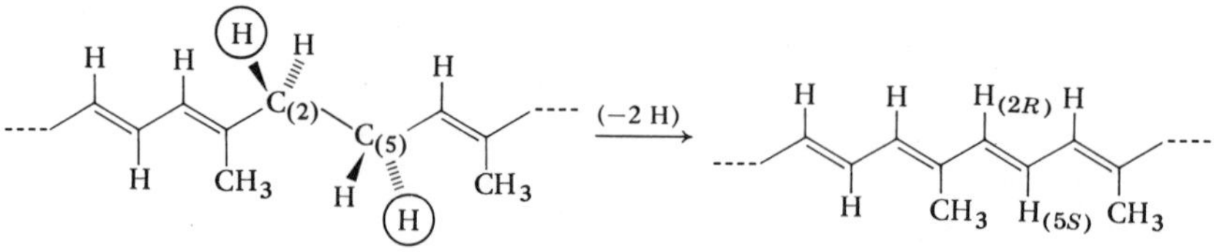

The pair of hydrogens eliminated (those ringed above) are the *pro-R*-hydrogen originating from $C_{(5)}$ of one MVA molecule and the *pro-S*-hydrogen originating from $C_{(2)}$ of another.

This work is covered more fully in Vol. II E (and its Supplement) and in the above-mentioned review by Goodwin (1971, loc. cit.). Taken in toto, it obviously provides further powerful evidence in favour of the scheme of lycopene biosynthesis outlined on p. 339 (Scheme 9).

Relatively little progress has been made on elucidating the means by which each of the steps in the sequence of reactions in Scheme 9 is effected. The dehydrogenation steps do not seem to require O_2 as such (cf. Porter and co-workers, 1970, loc. cit.) and though the co-factor requirements for the various steps may give a clue (cf. Porter's papers, pp. 337–340), detailed work may have to await isolation of the individual enzyme system(s) responsible. However, rather more attention has been devoted to the "dimerisation" step. Thus it should be mentioned that *opinion has changed as to the mechanism of the* $2 \times C_{15} \rightarrow C_{30}$ *condensation leading to squalene* [outlined in Vol. II B, 2nd Edn., p. 343, in connection with the formally similar $2 \times C_{20} \rightarrow C_{40}$ condensation leading to phytoene (Scheme 9)]. It is now thought that this condensation involves two molecules of farnesyl pyrophosphate rather than one of farnesyl pyrophosphate and one of its allylic rearrangement product, nerolidyl pyrophosphate*. An intermediate, *"presqualene pyrophosphate"* (LXV), has been identified (for leading refs., see R. H. Prince et al., Chem. and Ind., 1971, 720), and possible mechanisms for its formation and subsequent transformation into squalene have been suggested (cf. Prince et al.; Popják et al., J. Biol. Chem., 1971, 246: 6254; Beytia, Qureshi and Porter, 1973, loc. cit.).

An analogous C_{40} compound, *"prelycopersene pyrophosphate"*, has recently been isolated by Porter's group (ibid., 1972, 247: 6730; 1973, 248: 2755, 2768) from the two lycopersene-producing systems mentioned on p. 342 by depriving them of NADPH (thereby blocking the conversion to

---

* It now appears (Chichester et al., 1972, loc. cit.) that the analogous state of affairs also obtains for the corresponding $2 \times C_{20} \rightarrow C_{40}$ condensation.

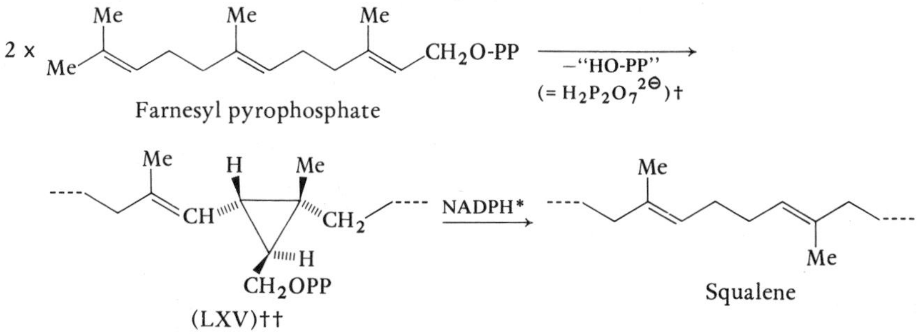

$$2 \times \quad \text{Me} \overset{\text{Me}}{\diagup} \diagdown \diagup \overset{\text{Me}}{\diagdown} \diagup \diagdown \diagup \overset{\text{Me}}{\diagdown} \diagup \text{CH}_2\text{O-PP} \quad \xrightarrow[\substack{(=\,H_2P_2O_7{}^{2\ominus})\dagger}]{-\text{"HO-PP"}}$$

Farnesyl pyrophosphate

(LXV)††

Squalene

* Nicotinamide adenine dinucleotide phosphate (old name, TPNH, used before); effects the reduction by donating one H as shown below:

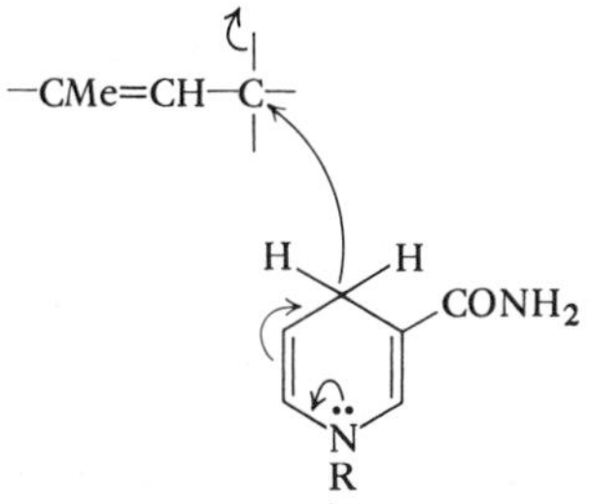

† For meaning of −PP, see note to Scheme 9.
†† Note—the *stereochemistry* shown here has recently been contested (by Popják, J. Amer. chem. Soc., 1973, 95: 2713).

lycopersene). The structure was deduced by hydride removal of the pyrophosphate residue to yield the corresponding alcohol, "prelycopersene alcohol", and m.s. study of the latter and its trimethylsilyl derivative, and of the hydrogenated product (only 7 moles H_2 absorbed, implying the presence of a ring, the detection of $M-C_{16}H_{33}$ and $M-C_{19}H_{39}$ fragments indicating its non-symmetrical position); and by comparison of this data, and the c.d. curve, with that previously obtained from presqualene pyrophosphate and alcohol (Popják et al., ibid., 1969, 244: 1897; H. C. Rilling, ibid., 1970, 245: 4597). A similar, probably identical, compound (*"prephytoene pyro-phosphate"*) was isolated by Rilling et al. (J. Amer. chem. Soc., 1972, 94: 3257) from a carotenoid-producing *Mycobacterium* extract dosed with labelled geranylgeranyl pyrophosphate. Here the structure elucidation included comparison with a *synthetic specimen* of the corresponding alcohol ("prephytoene alcohol") and ozonolytic degradation to a C_9 residue *containing the cyclopropane ring*, which was also synthesised (as the acetate) for comparison purposes. Both groups confirmed that their respective pyrophosphates could be converted into one or more of the subsequent members of the Porter—Lincoln series on incubation with the respective enzyme preparations in the presence of suitable co-factors (NADPH, etc.). "Prephytoene alcohol" has also been synthesised, from geranylgeraniol, by L. Crombie et al. (Chem. Comm., 1972, 1045).

The mechanism of the 2 x geranylgeranyl pyrophosphate → prelyco-persene (prephytoene) pyrophosphate → lycopersene (→ phytoene, etc.) series of reactions has been discussed by Porter's group (1973, loc. cit.), who infer it to be parallel to that suggested for the squalene series, above.

(iii) The cyclisation step

The level of dehydrogenation at which cyclisation of the end-groups of the series of compounds in the Porter—Lincoln pathway (cf. Scheme 9) starts occurring has long been in dispute. Originally, Porter and Lincoln (1950: Vol. II B, 2nd Edn., p. 338) suggested this occurs, in tomatoes, only at the lycopene stage. That is the β-carotene therein originates via lycopene → γ-carotene → β-carotene (pathway b, Scheme 13). However, subsequent studies by others of β-carotene formation in other systems indicated that in these systems it could not be arising solely in this manner (cf. 2nd Edn., p. 338; Simpson et al., Phytochem., 1970, 9: 1239). By 1966/1967 (2nd Edn.) it had been tentatively concluded that the β-carotene in tomatoes, and probably other organisms also, originates via one or more of at least two pathways (Scheme 13) the extent to which each is used depending on the particular organism. Further evidence has been presented in support of this thesis. Thus, several groups have studied the effect of treating an organism in such a way [temperature up or down*; no light; cultured in the presence of diphenylamine (cf. p. 341), etc.] that production of one or more of the carotenoids therein is hindered whilst synthesis of the others (which therefore, it can be inferred, must be on a different biosynthetic pathway) is unaffected. Further evidence can then be obtained by following the change in level of various carotenoids after removal of the hindering agent. Using such effects (Foppen et al., ibid., 1968, 7: 1605, for refs.), the following conclusions have been reached.

(a) The β-carotene in three varieties of tomato originates partly via pathway b (Scheme 13) and partly via a (using dimethylsulphoxide as selective inhibitor of lycopene formation; Simpson et al., 1970, loc. cit.) whilst that in various *Rhodotorula* species is formed entirely via pathway a or by both a and b depending on the species (Simpson et al., J. Bact., 1964, 88: 1688; Bonaly and Malenge, Biochim. Biophys. Acta, 1968, 164: 306).

---

* Cf. the early observation that the ratio of β-carotene to lycopene produced by ripening tomatoes depends on the ambient temperature: Goodwin and Jamikorn, Nature, 1952, 170: 104.

(b) The β-carotene in a *Mycobacterium* sp. and two other micro-organisms is formed via pathway b (nicotine inhibition of the cyclisation reaction caused the accumulation of lycopene (only); C. D. Howes and P. P. Batra, ibid., 1970, 222: 174; Goodwin, Biochem. J., 1971, 123: 302).

In addition, Suzue, Porter et al. (J. biol. Chem., 1970, 245: 4708) incubated [15,15'-^{3}H$_2$]lycopene (obtainable by ^{3}H$_2$/catalyst on the synthetic central-acetylenic analogue of lycopene) with the tomato enzyme system mentioned on p. 337, and isolated labelled β-, γ-, δ-, and α-carotenes (in order of decreasing activity; the assays were checked by gas—liquid radio-chromatography of perhydro-derivatives as on p. 341). Beeler, Porter et al. (ibid., 1969, 244: 3635) obtained similar results with "High-β" and "High-δ" tomatoes (bred to yield unusually large amounts of β- and δ-carotenes at the expense of the usual tomato carotenoid, lycopene) and a spinach system. They concluded (1970 paper) that in all these systems the lycopene was undergoing cyclisation in two different ways:

Lycopene ⟶ ┌─→ γ-carotene → β-carotene (= pathway b, below)
            └─→ δ-carotene → α-carotene (structures, 2nd Edn., pp. 273, 271).

Parallel experiments with labelled neurosporene, one step back in the biosynthetic sequence, are necessary before it can be said how much of the β-carotene in such systems normally arises as above and how much by pathway a (Scheme 13), or others.

The *mechanism* of the ring-closure reaction involved in the formation of cyclic carotenoids has been subjected to detailed study. The various schemes which have been put forward for the mode of formation of the β- and of the α-type of end-group from the acyclic precursor, whether it be lycopene or neurosporene, are outlined in Schemes 14, 15 and 16. The first of these involves formation of both types of end-group from a single carbonium-ion intermediate. Schemes 15 and 16 involve the initial formation of, respectively, the β- and the α-end-group, which then isomerises to give the other member of the pair.

Goodwin and his co-workers have now shown which of these schemes operates in nature by incorporating various doubly-labelled MVA's, of the type used on p. 346, into selected α- and β-carotene-producing systems and determining how many, and, by inference, which, of the labelled protons survive the cyclisation reaction. Thus addition of (4R)-[2-^{14}C,4-T]MVA (LVIII) to a carotenoid-producing system places tritium atoms at the C$_{(6)}$, C$_{(6')}$ positions of all acyclic carotenoids therein (and also at the 2, 2', 10,

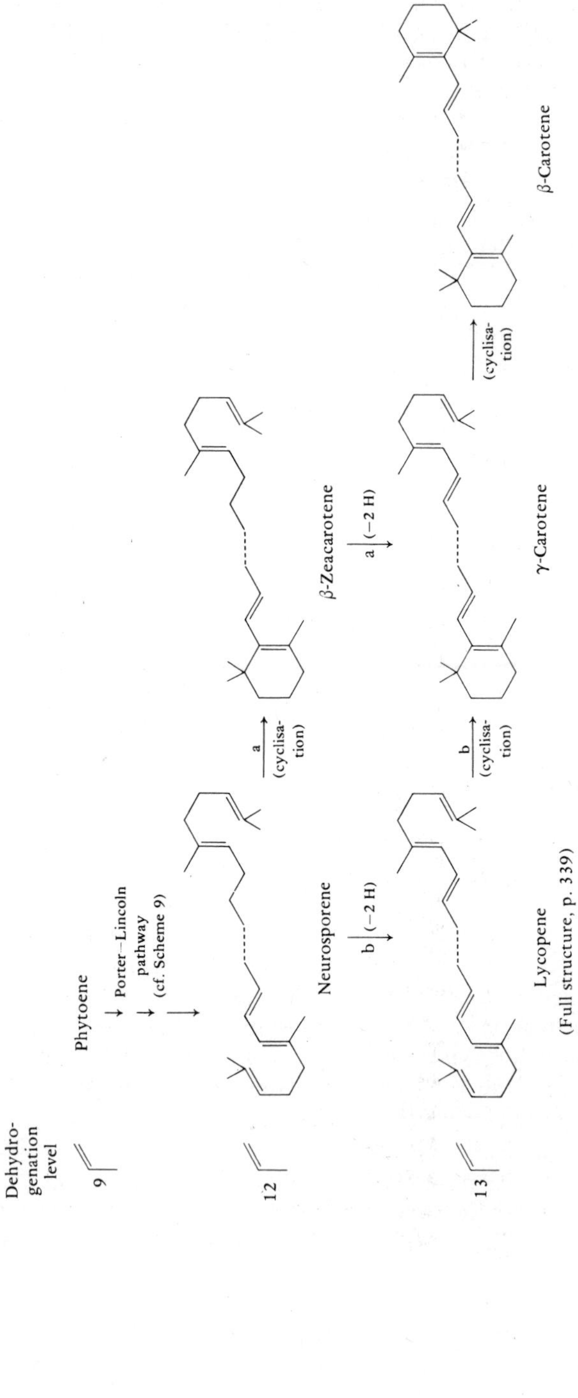
Dehydro-
genation
level
9
12
13
Phytoene
Porter–Lincoln
pathway
(cf. Scheme 9)
Neurosporene
b (−2 H)
Lycopene
(Full structure, p. 339)
a
(cyclisa-
tion)
b
(cyclisa-
tion)
β-Zeacarotene
a (−2 H)
γ-Carotene
(cyclisa-
tion)
β-Carotene

Scheme 13.

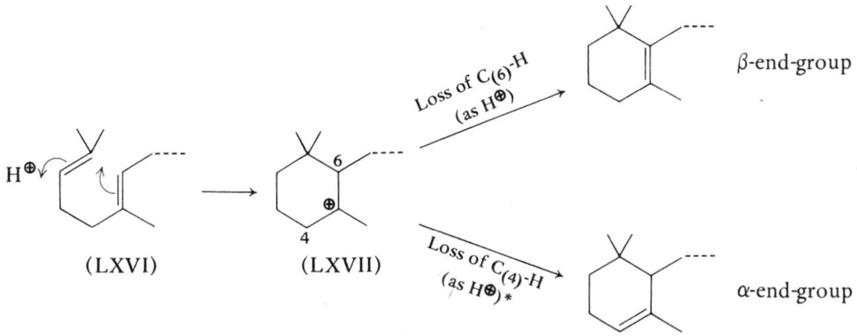

Scheme 14.

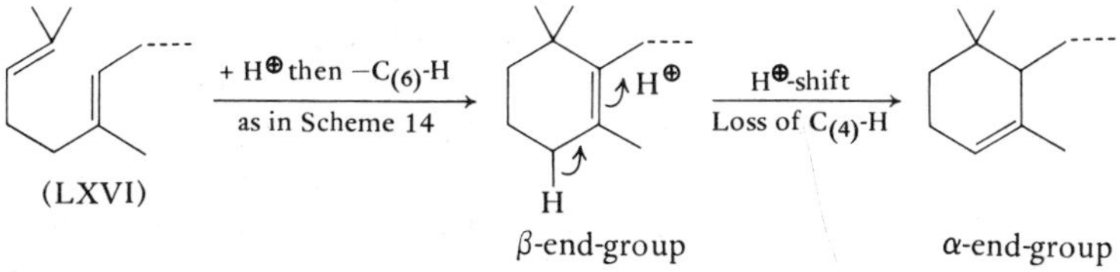

Scheme 15.

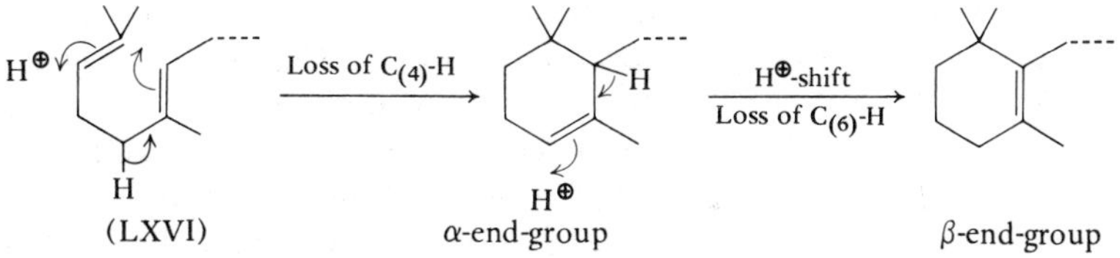

Scheme 16.

LXVI is the acyclic end-group of the precursor molecule in each case.

* The third possibility, loss of $H^{\oplus}$ from the methyl group, would lead to an exocyclic $=CH_2$ group. Carotenoids containing such a group have recently been detected in nature: cf. p. 224.

$10'$, 14, $14'$ positions, making a total of eight per molecule) but these would be lost on formation of a β-end-group if hypotheses of the above type were correct. This was found to be so. The β-carotene produced by a variety of such systems (carrot, tomato, fungi, maize seedlings: for refs. see Goodwin, 1971, loc. cit., p. 305) was found to contain only six tritium atoms per molecule (i.e. per eight ^{14}C atoms, the ^{14}C count being used as an internal standard as before: p. 344) corresponding to one tritium atom being lost from each end-group during the cyclisation process.

The above type of experiment was then carried out using $[2\text{-}^{14}C, 2\text{-}T_2]MVA$ which places a tritium atom at each of the $C_{(4)}$ and the $C_{(4')}$ positions of the acyclic precursor and a further eight elsewhere making a total of twelve (LXVIII). The $T/^{14}C$ ratio of the β-carotene so produced was

(LXVIII)

(* = ^{14}C label)

found to be 12:8. Thus none of the tritium had been lost on cyclisation of the acyclic precursor showing that *β-carotene arises via Scheme 14* and *not* via Scheme 16 (which involves loss of a proton from $C_{(4)}$ and $C_{(4')}$) (Williams, Britton and Goodwin, Biochem. J., 1967, 105: 99).

The α-carotene (one end-group of the α-type and one β-) formed simultaneously with the β-carotene in the above systems was also isolated and the $T/^{14}C$ ratio measured. This came to 7:8 using the $(4R)-[2-^{14}C,4-T]$ MVA (Williams, Britton and Goodwin, 1967, loc. cit.; Biochem. J., 1965, 97: 28C) corresponding to the loss of only one tritium atom at $C_{(6)}$ during cyclisation, this, presumably, being the one lost during formation of the β-end-group. Thus the tritium atom at $C_{(6)}$ at the other end of the acyclic precursor survives the cyclisation to the α-type end-group, thereby excluding the mechanism depicted in Scheme 15 for α-end-group formation. The $T/^{14}C$ ratios found for three of the other carotenoids (δ- and ε-carotenes; α-zeacarotene) in the 1967 (tomato) system showed that their α-end-groups were formed in the same way as that in α-carotene. The α-carotene formed from the $[2-^{14}C,2-T_2]$ MVA experiment contained one less tritium (T/^{14}C ratio = 11:8) than the β-carotene, but this is consistent with any of the above mechanisms.

Thus the β-end-group of both β- and α-carotenes arises according to Scheme 14 via carbonium ion LXVII with loss of the $C_{(6)}$-proton, whilst the α-end-group of α-carotene (and also of the δ- etc. carotenes, above) probably arises from the *same* carbonium ion but by loss of the $C_{(4)}$-proton (or via the first step of Scheme 16 which is the equivalent concerted reaction). From this, and knowing the configuration at the 6-position of the α-end-group of α-carotene (q.v.), it is possible to infer the manner in which the acyclic end-group must fold prior to cyclising (Goodwin, 1971, loc. cit., pp. 305—306).

The above is also relevant to the mode of biosynthesis of the β- and α-carotenes themselves. The $T/^{14}C$ ratios show that in these systems at least (tomato, carrot) the α-carotene does *not*, as was once thought (Vol. II B,

2nd Edn., p. 338), originate by isomerisation of β-carotene, nor does the β-originate from the α-. Instead each arises *independently of the other* from an (possibly the same) acyclic precursor.

(b) Biosynthesis of xanthophylls

(i) The insertion of oxygen functions

It is still thought that, in general, xanthophylls are probably formed by the direct oxidation (mechanism unknown) of the corresponding carotenes rather than by the stepwise dehydrogenation of hydroxy- etc. derivatives of phytoene, phytofluene, etc. (p. 338–9). However, there have been further reports of various oxygenated derivatives of the latter type occurring in nature, though usually only in trace amounts. These newly reported compounds are: phytoene-1,2-epoxide (LXIX) and possibly* the corresponding phytofluene and ζ-carotene derivatives, in tomatoes (Britton and Goodwin, Phytochem., 1969, 8: 2257); a methoxy-phytoene (LXX ?) and methoxy-phytofluene, and, probably, the corresponding hydroxy-derivatives of phytofluene, ζ-carotene, and unsymmetrical ζ-carotene, in various bacteria (Malhotra, et al., ibid., 1970, 9: 2369; Davies, Biochem. J., 1970, 116: 101; Weeks and Garner, Arch. Biochem. Biophys., 1967, 121: 35). The proposed structures were mainly inferred by showing the substances had a u.v. spectrum identical with the parent hydrocarbon (phytoene, etc.) but were more polar to an extent consistent with the presence of an oxygen function (e.g. an −OMe group) which was placed at $C_{(1)}$ by analogy with normal methoxy/hydroxy acyclic carotenoids. However, LXIX was isolated in an amount sufficient to examine its m.s. (molecular wt. and fragmentation

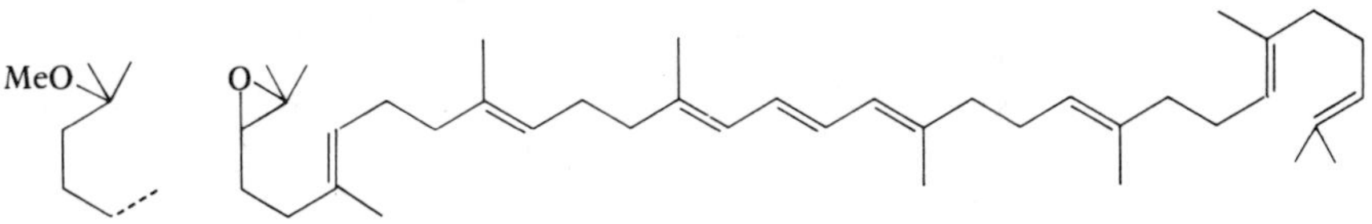

(LXX)   (LXIX)

pattern) and do small-scale reactions. Progress on the transformations labelled (i) and (ii) before (2nd Edn., p. 344) has been as follows:

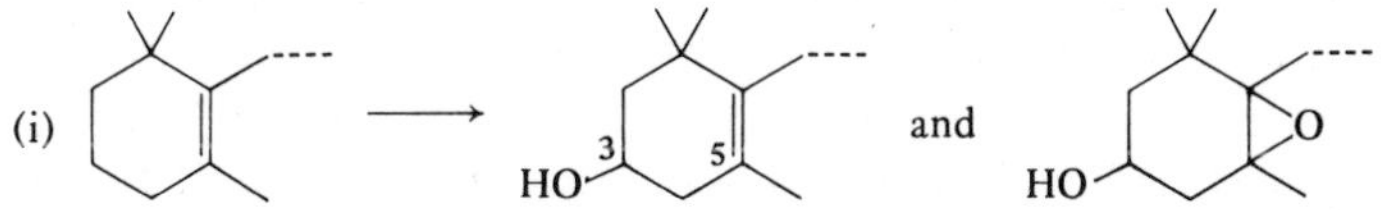

---

* Now confirmed (Goodwin et al., Phytochem., 1973, 12: 2759); but NB none of these compounds is *necessarily* a biogenetic intermediate; they may instead merely be oxidative degradation products.

Further work on the origin of the epoxide oxygen atom in the carotenoid 5,6-epoxides has shown that it is probably derived from molecular oxygen, rather than from water as earlier experiments had suggested. The origin of the hydroxyl oxygen is thereby rendered uncertain (Chichester, Yamamoto et al., Biochim. Biophys. Acta, 1965, 109: 303; 1968, 153: 459). This result is consistent with the finding (Vol. II B, 2nd Edn., p. 294) that on placing green leaves in oxygen, the zeaxanthin therein is converted into the corresponding 5,6-mono- and 5,6;5',6'-di-epoxides (antheraxanthin and violaxanthin). The $C_{(3)}$-position, the position most commonly hydroxylated in cyclic carotenoids, originates from the $C_{(5)}$ of MVA (cf. Section 8a). Measurement of the $T/^{14}C$ ratio (p. 344) of 3-hydroxy-β-carotene (β-cryptoxanthin) and of β-carotene biosynthesized by *Physalis* berries when steeped (a) in $[2-^{14}C,5-T_2]$MVA (LX) and (b) in $(5R)-[2-^{14}C,5-T_1]$MVA (LXI) has shown that hydroxylation involves stereospecific removal of only the "*pro-R*" $C_{(5)}$-protons (cf. p. 346). Therefore hydroxylation of the hydrocarbon end-group precursor does not proceed through the corresponding ketone (T. J. Walton, Britton and Goodwin, Biochem. J., 1969, 112: 383). Since the configuration of the product at $C_{(3)}$ is known to be R (p. 254), the overall reaction can be written as below:

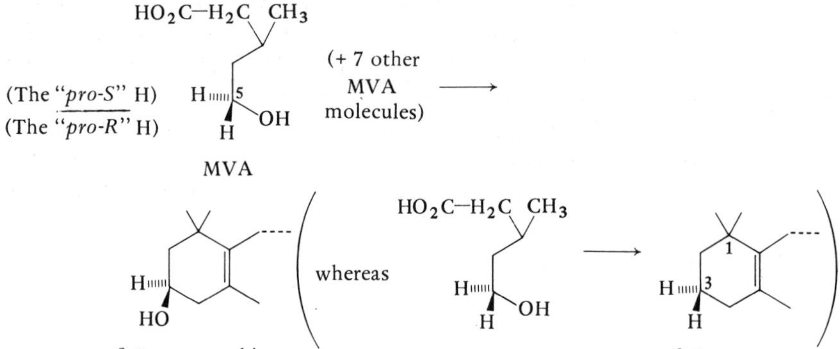

β-Cryptoxanthin   β-Carotene

Exactly parallel incorporation results were obtained for lutein (3,3'-dihydroxy-α-carotene, p. 249) synthesised by maize seedlings in the presence of the above labelled MVA's, in this case two *pro-R* H's from $C_{(5)}$ of MVA being lost for each molecule of lutein produced.

Goodwin et al. (Pure Appl. Chem., 1973, 35: 32) have shown, by a combination of diphenylamine-type inhibition studies (Vol. II B, 2nd Edn., p. 334) and ^{14}C-labelling, that the zeaxanthin in a *Flavobacterium* sp. is formed via the sequence: lycopene → β-carotene → zeaxanthin, O_2 being essential for the second (hydroxylation) stage.

There is some evidence that the keto-carotenoids are formed via the corresponding OH-compounds (e.g. rhodoxanthin, q.v.; Gribanovski-Sassu, Phytochem., 1972, 11: 3195).

(*ii*) $Me_2C(OH) \cdot CH_2 \cdots \rightarrow Me_2C(OMe) \cdot CH_2 \cdots$

The suggestion before (Vol. II B, 2nd Edn., p. 344) that acyclic methoxy-carotenoids (Section 7c) arise from the biological methylation of the corresponding hydroxy-compounds, has been confirmed by ^{14}C-incorporation studies, the methylating agent being identified as S-adenosyl-methionine (Singh et al., Biochem. J., 1973, 136: 413).

(*iii*) *Miscellaneous postulated pathways*

Support for the scheme proposed before (Vol. II B, 2nd Edn., p. 345) for the biosynthesis of the cyclopentane ring-system present in capsorubin, capsanthin, and kryptocapsin, has been provided by the measurement of the absolute configuration of capsorubin's proposed precursor, violaxanthin: the ring—OH has the same ("β"-type) configuration in each molecule (pp. 279, 252).

There has been further support for the proposal that torularhodin (q.v.) originates from γ-carotene by the series of dehydrogenation/oxidation reactions depicted before (Vol. II B, 2nd Edn., p. 345). The $C_{(2)}$-methyl of the isorenieratene-type end-group (cf. p. 331) originates from the methyl group attached to $C_{(3)}$ of MVA (p. 338), confirming (cf. 2nd Edn., p. 346) that formation of this end-group does involve a $C_{(1)} \rightarrow C_{(2)}$ methyl shift at some stage (D. J. Chapman et al., Biochem. J., 1973, 136: 395).

The possible mode of biogenesis of the characteristic structural feature of the allenic, acetylenic, C_{50} (and C_{45}), and C_{38} carotenoids are discussed at the end of the various sections (7f, 7g, 7h and 7j).

Chapter 8

The Cycloheptane, Cyclo-octane and Macrocyclic Groups

S. H. GRAHAM

Novel methods of synthesis have made some formerly inaccessible compounds readily available, and interest in cyclic compounds has been stimulated by their suitability for testing predictions of orbital symmetry theory. In eight-membered and larger rings both *cis* and *trans* double bonds may be encountered and the practice has developed of drawing formalised ring skeletons which allow differentiation of the two types. Thus I and II represent *cis*- and *trans*-cyclo-octene respectively, and III and IV represent *cis,cis*- and *cis,trans*-cyclodeca-1,5-diene:

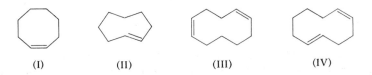

(I) 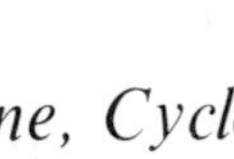   (II) 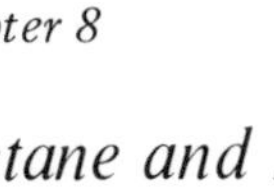   (III)    (IV)

1. Cycloheptane group

(a) Hydrocarbons

Cyclohepta-1,2-diene can be prepared by the action of potassium butoxide on 1-bromocycloheptene: it dimerises spontaneously but can be trapped as its adduct with 2,5-diphenyl-3,4-benzofuran (G. Wittig and J. Meske-Schuller, Ann., 1968, 711: 65). An improved route to **cycloheptyne** is the action of lead tetraacetate on 1-amino-4,5-pentamethylenetriazole (idem, ibid.).

Cyclohepta-1,3,5-triene can be prepared by pyrolysis of 7,7-dichloro-bicyclo[4,1,0]heptane, with only slight contamination by toluene, by correct choice of conditions (O. M. Nefedov and N. N. Noritskaya, Izvest.

Akad. Nauk, S.S.S.R., Ser Khim., 1965, 395). It is oxidised by ceric nitrate to benzaldehyde via, it is believed, the tropylium ion; some benzene is also formed (W. S. Trahanovsky et al., J. Amer. chem. Soc., 1968, 90: 7084). 3,7,7-*Trimethylcyclohepta*-1,3,5-*triene* is formed by pyrolysis of bornadiene (M. R. Willcott and C. J. Boriack, ibid., 1968, 90: 3287). The photochemical rearrangement of cycloheptatrienes leads to some bicyclo[3,2,0]hepta-dienes, as well as redistributing the substituents on the cycloheptatriene ring by a series of highly selective 1,7-methyl and 1,7-hydrogen shifts. In the rearrangement of 3,7,7-trimethylcycloheptatriene only the 1,3,7- and the 1,5,7-trimethylcycloheptatrienes are produced, so that only one of each of the possible methyl and hydrogen shifts is observed (L. B. Jones and V. R. Jones, ibid., 1967, 89: 1880; 1968, 90: 1540):

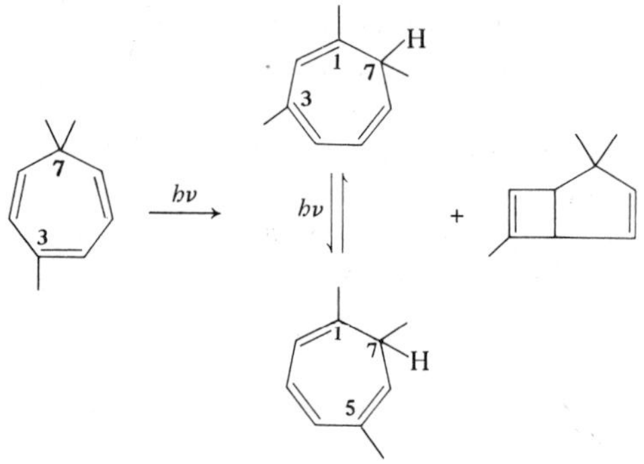

The thermal rearrangements of cycloheptatrienes involve, of course, 1,5-hydrogen shifts: when this is not possible in the 7,7-disubstituted cycloheptatrienes a skeletal rearrangement, which involves successive electro-cyclic ring closure, [1,5] sigmatropic rearrangement of the bicycle, and electrocyclic ring opening, can be observed. That it is a true skeletal re-arrangement rather than a succession of methyl shifts is shown by deuterium labelling: alkyl groups — other than those on $C_{(7)}$ — do not move relative to the label (J. A. Berson and M. R. Willcott, ibid., 1966, 88: 2494). Cycloheptatriene forms both [4π + 2π] and [6π + 4π] cycloadducts with 2,6-dimethyl-3,5-diphenyl-cyclopentadiene (K. N. Houk and R. B. Wood-ward, ibid., 1970, 92: 4143 and 4145):

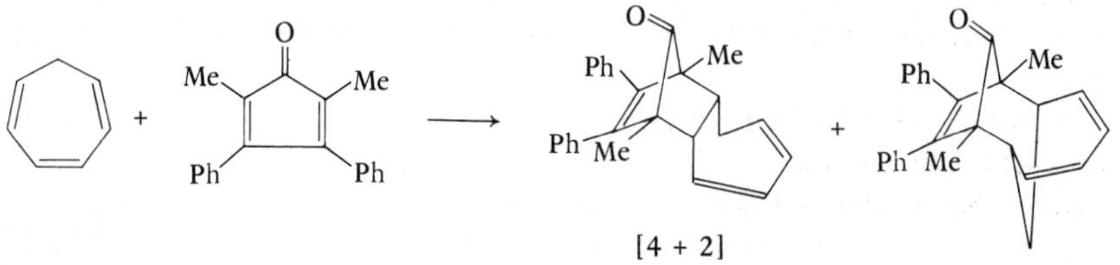

[4 + 2]

[6 + 4]

Thermal hydride shifts are also reported in the cycloheptadiene series (V. A. Mironov et al., Tetrahedron Letters, 1969, 499).

(b) Alcohols

trans-Cycloheptane-1,3-diol is obtained, via its diacetate, by the action of lead tetra-acetate on bicyclo[4,1,0]heptane (Sung Moon, J. org. Chem., 1964, 29: 2718). Also formed is the acetate of cyclohept-2-enol and, in general, derivatives of this alcohol are accessible by solvolysis of 8-halogeno-bicyclo[4,1,0]heptanes (P. S. Skell and S. R. Sandler, J. Amer. chem. Soc., 1958, 80: 2024). Since the ring opening, which is necessarily disrotatory, is constrained to give derivatives of the *cis*-enol, the *endo*-halogen is selectively lost (S. F. Cristol, R. M. Segueira and C. D. de Puy, ibid., 1965, 87: 4007).

Acetolysis of 4-cycloheptenylmethyl brosylate gives almost exclusively *endo*-bicyclo[3,2,1]octyl acetate (G. le Ny, Compt. Rend., 1960, 251: 1256). Since the rate of the solvolysis shows a thirty-fold acceleration over that of the dihydro derivatives, assistance by the π-electrons is postulated, Even when the brosylate function is replaced by the poorer leaving group, dimethylsulphonium, the reaction still proceeds, and this is the first recorded example of non-enzymatic alkylation of a double bond by a sulphonium salt . (C. Chuit and H. Felkin, ibid., 1967, 264: 1412).

(c) Ketones

Cycloheptanone can be prepared from the acid chloride of 7-heptenoic acid by cyclisation (aluminium chloride catalyst) and reduction with triphenyltin hydride (W. S. Trahanovsky et al., J. org. Chem., 1969, 34: 3679). Cycloheptane-1,3-dione is prepared from 2,7-dimethoxycycloheptatriene by reduction and hydrolysis (T. Maclean and R. P. A. Sneeden, Tetrahedron, 1965, 21: 31).

(d) Carboxylic acids

4-Cycloheptenylcarboxylic acid is obtained in good yield by the action of alkali on the methiodide of 2-(N-pyrrolidyl)bicyclo[3,2,1]octan-8-one (G. Stork and H. K. Landesman, J. Amer. chem. Soc., 1956, 78: 5129).

2. Tropylium salts, tropone

Tropone can be prepared by air oxidation of tropylium azide in a benzene–alumina slurry (L. N. McCullagh and D. S. Wulfman, Synthesis, 1972, 422). It is better regarded as an unsaturated ketone rather than as a quasi-aromatic structure. Comparison with appropriate models and molecular orbital cal-culations show that the dipole moment is not consistent with much contribution from dipolar structures. Analysis of the n.m.r. spectrum leads to the same conclusion (D. J. Bertelli and T. G. Andrews, Tetrahedron Letters, 1967, 4467; J. Amer. chem. Soc., 1969, 91: 5280, 5286). Similarly, X-ray diffraction studies on 2-chlorotropone show that while the bonds, other than the $C_{(1)}-C_{(2)}$ bond, are intermediate between single and double bonds, there is alternation of bond length so that delocalisation is only partial (E. J. Forbes et al., Chem. Comm., 1966, 114; D. J. Watkin and T. A. Hamer, J. chem. Soc., B, 1971, 2167).

(a) Tropylium salts

Tropylium fluoborate is prepared by the reaction of phosphorus pentachloride with cycloheptatriene to give the double salt $(C_7H_7^{\oplus}PCl_6^{\ominus})(C_7H_7^{\oplus}Cl^{\ominus})$, which is then treated with fluoroboric acid (K. Conrow, Org. Synth., 1963, 43: 101). Oxidation of cycloheptatriene to the tropylium ion by the ion radical $Ar_3\dot{N}^{\oplus}$ has also been reported (P. Beresford and A. Ledwith, Chem. Comm., 1970, 15).

Tricarbonyltropyliumchromium salts may be prepared from tricarbonyl-cycloheptatrienechromium (J. D. Munro and P. L. Pauson, J. chem. Soc., 1961, 3475). Their reactions with nucleophiles are described in a series of papers (P. L. Pauson and K. H. Todd, J. chem. Soc., C, 1970, 2638 and earlier papers).

(b) Tropone

The study of the cycloaddition reactions of tropone was prompted by the predictions of Woodward and Hoffmann. With cyclopentadiene a $[6\pi + 4\pi]$ ·adduct V is obtained (R. C. Cookson et al., Chem. Comm., 1966, 15). However 2-chlorotropone and cyclopentadiene form the $[4\pi + 2\pi]$ adducts VI and VII, and similarly diethyl azodicarboxylate adds to give VIII (Y. Kitahara et al., Tetrahedron Letters, 1967, 3003). $[6\pi + 4\pi]$ Adducts are also formed with butadiene, isoprene and piperylene (Sho Ito et al., ibid., 1972, 2223). Further examples are given by Woodward and Hoffmann ("The Conservation of Orbital Symmetry", Verlag Chemie, Academic Press, 1970, pp. 84—85). Photodimerisation of 2-chlorotropone gives the $[6\pi + 6\pi]$ adduct IX, as well as X and XI formed by $[4\pi + 2\pi]$ addition (H. Watanabe, Tetrahedron Letters, 1968, 4065). Tropone itself gives only the $[6\pi + 6\pi]$ adduct XII when irradiated in sulphuric acid, but irradiation of an aceto-nitrile solution gives the adducts XIII, XIV and XV (T. Mukai, T. Tezuka and Y. Akazubi, J. Amer. chem. Soc., 1966, 88: 5025; A. S. Kende, ibid., 1966, 88: 5026; 1967, 89: 5283). Diphenylketene and enamines of cyclic ketones add to tropone to give XVI and XVII respectively (R. Gomper, A. Studener and W. Elsner, Tetrahedron Letters, 1968, 1019; Y. Kitahara et al., Chem. Comm., 1969, 737).

Nucleophilic displacements can be observed with 3-substituted tropones (Schuichi Seto, ibid., 1968, 562; K. Takase, Bull. chem. Soc., Japan, 1964, 37: 1288).

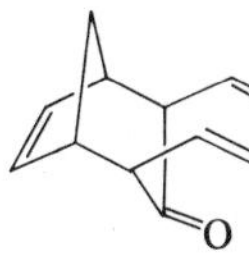

(V)

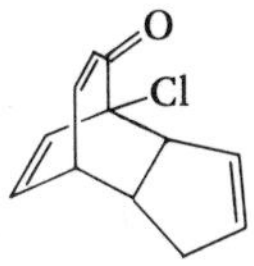

(VI) *exo*
(VII) *endo*

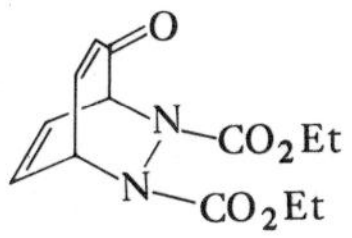

(VIII)

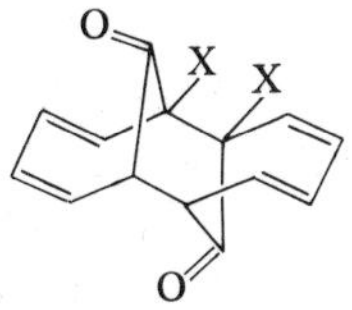

(IX) X = Cl
(XII) X = H

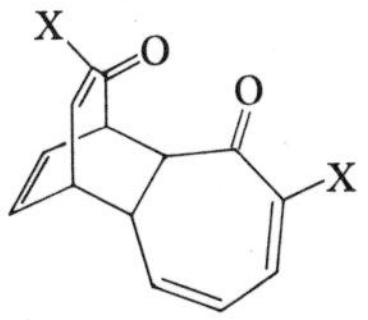

(X) X = Cl
(XIII) X = H

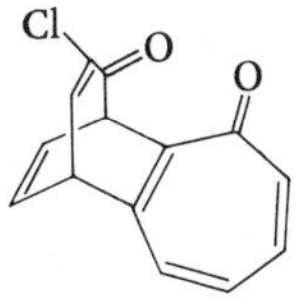

(XI)

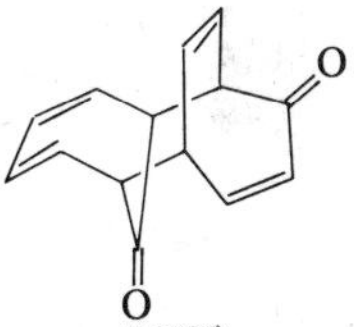

(XIV)

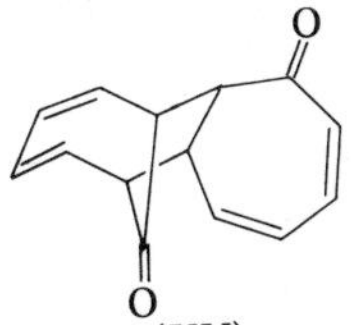

(XV)

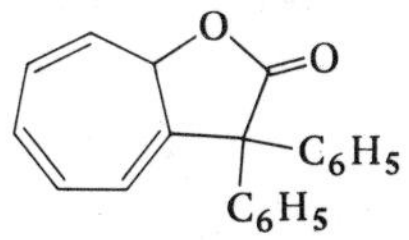

(XVI)

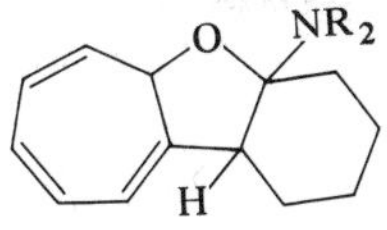

(XVII)

(c) *Tropolones*

A new synthesis of tropolones involves addition (e.g. of acrylonitrile) to the betaine from 3-hydroxy-N-methylpyridinium hydroxide to give XVIII; pyrolysis of the methohydroxide of this gives a dihydrotropone which is oxidised under the conditions of the experiment to a dimethylaminotropone (hydrolysable to a tropolone) (A. R. Katritzky and Y. Takeuchi, J. Amer. chem. Soc., 1970, 92: 4135):

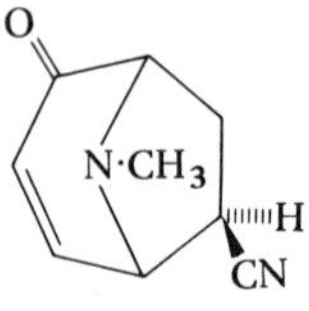

(XVIII)

γ-*Tropolone* is readily prepared from the photoadduct of 1,2-dichloro-ethylene and 3-acetoxycyclopentenone by brief treatment with alkali: the method has been extended to the preparation of stipitatic and stipitatonic acids (P. de Mayo et al., Chem. Comm., 1967, 704; J. org. Chem., 1969, 34: 794). Cycloheptane-1,2-dione, an intermediate in the original tropolone synthesis, is available from the mixed ester condensation of diethyl oxalate and a pimelic ester (S. D. Saraf, Austral. J. Chem., 1969, 22: 2025).

The Mannich reaction affords a convenient means of introducing carbon side-chains into the tropolone ring (Pauson et al., J. chem. Soc., C, 1970, 1323 and earlier work there cited).

(d) Rearrangements of tropones

As well as the better known base-catalysed rearrangement to benzoic acids, 2-halotropones may rearrange to salicylaldehydes, and the balance between the competing processes is affected by the alkali concentration: more dilute alkali favours hydroxylaldehyde formation (E. J. Forbes, D. C. Warrell and W. J. Fry, J. chem. Soc., C, 1967, 1693). Labelling experiments show that the aldehydic carbon derives from $C_{(3)}$ of the tropone:

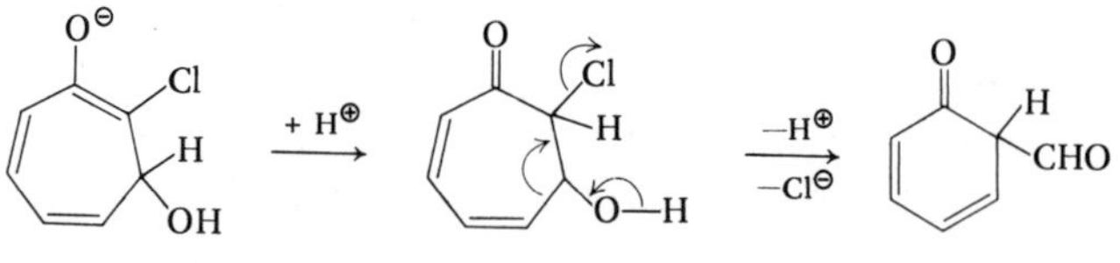

The protonation of the initial anion by water will not occur in more concentrated alkali and hence formation of benzoic acid, which involves no protonation, supervenes.

The formation of benzoic acids probably involves one or other of the intermediates XIX or XX, though it has been objected that when two potential leaving groups are available it is the stronger base which is ejected.

Should ring contraction precede expulsion, giving XXI or XXII, then the reaction path would be dictated by the capacity of substituents to delocalise the anionic charge. Electron-withdrawing substituents do have the orienting influence which would then be predicted, but attempts to demonstrate directly the participation by ions such as XXI or XXII failed. For simple tropones, XIX and XX are best regarded as the most probable intermediates: the weaker base may be retained because it is better able to stabilise anionic charge in the transition state leading to final product (R. M. Magid, C. R. Grayson and D. R. Cousar, Tetrahedron Letters, 1968, 4819 and refs. cited therein).

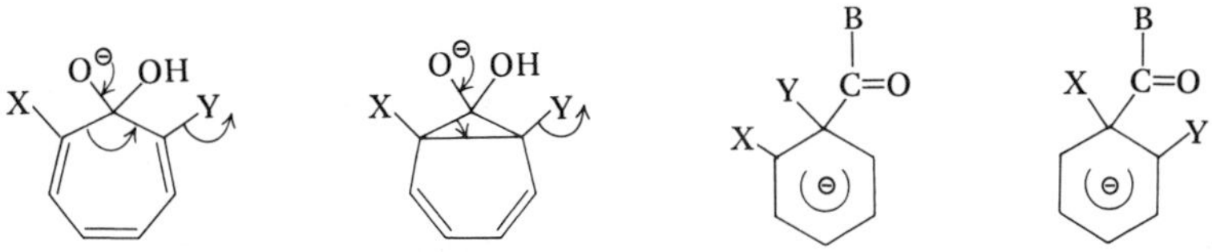

(XIX)   (XX)   (XXI)   (XXII)

Various nitro-2-chlorotropones rearrange in the presence of mild oxidising agents (e.g. silver nitrate) to m-hydroxybenzaldehydes, indeed the process is so facile that some rearrangement occurs even in the absence of specifically added oxidising agent (Forbes, M. J. Gregory and D. C. Warrell, J. chem. Soc., C, 1968, 1969). Labelling experiments show that the initial attack of nucleophile is on $C_{(6)}$:

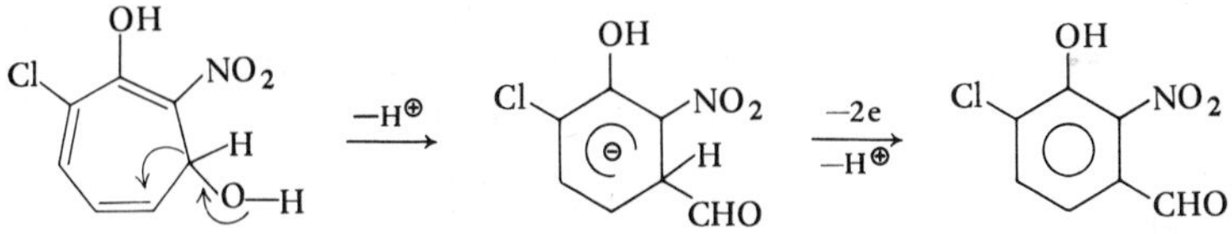

Hexafluorotropone (plates, m.p. 102–103°), which is obtained by hydrolysis of octafluorocycloheptatriene, shows little tendency to rearrange to a benzenoid structure. It is converted by methoxide ion into a symmetrical tetrafluorodimethoxytropone (J. C. Tatlow et al., Chem. Comm., 1972, 803).

(e) Photochemistry of tropolones

Irradiation of an aqueous solution of α-tropolone gives the bicyclic ketol XXIII as well as 3-oxocyclopent-4-enylacetic acid. The ketol XXIII is isomerised to XXIV at 150° (the process is of interest as being a possible example of an *antara-antara*[3,3] sigmatropic rearrangement); some tropolone is also formed and it can be obtained by the action of base on XXIII (A. C. Day and M. A. Ledlie, Chem. Comm., 1970, 1265). Tropone adds on oxygen when irradiated in the presence of a sensitiser: the resulting 2,5-peroxycycloheptadienone is reduced to tropolone with thiourea and isomerised to 5-hydroxytropolone with triethylamine (M. Oda and Y. Kitahara, Tetrahedron Letters, 1969, 3295). 2-Methoxytropone gives a similar peroxide (reducible to 5-hydroxytropolone) which reacts with alcohols to give esters $RO_2C \cdot CH : CH \cdot CHOH \cdot CH : CH \cdot CO_2R$ (E. J. Forbes and J. Griffiths, J. chem. Soc., C, 1968, 575).

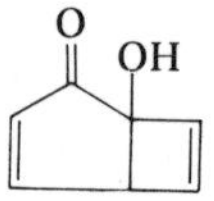

(XXIII) 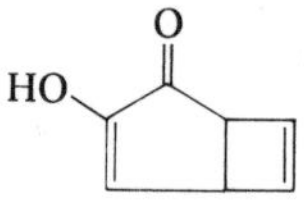   (XXIV)

(f) Thia analogues of tropolone

Bis-thiotropolone is formed from sodium hydrogen sulphide and 1,2-diethoxytropylium fluoborate (C. E. Forbes and R. H. Holm, J. Amer. chem. Soc., 1968, 90: 6884). It is unstable to air but various metal complexes are stable to air and moisture.

3. Macrocyclic compounds

(a) Hydrocarbons

Macrocyclic olefins are now often the best starting materials for the synthesis of other macrocycles, since they are available from the cyclo-oligomerisation of olefins (for review see G. Wilke, Angew. Chem., intern. Edn., 1963, 2: 105, and for specific examples see P. Heimbach et al., ibid., 1969, 8: 753; Ann., 1969, 727: 143 et seq.). The self-condensation of allylic halides in the presence of nickel carbonyl is also of value, for example the preparation of 1,4,7-trismethylenecyclononane from 1,1-bischloromethylethylene (E. J. Corey and M. Semmelhack, Tetrahedron Letters, 1966, 6237), and of various cyclic 1,5-dienes (E. J. Corey and E. K. W. Wat, J. Amer. chem. Soc., 1967, 89: 2757). For a review of synthetic methods for cycloalkynes see H. Meier (Synthesis, 1972, 235–253).

(*i*) *Cyclic olefins*

trans-**Cyclo-octene** can be obtained by photoisomerisation of *cis*-cyclo-octene (J. S. Swenton, J. org. Chem., 1969, 3217) or from dimethylamino-cyclo-octane (A. C. Cope and R. D. Bach, Org. Synth., 1969, 49: 39): in both cases it is separated from accompanying *cis*-isomer by its capacity to add silver nitrate. It can be prepared from *trans*-cyclo-octane-1,2-diol by the action of butyl lithium on its benzaldehyde acetal, a stereospecific route which is simpler than the thionocarbonate procedure (G. H. Whitham et al., Chem. Comm., 1968, 1593; see also E. Vedejs and P. L. Fuchs, J. Amer. chem. Soc., 1971, 93: 4070). Its absolute configuration, and conformation in the crystal lattice, have been determined by X-ray diffraction methods (D. P. Shoemaker et al., ibid., 1970, 92: 5260).

Cyclo-octyne forms stable complexes with cuprous bromide and gold bromide, complexing with other metals can convert it into oligomers (G. Wittig and S. Fischer, Ber., 1972, 105: 3542).

Cyclo-octa-1,2-diene can be obtained from cyclo-octyne by base-catalysed isomerisation; it is trapped as its tetracyclone adduct as it is unstable (Wittig et al., Ann., 1968, 711: 55). cis,cis-*Cyclo-octa*-1,4-*diene* can be prepared from the 1,3-diene by reaction with *N*-bromosuccinimide and reduction of the resulting mixture of bromides (Sung Moon and C. R. Ganz, J. org. Chem., 1969, 34: 465).

Dimerisation of isoprene with a nickel catalyst gives mainly 1,5-dimethyl-*cis,cis*-cyclo-octa-1,5-diene by head-to-tail dimerisation (W. Heggie and J. K. Sutherland, Chem. Comm., 1972, 957).

trans,trans-*Cyclo-octa*-1,5-*diene* is formed by irradiation of the cuprous chloride complex of the *cis,cis*-isomer. It is very unstable, although models indicate a lesser degree of strain than in the *cis,trans*-diene. It has a remarkable ultraviolet absorption (λ_{max} 246 nm, $\epsilon \sim$ 1500), and this seems to be due to genuine interaction between non-conjugated π-orbitals and not to ring-strain (G. M. Whitesides, G. L. Coe and A. C. Cope, J. Amer. chem. Soc., 1969, 91: 2608). Similar absorption has been reported for 1,4-bis-methylenecyclo-octane (C. S. Dean, D. A. Jonas and S. H. Graham, J. chem. Soc., C, 1968, 3045) as well as for the sesquiterpene germacratriene.

Additions to *cis,cis*-cyclo-octa-1,5-diene may give tetra-substituted cyclo-octanes, as in the addition of bromine (R. A. Raphael et al., ibid., 1969, 474) or of peracids (Cope et al., J. org. Chem., 1969, 34: 2231): alternatively bridged bicyclic products are obtained with mercuric acetate (C. Gunter et al., Helv., 1970, 53: 1619), nickel carbonyl (B. Fell, W. Seide and F. Asinger, Tetrahedron Letters, 1968, 1003) and sulphur dichloride (Corey and E. Block, J. org. Chem., 1966, 31: 1663; E. D. Weil et al., ibid., 1669):

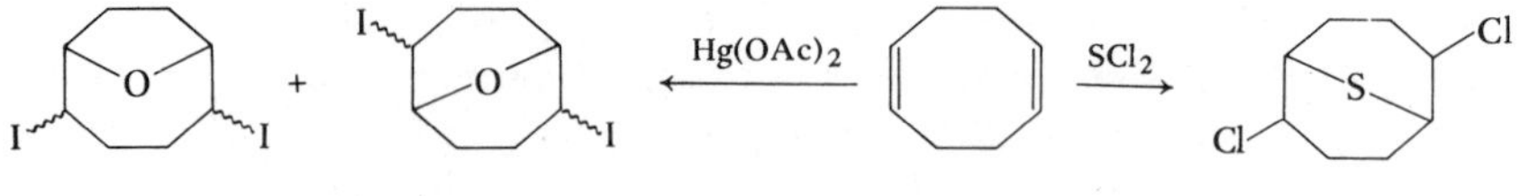

Molecules with exocyclic double bonds tend more to give bridged bicyclic adducts, as in addition of bromine to 1,5-bismethylenecyclo-octane (Graham et al., Tetrahedron, 1968, 24: 3445).

Cyclo-octa-1,3,5-triene is known as the all-*cis* isomer: it has been suggested that the *cis,trans,cis* isomer may be an intermediate in the degenerate electrocyclic rearrangement of bicyclo[4,2,0]octadiene which is detected by the scrambling of a deuterium label:

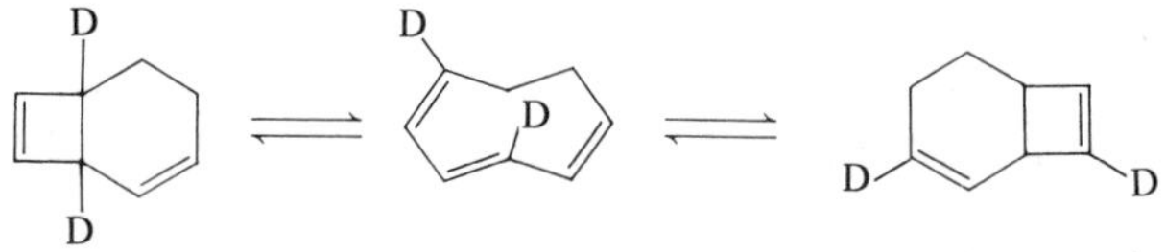

The thermal opening and closing of the cyclobutene ring must be conrotatory, necessitating *trans*-geometry in the intermediate (J. E. Baldwin and M. S. Kaplan, J. Amer. chem. Soc., 1972, 94: 4696).

Addition of sulphur dioxide to *cis,cis,cis*-cyclo-octatriene is governed by orbital symmetry requirements: it is necessarily suprafacial and hence forms XXV rather than XXVI. The adduct XXV dissociates at 100–120°, but XXVI only at 200°, and then less cleanly (W. L. Mock, ibid., 1970, 92: 3807):

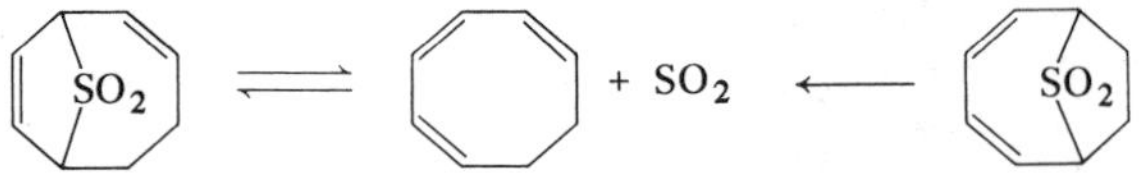

(XXV)   (XXVI)

Cyclo-octatrienes are connected by thermally permitted pathways with octatetraenes on the one hand and bicyclo[4,2,0]octadienes on the other, ring closure of the octatetraene is a conrotatory process, the formation of a bicyclic diene is disrotatory (hence leading to a *cis* ring-fusion). The *trans,cis,cis,trans*-decatetraene cyclises smoothly to triene at a low temperature, so that formation of the bicycle may be seen as a separate stage. The all-*cis*-tetraene, because of steric hindrance in the transition state for cyclisation, requires a much higher temperature and only the bicycle is obtained. The mono-*trans*-tetraene behaves in the anticipated intermediate fashion (R. Huisgen et al., ibid., 1967, 89: 7130):

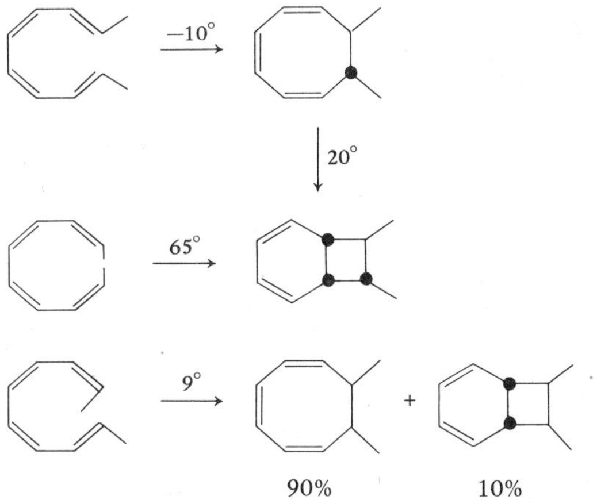

A related observation is that carbonation of cyclo-octa-tetraene dianion gives *trans,cis,cis,trans*-dicarboxyoctatetraene, which is thermally isomerised to *trans*-7,8-dicarboxybicyclo[4,2,0]octadiene (T. S. Cantrell, Tetrahedron Letters, 1968, 5635). The formation of cyclo-octatriene by isomerisation of octa-1,7-diyne presumably involves a similar ring-closure (G. Eglinton, R. A. Raphael and J. A. Zabkiewicz, J. chem. Soc., C, 1969, 469).

Cyclonona-1,2-diene (L. Skattebøl and S. Solomon, Org. Synth., 1969, 49: 35) can be resolved into chiral forms (A. C. Cope et al., J. Amer. chem. Soc., 1970, 92: 1243) or prepared in a state of high optical purity from chiral *trans*-cyclo-octene via its dibromocarbene adduct (Cope, loc. cit.), or diazomethane adduct (Corey and J. I. Shulman, Tetrahedron Letters, 1968, 3655). The chiral allene reacts with mercury salts in ethanol to give 3-ethoxycyclononenes with a degree of stereospecificity depending on the salt used (R. D. Bach, J. Amer. chem. Soc., 1969, 91: 1771). The dimerisation of cyclonona-1,2-diene to give the tricyclic dienes XXVII, XXVIII and XXIX is of interest as the product distribution varies accordingly as chiral or racemic allene is used: two allene molecules of the same chirality cannot generate XXVIII as efficiently as can two of opposite chirality. Although it has not been established that the dimerisation is a concerted process, these results are in accord with the predictions of

Woodward and Hoffmann (W. R. Moore, Bach and T. M. Ozretich, ibid., 1969, 91: 5918):

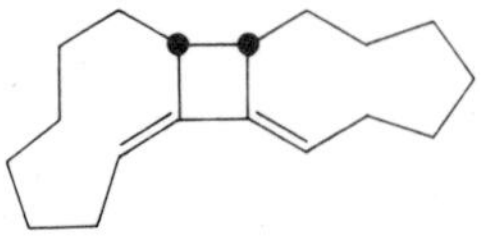

(XXVII) 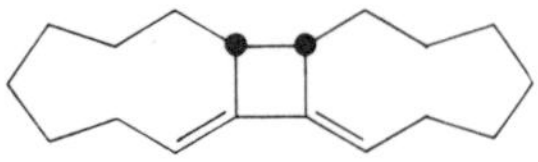   (XXVIII)

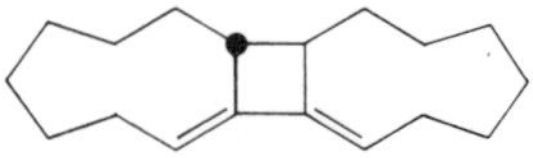

(XXIX)

	XXVII	XXVIII	XXIX
Yield from chiral allene	0.4%	11.8%	79.8%
Yield from racemic allene	6.3%	62.5%	31.2%

Benzene-sensitised photolysis of cyclonona-1,2-diene furnishes the first example of an allene → cyclopropylidene change, the reverse of the usual process.

The intermediate cyclopropylidene forms tricyclo$[3,3,0,0^{2,9}]$nonane by an insertion reaction (H. R. Ward and E. Karafiath, ibid., 1968, 90: 2193):

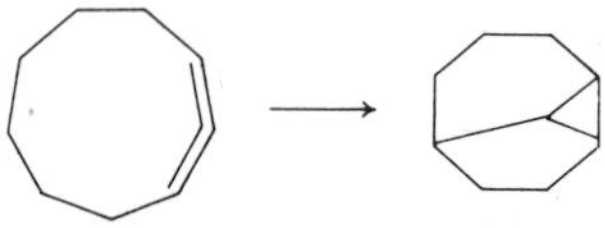

Cyclononatetraene is more stable than might have been anticipated, it is formed from the aromatic cyclononatetraenide ion by quenching in water and has a half-life of 50 min at 23° (P. Radlick and G. Alford, ibid., 1969, 91: 6529; S. Masamune et al., Chem. Comm., 1969, 1203):

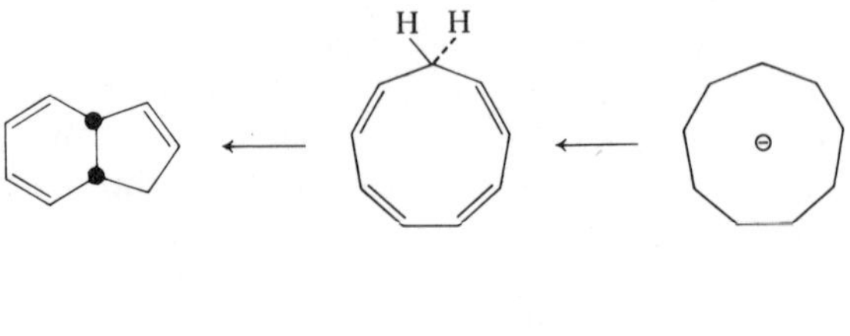

The formation of the cyclononatetraenide ion from carbene adducts of cyclo-octatetraene gives first the mono-*trans* ion, as required by orbital symmetry rules, and this rearranges to the all-*cis* (planar) iron (G. Boche, D. Martons and W. Danzer, Angew. Chem., intern. Edn., 1969, 8: 984). Nonafulvenes (methylenecyclononatetraenes) can exist if stabilised with electron-donating substituents (K. Hafner and H. Tappe, ibid., 1969, 8: 593). They undergo electrocyclic ring-closure to dehydro-indenes.

Cyclodecene is made by dehydrochlorination of chlorocyclodecane (J. G. Traynham et al., J. org. Chem., 1967, 32: 510). Use of potassium butoxide for the reaction gives almost pure *cis*-olefin, while with lithium dicyclohexylamide the *trans*-olefin is obtained (see p. 382, see also Corey and T. Durst, J. Amer. chem. Soc., 1968, 90: 5553 for a stereospecific route to *trans*-cyclodecene).

cis,trans-Cyclodeca-1,5-diene is available from co-polymerisation of butadiene and ethylene in the presence of a complex nickel catalyst (G. Wilke and P. Heimbach, Belg. P. 630,046/1958). It is isomerised thermally to *cis*-1,2-divinylcyclohexane (Heimbach, Angew. Chem., intern. Edn., 1964, 3: 702) or, by irradiation in the presence of iron pentacarbonyl into a mixture of the cyclodeca-1,6-dienes in which the *cis,cis*-isomer predominates (idem, ibid., 1966, 5: 594). Most additions occur preferentially at the *trans* double bond, including reduction with di-imide (Traynham et al., J. org. Chem., 1967, 32: 3285).

cis,cis-Cyclodeca-1,6-diene is the most stable of all the isomeric cyclodecadienes and is obtained from the others by isomerisation with potassium deposited on alumina (A. J. Hubert and J. Dale, J. chem. Soc., 1963, 4091): it cannot therefore be hydrogenated to mono-ene over this catalyst as can larger cyclic dienes, for hydrogenation proceeds by reductive removal of the conjugated isomer from an equilibrium (Hubert, ibid., C, 1967, 3222). Substituted cyclodeca-1,6-dienes are reported to be unreactive towards normal olefinic additive reagents (H. W. Guin et al., J. Amer. chem. Soc., 1966, 88: 5366).

For improved preparations of cyclododecyne and cyclododecene see W. Ziegenbein and W. M. Schneider (Ber., 1965, 98: 825, also M. Ohno and M. Okamoto, Org. Synth., 1969, 49: 30).

(*ii*) *Annulenes*

Cyclo-octa-1,3,5,7-tetraene has alternating single and double bonds: Kekulé type isomers, XXX and XXXI have been isolated. They interconvert at −12° (ΔF^* 78.65 kJ · mole^{-1}); at equilibrium about 5% of XXXI is present, but its proportion can be increased by photo-irradiation (F. A. L.

Anet and L. A. Bock, J. Amer. chem. Soc., 1968, 90: 7130). 1,2-Dicarboxy-cyclo-octatetraene apparently exists exclusively as XXXII, it will not form an anhydride. Attempts to resolve it were unsuccessful (Cope and J. E. Meili, ibid., 1967, 89: 1883).

Photo-irradiation of 1,2,4,7-tetraphenylcyclo-octatetraene isomerises the normal all-*cis* olefin to the strained mono-*trans* olefin (E. W. White et al., ibid., 1969, 91: 523):

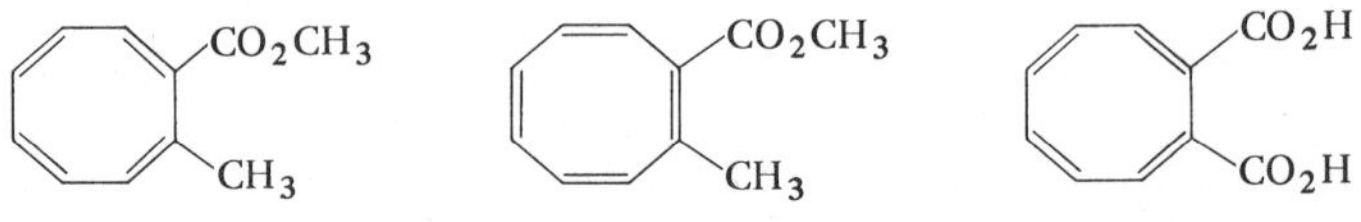

(XXX) (XXXI) (XXXII)

The equilibrium between cyclo-octatetraene and bicyclo[4,2,0]octatriene lies entirely towards the monocycle. Bridging positions 1 and 4 with another ring displaces the equilibrium towards the triene (L. A. Paquette and J. C. Phillips, Chem. Comm., 1969, 680). Dienophiles normally form only the adducts of the bicyclic triene, but the more reactive dienophile azo-dicarboxylic-*N*-phenylimide forms XXXIII as well as XXXIV (A. B. Evrin, Tetrahedron Letters, 1968, 5863). Similarly it reacts with chlorocyclo-octatetraene to give the adduct of both 1-chloro- and of 7-chloro-bicyclo[4,2,0]octatriene showing that both bicycles are present at equilibrium – through the latter is more reactive (R. Huisgen et al., J. Amer. chem. Soc., 1970, 92: 4105). Cyclo-octatetraene itself behaves as a dieno-phile with 5,5-dimethoxytetrachlorocyclopentadiene (G. I. Fray and D. P. S. Smith, J. chem. Soc., C, 1969, 2710):

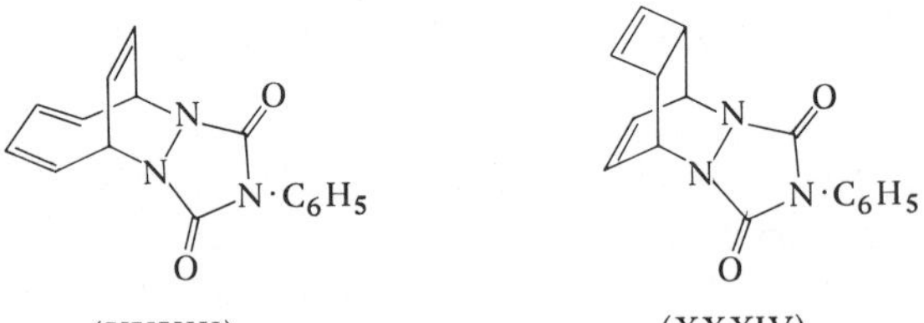

(XXXIII) (XXXIV)

The 1-bromobicyclo[4,2,0]octatrienes are intermediates in the stereospecific rearrangement of bromocyclo-octatetraene to *trans*-β-bromostyrene. By using di-substituted compounds it has been shown that a 1,3-shift of bromine occurs and the reaction is considered to occur within an ion pair, accounting for the stereospecificity. The final ring-opening is conrotatory (Huisgen and W. E. Konz, J. Amer. chem. Soc., 1970, 92: 4102):

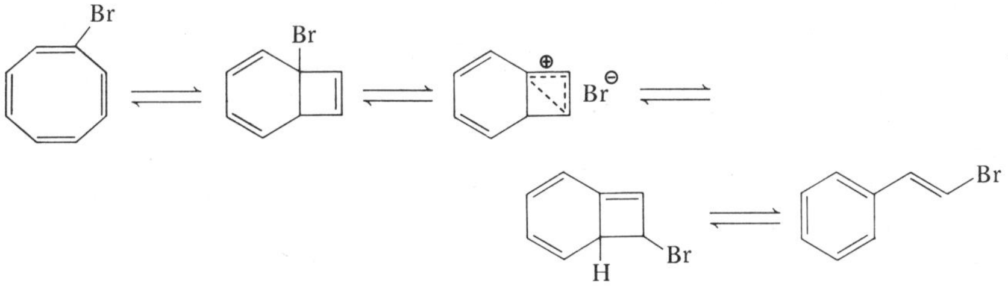

The cyclo-octatetraene 1,2-dichlorides react with Lewis acids in aprotic solvents to give *exo*-8-chlorohomotropylium salts, and with fluorosulphonic acid to give the *endo* isomer. These salts are intermediates in the inter-conversion of the *cis* and the *trans* isomers of the dihalides, and in the addition of chlorine to cyclo-octatetraene (Huisgen and J. Gasteiger, Angew. Chem., intern. Edn., 1972, 11: 1104; Tetrahedron Letters, 1972, 3665, where earlier refs. are given).

As well as the iron tricarbonyl complex, in which the metal is bonded to one butadiene unit of the cyclic olefin, cyclo-octatetraene and its tetra-methyl homologue form a series of complexes with other metal carbonyls in which the metal is bonded to a hexatriene unit, leaving one isolated double bond (M. J. Bennett, F. A. Cotton and J. Takats, J. Amer. chem. Soc., 1968, 90: 903, where earlier papers are cited), while in *bis*-cyclo-octa-tetraene iron (a selective catalyst for the co-dimerisation of butadiene and ethylene) the metal is bonded to four carbon atoms of one ring and six of the other (G. Allegra et al., ibid., 1968, 90: 4455). (For the structure of cyclo-octatetraene di-iron pentacarbonyl see R. Pettit et al., ibid., 1966, 88: 3158; Cotton and M. D. La Prade, ibid., 1968, 90: 2026). These compounds at ambient temperatures all show a single n.m.r. absorption due to valence tautomerism within the complex. Alternative interpretations have been made of the temperature dependence of their n.m.r. spectra (M. Green et al., Chem. Comm., 1967, 523 and refs. cited therein).

Cyclo-octatetraene iron tricarbonyl will undergo electrophilic substitu-tion, e.g. the introduction of the aldehyde group (J. Lewis et al., ibid., 1969, 595).

Higher annulenes. In the study of these molecules, n.m.r. spectroscopy is the key tool. While the $4n + 2$ annulenes can sustain a diamagnetic ring current, the $4n$ annulenes sustain a paramagnetic current (for a theoretical explanation see H. C. Longuet Higgins, Special Publication No. 21, The

Chemical Society, London, 1967, p. 109). Consequently while the exterior protons in both cases resonate at about $\tau 1 - \tau 4$ the interior protons in the $4n + 2$ annulenes are shielded (to $> \tau 10$) and deshielded (to $< \tau 0$) in the case of the $4n$ annulenes. The ratio of interior to exterior protons may help to decide between alternative configurations for a particular annulene. These effects are commonly seen only in low-temperature spectra as site-averaging may be rapid at room temperature.

The Biot—Savart shielding law has been used to set up relationships between chemical shift and ring current, and hence to compare the aromatic character of related annulenes (R. C. Haddon, Tetrahedron, 1972, 28: 3613).

For a review of annulene chemistry see F. Sondheimer (Accts. Chem. Res., 1972, 5: 81), and for a review of conformational mobility and bond shift in annulenes see J. F. M. Oth (Pure and Appl. Chem., 1971, 25: 573), where a system of notation for specifying configurational isomers is also described.

It has been estimated that the resonance energy of [18]annulene (the only macrocyclic annulene of reasonable stability) cannot exceed 78 kJ · mole^{-1} (J.-M. Gilles, Oth, Sondheimer and E. P. Woo, J. chem. Soc., B, 1971, 2177). On the basis of the n.m.r. studies referred to above a tetradehydro[18]annulene would be more aromatic than [18]annulene itself, though the presence of the triple bonds (while releasing the steric strains due to internal protons) might have been expected to suppress bond delocalisation, for the Kekulé canonical forms would no longer be equivalent. [18]Annulene undergoes rapid conformational interconversion which renders all protons equivalent above 40°, but 1,2-dihalogeno derivatives exist as single conformations with temperature independent n.m.r. spectra (Oth, Angew. Chem., intern. Edn., 1972, 11: 51).

[10]Annulene can be prepared by low temperature photolysis of cis-9,10-dihydronaphthalene: earlier claims that it could similarly be made from the trans-isomer have been challenged (S. Masamune and R. T. Seidner, Chem. Comm., 1969, 542, but cf. E. E. van Tamelen and R. H. Greeley, ibid., 1971, 601). Two isomers of the annulene are generated and both can be obtained crystalline. One collapses thermally at $-25°$ to cis-9,10-dihydronaphthalene and the other at $-10°$ to the trans-isomer. Since the ring closures must be disrotatory the two annulenes can be identified as the mono-trans and the all-cis isomer respectively. Their n.m.r. spectra, as well as their low stability, show that neither is aromatic (Masamune et al., J. Amer. chem. Soc., 1971, 93, 4966). Cyclodeca-2,4,8-triene-1,6-dione shows no tendency to tautomerise to a dihydroxy[10]annulene (P. J. Mulligan and Sondheimer, ibid., 1967, 89: 7118).

[12]Annulene is thermally unstable, but its dianion is much more stable

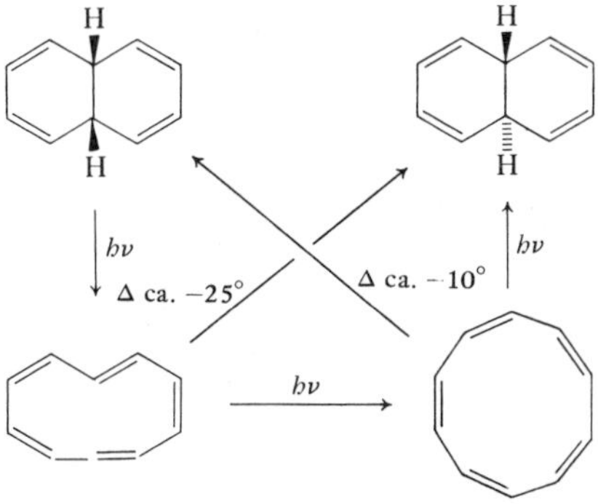

Mono-*trans* [10] annulene   All-*cis* [10] annulene

and supports a diamagnetic ring current. The annulene undergoes rapid conformational isomerisation but, unlike larger annulenes, does not undergo bond shift. Presumably steric repulsions of internal protons prevent it becoming co-planar (Oth and G. Schröder, J. chem. Soc., B, 1971, 904).

Two different syntheses of [16] annulene both give the same isomer which should be predominantly XXXV, since it has four interior protons. This is the structure found by X-ray diffraction (S. M. Johnson, I. C. Paul and G. S. D. King, ibid., B, 1970, 643) but detailed analysis of the n.m.r. spectrum cannot be reconciled with its being the sole configuration in solution and the presence of 25% of XXXVI, with a relatively slow interconversion rate, is postulated (Oth and Gilles, Tetrahedron Letters, 1968, 6259). The annulene undergoes both photochemical and thermal ring closure by double conrotatory and double disrotatory modes respectively (Schröder, W. Martin and Oth, Angew. Chem., intern. Edn., 1967, 6: 870).

It reacts with potassium to give an aromatic dianion, with the inner protons resonating at $\tau 18.17$ and the outer between $\tau 1$ and $\tau 3$ (Oth, G. Anthoine and Gilles, Tetrahedron Letters, 1968, 6265):

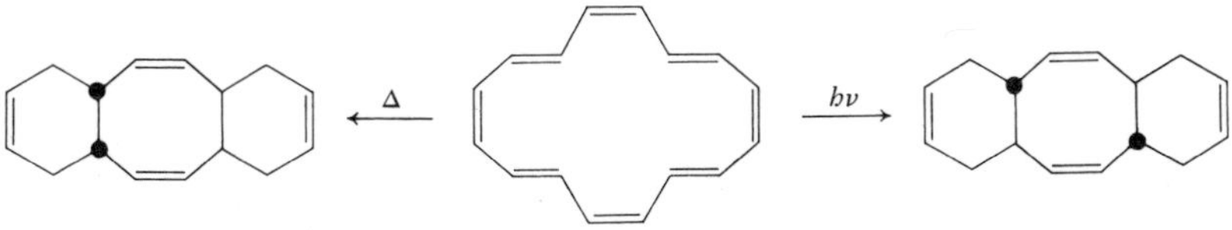

(XXXV)

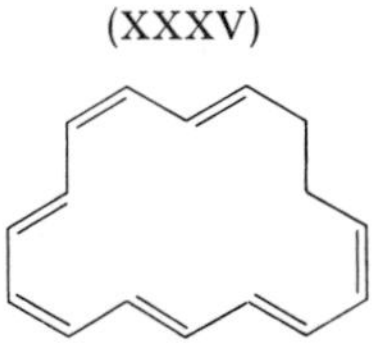

(XXXVI)

[22] Annulene exists as unstable dark purple crystals which decompose on heating. It is aromatic (i.e. sustains a diamagnetic ring current) and seems to be a conformational mixture (Sondheimer et al., Chem. Comm., 1971, 338).

Much work has been carried out on dehydroannulenes and an octa-dehydro [24] annulene has been prepared (Sondheimer et al., J. Amer. chem. Soc., 1970, 92: 6682, where refs. to earlier work are given). The compound is unstable, it reacts with potassium to give first a radical ion, then a dianion which is a [26] annulene and aromatic. A tetradehydro [18] annulene has also been shown to be aromatic (J. Ojima et al., Tetrahedron Letters, 1968, 1115). [17] Annulenone, a higher homologue of tropone, is a dark red unstable solid: the inner protons resonate at τ—0.31 and the outer at ca. τ4.8 (G. W. Brown and Sondheimer, J. Amer. chem. Soc., 1969, 71: 760).

Bridged annulenes. Improved yields of 1,6-methano [10] annulene (XXXVII) can be obtained by dehydrogenation of the precursor (XXXVIII) with dichlorodicyanoquinone (P. H. Nelson and K. G. Untch, Tetrahedron Letters, 1969, 4475). The enthalpy of formation of the annulene is 315 kJ · mol^{-1} (E. Vogel et al., Helv., 1969, 52: 418) and this has been analysed in terms of resonance stabilisation.

(XXXVII)   (XXXVIII)   (XXXIX)

2-Amino-1,6-methano [10] annulene exists in solution free of imino tautomer, but the 2-hydroxy compound exists in a solvent-dependent equilibrium with its keto tautomer (XXXIX): the keto form predominates in benzene solutions but the enol in dimethylsulphoxide (Vogel, W. Schrock and W. A. Boll, Angew. Chem., intern. Edn., 1966, 5: 732). 1,6-Methano-[10] annulene forms a complex with chromium tricarbonyl, in which the metal is bonded to atoms 2,3,4 and 5 (Vogel et al., ibid., 1966, 5: 518). However the π-electron system is delocalised over the whole ring as shown by X-ray measurements and n.m.r. studies (P. E. Baikie and D. S. Mills, J. chem. Soc., A, 1969, 328; H. Gunther, R. Wenzl and W. Grimme, J. Amer. chem. Soc., 1969, 91: 3808; C. M. Grammacioli and M. Simonetta, Tetrahedron Letters, 1971, 173). A number of bridged [14] annulenes have been synthesised (Vogel et al., Angew. Chem., intern. Edn., 1970, 9: 513) and shown to be aromatic: X-ray diffraction studies show the annulene ring to be not quite planar with slight bond length alternation (M. Simonetta et al., ibid., 1970, 9: 519; Chem. Comm., 1971, 973). For an analysis of the n.m.r. spectrum see J. D. Roberts et. al., Tetrahedron Letters, 1969, 4307). The bridged [14] annulenes have weaker ring currents as they are forced to depart from planarity (Vogel and H. Reed, J. Amer. chem. Soc., 1972, 94: 4388). A bridged analogue of tropolone is also known (J. Reisdorff and Vogel, Angew. Chem., intern. Edn., 1972, 11: 218).

An alternative approach to the synthesis of bridged [14] annulenes has been the preparation of both *cis* and *trans*-15,16-dimethyldihydropyrene, (XL and XLI) (R. H. Mitchell and V. Boekelheide, Chem. Comm., 1970, 1555, where earlier refs. are given). The compounds sustain a diamagnetic ring current and the methyl absorptions appear at high field, $\tau 12.06$ (*cis*) and $\tau 14.25$ (*trans*). The *cis*-isomer is not quite planar and so the methyl groups are less deeply embedded in the π-electron cloud. A bridged [16] annulene has been made in similar fashion and forms a dianion which sustains a diamagnetic ring current (Mitchell and Boekelheide, ibid., 1970, 1557):

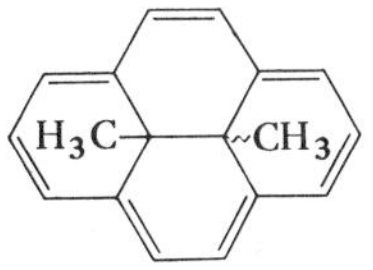

(XL) *cis*
(XLI) *trans*

(b) Functionally substituted macrocycles

These are often most accessible from a suitable cyclic olefin, for example the preparation of cyclodecanol and cyclodecanone from *cis,trans*-cyclodeca-1,5-diene (K. Schank and J. H. Felzmann, Ber., 1967, 100: 3835). Additions to cyclic olefins, particularly to cyclodecene, may be complicated by trans-annular reactions. This can be avoided by addition to a diene and reduction of the surviving double bond (M. Havel, M. Svoboda and J. Sicher, Coll. Czech. chem. Comm., 1969, 34: 340 for a preparation of 1,2-dibromocyclo-decane). However cyclodecene adds chlorine normally to give the *trans*-1,2-dichloride (J. G. Traynham and D. W. B. Stone, J. org. Chem., 1970, 35: 2025).

Various procedures will expand or contract rings of accessible size to the less accessible ones, for example cyclododecanone may be converted into cyclo-undecanone (E. W. Garbisch and J. Wohllebe, ibid., 1968, 33: 2157):

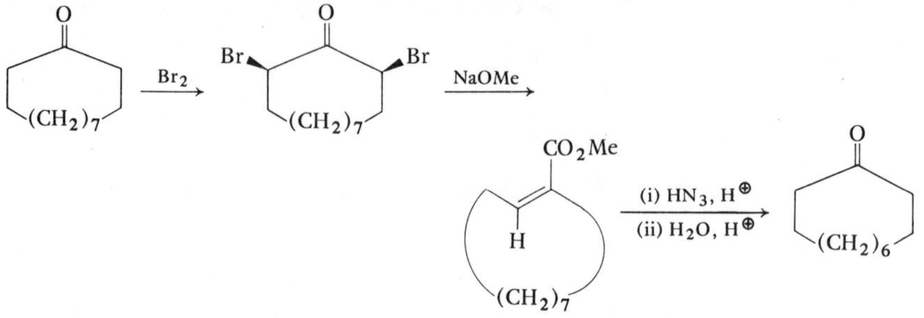

The same ketone may also be homologated to 14-membered or to 16-membered rings (R. C. Cookson and Prithipal Singh, J. chem. Soc., C, 1971, 1477):

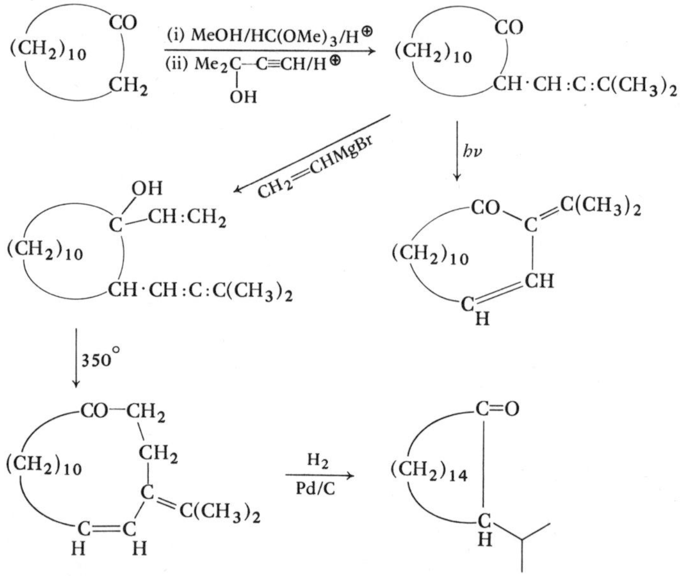

Cope rearrangement of the silyl ethers of suitable bis-olefinic cyclic alcohols results in bis-homologation of the ring. The use of the silyl ether (siloxy-Cope rearrangement) suppresses competing processes (R. W. Thies, J. Amer. chem. Soc., 1972, 94: 7074):

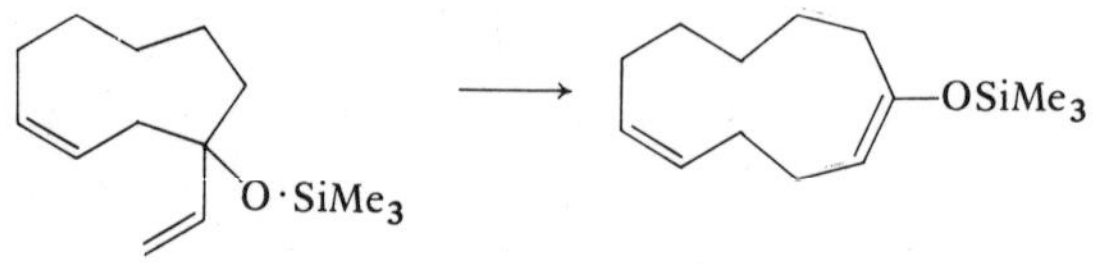

The enamine of cyclo-octanone may be bis-homologated to α-carbomethoxy-cyclodecanone via its [2 + 2] adduct with methyl acetylenecarboxylate (R. D. Burpitt and J. G. Thweatt, Org. Synth., 1968, 48: 56):

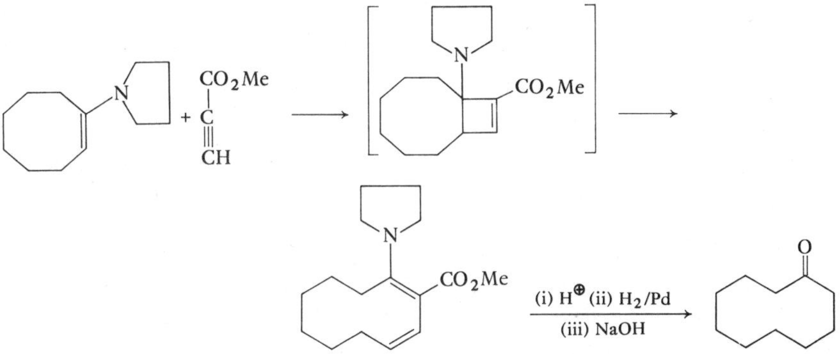

Cyclododecane-1,6-dione is formed by Cope rearrangement of 1,2-bisvinyl-cyclo-octane-1,2-diol (E. N. Marvel and T. Tao, Tetrahedron Letters, 1969, 1341). The modified acyloin synthesis using chlorotrimethylsilane may be applied to esters of cycloalkane-1,2-dicarboxylic acids: the bis-trimethylsilyl esters of the annellated cyclobutenediols which are formed are rearranged and hydrolysed to bis-homologous cycloalkane-1,2-diones (T. Mori, T. Nakahara and H. Nozacki, Canad. J. Chem., 1969, 47: 3266).

Muscone has been made from dodecatrienyl nickel by reaction with allene and carbon monoxide, followed by hydrogenation (R. Baker, B. N. Blackett and R. C. Cookson, Chem. Comm., 1972, 802):

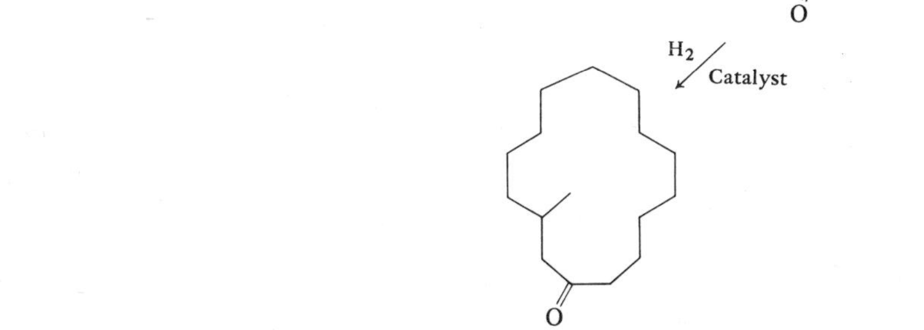

Muscone

Cycloalken-3-ols, or their derivatives, are obtainable by solvolysis of the halocarbene adducts of cycloalkenes (G. H. Whitham and M. Wright, ibid., 1967, 294: C. B. Reese and A. Shaw, J. Amer. chem. Soc., 1970, 92: 2566). Loss of an *exo*-halogen is more rapid than loss of an *endo*-halogen (there is less steric hindrance) and since the ring-opening is disrotatory a *trans*-olefin is obtained from a di-halocarbene adduct. Since the *trans*-cycloalkenes are chiral, diastereoisomeric alcohols could be produced. The diastereoisomers of 3-substituted cyclononenes and cyclodecenes of course interconvert rapidly, but those of *trans*-cyclo-octen-3-ols (and *cis,trans*-cyclononadienols) do not: the product obtained is that resulting from nucleophilic attack on the same side of the molecule as the departing anion (Reese and Shaw, Chem. Comm., 1970, 1365, 1367; Whitham et al., ibid., 1968, 1593). (1*RS*,2*RS*)-*trans*-cyclo-oct-2-en-1-ol has been synthesised (Whitham and Wright, J. chem. Soc., C, 1971, 883). It isomerises to the *cis*-isomer when heated with silver nitrate in aqueous dioxan:

trans-Cyclo-oct-2-en-1-ol

In this way, substituted derivatives of the previously unknown *cis,trans*-cyclo-octa-1,4-diene have been prepared (M. S. Baird and Reese, Chem. Comm., 1970, 1644).

In a similarly stereospecific process, solvolysis of the *p*-nitrobenzoate of *endo*-bicyclo[6,1,0]nonan-2-ol gives derivatives of *cis*-cyclononen-4-ol, together with bicyclic products of retained configuration (C. D. Poulter, E. C. Friedrich and S. Winstein, J. Amer. chem. Soc., 1970, 92: 4274). Only bicyclic products are formed from esters of the *exo*-bicyclic alcohol, but from *exo*-bicyclo[7,1,0]decan-2-yl esters some *trans*-cyclodecen-4-ol is obtained (Poulter and Winstein, ibid., 4281).

α,ω-Dihalides alkylate α,α,ω,ω-tetracarboxylic esters to give macrocyclic tetra-esters. Cyclic diesters are also formed when the dihalide chain is at least fourteen carbons in length (G. Schill, K. Rothmaier and H. Ortlich, Synthesis, 1972, 426).

Fragmentation reactions (for a review see C. A. Grob and P. W. Schiess, Angew. Chem., intern. Edn., 1967, 6: 3) are often the best routes to particular derivatives, for example the formation of 5-carboxycyclo-octene from 2-(*N*-pyrrolidyl)-bicyclo[3,3,1]nonane (G. Stork and H. Landesman, J.

Amer. chem. Soc., 1956, 78: 5129; S. H. Graham et al., J. chem. Soc., C, 1968, 1491). Carbonyl bridges may be opened photochemically, as in a synthesis of cyclopentadecanone via the bicyclic ketone XLII (H. Nozaki et al., Tetrahedron Letters, 1967, 779):

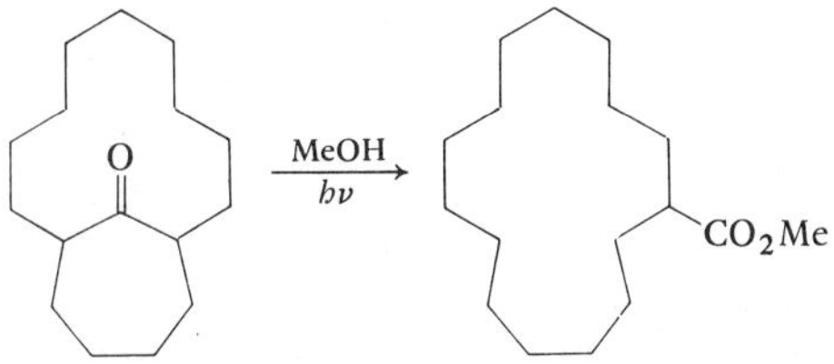

(XLII)

Transannular hydride shifts characteristically accompany reactions in which a positively charged carbon is generated in an eight to eleven-membered ring (for a review see A. C. Cope, M. M. Martin and M. A. McKervey, Quart. Reviews, 1966, 20: 119). Transannular transfer of hydrogen in a free-radical reaction has also been observed (Cope and J. E. Engelhart, J. Amer. chem. Soc., 1968, 90: 7092; M. Fisch and G. Ourisson, Chem. Comm., 1965, 407), and transannular reaction accompanies Simmons–Smith addition to cyclo-octyne (G. Wittig and J. J. Hutchinson, Ann., 1970, 741: 79):

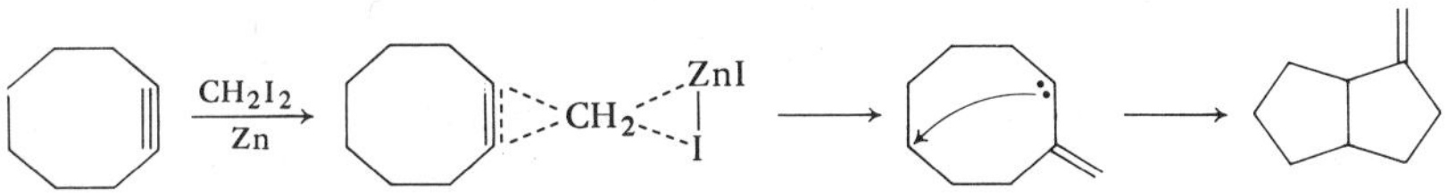

Transannular insertion reactions have been observed in reactions of nitrenium ions (O. E. Edwards et al., J. Amer. chem. Soc., 1965, 87: 678) and in reactions of carbenes (Cope and S. S. Hecht, ibid., 1967, 89: 6920). Transannular participation by π-electrons, as in the solvolysis of 4-cyclo-octenylmethyl tosylate (Graham et al., Tetrahedron Letters, 1967, 23: 299, and refs. cited therein), can lead to bridged bicyclic alcohols with predominantly *endo*-geometry. By contrast the silver perchlorate assisted rearrangement of 5-(N-chloro-N-methylamino)cyclo-octene gives products of *exo*-geometry; the halide ion never becomes completely free, as shown by the failure of silver ion to capture it (J. D. Hobson and W. D. Riddell, Chem. Comm., 1968, 1178).

   It has long been known that elimination reactions of medium-ring cyclo-alkyl halides, esters and ammonium salts gave rise to mixtures of the *cis* and the *trans* olefins in ratios which varied with the nature of the leaving group and the basic reagent used. Since this was true even where the *cis*-olefins were of much lower energy (ten-membered rings and smaller), the transition state did not closely resemble the final olefin, and explanations were offered on the basis of special conformational properties of the rings, all sharing the commonly held assumption that the transition state for ionic elimination possessed *anti-periplanar* geometry. However Sicher and his school have shown that the plot of rate of *trans*-olefin formation as a function of ring size is the same for an ionic elimination, Hofmann pyrolysis of an ammonium hydroxide, as for a process undoubtedly involving *syn-periplanar* geometry in the transition state, pyrolysis of an amine N-oxide. The same was not true for production of *cis*-olefins, the plot of the rate of olefin formation against ring size was quite different for ionic and pyrolytic techniques. They therefore concluded that formation of *trans*-olefins by Hofmann elimination was a *syn*-process, and formation of *cis*-olefins was an *anti*-process (Sicher et al., Tetrahedron Letters, 1966, 1619, 1627). A characteristic of the *syn*-elimination is that rates remain high throughout the range of eight to twelve-membered rings, normally exceeding the rates for corresponding open-chain molecules, while for *anti*-eliminations the rates throughout this range are below those for open-chain molecules and pass through a minimum about the twelve-membered ring. The same *syn–anti* dichotomy has been found to apply also to elimination reactions of cycloalkyl halides and tosylates when non-dissociating solvents (butanol or benzene) are used. In dimethylformamide both the *cis*- and the *trans*-olefins are formed by *anti*-elimination, and little *trans*-olefin is formed when the ring is smaller than thirteen-membered. Thus for nine-, ten-, and eleven-membered rings choice of base and solvent system drastically affects the course of the reaction: cyclodecyl bromide can give as little as 11% *cis*-olefin (in benzene) or as much as 94% (in dimethylformamide) (J. Závada, J. Krupička and Sicher, Coll. Czech. chem. Comm., 1968, 33: 1393, where earlier refs. are given; cf. p. 371).

It has been shown in eliminations from 1,1,4,4-tetramethylcyclodecyl-7-tosylate (XLIII) and the corresponding trimethylammonium hydroxide to give the olefins XLIV and XLV that deuterium replacement of the *cis*-protons on $C_{(6)}$ or $C_{(8)}$ does not alter the ratio in which the two olefins are produced, but replacement of the *trans*-proton on $C_{(8)}$ raises the proportion of XLIV, and similarly *trans*-deuteriation on $C_{(6)}$ raises the proportion of XLV due to the operation of the primary isotope effect: *cis*-deuterium label is retained in the olefinic products and *trans*-label is lost (Závada, Svoboda and Sicher, ibid., 1968, 33: 1415, 4027). Analogous experiments with deuteriated cyclo-octyltrimethylammonium hydroxide confirmed that *trans*-cyclo-octene is formed by a *syn*-elimination but showed that a proportion of the *cis*-cyclo-octene was also formed by *syn*-elimination (J. L.

Coke, M. P. Cooke and M. C. Mourning, Tetrahedron Letters, 1968, 2253):

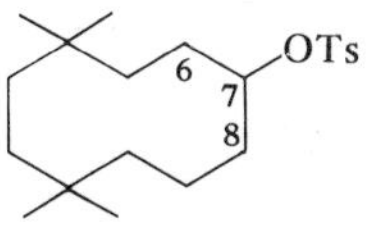

(XLIII) 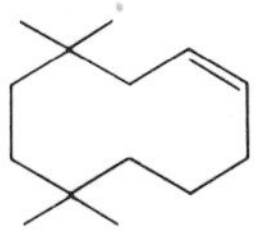   (XLIV) 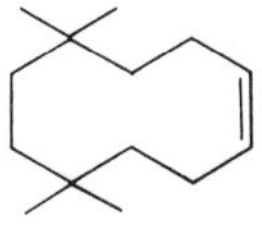   (XLV)

Vicinal dihalogeno cyclodecanes and cyclododecanes eliminate preferentially via a *syn*-mechanism, using zinc and methanol, but via an *anti*-mechanism using iodide ion (Sicher, M. Havel and Svoboda, ibid., 1968, 4269). For a review on stereochemistry and mechanism of elimination reactions see Sicher (Pure and Appl. Chem., 1971, 25: 655).

4. The nonadrides

These are a group of mould metabolites, *glauconic acid* (XLVI), the related deoxy compound *glaucanic acid* (XLVII) (N. Wijkman, Ann., 1931, 485: 61), *byssochlamic acid* (XLVIII) (H. Raistrick and G. Smith, Biochem. J., 1933, 27: 1814), and the two *rubratoxins*, A (XLIX) and B (L) (M. O. Moss et al., J. chem. Soc., C, 1971, 619; Tetrahedron Letters, 1969, 367; G. Buchi

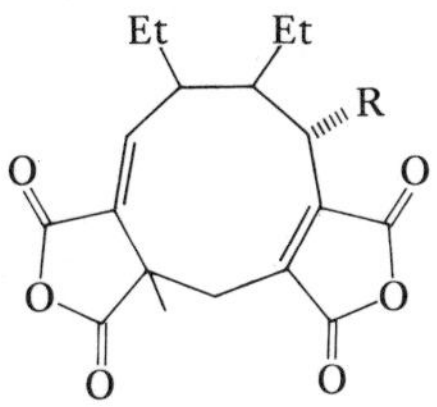

(XLVI) R = OH
(XLVII) R = H

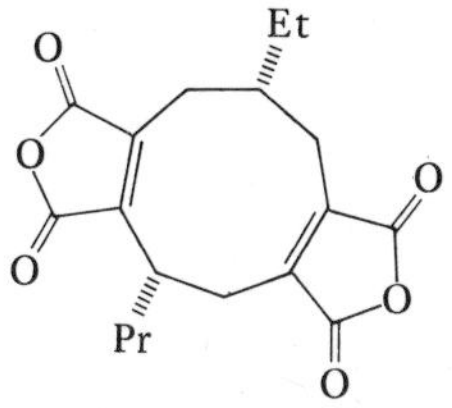

(XLVIII)

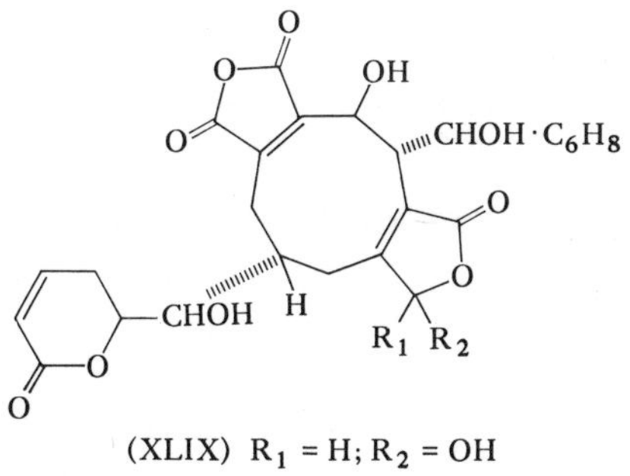

(XLIX) R$_1$ = H; R$_2$ = OH
(L) R$_1$, R$_2$ = O

et al., J. Amer. chem. Soc., 1970, 92: 6638) all share the common feature of being bis-anhydrides of cyclononadienetetracarboxylic acids: the free acids dehydrate spontaneously, hence cannot be isolated, but the anhydrides dissolve in aqueous alkali to give solutions of their salts.

The determination of these structures leaned heavily on X-ray diffraction studies on the bis-*p*-bromophenyl-hydrazides of the acids (J. M. Robertson et al., Experientia, 1962, 18: 352; J. chem. Soc., 1963, 5502): chemical degradation was also used (D. H. R. Barton, J. K. Sutherland et al., Experientia 1962, 18: 345; J. chem. Soc., 1965, 1769 et seq.). Some of the early observations, notably the pyrolysis of glauconic acid to glauconin (LI), and α,β-diethylacraldehyde gave results which were difficult to interpret. It is now known that the pyrolysis involves a Cope rearrangement via LII to LIII. A retro-Michael reaction then splits LIII to glauconin and the aldehyde:

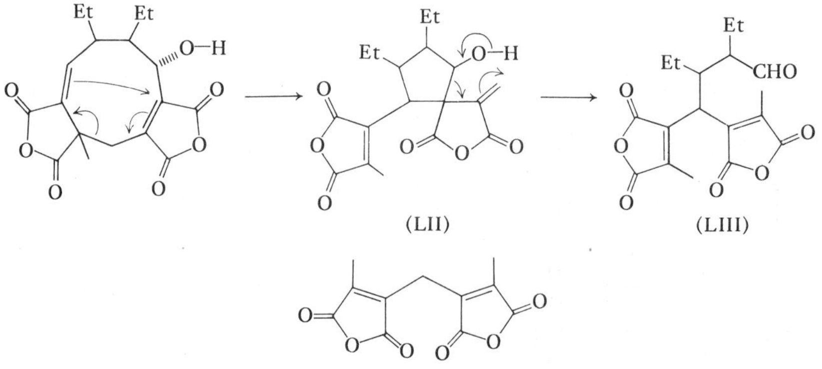

(LII)   (LIII)

(LI)

Both double bonds of byssochlamic acid are very unreactive, but one of the anhydride functions could be converted into an *N*-hydroxyimide, and the action of base on the toluenesulphonyl derivative of this gave the ketone LIV: nitrosation of the reactive methylene of this ketone and Beckmann rearrangement of the resulting oximinoketone gave the nitrile acid LV which could be oxidised to a mixture of *S*-*n*-propylsuccinic acid and β-ethylglutaric acid:

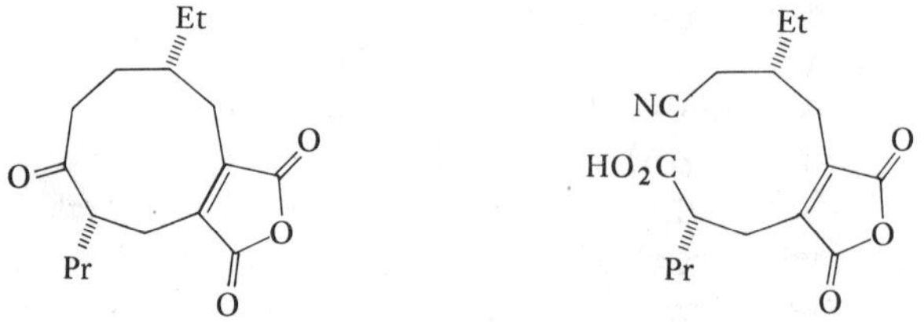

(LIV)   (LV)

A suggestion that these materials arose by dimerisation of two C_9 units was confirmed by feeding the labelled anhydride LVI to a culture of *Penicillium purpurogenum*: the label was efficiently incorporated into the glauconic and glaucanic acids (Sutherland et al., J. chem. Soc., Perkin I, 1972, 2584). The cycloaddition of LVI to the derived anion LVII would be thermally permitted ($4\pi + 6\pi$), but an attempt to realise this process in vitro, by treatment of LVI with sodium hydride and triethylamine, led to a cyclic product, LIX, with "unnatural" geometry. This was ascribed to isomerisation of LVII to LVIII prior to addition (Sutherland et al., Chem. Comm., 1968, 1192):

(LVI)　　　　　(LVII)　　　　　(LVIII)　　　　　(LIX)

5. Conformations of larger rings

A comprehensive review, written from one specific standpoint, is available (J. Dale, Angew. Chem., intern. Edn., 1966, 5: 1000; see also Pure and appl. Chem., 1971, 25: 469) and also a brief review of the special case of the ten-membered ring (Sicher, Ind. chim. Belge, 1967, 331). For a review of X-ray diffraction techniques as applied to medium-sized rings see J. Dunitz (Pure and appl. Chem., 1971, 25: 495). The thesis advanced by Dale is that cyclic structures are derived by tracing strain-free paths along the diamond lattice which incorporate the requisite number of carbon atoms. For the ten-membered ring this involves placement of some of the hydrogens in "intra-annular" positions with resulting strain, reflected in enlarged internal bond angles: however this conformation has been solidly established by experiment and is also that predicted by absolute calculation of strain energy (J. B. Hendrickson, J. Amer. chem. Soc., 1964, 86: 4854). However the "saddle" conformation LX predicted for cyclo-octane on the diamond lattice approach is not in fact populated. Careful n.m.r. studies on deuterated and fluorinated cyclo-octanes are best interpreted on the basis of the boat–chair conformation LXI (F. A. L. Anet and M. St. Jaques, ibid., 1966, 88: 2585, but cf. 1968, 90: 5243; J. D. Roberts et al., ibid., 1969, 91: 1386). The same conformation has been found in a number of X-ray diffraction studies (H. B. Burgi and J. D. Dunitz, Helv., 1968, 51: 1514 and refs. cited therein). However absolute-energy calculations show that two or three conformations of cyclo-octane are competitive in energy (Hendrickson, J. Amer. chem. Soc., 1967, 89: 7036), and a distorted crown conformation, (LXII), has been observed in *trans,syn,trans*-1,2,5,6-tetrabromocyclo-octane (R. A. Raphael, G. Ferguson et al., Chem. Comm., 1968, 103). The cyclo-octane ring can thus be expected to be able to populate varying conformations to suit the circumstances of substitution pattern and environment, though the observation (N. L. Allinger et al., J. org. chem., 1970, 35: 1255) that reduction of 3-*tert*-butylcyclo-octanone gives only one alcohol, and that this in turn selectively eliminates towards $C_{(4)}$ suggest that in some cases, conformational preferences may still be quite marked.

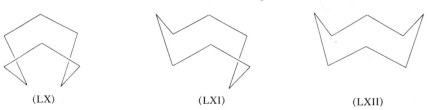

(LX)   (LXI)   (LXII)

For the larger rings, the derivation of ring skeletons from the diamond lattice must be guided by further considerations. Two parallel polymethylene chains, linked at each end by ethane bridges, are preferable to more "open" conformations, but this structure can only be realised for the case $(CH_2)_{4n+2}$. Such cyclic paraffins may populate substantially one conformation in the liquid phase with the consequence that they possess a relatively small entropy of fusion and therefore a high melting point. Cycloalkanes of general formula $(CH_2)_{4n}$ would have to possess methylene bridges (with resulting transannular strain) or three-carbon bridges (with a more open ring). They have lower melting points and greater entropies of fusion, which may be attributed to their being conformationally inhomogeneous in the liquid phase.

A feature common to most of these cyclic systems (but not encountered in chair cyclohexane) is the phenomenon of pseudo-rotation, whereby ring positions are interchanged by a process involving no angle distortion, and hence with a low energy of activation (Hendrickson, J. Amer. chem. Soc., 1967, 89: 7047). Thus, although the boat—chair conformation for cyclo-octane possesses several different types of proton, site-averaging by pseudo-rotation reduces these to two groups only (like protons are always *trans* to each other) and only two n.m.r. signals are seen at low temperatures: at ambient temperatures, ring inversion further interchanges these two groups to render all protons equivalent. Similarly although three types of ring position may be distinguished in cyclodecane, all methylenes are equivalent on the n.m.r. time scale.

This process of pseudo-rotation can be frozen by *gem*-disubstitution; thus 1,1,4,4-tetramethylcyclodecane has all its sites necessarily fixed, since the methyls cannot be accommodated at "intra-annular" positions, and ring inversion is the only site-averaging process possible. Consequently deamination of 7-amino-1,1,4,4-tetramethylcyclodecane proceeds without hydride shift, and similarly no hydride shift is observed in addition reactions of 5,5,8,8-tetramethylcyclodecene or in solvolysis of its epoxide. In the accepted conformation for the ten-membered ring, the 6,7-bond of 1,1,4,4-tetramethylcyclodecane is *cisoid*, and the 7,8-bond is *transoid*. Measurement of heats of hydrogenation show that *cis*-4,4,7,7-tetramethyl-cyclodecene has in fact a lower enthalpy than its *trans*-isomer and that the reverse is true for the 5,5,8,8-tetramethyl isomer (Sicher, R. B. Turner et al., Tetrahedron, 1966, 22: 659).

4,4,7,7-Tetramethylcyclononanone and several of its derivatives have unexpectely high melting points (suggestive of conformational homogeneity) and also relatively high barriers to ring inversion (two methyl signals in the n.m.r. spectrum coalesce only at near-ambient temperatures) (G. Borgen and Dale, Chem. Comm., 1970, 1105: see also A. T. Blomquist and R. D. Miller, J. Amer. chem. Soc., 1968, 90: 3233). The substituents are suppressing site averaging by pseudo-rotation and thereby allowing observation of a process with higher energy of activation.

In larger rings *gem*-dimethyl substitution can play another role: such substituent pairs are best accommodated at the "corners" of the molecules, and their siting can affect the conformation. Cyclohexadecane is conformationally inhomogeneous, it has to choose between a diamond lattice conformation which is "open" or a more compact conformation which is not strain-free. 3,3,7,7,11,11,15,15-Octamethylcyclohexadecan-1,9-dione however is conformationally homogeneous, and populates the "open" conformation with alkyl substituents at the corners. Similarly an octamethylcyclo-eicosanedione tenaciously retains solvent of crystallisation, being constrained to have an "open" ring, while the parent dione does not (Borgen and Dale, Chem. Comm., 1970, 1340, 1342).

Some cyclic olefins present simpler conformational problems than the cycloalkanes. Cycloheptene and cyclo-octa-1,5-diene can both populate chair and flexible conformations like cyclohexane, but an examination of models suggests that the relative energy of the chair should rise along this series, for the complete absence of eclipsing which characterises the cyclohexane chair is replaced by completely eclipsed bonds in the cyclo-octadiene chair. Dipole moment and n.m.r. studies indicate from 8% to under 2% of boat conformer in benz-cycloheptenes (N. L. Allinger and W. Szkrybalo, J. org. Chem., 1962, 22: 722; G. L. Buchanan and J. M. McCrae, Tetrahedron, 1967, 279; S. Kabuss, H. Friebolin and H. Schmid, Tetrahedron Letters, 1965, 469), though calculations have suggested the boat conformer might be of lower energy (N. L. Allinger et al., J. Amer. chem. Soc., 1968, 90: 5773). For cyclo-octa-1,5-diene the position is obscure; the n.m.r. spectrum is not informative (M. St. Jaques, M. A. Brown and Anet, Tetrahedron Letters, 1966, 5947), studies on molecular beam deflection in an electric field are reported to fit best the tub conformation (W. Klemperer et al., J. Amer. chem. Soc., 1970, 92: 6325), while dibenzcyclo-octadiene exists as a chair conformation in the crystalline state (W. Baker et al., J. chem. Soc., 1945, 27). A dibromocyclo-octadiene, however, exists in a twist conformation (D. D. MacNicol, R. A. Raphael et al., J. chem. Soc., Perkin II, 1972, 1632).

cis,cis-Cyclodeca-1,6-diene exists in the chair conformation LXIII as shown by X-ray and n.m.r. studies (B. W. Roberts et al., J. Amer. chem. Soc., 1968, 90: 5263, 5264; Dale, T. Ekeland and J. Schaug, Chem. Comm., 1968, 1477). This conformation may be populated preferentially as it avoids repulsion between two π-electron clouds, as occurs in the otherwise apparently strain-free conformation LXIV of its trans,trans-isomer. At all events conformation LXIII seems to be in a considerable energy trough as the cis,cis-1,6-diene is not only much more stable than the trans,trans-, but also much more stable than any of its other isomers (for equilibration of geometrical isomers of dienes by photo-irradiation in the presence of a sensitiser see Dale and C. Moussebois, J. chem. Soc., C, 1966, 264). This conformational preference has been suggested as the reason for ketone LXV failing to undergo a concerted 2 + 2 cycloaddition on photo-irradiation; it gives the cis,anti,cis-tricyclo[5,3,0,0^{2,6}]decadione in a stepwise process (J. R. Scheffer and M. R. Lungle, Tetrahedron Letters, 1969, 845).

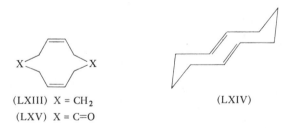

(LXIII) X = CH$_2$ (LXIV)
(LXV)   X = C=O

However a stereospecific addition to a cis,cis-cyclodeca-1,6-diene, which apparently proceeds via a boat conformation, has been reported (H. W. Guin et al., J. Amer. chem. Soc., 1966, 88: 5366).

6. Catenanes and very large rings

Mass spectrometry is a valuable technique for exploring the structure of catenanes: the normal fragmentation products of both cyclic components can be observed with the mass augmented by the second cycle. It has also been possible to detect hydrogen transfer (during fragmentation) from one catenated ring to the other (W. Vetter and G. Schill, Tetrahedron, 1967, 23: 3079).

Rational syntheses of catenanes are of two types. Ziegler cyclisation of LXVI can give two ketones, one of these possessing linked macrocycles — the so-called statistical approach (A. Luttringhaus and G. Isele, Angew. Chem., intern. Edn., 1967, 6: 956). Intramolecular formation of the tertiary amine from LXVII necessarily results in production of linked macrocycles (G. Schill, Ber., 1967, 100: 2021; Schill, K. Murjaha and Vetter, Ann., 1970, 740: 18). In the first case oxidative hydrolysis separates the heterocyclic and aromatic rings to give the catenane, in the second, two stages are required to free the catenated ring from the aromatic template. A similar technique has been used to synthesise a "rotaxane", a molecule in which a chain

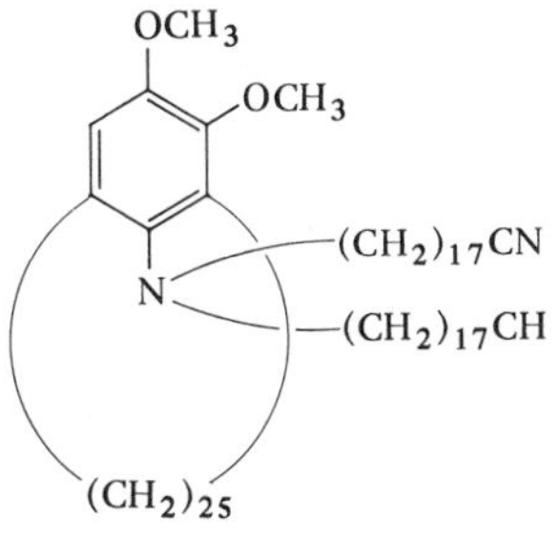

(LXVI)

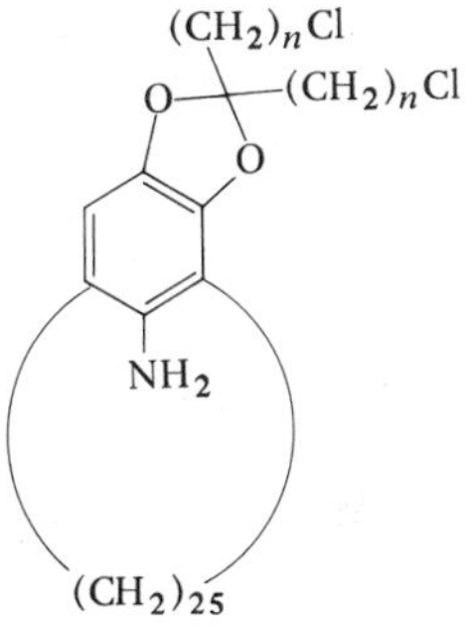

(LXVII)

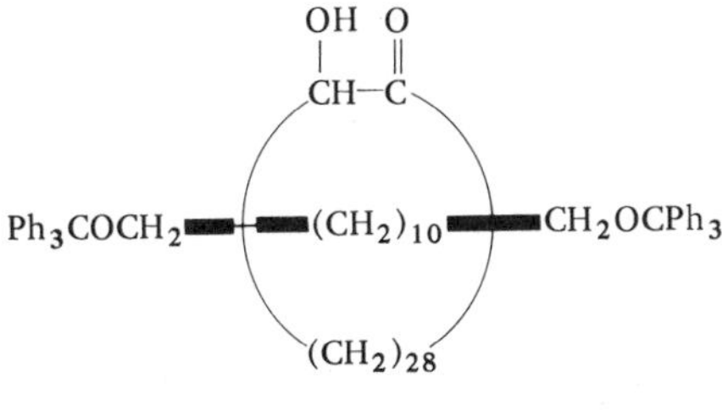

(LXVIII)

terminated by bulky substituents is threaded through a macrocycle (Schill and H. Zollenkopf, Ann., 1969, 721: 53). An example of this is LXVIII, which was prepared by tritylation of dodecane-1,10-diol in the presence of a 30-membered cyclic acyloin. By equilibrating the trityl ether of dodecane-diol with a mixture of all the cyclo-alkanes from $C_{12}H_{24}$ to $C_{40}H_{80}$, it could be shown that a rotaxane was formed only from $C_{29}H_{58}$, and this dissociated on heating. Smaller rings could not admit the triphenylmethyl groups and the larger rings could not retain them. By preparing the ether from the diol in the presence of the same cyclo-alkane mixture, rotaxanes were formed from all the rings from $C_{25}H_{50}$ to $C_{29}H_{58}$ inclusive (T. Harrison, Chem. Comm., 1972, 231).

The production of very large rings may be approached by cyclo-oligomeri-sation of a cycloalkyne (or cycloalkadiyne) with butadiene over a zero-valent nickel catalyst. This builds on an annellated cyclodecatriene ring: hydro-genation of the di-substituted olefinic bonds and oxidative fission of the remaining double bonds gives a cyclic diketone with eight (or sixteen) more methylenes than the original cycle (P. Heimbach and W. Brenner, Angew. Chem., intern. Edn., 1966, 5: 961). A cycloalkene may be polymerised as shown below over a suitable catalyst — $WCl_6 \cdot EtAlCl_2 \cdot EtOH$ (E. Wasserman, D. A. Ben Efraim and R. Wolovsky, J. Amer. chem. Soc., 1968, 90: 3286). The production of C_{16}, C_{24} and C_{32} monocycles from cyclo-octene has been definitely established, and evidence obtained for the formation of monocycles up to C_{120}:

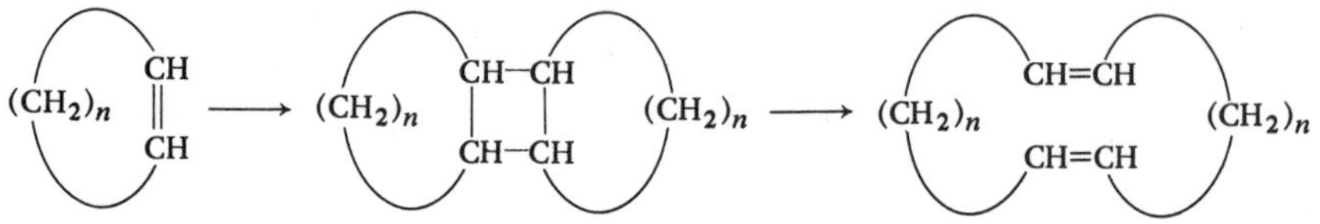

Autoxidation of bifunctional alkylidenephosphoranes can give macrocyclic olefins (medium-sized rings are not formed in this way). Mass spectral evidence of the formation of $C_{60}H_{108}$ has been obtained (H. J. Bestmann and H. Pfüller, Angew. Chem., intern. Edn., 1972, 11: 508).

GUIDE TO THE INDEX

This index is constructed in a similar manner to the volume indexes of the first edition of the Chemistry of Carbon Compounds. However, to make the index easier to use, more descriptive entries have been made for the commonly occurring individual, and groups of chemicals.

The indexes cover primarily the chemical compounds mentioned in the text, and also include reactions and techniques, where named, and some sources of chemical compounds such as plant and animal species, oils, etc.

Chemical compounds have been indexed alphabetically under the names used by authors, editing being restricted to ensuring uniformity of entries under the same heading. In view of the alternative nomenclature that can often be used, a limited amount of cross-referencing has been done where it is considered to be helpful, but attention is particularly drawn to Convention 2 below.

For this and the succeeding volumes, the indexing conventions listed below have been adopted.

1. Alphabetisation

(a) The following prefixes have not been counted for alphabetising:

n-	*o-*	*as-*	*meso-*	D	*C-*
sec-	*m-*	*sym-*	*cis-*	DL	*O-*
tert-	*p-*	*gem-*	*trans-*	L	*N-*
	vic-				*S-*
		lin-			*Bz-*
					Py-

Some prefixes and numbering have been omitted in the index, where they do not usefully contribute to the reference.

(b) The following prefixes have been alphabetised:

Allo	Epi	Neo
Anti	Hetero	Nor
Cyclo	Homo	Pseudo
	Iso	

[391]

(c) A letter by letter alphabetical sequence is followed for entries, firstly for the main entry, followed by the descriptive entry. The only exception to this sequence is the placing of plural entries in front of the corresponding individual entries to prevent these being overlooked by a strict alphabetical sequence which could lead to a considerable separation of plural from individual entries. Thus "butanes" will come before n-butane, "butenes" before 1-butene, and 2-butene, etc.

2. Cross references

In view of the many alternative trivial and systematic names for chemical compounds, the indexes should be searched under any alternative names which may be indicated in the main body of the text. Only a limited amount of cross-referencing has been carried out, where it is considered that it would be helpful to the user.

3. Esters

In the case of lower alcohols esters are indexed only under the acid, e.g. propionic methyl ester, not methyl propionate. Ethyl is normally omitted e.g. acetic ester.

4. Derivatives

Simple derivatives are not normally indexed if they follow in the same short section of the text.

5. Collective and plural entries

In place of "— derivatives" or "— compounds" the plural entry has normally been used. Plural entries have occasionally been used where compounds of the same name but differing numbering appear in the same section of the text.

6. Main entries

The main entry of the more common individual compounds is indicated by heavy type. Where entries relate to sections of three pages or more, the page number is followed by "ff".

Index